Student Solutions Ma

General, Organic, and Biochemistry
An Applied Approach

James Armstring
City College of San Francisco

Prepared by

Lenore K. Hoyt
University of Louisville

Australia • Brazil • Japan • Korea • Mexico • Singapore • Spain • United Kingdom • United States

ISBN-13: 978-0-534-49352-3
ISBN-10: 0-534-49352-1

Brooks/Cole
20 Davis Drive
Belmont, CA 94002-3098
USA

Cengage Learning is a leading provider of customized learning solutions with office locations around the globe, including Singapore, the United Kingdom, Australia, Mexico, Brazil, and Japan. Locate your local office at: **www.cengage.com/global**

Cengage Learning products are represented in Canada by Nelson Education, Ltd.

To learn more about Brooks/Cole, visit **www.cengage.com/brookscole**

Purchase any of our products at your local college store or at our preferred online store **www.cengagebrain.com**

Printed in the United States of America
1 2 3 4 5 6 7 15 14 13 12 11

Table of Contents

Chapter 1

1

Measurements in Science and Medicine

Solutions to Section 1.1 Core Problems

1.1 *State whether each of the following is a unit of distance, volume, or mass.*

Remember in each of these, the prefix is irrelevant to what *kind* of unit it is; we're only looking at whether the root is meter, liter or gram.

a) kilogram **mass** (root unit is gram)

b) deciliter **volume** (root unit is liter)

c) centimeter **distance** (root unit is meter)

1.3 *Give the standard abbreviation for each of the following units.* See Tables 1.2 and 1.3.

a) liter **L** *b) centimeter* **cm** *c) milligram* **mg** *d) microliter* **µL**

1.5 *Write the full name for each of the following metric units.*

See Tables 1.2 and 1.3. If there are two letters, the first letter is the Table 1.3 prefix, and the second is the Table 1.2 unit. If there's only one letter, it can only be the Table 1.2 unit.

a) m **meter**

NOTE:
this can't be "milli" because the prefixes in Table 1.3 never stand by themselves.

b) dL **deciliter**

c) kg **kilogram**

d) µm **micrometer**

1.7 *The following three measurements describe properties of a television. Which of these is the mass of the television, which is its volume, and which is its height?*

a) 50 cm This is a distance unit (based on the meter) and must be the **height**.

b) 18 kg This is a **mass** unit (based on the gram).

c) 47 L This is a **volume** unit.

1.9 *a) Which is a more appropriate unit to express the mass of a piano: grams or kilograms?*

Of the two, the **kilogram** is closer to the weight of the piano (it would weigh many kg, but a thousand times greater in grams.) We generally choose the common unit closest to the mass of the object.

b) Which is a more appropriate unit to express the volume of a drinking cup: microliters or milliliters?

The volume of a drinking cup is many **millilters**, but would be many thousand microliters. Of the two, a milliliter is closer to the actual volume.

c) Which is a more appropriate unit to express the width of this page: centimeters or meters?

In this case, either unit is in range, but we'd probably express the distance more conveniently as several **centimeters** rather than as a fraction of a meter (22 cm vs. 0.22 m). Given a choice between a number in the tens or hundreds, versus a number that starts with a decimal, we usually take the larger number with the smaller unit. Neither unit is unreasonable, though.

1.11 *From the list below, select all of the statements that cannot possibly be correct based on the size of the measurement.*

If you need to, refer back to Tables 1.4-1.6. You should get comfortable enough with those "rough descriptions" that problems like this are easy for you.

a) "The mass of this pebble is 8.2 grams."

Reasonable—It's easy to imagine a pebble with the same mass as 8 or 9 paperclips.

b) "My kitchen sink has a capacity of 22 microliters of water."

Not possible. A microliter is a very tiny volume of matter. A kitchen sink has a volume measured in liters. (Even the kitchen sink in a doll's house would have a volume much greater than 22 microliters!)

c) "The distance between Jill's house and her office is 6.3 kilometers."

Reasonable—this distance is in the range you would expect for a drive partway across a town.

d) "This plastic drinking cup weighs 85 kilograms."

Not possible. 85 kilograms is more like the weight of a large person. A plastic drinking cup should have a mass measured in grams, at least much less than a kilogram.

e) "My coffee cup holds 150 milliliters."

Reasonable—this is a volume you'd expect for a coffee cup (remember one deciliter is about half a cup, and this is 1.5 dL—fine for a medium-small coffee cup.)

f) "The width of Bob's classroom is 8.1 millimeters."

Not possible. This is a tiny distance (1 mm is about the thickness of a dime). The width of a classroom would be on the order of meters.

g) "A mouse is 9 centimeters long."

Reasonable—one cm is around the width of a person's pinkie finger. A mouse could easily be about 9 of those units long.

Solutions to Section 1.2 Core Problems

1.13 *Which digit (if any) is uncertain in the measurement below?*

The last digit in any measurement is assumed to be uncertain.

"One cup of sliced peaches contains 1.19 mg of Vitamin E." **The 9 in 1.19 mg is uncertain.**

1.15 *Using the ruler pictured below, measure the length of the nail as precisely as you can. Express your answer in centimeters, to an appropriate number of decimal places.*

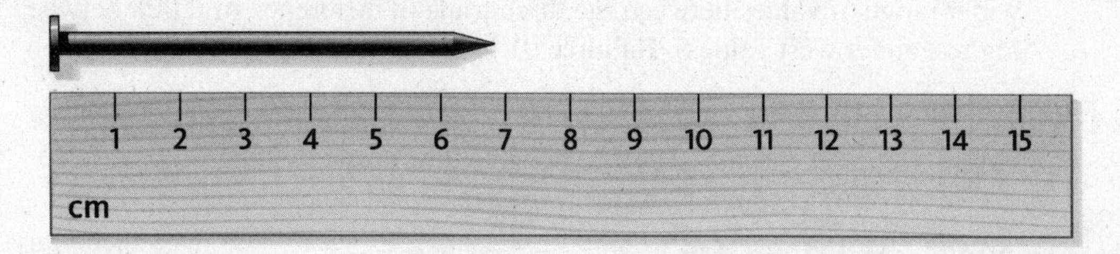

The tip of the nail reaches about 4/5 of the way between the 6 and the 7, so we would record the length as **6.8 cm**. 6.7 or 6.9 cm would also be acceptable measurements. We know it's more than 6 cm, close to 7 cm, but without marks between 6 and 7 we have to estimate as accurately as possible. This is why the last digit is always assumed to be uncertain—we always read the marked values, then estimate as best we can the next digit, between the marks.

1.17 *What is the volume of water in the following graduated cylinder? (Note that each small graduation is 0.5 mL.)*

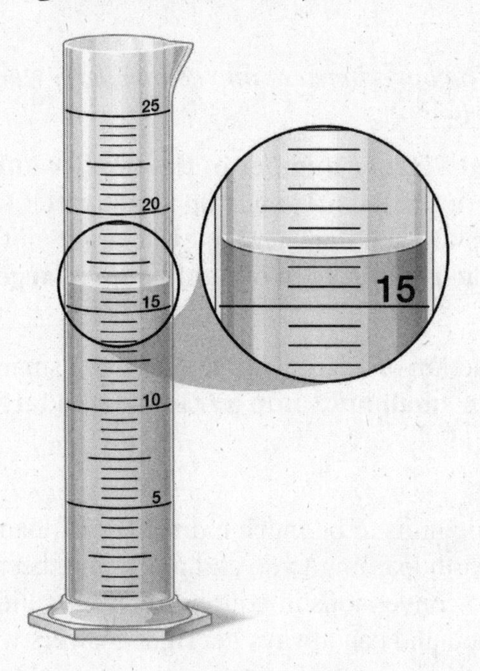

The volume is 16.2 mL. The meniscus of the water is just over two marks (1 mL) above the 15-mL mark, so we know for sure it's a little over 16 mL. We estimate the last digit: about 2/5 of the distance between the 16-mL mark and the 16.5-mL mark.

1.19 *A piece of metal weighing 26.386 g is used to check two different balances. The piece of metal is weighed three times on each balance, with the following results:*

Balance 1:	*26.375 g*	*26.377 g*	*26.378 g*
Balance 2:	*26.389 g*	*26.381 g*	*26.385 g*

a) *Which balance is more precise? Explain your reasoning.*

"More precise" means "giving the same value or very similar values in repeat measurements." Balance #1 gave three readings that were only slightly different (a difference of 0.003 between the highest and lowest values), while balance #2 has a wider range of values between the three trials (a difference of 0.008 between the highest and lowest values). **Balance #1 is more precise.**

b) *Which balance is more accurate? Explain your reasoning.*

"More accurate" means "closer to the actual or correct value."
The average of the three readings from balance #1 is:

$$\mathbf{\frac{26.375 + 26.377 + 26.378}{3} = 26.377\ g}$$

The average of the three readings from balance #2 is:

$$\mathbf{\frac{26.389 + 26.381 + 26.385}{3} = 26.385\ g}$$

While balance #2 was less precise (wider range of values), all of these values—and the average of the three—are closer to the actual value, 26.836 g, than any of the readings taken from balance #1. **Balance #2 is more accurate.**

Solutions to Section 1.3 Core Problems

1.21 *Complete the following statements. Example: To convert meters into centimeters, you must move the decimal point* <u>*two*</u> *places to the* <u>*right*</u>*.*

In each case, remember that we're going to need a **larger number of the smaller unit.** For example, it takes a lot (100) of centimeters (a small unit) to add up to one meter (a larger unit), and a lot (1000) of meters to add up to one kilometer (an even larger unit). Therefore **we'll always move the decimal to the right to turn big units into a larger number of small units**.

The reverse of this is that we'll need a small fraction of a large unit to equal one small unit (**we'll move the decimal to the left to turn small units into a fraction of a larger unit**.)

NOTE:

Some students find this way of converting metric units to be much more difficult than setting up the conversion factor. It's actually worth learning to do and practicing, because it's a much faster way, and makes it easier to do conversions in your head once you learn it. However, if you are fine with the other method and can always get right answers with it, and setting up this kind of statement seems much more difficult to you, you might ask if your instructor cares *how* you get the right answer, or only *that* you get the right answer in unit conversion problems.

a) *To convert grams into milligrams, you must move the decimal point* **3** *places to the* **right**.

Milligrams are smaller than grams. We'll need a larger number of milligrams, so we'll move the decimal to the **right** to make a larger number. Since 1 g = 1000 mg or 10^3 mg, we'll need to move the decimal **3 places** (three powers of ten) **to the right.**

Another way of solving this: rewrite the question as "1 g = ? mg," work the conversion as usual, and find that 1 g = 1000 mg (the decimal has moved **3 places to the right.**)

b) *To convert milliliters into deciliters, you must move the decimal point* **2** *places to the* **left**.

dL are larger than mL. We'll need a smaller number of dL, so we'll move the decimal to the **left** to make a smaller number. Since 1 L = 10 dL (10^1 dL), and 1 L = 1000 mL (10^3 mL), the difference between mL and dL is 10^2. We'll need to move the decimal **2 places** (two powers of ten) **to the left.**

Alternatively, rewrite the question as "1 mL = ? dL," work the conversion as usual, and find that 1 mL = 0.01 dL (the decimal has moved **2 places to the left.**)

c) *To convert centimeters into millimeters, you must move the decimal point* **1** *places to the* **right**.

mm are smaller than cm. We'll need a larger number of mm, so we'll move the decimal to the **right** to make a larger number. Since 1 m = 100 cm (10^2 cm), and 1 m = 1000 mm (10^3 mm), the difference between mm and cm is 10^1 or just 10. We'll need to move the decimal **1 place** (one power of ten) **to the right.**

Or: rewrite the question as "1 cm = ? mm," work the conversion as usual, and find that 1 cm = 10 mm (the decimal has moved **1 place to the right.**)

d) *To convert kilograms into micrograms, you must move the decimal point* **9** *places to the* **right**.

μg are smaller than kg. We'll need a larger number of μg, so we'll move the decimal to the **right** to make a larger number. 10^3 g = 1 kg, so that's three spaces, and then 10^6 μg = 1 g, so that's six more places. We'll need to move the decimal **9 places** (nine powers of ten) **to the right.**

Or: rewrite the question as "1 kg = ? μg," work the conversion as usual, and find that 1 kg = 10^9 μg (the decimal has moved **9 places to the right.**)

1.23 *A textbook is 27.2 cm tall. Express this measurement in each of the following units:*

a) *meters*

$$\mathbf{27.2\ cm \times \frac{1\ m}{10^2\ cm} = 0.272\ m}$$

b) *millimeters*

Since 1 m = 100 cm and 1 m = 1000 mm,
we can also say that 100 cm = 1000 mm.

We can also write this in exponential (power of 10) notation: 10^2 cm = 10^3 mm
Then we use this relationship to make a conversion factor:

$$\mathbf{27.2\ cm \times \frac{10^3\ mm}{10^2\ cm} = 272\ mm}$$

c) *micrometers (use scientific notation)* since 1 m = 10^2 cm and 1 m = 10^6 µm, 10^2 cm = 10^6 µm

$$\mathbf{27.2\ cm \times \frac{10^6\ \mu m}{10^2\ cm} = 27.2 \times 10^4\ \mu m = 2.72 \times 10^5\ \mu m}$$

1.25 *Write a relationship between each of the following units.*

There is usually more than one correct way to write the relationship. You also have the option of either writing out the entire word or using the abbreviation for each unit. (Abbreviations are used in the answers below.)

a) *liters and deciliters* — Since "deci" means 1/10, **1 dL = 1/10 L, or 10 dL = 1 L.**

b) *kilometers and meters* — Since "kilo" means 1000 or 10^3, **1 km = 1000 m or 1 km = 10^3 m**

c) *grams and micrograms* — "Micro" means 1/1,000,000, and can also be written as $1/10^6$ or 10^{-6}. The following are all mathematically equivalent ways of writing the relationship:
1/1,000,000 g = 1 µg or 1 g = 1,000,000 µg
$1/10^6$ g = 1 µg or 1 g = 10^6 µg
10^{-6} g = 1 µg

d) *millimeters and centimeters* — The simplest solution here is to write the relationships between each of these units and the root unit, meters:

1000 mm = 1 m and 100 cm = 1 m, so **1000 mm = 100 cm** (or, in scientific notation, **10^3 mm = 10^2 cm**.) We can also simplify this to show the relationship of the smaller unit (mm) to **one** of the larger unit (cm), a more familiar representation of the relationship, by dividing both sides by 100:
10 mm = 1 cm

Solutions to Section 1.4 Core Problems

1.27 *Express each of the following relationships as a pair of conversion factors.*

Remember that for any statement that two quantities are equal, you can write two possible ratios to use as conversion factors.

a) *There are 1.609 kilometers in a mile.*

$$\mathbf{\frac{1.609\ km}{1\ mi}\ \text{or}\ \frac{1\ mi}{1.609\ km}}$$

b) *An object that weighs one kilogram weighs 2.205 pounds.*

$$\frac{\textbf{1 kg}}{\textbf{2.205 pounds}} \textbf{ or } \frac{\textbf{2.205 pounds}}{\textbf{1 kg}}$$

c) *A deciliter equals 100 milliliters.*

$$\frac{\textbf{1 dL}}{\textbf{100 mL}} \textbf{ or } \frac{\textbf{100 mL}}{\textbf{1 dL}}$$

1.29 *What (if anything) is wrong with each of the following conversion factor setups?*

a) $5.3\ \cancel{\text{cm}} \times \frac{100\ \text{m}}{1\ \cancel{\text{cm}}} = 5300\ \text{m}$ **The conversion factor is wrong. cm are smaller than m, so it's 100 cm = 1 m, not 100m = 1 cm.**

b) $5.3\ \cancel{\text{cm}} \times \frac{100\ \cancel{\text{cm}}}{1\ \text{m}} = 5300\ \text{m}$ **The conversion factor has the correct relationship (it's true that 100 cm = 1 m), but the units don't cancel properly. This problem would give an answer in cm^2/m, not m. The conversion factor needs to be flipped to 1 m/100 cm.**

1.31 *Carry out each of the following unit conversions, using only the information given in the problem. Use a conversion factor to solve each problem.*

a) *There are 32 fluid ounces in a quart. If a beverage container holds 19.7 fluid ounces of juice, how many quarts of juice does it hold?*

We're told that 32 fl oz = 1 qt, so this will be the source for our conversion factor. In this and all problems that follow, the abbreviation "sf" means "significant figures."

$$\textbf{19.7 fl oz} \times \frac{\textbf{1 qt}}{\textbf{32 fl oz}} = \textbf{0.615625 qt (calculator answer)} = \textbf{0.616 qt (rounded to 3 sf)}$$

b) *One ounce equals 28.35 g. If a beaker weighs 4.88 ounces, how many grams does it weigh?*

$$\textbf{4.88 oz} \times \frac{\textbf{28.35 g}}{\textbf{1 oz}} = \textbf{138.348 g (calculator answer)} = \textbf{138 g (rounded to 3 sf)}$$

c) *There are 236.6 mL in one cup. If a flask holds 247 mL of water, how many cups of water does it hold?*

$$\textbf{247 mL} \times \frac{\textbf{1 cup}}{\textbf{236.6 mL}} = \textbf{1.043956 cup (calculator answer)} = \textbf{1.04 cup (rounded to 3 sf)}$$

d) *There are 39.37 inches in one meter. If a woman is 1.52 m tall, what is her height in inches?*

$$\textbf{1.52 m} \times \frac{\textbf{39.37 in}}{\textbf{1 m}} = \textbf{59.8424 in (calculator answer)} = \textbf{59.8 in (rounded to 3 sf)}$$

e) A typical member of the species Vibrio cholerae (the bacterium that causes cholera) is 8.8×10^{-6} m long. How long is this in inches, given that there are 39.37 inches in a meter?

$$8.8 \times 10^{-6} \text{ m} \times \frac{39.37 \text{ in}}{1 \text{ m}} = 3.46456 \times 10^{-4} \text{ in (calculator answer)}$$

$$= 3.5 \times 10^{-4} \text{ in (rounded to 2 sf)}$$

Solutions to Section 1.5 Core Problems

1.33 *Fred does not know how many ounces are in a ton, but he knows how each of these units is related to a pound. Propose a strategy that Fred could use to convert a weight in ounces into the corresponding weight in tons.*

He could convert the weight in ounces into pounds (using the relationship between ounces and pounds), then convert the weight in pounds into tons (using the relationship between pounds and tons).

1.35 *Carry out each of the following conversions, using a combination of two conversion factors for each conversion.*

In each case, we find that we can't go directly from the given unit to the final unit, and we must convert to an intermediate unit first. We use each conversion factor in such a way that the units cancel properly. Then we round our answer appropriately (the abbreviation "sf" means "significant figures.")

a) Convert 0.235 km into feet, using the fact that one km equals 0.621 mile and one mile equals 5280 feet.

$$0.235 \text{ km} \times \frac{0.621 \text{ mi}}{1 \text{ km}} \times \frac{5280 \text{ feet}}{1 \text{ mile}} = 770.5368 \text{ ft (calculator answer)}$$

$$= 771 \text{ ft (rounded to 3 sf)}$$

b) Convert 0.175 g into grains, using the fact that one ounce equals 28.35 g and one ounce equals 437.5 grains.

$$0.175 \text{ g} \times \frac{1 \text{ ounce}}{28.35 \text{ g}} \times \frac{437.5 \text{ grains}}{1 \text{ ounce}} = 2.7006173 \text{ grains (calculator answer)}$$

$$= 2.70 \text{ grains (3 sf)}$$

There's another option in this particular problem: since both conversion factors are in terms of one ounce, we can also conveniently say that 28.35 g = 437.5 grains. (This is less convenient when the two conversion factors are in terms of different amounts of the intermediate unit.)

$$0.175 \text{ g} \times \frac{437.5 \text{ grains}}{28.35 \text{ g}} = 2.70 \text{ grains (3 sf)}$$

c) Convert 25 teaspoons into fluid ounces, using only the relationships in Table 1.5.

Table 1.5 gives the relationship between each of these units and mL, so mL is our intermediate unit.

$$25 \text{ teaspoons} \times \frac{4.93 \text{ mL}}{1 \text{ teaspoon}} \times \frac{1 \text{ fl oz}}{29.57 \text{ mL}} = 4.168076 \text{ fl oz (calculator answer)}$$

$$= 4.2 \text{ fl oz (2 sf)}$$

d) *Convert* 5.1×10^7 *kg into tons, using the fact that there are 2.205 pounds in a kilogram and 2000 pounds in a ton.*

$$5.1 \times 10^7 \text{ kg} \times \frac{2.205 \text{ pounds}}{1 \text{ kg}} \times \frac{1 \text{ ton}}{2000 \text{ pounds}}$$

$$= 5.62275 \times 10^4 \text{ tons (calculator answer)} = 5.6 \times 10^4 \text{ tons (2 sf)}$$

1.37 *A quatern is an English unit of volume. There are eight quaterns in one quart. Using this fact, the information in Table 1.5, and your knowledge of metric units, convert 0.125 L into quaterns.*

Table 1.5 has the relationship between quarts and liters:

$$0.125 \text{ L} \times \frac{1.057 \text{ quart}}{1 \text{ L}} \times \frac{8 \text{ quaterns}}{1 \text{ quart}} = 1.057 \text{ quaterns (calculator answer)}$$

$$= 1.06 \text{ quaterns (3 sf)}$$

Solutions to Section 1.6 Core Problems

1.39 *Write each of the following as a pair of conversion factors.*

a) *A pound of watermelon costs 22¢.*

$$\frac{1 \text{ pound}}{22 \text{ cents}} \text{ or } \frac{22 \text{ cents}}{1 \text{ pound}}$$

b) *There is 0.8 mg of antihistamine in each teaspoon of cold medicine.*

$$\frac{0.8 \text{ mg drug}}{1 \text{ teaspoon cold medicine}} \text{ or } \frac{1 \text{ teaspoon cold medicine}}{0.8 \text{ mg drug}}$$

c) *The flow rate of an intravenous solution is 65 mL per hour.*

$$\frac{65 \text{ mL}}{1 \text{ hour}} \text{ or } \frac{1 \text{ hour}}{65 \text{ mL}}$$

d) *The density of lead is 11.3 g/mL.*

$$\frac{11.3 \text{ g lead}}{1 \text{ mL lead}} \text{ or } \frac{1 \text{ mL lead}}{11.3 \text{ g lead}}$$

1.41 *Orange juice contains 2.10 mg of potassium per mL of juice.*

A compound unit given in a problem usually becomes a conversion factor.

Our two options here are $\frac{2.10 \text{ mg potassium}}{1 \text{ mL juice}}$ or $\frac{1 \text{ mL juice}}{2.10 \text{ mg potassium}}$

a) *If you want to get your daily recommended amount of potassium (3500 mg) from orange juice, how many milliliters of juice must you drink?*
To get 3500 mg of potassium,

$$\textbf{3500 mg potassium} \times \frac{\textbf{1 mL juice}}{\textbf{2.10 mg potassium}} = \textbf{1666.667 mL (calculator)}$$

$$= \textbf{1700 mL (2 sf)}$$

b) *If you drink one cup (236.5 mL) of orange juice, how many milligrams of potassium will you consume?*
It's convenient that the problem tells us how many mL are in a cup, or we would have to do that conversion as well. As it is, we can start from a unit that appears in our conversion factor:

$$\textbf{236.5 mL juice} \times \frac{\textbf{2.10 mg potassium}}{\textbf{1 mL juice}} = \textbf{496.65 mg potassium (calculator)}$$

$$= \textbf{497 mg (3 sf)}$$

NOTE:
Because the conversion factor is not exact, and is only given with 3 significant figures, our answer is limited to 3 sf, even though the given measurement has 4.

1.43 *A car gets 31 miles per gallon of gasoline.*

Again, our compound unit will probably be used as a conversion factor. In this problem, the conversion factor only has two significant figures, so that could limit the sf in some of our answers.

Our two options are: $\frac{\text{31 miles}}{\text{1 gallon}}$ or $\frac{\text{1 gallon}}{\text{31 miles}}$

a) *How far can you travel on 8.2 gallons of gasoline?*

$$\textbf{8.2 gallons} \times \frac{\textbf{31 miles}}{\textbf{1 gallon}} = \textbf{254.2 miles (calculator)} = \textbf{250 miles (2 sf)}$$

b) *How many gallons of gasoline do you need to go 186 miles?*

$$\textbf{186 miles} \times \frac{\textbf{1 gallon}}{\textbf{31 miles}} = \textbf{6 gallons (calculator)} = \textbf{6.0 gallons (2 sf)}$$

1.45 *A doctor prescribes 0.2 g of carbamazepine to be taken three times daily. The pharmacy supplies tablets that contain 100 mg of carbamazepine per tablet. How many tablets should be taken per dose?*

Our two conversion factors are $\frac{\text{100 mg}}{\text{1 tablet}}$ or $\frac{\text{1 tablet}}{\text{100 mg}}$

One dose is 0.2 g; we'll have to convert that to mg in our calculation. (We could, alternatively, have changed the conversion factors into g.)

$$\mathbf{0.2\ g \times \frac{1000\ mg}{g} \times \frac{1\ tablet}{100\ mg} = 2\ tablets}$$

NOTE:
About significant figures (sf) in this problem: in this case, both the dose and the conversion factor, as given, have only 1 sf, and tablets are whole-number countable objects anyway. If the calculation had given us a decimal value, we would have rounded to the nearest tablet (or possibly ½ tablet, since some tablets can be cut in half.) In reality, though, pharmaceutical preparations are much more tightly controlled than this; while the number given (100 mg) has only one sf, it's probably correct to at least three, and possibly four, sf (that is, it may be 100.0 mg in reality).

1.47 *A doctor prescribes 300 mg of Pediazole to be taken every six hours. The pharmacy supplies a solution that contains 200 mg of Pediazole per 5 mL of solution. How many mL of solution should be taken per dose?*

Conversion factors: $\frac{200\ mg}{5\ mL}$ or $\frac{5\ mL}{200\ mg}$

One dose is 300 mg: $\mathbf{300\ mg \times \frac{5\ mL}{200\ mg} = 7.5\ mL}$

While the answer should technically be rounded to 1 sf, there's no reason not to measure the dose to 7.5 mL instead of rounding. (As in problem 1.45, both the dose and the conversion factor, as given, have only 1 sf. In reality, though, pharmaceutical preparations are much more tightly controlled than this.)

1.49 *An intravenous solution is infused at a rate of 135 mL per hour. At this rate, how long will it take to infuse 500 mL of solution? (Give your answer to three significant figures.)*

Conversion factors: $\frac{135\ mL}{1\ hour}$ or $\frac{1\ hour}{135\ mL}$

$$500\ mL \times \frac{1\ hour}{135\ mL} = 3.70\ hours$$

Since we don't normally express parts of hours as decimals, but as minutes, we take the 0.70 hours and convert to minutes:

$$0.70\ hours \times \frac{60\ minutes}{1\ hour} = 42\ minutes$$

So the answer, to three sf, is **3 hours 42 minutes.**

1.51 *A doctor prescribes chloramphenicol at a dosage of 25 mg/kg every 12 hours for an infant who weighs 3160 g. This medication is available as a solution containing 500 mg in each 5 mL of liquid. How many milliliters of liquid should the infant receive every 12 hours?*

This problem is more involved. First, let's go back to basics and set up the problem: "How many mL should the infant receive?" (The 12 hours is a distracter—once we realize that the dosage is also given for every 12 hours, we can ignore that.) The only other measurement given in the problem is the mass of the infant, so let's start with that and see where it goes:

3160 g body weight = ? mL liquid

Then, let's pick out the unit relationships given in the problem. We have two sets of conversion factors (abbreviating chloramphenicol as chlor.):

$$\frac{25\text{ mg chlor.}}{1\text{ kg body weight}} \text{ or } \frac{1\text{ kg body weight}}{25\text{ mg chlor.}} \quad \text{and} \quad \frac{500\text{ mg chlor.}}{5\text{ mL liquid}} \text{ or } \frac{5\text{ mL liquid}}{500\text{ mg chlor.}}$$

The remaining piece of information is the weight of the infant. Notice that the dosage is given in mg chlor./kg (of body weight), so we'll have to convert the infant's weight to kg:

$$\mathbf{3160\text{ g body weight} \times \frac{1\text{ kg}}{1000\text{ g}} \times \frac{25\text{ mg chlor.}}{1\text{ kg body weight}} \times \frac{5\text{ mL liquid}}{500\text{ mg chlor.}} = 0.79\text{ mL liquid}}$$

The answer is rounded to 2 significant figures to be consistent with the 25 mg/kg dosage given in the problem.

1.53 *A piece of cork weighs 7.545 g and has a volume of 31.7 mL.*

a) *Calculate the density of cork.* Density = mass/volume, so

$$\mathbf{\frac{7.545\text{ g}}{31.7\text{ mL}} = 0.238013\text{ g/mL (calculator answer)} = 0.238\text{ g/mL (3 sf)}}$$

b) *What is the specific gravity of cork, based on your answer to part a?*

Specific gravity is the ratio of the density of a substance to the density of water (assuming that the density of water is very close to 1 g/mL and doesn't limit significant figures):

$$\mathbf{\frac{0.238\text{ g/mL}}{1\text{ g/mL}} = 0.238}$$

Remember that in practice, the only difference between density and specific gravity is that density has units (g/mL) and specific gravity doesn't. The two terms tend to be used interchangeably.

1.55 *The density of concrete is 2.8 g/mL.*

When the density of a substance is given in the problem, it's usually used as a conversion factor.

In this case our two conversion factors are: $\frac{2.8\text{ g}}{1\text{ mL}}$ or $\frac{1\text{ mL}}{2.8\text{ g}}$

a) *If the volume of a piece of concrete is 235 mL, what is its mass?*

$$\mathbf{235\text{ mL} \times \frac{2.8\text{ g}}{1\text{ mL}} = 658\text{ g (calculator answer)} = 660\text{ g (2 sf)}}$$

b) *If the mass of a piece of concrete is 1600 g, what is its volume?*

$$\mathbf{1600\ g \times \frac{1\ mL}{2.8\ g} = 571.42857\ mL\ (calculator) = 570\ mL\ (2\ sf)}$$

Solutions to Section 1.7 Core Problems

1.57 a) *When aspirin is heated, it melts at 275°F. Convert this to a Celsius temperature.*

$$(°F - 32) \div 1.8 = °C$$

$$(275°F - 32) \div 1.8 = \mathbf{135°C}$$

b) *Isopropyl alcohol boils when it is heated to 82.5°C. Convert this to a Fahrenheit temperature.*

$$(°C \times 1.8) + 32 = °F$$

$$(82.5°C \times 1.8) + 32 = \mathbf{181°F}$$

c) *Convert the boiling temperature of isopropyl alcohol to a Kelvin temperature.*

$$82.5°C + 273 = \mathbf{356\ K}$$

NOTE:
By convention, K doesn't get a degree sign like °F and °C.

1.59 *What is the freezing point of water in Celsius and in Fahrenheit?*

These are important values to memorize, because you can often make a good estimate of a temperature conversion without doing the math if you know the four values in each system in Figure 1.15 (this will often be enough to let you choose the right answer in a multiple-choice problem without using the equations). The freezing point of water is **0°C or 32°F**.

Solutions to Concept Questions

* indicates more challenging problems.

1.61 *You are getting a new washing machine. Each of the people below asks you, "How large is it?" In each case, tell whether the person is likely to be asking you for a mass, a volume, or a distance (length, width, or height), and explain your answer.*

a) *The person who is calculating the shipping charge.*

Shipping charges are usually calculated by weight, so this person is probably most interested in the weight or mass of the package. (However shippers usually also have size limits that specify the maximum length in the largest direction, or the maximum sum of length + height + width, so they might also ask for those dimensions as well.)

b) *The person who is designing the arrangement of appliances in the laundry room.*

This person is going to be most interested in the width (across the machine), but may also be interested in the distance from front to back (if trying to make sure there will be plenty of room to stand in front of the machine and sort laundry) and the height (if installing a countertop to be level with the top of the machine.) Since the floor will hold the weight of the machine no matter what, the weight isn't important.

c) *The person who will be using the machine to wash clothing.*

This person is likely to be interested in, not the volume occupied by the entire machine, but the volume of clothes the machine can wash at one time (the capacity), the volume of water it will use for each load, or the volume of the tub.

d) *The people who have to carry the washing machine up a flight of stairs.*

These people are going to be most interested in the weight, so they know how many people and what equipment will be appropriate for the job. Length, width and height dimensions may also be important – for example, to be sure the machine will fit through a specific doorway or turn on a stairway.

1.63 *What is a unit, and why are units important in health care?*

From Section 1.1, **a unit is a measurement whose size everyone agrees upon, so it can be used as a basis for other measurements.** Units are important in health care because we express so many things in terms of numbers—patients' height, weight, age, temperature, heart rate, blood pressure, etc; lab results; dosages of medications; etc. Unless two people are using the same units when they compare or communicate these things, the numbers are meaningless.

1.65 *The most common mass units in the metric system are the gram, the kilogram, and the milligram. The mass of a calculator could be expressed using any of these units, but in practice it would normally be expressed in grams. Why is this?*

The mass of a calculator is closer to a gram than it is to a kilogram or a milligram, so giving the mass in terms of the number of grams is more convenient than using other units. The mass of a calculator would be only a fraction of a kg, and would be tens or hundreds of thousands of milligrams. A rough rule is that we are most likely to use the unit that gives us a number between 1 and 1000 or so, just because it keeps us from having to use either a decimal or an inconveniently large value.

1.67 *At her annual checkup on May 2nd, Katie was 127.2 cm tall. On May 6th, Katie went back to her pediatrician and was measured again. This time, she was 127.4 cm tall. Is it safe to conclude that Katie grew between the two visits to the doctor? Why or why not?*

The last digit of a measurement is uncertain, the best estimate we can make at the time using the particular tool at our disposal. **We can't really conclude that Katie's height increased between the two visits, because both measurements have an uncertainty of ±1 in the last digit (in this problem, ± 0.1 cm)—it's possible that her true height at the first measurement was 127.3 (or 127.1) and that her true height at the second measurement was 127.3 (or 127.5).** It's possible that she grew by 0.2 cm in four days, or even by as much as 0.4 cm, or not at all. We can't tell for sure from these measurements.

1.69 *A friend who is not yet familiar with the metric system tells you that 5 L is the same volume as 0.005 mL. How would you explain to your friend that this is incorrect?*

First of all, most people in the US are familiar enough with the metric system to have a pretty good idea of what 2 L looks like (a two-liter pop bottle) and what a cubic centimeter (cc or cm^3) looks like (a common unit used for doses of medications). You could explain to your friend that one mL is the same as one cc, so **mL are small and L are large (relative to each other) and it doesn't make sense that a tiny fraction of a tiny mL would be equal to 5 L (that is, two and a half pop bottles).**

Another way of explaining this is to learn what the prefixes mean. The "m" in mL stands for 1/1000 of the base unit, so 1 mL = 1/1000 L, which means (through a tiny bit of algebra) that 1000 mL = 1 L.

Then you can show the math: $$\mathbf{5\,L \times \frac{1000\,mL}{1\,L} = 5000\,mL}$$

1.71 *You have two pieces of aluminum, one larger than the other. Which of the following will be the same for both pieces, and which will be different?*

a) the mass — Mass is an extensive property. Since the mass of a sample depends on how much of the substance is present, **the bigger sample will have more mass.**

b) the volume — Volume is an extensive property. Since the volume of a sample depends on how much of the substance is present, **the bigger sample will have a greater volume.**

c) the density — Density is an intensive property. The bigger piece will have both more mass and a greater volume, but **density, the ratio of mass to volume, will be the same for both pieces.**

d) the specific gravity — Specific gravity is calculated from (and used almost interchangeably with) density, so it's also an intensive property. **Both pieces will have the same specific gravity.**

1.73 *Mr. Huynh goes out for a long walk on a hot summer day. On the way home, he stops at his doctor's office for a routine urinalysis. The lab finds that the specific gravity of his urine is above the normal range. Suggest a reasonable explanation for this.*

Dehydration can cause the urine to be more concentrated than usual. Generally, the more concentrated a solution is, the greater its density/specific gravity. Probably, if Mr. Huynh drank some water to replace what he may have sweated away, his urine would be in the normal range again.

1.75 *In each of the pictures that follow, an object has been placed into a container of alcohol. The density of alcohol is 0.78 g/mL. Match each picture with one of the following descriptions.*

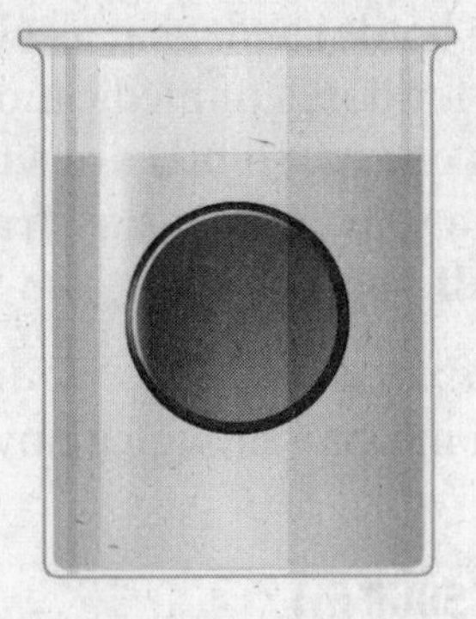

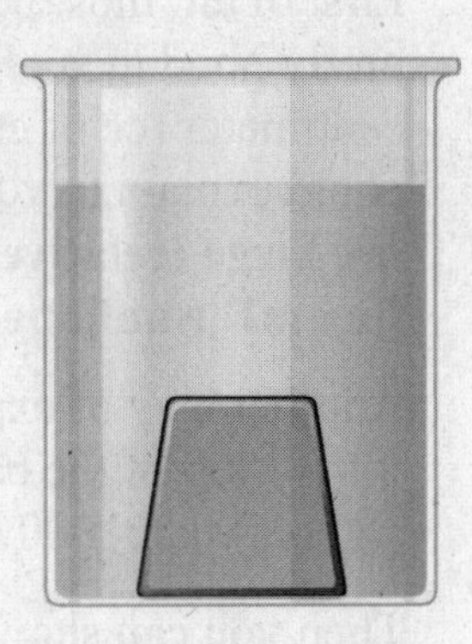

In any combination of two liquids or a liquid and a solid (as long as the two are not soluble in each other), the less dense of the two will rise to the top and the more dense will go to the bottom. A solid object with a greater density than alcohol will sink to the bottom, and a solid object that is less dense than alcohol will float to the top. If the two have the same density, the object will essentially stay wherever it is placed in the liquid (the technical term for this is that it will be "neutrally buoyant.")

a) *The density of the object is 0.89 g/mL.* **Picture #3: The object will sink.**

b) *The density of the object is 0.71 g/mL.* **Picture #2: The object will float.**

c) *The density of the object is 0.78 g/mL.* **Picture #1: The object will stay where it is placed and will neither sink nor float to the top on its own.**

Solutions to Summary and Challenge Problems

* indicates more challenging problems.

1.77 *A company has built a new device that can monitor blood sugar levels. The company tests the device by measuring the blood sugar level three times using the same blood sample, and obtains the following results:*

Test 1: 77.3 mg/dL *Test 2: 79.7 mg/dL* *Test 3: 78.9 mg/dL*

Calculate the average of these three tests, and round your answer to a reasonable number of decimal places. Remember that the answer should have only one uncertain digit.

To calculate the average, add the three values together and divide by three:
(77.3 + 79.7 + 78.9)/3 = 78.63333333…
The highest and lowest values are 2.4 mg/dL apart, so there is uncertainty in the first digit to the left of the decimal place. Therefore we round the average to **79 mg/dL.**

1.79 *Use the information in Tables 1.4 through 1.6 and your knowledge of metric-to-metric conversions to carry out each of the following unit conversions.*

a) *A rod is an old English unit of distance. There are 320 rods in one mile. If a road is 618 rods long, how long is it in kilometers?*

Our original measurement is 618 rods, the length of the road. We're asked to figure out the distance in km, so our setup becomes:
618 rods = ? km
We don't have a relationship between rods and km, but we're told that there are 320 rods in one mile, and we can look up the relationship between km and miles (1 mile = 1.609 km). We can write two conversion factors for each relationship:

$$\frac{320\text{ rods}}{1\text{ mile}} \quad \text{and} \quad \frac{1\text{ mile}}{320\text{ rods}}$$

$$\frac{1\text{ mile}}{1.609\text{ km}} \quad \text{and} \quad \frac{1.609\text{ km}}{1\text{ mile}}$$

Our original measurement is in rods, so we will first use the conversion factor with rods in the denominator; then, to cancel out miles and end up with kilometers, we'll use the conversion factor with miles in the denominator and kilometers in the numerator.

$$\mathbf{618\text{ rods} \times \frac{1\text{ mile}}{320\text{ rods}} \times \frac{1.609\text{ km}}{1\text{ mile}} = 3.10738125\text{ km (calculator answer)}}$$

The original measurement has 3 significant figures, so we round our answer to **3.11 km** (the value 320 rods = 1 mile is a definition and therefore does not affect significant figures.)

b) *A dram is an old English unit of weight. There are 16 drams in one ounce. If a bottle of medicine weighs 3.26 drams, what is its mass in grams?*

The measurement in the problem is 3.26 drams. We're asked to figure out the mass in grams:
3.26 drams = ? g
The problem tells us that 1 ounce = 16 drams, and we can look up the relationship between ounces and grams. We can write two conversion factors for each relationship:

$$\frac{16\text{ drams}}{1\text{ ounce}} \quad \text{and} \quad \frac{1\text{ ounce}}{16\text{ drams}}$$

$$\frac{1\text{ ounce}}{28.35\text{ g}} \quad \text{and} \quad \frac{28.35\text{ g}}{1\text{ ounce}}$$

We select the conversion factors that cancel out drams and ounces, leaving an answer in grams:

$$\mathbf{3.26\text{ drams} \times \frac{1\text{ ounce}}{16\text{ drams}} \times \frac{28.35\text{ g}}{1\text{ ounce}} = 5.7763125\text{ g (calculator answer)}}$$

The original measurement has 3 significant figures, so we round our answer to **5.78 g** (the relationship 16 drams = 1 ounce is a definition and therefore does not affect significant figures.)

c) *A troy ounce is a unit of weight, used to measure the weight of gold and other precious metals. One troy ounce equals 1.097 "normal" ounces. Convert 31.5 grams into troy ounces. (The "normal" weight units are called avoirdupois weights.)*

The measurement in the problem is 31.5 grams. We're asked to figure out weight in troy ounces:
31.5 g = ? troy ounces
The problem tells us that 1 troy ounce = 1.097 ounces, and we can look up the relationship between ounces and grams. We can write two conversion factors for each relationship:

$$\frac{1 \text{ troy ounce}}{1.097 \text{ ounce}} \quad \text{and} \quad \frac{1.097 \text{ ounce}}{1 \text{ troy ounce}}$$

$$\frac{1 \text{ ounce}}{28.35 \text{ g}} \quad \text{and} \quad \frac{28.35 \text{ g}}{1 \text{ ounce}}$$

We select the conversion factors that cancel out grams and ounces, leaving an answer in troy ounces and then rounding to 3 significant figures:

$$31.5 \text{ g} \times \frac{1 \text{ ounce}}{28.35 \text{ g}} \times \frac{1 \text{ troy ounce}}{1.097 \text{ ounce}} = 1.012863365 \text{ troy ounces (calculator answer)}$$

$$= \mathbf{1.01 \text{ troy ounces}} \text{ (rounded)}$$

d) *Until recently, large volumes were measured in British gallons in the United Kingdom. One British gallon equals 1.201 "normal" (American) gallons. Convert 26.5 liters into British gallons.*

We don't have a conversion factor between liters and British gallons or between liters and gallons. From Table 1.5, we know that 1 L = 1.057 quarts, and we also know that 4 quarts = 1 American gallon. Using the appropriate conversion factors from each of these equalities, we convert from L to British gallons and round appropriately (3 sf):

$$\mathbf{26.5 \text{ L} \times \frac{1.057 \text{ quart}}{1 \text{ L}} \times \frac{1 \text{ gallon}}{4 \text{ quarts}} \times \frac{1 \text{ British gallon}}{1.201 \text{ gallon}}}$$

$$= \mathbf{5.830661948 \text{ British gallons (calculator answer)}}$$

$$= \mathbf{5.83 \text{ British gallons}} \text{ (rounded)}$$

e) *Using the fact that one teaspoon equals 4.93 milliliters, convert 15 teaspoons into deciliters.*

We're given 4.93 mL = 1 tsp. We write separate relationships for mL and dL with L:
1000 mL = 1 L and 10 dL = 1 L
We can shorten the problem by combining those two relationships into one:

1000 mL = 10 dL

Then $$\mathbf{15\ tsp \times \frac{4.93\ mL}{1\ tsp} \times \frac{10\ dL}{1000\ mL} = 0.7395\ dL\ (calculator\ answer)}$$

$$\mathbf{= 0.74\ dL\ (rounded)}$$

f) *Using the fact that one cup equals 2.366 dL, convert 5 cups into liters.*

We're given 1 cup = 2.366 dL, and we know that 10 dL = 1 L:

$$\mathbf{5\ cups \times \frac{2.366\ dL}{1\ cup} \times \frac{1\ L}{10\ dL} = 1.183\ (calculator\ answer) = 1\ L\ (rounded\ to\ one\ sf)}$$

1.81 *The total mass of living organisms on Earth is estimated to be 3.6×10^{14} kg. Convert this mass into tons. (1 ton = 907 kg).*

$$3.6 \times 10^{14}\ \text{kg} \times \frac{1\ \text{ton}}{907\ \text{kg}} = 3.969128997 \times 10^{11}\ \text{tons (calculator answer)}$$

$$\mathbf{= 4.0 \times 10^{11}\ tons\ (rounded\ to\ two\ sf)}$$

1.83 *The store is advertising 3 pounds of oranges for $1.00.*

a) *If you buy 8.5 pounds of oranges, how much money will you spend?*

First, rewrite the relationship between pounds and cost to make it look more like our other conversion factors (this isn't necessary but might make it easier to see the setup):

3 pounds = 1.00 dollars

Then,

$$\mathbf{8.5\ pounds \times \frac{1.00\ dollars}{3\ pounds} = 2.8333333333\ dollars\ (calculator\ answer)}$$

Since prices are almost always rounded to the nearest penny, regardless of the number of sf in the original measurement, we might write this as **$2.83**. However, since there were only 2 sf in 8.5 pounds, we might find at the checkout that we were actually spending a few cents more or less (once the oranges are weighed on a more precise scale).

b) *If you spend $2.75 on oranges, how many pounds of oranges will you buy?*

NOTE:

We need to use the other form of the conversion factor this time (pounds on top, rather than on the bottom as in the last problem):

$$\mathbf{2.75\ dollars \times \frac{3\ pounds}{1.00\ dollar} = 8.25\ pounds\ (calculator\ answer,\ already\ 3\ sf)}$$

*1.85 *An intravenous solution is infused at a rate of 0.75 mL per minute. At this rate, how many hours will it take to infuse a total of 100 mL of solution?*

Even though this looks like a problem we haven't dealt with before, it's the same process.

100 mL = ? hours

We're given the relationship between mL and minutes (0.75 mL = 1 min), and we know that 60 min = 1 hour. So our setup becomes:

$$\mathbf{100\ mL \times \frac{1\ min}{0.75\ mL} \times \frac{1\ hour}{60\ min} = 2.2222222222\ hours\ (calculator\ answer)}$$

The original measurement (100 mL) has only 1 sf, so we round our answer to **2 hours** (and we aren't surprised if it takes a little more or less.)

*1.87 *Chicken broth contains 6.27 mg of sodium per milliliter. The U.S. government recommends that adults consume no more than 2400 mg of sodium per day. How many cups of chicken broth would you need to eat to consume 2400 mg of sodium? (See Table 1.5 for useful information.)*

The problem gives us two possible conversion factors, 6.27 mg/mL and 2400 mg/day.
Our setup is: 2400 mg = ? cups
From Table 1.5, 1 cup = 2.366 dL, and we know that 10 dL = 1000 mL. So:

$$\mathbf{2400\ mg \times \frac{1\ mL}{6.27\ mg} \times \frac{10\ dL}{1000\ mL} \times \frac{1\ cup}{2.366\ dL} = 1.617815383\ cups\ (calculator\ answer)}$$

$$\mathbf{= 1.6\ cups\ (rounded\ to\ 2\ sf)}$$

So just a cup and a half of chicken broth contains an entire day's limit of sodium, without eating any other foods that contain sodium.

*1.89 *At the local store, a gallon of milk costs $3.29, a 0.75 L bottle of wine costs $9.49, a box containing twelve cans of cola costs $5.89, and a box containing 24 bottles of drinking water costs $5.99. (A can of cola contains 12 fluid ounces and a bottle of water contains 500 mL.) Calculate the cost of one fluid ounce of each beverage, and then rank the beverages in order from lowest cost to highest cost per fluid ounce.*

For each beverage, we calculate the cost of one fluid ounce (1 fl oz). There are shorter pathways for some of these calculations, but let's use the relationships in Table 1.5, assuming those are the ones we know.
Our setup in each case is: 1 fl oz = ? dollars
Milk (for this one we need to know also that 4 quarts = 1 gallon):

$$\mathbf{1\ fl\ oz \times \frac{29.57\ mL}{1\ fl\ oz} \times \frac{1\ quart}{946.4\ mL} \times \frac{1\ gallon}{4\ quarts} \times \frac{3.29\ dollars}{1\ gallon}}$$

$$\mathbf{= 0.026\ dollars\ (or\ 2.6\ cents\ per\ fl\ oz)}$$

Wine:

$$1\ \text{fl oz} \times \frac{29.57\ \text{mL}}{1\ \text{fl oz}} \times \frac{1\ \text{L}}{1000\ \text{mL}} \times \frac{9.49\ \text{dollars}}{0.75\ \text{L}} = 0.374\ \text{dollars (or 37.4 cents per fl oz)}$$

Cola:

$$1\ \text{fl oz} \times \frac{1\ \text{can}}{12\ \text{fl oz}} \times \frac{5.89\ \text{dollars}}{12\ \text{cans}} = 0.041\ \text{dollars (or 4.1 cents per fl oz)}$$

Water:

$$1\text{ fl oz} \times \frac{29.57\text{ mL}}{1\text{ fl oz}} \times \frac{1\text{ bottle}}{500\text{ mL}} \times \frac{5.99\text{ dollars}}{24\text{ bottles}} = 0.015\text{ dollars (or 1.5 cents per fl oz)}$$

Ranking in order, from lowest to highest cost per fl oz: **water < milk < cola < wine**.
There is a second approach to this that is mathematically the same, though it looks a little different. Instead of setting up the problem as "1 fl oz = ? dollars," we can set each problem up to get the answer in the units of dollars/fl oz. To do this, we see that the final units have a cost over a volume, and we start each problem with the cost/volume for that beverage.
Here's the setup for milk:

$$\frac{3.29\text{ dollars}}{1\text{ gallon}} \times \frac{1\text{ gallon}}{4\text{ quarts}} \times \frac{1\text{ quart}}{946.2\text{ mL}} \times \frac{29.57\text{ mL}}{1\text{ fl oz}} = 0.026\text{ dollars/fl oz}$$

In this case the unit we're trying to convert, the volume unit, is on the bottom of the ratio. In the calculation, we cancel gallons, quarts and mL, while keeping dollars. The only difference between this approach and the other one is the step where we multiply by 1 fl oz to get an amount in dollars alone. Mathematically the two are equivalent.

NOTE:
This problem specifically asked us to calculate the cost per fluid ounce, but if we only needed the ranking, we could actually have calculated the cost per any unit volume and come up with the same order (that is, the most expensive on a per-ounce basis is also going to be the most expensive per mL, L, gallon, etc.)

1.91 *Sharon has inherited a set of tableware. She thinks that it is made of silver (density = 10.5 g/mL), but she knows that it could also be made from aluminum (density = 2.7 g/mL), stainless steel (density = 8.0 g/mL), or nickel (density = 8.9 g/mL). Sharon carries out the following measurements on a fork from the set. From her data, help her identify what metal the tableware is made from.*

Mass of fork = 45.718 g *Volume of fork = 5.7 mL*

$$\text{Density} = \frac{\text{mass}}{\text{volume}} = \frac{45.718\text{ g}}{5.7\text{ mL}} = 8.0\text{ g/mL}$$

The tableware is probably stainless steel (and certainly cannot be pure silver, aluminum or nickel).

1.93 *Potential blood donors are screened for iron deficiency. If a donor's blood density is below 1.05 g/mL, his or her iron level is too low. A 1.50 mL sample of blood from a prospective donor weighs 1.593 g. Would this person be accepted as a blood donor?*

$$\text{Density} = \frac{\text{mass}}{\text{volume}} = \frac{1.593\text{ g}}{1.50\text{ mL}} = 1.06\text{ g/mL (rounded to 3 sf)}$$

This person is eligible to donate based on blood density.

*1.95 *A bottle contains 22.6 fluid drams of alcohol. Calculate the mass of alcohol in the bottle using the following information:*

1 fluid ounce = 8 fluid drams *density of alcohol = 0.79 g/mL*

1 fluid ounce = 29.57 mL

The desired unit for mass is not specified, so we'll go for grams:
22.6 fluid drams = ? g

$$\mathbf{22.6\ fl\ dr} \times \frac{\mathbf{1\ fl\ oz}}{\mathbf{8\ fl\ dr}} \times \frac{\mathbf{29.57\ mL}}{\mathbf{1\ fl\ oz}} \times \frac{\mathbf{0.79\ g}}{\mathbf{1\ mL}} = \mathbf{65.9928475\ g\ (calculator\ answer)}$$

$$= \mathbf{66\ g\ (rounded\ to\ 2\ sf)}$$

Chapter 2

2

Atoms, Elements, and Compounds

Solutions to Section 2.1 Core Problems

2.1 *Which of the following will make a homogeneous mixture, and which will make a heterogeneous mixture? Assume that the two components are stirred well.*

a) A teaspoon of salt and a cup of water

Homogeneous mixture: the salt will dissolve completely in the water to form a solution, and the composition of the solution will be the same throughout.

b) A teaspoon of pepper and a cup of water

Heterogeneous mixture: the pepper will not dissolve in the water, but will instead remain as flakes of pepper. The composition of the mixture will be different in different places.

2.3 *Give an example of a substance (other than salt) that could form a homogeneous mixture if it were mixed with water.*

We're looking for something that will dissolve cleanly in water. Some examples are **alcohol, food coloring, bleach, vinegar, baking soda, hydrogen peroxide.**

2.5 *Which of the following statements describe an intensive property and which describe an extensive property?*

a) The glass in my front window is transparent.

Intensive: the glass is transparent regardless of how big the window is. This is a property of the substance (the glass), not just of this object (this window).

b) The glass in my front window weighs 1.2 kg.

Extensive: smaller pieces of glass would weigh less, larger pieces would weigh more. This property is specific to this object.

c) The glass in my front window is 2 mm thick.

Extensive: windows made of the same glass could be thicker or thinner. This property is specific to this object.

d) The glass in my front window does not dissolve in water.

Intensive: the glass does not dissolve in water. This is a property of the substance, not just of this object.

2.7 *Chalk is a naturally occurring mineral. In ancient times, people discovered that if chalk is heated, it breaks down into a fluffy white powder (called quicklime) and a colorless, odorless gas (called fixed air). A sample of chalk from anywhere on Earth will break down to produce 56% quicklime and 44% fixed air.*

a) *Based only on the information in this question, is chalk an element, a compound, or a mixture? Or can you tell? Explain your reasoning.*

Chalk can break down into other substances, so it can't be an element. But it also has a fixed composition (always gives the same proportions of those substances), so it must be a **compound**. If it were a mixture, then the proportions of the component substances would be variable.

b) *Based on this information, is quicklime an element, a compound, or a mixture? Or can you tell? Explain your reasoning.*

Quicklime is produced from the breakdown of a pure substance (chalk), so it's likely to be a pure substance and not a mixture (though in some cases a compound can break down into a mixture of compounds or elements). However, we don't have any information on whether quicklime itself can be broken down into yet other substances or not, so **we can't decide whether it is an element or a compound.**

2.9 *Because pure gold is rather soft, gold-colored jewelry is often made out of 14-carat gold, which contains 58% gold and 42% silver. Gold-colored jewelry can also be made from 16-carat gold, which contains 67% gold and 33% silver. Based on this information, do gold and silver form a compound when they are mixed, or do they form a mixture? Explain your reasoning.*

Since gold and silver can be combined in several different ratios, they must form a mixture, not a compound. Also, like other mixtures, the mixture of the two metals resembles the component metals: it has a metallic sheen, can be worked into different shapes, and has a color very much like that of pure gold.

2.11 *Sulfur and oxygen are both nontoxic. When these substances are mixed and heated, they can combine to make a poisonous gas. Based on this information, is this combination likely to be a compound, or is it probably a mixture? Explain your reasoning.*

Mixtures of substances have physiological effects similar to those of the component substances. Since the individual substances are non-toxic, their mixtures are also expected to be non-toxic. **Compounds, on the other hand, usually have very different properties, including toxicity, than their component elements. Therefore, the combination described must produce a compound, not a mixture.**

Solutions to Section 2.2 Core Problems

2.13 *Write the name of each of the following elements.* No way around this but to memorize; these elements are listed in Table 2.5.

a) C **carbon** *b) N* **nitrogen** *c) Cl* **chlorine** *d) Mg* **magnesium**
e) Co **cobalt** *f) Se* **selenium**

2.15 *Each of the following elements has a symbol that does not match its English name. What is the symbol for each of these?* Again, these must be memorized from Table 2.5.

a) iron **Fe** *b) sodium* **Na** *c) silver* **Ag** *d) lead* **Pb**

2.17 *Using the periodic table, find elements that match each of the following descriptions:*

a) A nonmetal in Group 6A

Nonmetals are on the right of the heavy "stair step" line (the blue elements in Figure 2.3), and Group 6A is the Oxygen family. **O, S, and Se** are all nonmetals in Group 6A. (Te and Po are also in Group 6A, but are metalloids.)

b) A metal in the same group as carbon

Carbon is C, element #6. Metals (the pink elements in Figure 2.3) are to the left of the heavy "stair step" line, and groups are vertical columns. Si and Ge are metalloids in the same group as carbon. The metals in the group are **Sn and Pb.**

c) A metalloid in the same period as calcium

Periods are horizontal rows; calcium is in Period 4. The metalloids in this period are **Ge and As.**

d) An element that is shiny and conducts electricity

These are characteristics of metals, so we can choose **any metal** in the entire periodic table (any element in pink in Figure 2.3): **Li, Be, Al, Sn, Bi and any elements under them, as well as any element from a B group.**

e) A halogen in Period 3

Periods are horizontal rows. Halogens are elements in Group 7A (except H). There is only one possible answer for this, where Period 3 and Group 7A intersect: **Cl**.

f) An alkaline earth metal

The alkaline earth metals are Group 2A: **Be, Mg, Ca, Sr, Ba, Ra.**

2.19 *Columbium is a name that was once used for one of the chemical elements. Columbium is a shiny, silvery solid that conducts electricity well and can bend without shattering. Is columbium a metal, a metalloid, or a nonmetal?*

The properties listed are characteristics of a **metal**. Columbium is now called Niobium; the name columbium was derived from the mineral source, columbite.

Solutions to Section 2.3 Core Problems

2.21 *Only one of the following statements could possibly be true. Which one is it?*

Understanding the answers to this sort of problem depends on understanding two things: (1) that matter is made up of atoms, and that the smallest possible sample of an element is a single atom (fractions of atoms of elements are not possible); (2) that atoms are very, very small, and it takes huge numbers of them to make up a sample even large enough to see with a microscope, let alone to make objects that can be handled.

a) *A gold coin contains 0.000000000000000000000252 atoms of gold.*

False. This number is a very small fraction of one atom. It is not possible to have a fraction of an atom of gold.

b) *A gold coin contains 528 atoms of gold.*

False. Atoms are extremely small; a sample of gold that only contains 528 atoms would be much too small to see, even with most microscopes.

c) *A gold coin contains 0.25 atoms of gold.*

False. This number is a fraction of one atom. It is not possible to have a fraction of an atom of gold.

d) *A gold coin contains 86,800,000,000,000,000,000,000 atoms of gold.*

While we don't know how many atoms of gold it would take to make a gold coin, this is the only one of the four statements that could possibly be **true**. (In chapter 5, we will look at how to actually calculate the mass of this many gold atoms and see if it's reasonable for a gold coin.)

2.23 *Only one of the following could possibly be a true statement. Which one is it?*

This question expects you to understand the relationship between amu, a tiny, tiny unit of mass used to describe masses of individual atoms, and grams, the mass unit we use in relation to objects like paper clips and coins. While we are not required to memorize the numerical relationship between an amu and a gram, you should have an idea of the enormous difference in size between the two:

$$1 \text{ amu} = 0.000000000000000000000001661 \text{ g} \quad (1.661 \times 10^{-24} \text{ g})$$

a) *A penny weighs 0.000000000000000000000827 amu*

False. An amu is a unit used to measure masses of individual atoms. Even a single proton weighs 1 amu, while this is a small fraction of an amu.

b) *A penny weighs 3,680 amu*

False. A penny's mass would be measured on the order of grams. This mass is much too small to add up to the mass of a penny.

c) *A penny weighs 1,810,000,000,000,000,000,000,000 amu*

This is the only one of the statements that could possibly be **true.** We can actually do the calculation here to check (the given value, put into scientific notation, is 1.81×10^{24} amu):

$$\mathbf{1.81 \times 10^{24}\,amu \times \frac{1.661 \times 10^{-24}\,g}{1\,amu} = 3.01\,g}$$

Three grams is a reasonable estimate for the mass of a penny (see Table 1.6 for a refresher on masses.)

2.25 *Proteins are made from the elements carbon, oxygen, hydrogen, nitrogen, and sulfur. How many different types of atoms are there in proteins?*

There are five elements in proteins, which means there are **five types of atoms**. When we say "type" of atom, we are referring to its element identity.

Solutions to Section 2.4 Core Problems

2.27 *Fill in the blanks in the following statements with the names of the appropriate subatomic particles.*

Our options for each blank are **proton, neutron, electron** and we know the properties of each (Table 2.6).

a) **Electrons** *have a negative charge.*

Protons are positive and neutrons are neutral.

b) **Protons** *and* **neutrons** *have roughly the same mass.*

Each of these particles has a mass of about 1 amu, while the mass of an electron is much smaller.

c) *The number of* **protons** *determines which element an atom is.*

The element identity of an atom is determined by the number of protons in its nucleus.

d) *To find the charge on an atom, you must know the numbers of* **protons** *and* **electrons**.

These are the two charged particles, and the difference between them determines the charge of an atom or ion.

2.29 *You have an atom that contains 16 protons, 17 neutrons, and 18 electrons.*

a) *What element is this?*

The element identity of an atom is determined by the number of protons (16 in this atom). Element #16 is **Sulfur.**

b) *What is the mass number of this atom?*

The mass number is the sum of the neutrons and protons: 16 + 17 = **33**.

c) *What is the atomic number of this atom?*

The atomic number is the number of protons (**16** in this atom.)

d) *What is the charge on this atom?*

The charge is determined by the number of protons (+1 each) and electrons (-1 each). This atom has two more electrons than protons, so it has a charge of **-2**:

total charge of the protons	=	+16
total charge of the electrons	=	–18
Total charge of this ion	=	**–2**

e) *What is the approximate mass of this atom in amu?*

The mass number is the approximate mass of the atom in amu, **33 amu** for this atom.

2.31 *Many silver atoms have a mass number of 107.*

a) *How many protons are there in an atom of silver-107?*

Silver is Ag, element 47. The atomic number is the number of protons, so this atom has **47 protons**.

b) *How many neutrons are there in an atom of silver-107?*

To get the number of neutrons, subtract the number of protons from the mass number:

107 – 47 protons = 60 neutrons.

c) *How many electrons are there in an electrically neutral atom of silver-107?*

In any neutral atom, the number of electrons and the number of protons must be equal: **47 electrons** for a neutral atom of silver.

d) *Silver commonly has a charge of +1. How many electrons are there in a silver-107 atom that has this charge?*

For an atom to have a +1 charge, the neutral atom must lose one electron (the ion must have one less electron than the number of protons.) The Ag^+ ion must have **46 electrons.**

2.33 What is the significance of the number 16 in the symbol ^{16}O?

The upper-left position is reserved for the **mass number (number of neutrons + number of protons).**

Solutions to Section 2.5 Core Problems

2.35 *Which of the following pairs of atoms are isotopes?*

Isotopes have the same atomic number (# protons), and are therefore the same element, but have different mass numbers.

a) ^{39}K *and* ^{40}K

Isotopes: both atoms are the same element but they have different mass numbers.

b) *An atom with seven protons and eight neutrons, and an atom with 8 protons and 8 neutrons.*

Not isotopes: The atoms have different mass numbers (15 and 16), but they are also different elements because they have different numbers of protons (one is N, the other is O).

2.37 *A beaker is filled with a large number of atoms, all of which have 34 protons and 44 neutrons. Which of the following would look and behave similarly to this collection of atoms?*

a) A group of atoms that have 33 protons and 45 neutrons.
b) A group of atoms that have 35 protons and 44 neutrons.
c) A group of atoms that have 34 protons and 45 neutrons.

Chemical behavior is determined by the number of protons (atomic number), regardless of the number of neutrons and the mass number. The only group of atoms that would behave similarly is **(c), the group that also has 34 protons in its nuclei.** (Another way of looking at this is that the original sample is a sample of Se, and the other two samples are a) As and b) Br.

2.39 *The atomic weight of gold (from the periodic table) is 197.0 amu. Based on this fact, which of the following conclusions can be drawn?*

The atomic weight is an average of all naturally-occurring atoms of an element. It doesn't tell us anything specific about which isotopes exist or in what proportions, but it does give us an idea of what mass numbers are likely to exist.

a) All gold atoms weigh 197 amu.

We cannot conclude this; for example, a mixture of 50% ^{196}Au and ^{198} Au would give the same average.

b) Most gold atoms weigh 197 amu.

We cannot conclude this; there don't have to be any atoms at all with a mass of 197 amu.

c) More gold atoms weigh 197 amu than any other mass.

We cannot conclude this; there don't have to be any atoms at all with a mass of 197 amu.

d) We cannot conclude anything about the masses of individual gold atoms.

This is the only safe answer (though we could predict that gold atoms with masses very different from 197 are unlikely to exist—we would not expect to see atoms of gold-160 or gold-220, for example.)

Solutions to Section 2.6 Core Problems

2.41 *A formula unit of table sugar (sucrose) contains 12 carbon atoms, 22 hydrogen atoms, and 11 oxygen atoms. Write the chemical formula of table sugar.*

We write the element symbols for the elements in the order given, and put a subscript after each symbol to show how many atoms of that element are present: $\mathbf{C_{12}H_{22}O_{11}}$.

2.43 *Monosodium glutamate (MSG) is a commonly-used food additive that has the chemical formula $NaC_5H_8NO_4$.*

a) *How many atoms of each element are there in one formula unit of MSG?*

Remember that any element listed without a number is assumed to have the subscript "1." There are **one atom of sodium (Na), five atoms of carbon (C), eight atoms of hydrogen (H), one atom of nitrogen (N), and four atoms of oxygen (O).**

b) *If you have three formula units of MSG, how many carbon atoms do you have?*

Each formula unit contains five atoms of carbon, so **three formula units contain a total of $5 \times 3 = 15$ atoms of carbon.**

Solutions to Section 2.7 Core Problems

2.45 *Which of the following is a true statement? Refer to Figure 2.9 as you answer this question.*

Figure 2.9 illustrates the electron shells for an atom of argon (Ar). Argon has 18 electrons, arranged in three shells. The shells represent the area where the electrons are most likely to be found, but in reality the electrons do not stay on the surface of the electron shell; they can go anywhere within the shell, and can even go outside the surface.

a) *Electrons in the first shell must remain inside the region labeled shell 1.*

Not true; the shell isn't a cage, it's an area of probability of finding an electron.

b) *Electrons in the first shell must remain on the surface of the shell.*

Definitely **not true.** Electrons can go anywhere inside or outside the shell.

c) *Electrons in the first shell can sometimes be outside the region labeled shell 1.*

This is the only true statement.

2.47 Write the electron arrangement for each of the following elements.

In a neutral atom the number of electrons = the number of protons = the atomic number, so we can easily find out how many electrons each element has in its atoms. These elements are both among the first 18 elements, so their electrons occupy the lowest possible shells.

The shells are keyed to the periodic table: we can remember that shell 1 holds 2 electrons because He, the first noble gas, has 2 electrons. We can remember that shell 2 holds 8 electrons because Ne, the next noble gas after He, has 8 more electrons than He ($10 - 2 = 8$.)

a) *nitrogen*

Each N atom has a total of 7 electrons, filling the shells systematically.
shell 1: 2 electrons **shell 2: 5 electrons**
This uses up all the electrons, so no further shells are needed.

b) *magnesium*

Each Mg atom has a total of 12 electrons, filling the shells systematically.
shell 1: 2 electrons **shell 2: 8 electrons** **shell 3: 2 electrons**
This uses up all the electrons, so no further shells are needed.

2.49 *An atom has the following electron arrangement. What element is it?*

shell 1: 2 electrons *shell 2: 8 electrons* *shell 3: 8 electrons*

The total number of electrons in this atom is 18, so (assuming it is a neutral atom) its atomic number is 18. **The element is Ar, argon.**

Solutions to Section 2.8 Core Problems

2.51 *Beryllium combines with nitrogen to form the compound Be_3N_2.*

a) *Magnesium and beryllium show similar chemical behavior. Write the chemical formula for the compound that is formed by magnesium and nitrogen.*

Mg and Be are in the same group of the periodic table and are expected to combine with other elements in the same ratios in the compounds they form. The compound formed by magnesium and nitrogen is **Mg_3N_2.**

b) *Phosphorus and nitrogen show similar chemical behavior. Write the chemical formula for the compound that is formed by beryllium and phosphorus.*

P and N are in the same group of the periodic table and are expected to combine with other elements in the same ratios in the compounds they form. The compound formed by beryllium and phosphorus is **Be_3P_2.**

NOTE:
We could even carry this one step further and predict the formula of the compound formed by magnesium and phosphorus: Mg_3P_2.

2.53 *The electron arrangement of selenium (Se) is*

shell 1: 2 electrons *shell 2: 8 electrons* *shell 3: 18 electrons* *shell 4: 6 electrons*

a) *How many valence electrons does an atom of selenium contain?*

By definition, valence electrons are the electrons in the outermost shell (in this case, shell 4). Selenium has **6 valence electrons.** (As explained in Section 2.9, we could also have determined this by seeing that Se is in Group 6A.)

b) *Draw the Lewis structure of a selenium atom.*

The Lewis structure includes the element symbol and the valence electrons, represented as dots. (The paired and single electrons can be shown in any of the four positions around the atom.)

2.55 *Draw the Lewis structure of a carbon atom. (Hint: first figure out the electron arrangement.)* Carbon's electron arrangement is

shell 1: 2 electrons shell 2: 4 electrons

Carbon has four valence electrons. Its Lewis structure is: $\cdot\dot{\underset{\cdot}{C}}\cdot$

2.57 *Lead (Pb) has four valence electrons. What element from Period 2 (Li through Ne) should show similar chemical behavior to lead, based on its number of valence electrons?*

The Period 2 element with four valence electrons is **Carbon.**

Solutions to Section 2.9 Core Problems

2.59 *Based on its position on the periodic table, how many valence electrons does an atom of selenium (Se) contain? Which shell do they occupy?*

For representative elements, the group number is the number of valence electrons (for Se, Group 6A). Also for representative elements, the period number is the valence shell (for Se, Period 4). So Selenium has **6 valence electrons in shell 4.**

2.61 *A representative element has the following electron arrangement:*

shell 1: 2 electrons *shell 2: 8 electrons* *shell 3: 18 electrons*
shell 4: 8 electrons *shell 5: 1 electron*

Without looking at the periodic table, answer the following questions:

a) *How many valence electrons does this element have?*

The outermost shell is shell 5, with **1 valence electron**.

b) *What is the group number for this element?*

By definition, representative elements are those in the A groups. For representative elements, the group number = the number of valence electrons. **This element is in Group 1A (the alkali metals).**

c) *Which period is this element in?*

For representative elements, the period number is the valence shell. **This element is in Period 5.**

2.63 *Draw Lewis structures for each of the following atoms. You may use the periodic table.*

Lewis structures represent valence electrons as dots. For each element, we'll use the group number to determine the number of valence electrons.

a) *K* Group 1A, one valence electron: $K\cdot$

b) *Pb* Group 4A, four valence electrons: $\cdot\dot{\underset{\cdot}{Pb}}\cdot$

c) *Br* Group 7A, seven valence electrons: $:\ddot{\underset{\cdot\cdot}{Br}}\cdot$

2.65 *Using the periodic table, give an example of each of the following:*

a) *A metal that has five valence electrons.*

Looking for a metal (the pink elements in Figure 2.3) in Group 5A: **Bi.**

b) *A representative element that has its valence electrons in the sixth shell.*

For representative elements, the period number and the valence shell number are the same, so any representative element in Period 6: **Cs, Ba, Tl, Pb, Bi, Po, At, Rn.**

c) *A transition element that is in the fourth period.*

The fourth period is the fourth row of the periodic table, and the transition elements (or transition metals) are the B groups: **Sc, Ti, V, Cr, Mn, Fe, Co, Ni, Cu, Zn.**

d) *An element that should show similar chemical behavior to oxygen.*

Elements in the same group show similar chemical behavior (and the closer two elements are within a group, the more similar their chemical behavior will be). The element that should have the most similar chemical behavior to oxygen is **sulfur, S,** but other Group 6A elements should also show similar chemical behavior: **Se, Te, Po.** (The last three, being metalloids, are likely to be less similar to O than S, since O and S are both nonmetals.)

Solutions to Concept Questions

* indicates more challenging problems.

2.67 a) *What is the fundamental difference between a mixture and a compound?*

Both mixtures and compounds are made up of two or more other substances, but **mixtures have variable composition** (any proportions of the other substances can be mixed, within certain limits) **while compounds have constant composition** (only certain very specific amounts of two or more elements will combine to form a compound).

b) *What is the fundamental difference between an element and a compound?*

Both elements and compounds are pure substances (cannot be separated into two or more other substances by **physical** processes). However, **compounds can be broken down into two or more other substances by chemical processes, while elements cannot.**

Another way to express the fundamental difference between compounds and elements is that **in an element, every atom has the same number of protons, while a compound must by definition by composed of at least two different elements (atoms with different numbers of protons)**.

2.69 *Physiological saline is a mixture of salt and water that can be used in intravenous injections. It always contains 0.9% salt and 99.1% water. Is physiological saline a compound? Explain your answer.*

Start by looking at the way "compound" and "mixture" are defined, and the characteristics of each. Physiological saline is a mixture, not a compound. Here are four possible ways to explain or rationalize this:

(1) **While the term "physiological saline" is only applied to a mixture that has the composition above, those percentages are agreed upon by humans, they aren't a law of nature.** (It's essentially a labeling law—similar to the rule for the maximum amount of fat a product could contain and still be labeled "low-fat.") Salt and water can be mixed in any proportions to make saltwater.

(2) **The salt and water are not chemically bonded to each other; they can still be easily separated by physical methods** (for example, letting the water evaporate away from the salt). This is a characteristic of mixtures.

(3) **The mixture of salt and water looks like one of its components (water) and has properties that are a combination of those of the components, a common characteristic of a mixture.** Compounds typically have properties that are very different from those of the component elements.

2.71 *A student says, "A carbon atom is not the smallest possible amount of carbon, because atoms are built from smaller particles." How would you respond to this student?*

While it's true that atoms are built from smaller particles, these smaller particles are not "carbon," just as a brick is not a house, a piece of fingernail is not a hand, and a grain of sand is not a beach. **If you take apart an atom of carbon into smaller particles, it's not carbon anymore.**

2.73 *Why do chemists express the masses of atoms using the atomic mass units, rather than a normal metric unit such as grams or milligrams?*

We often choose units based on their convenience for a given object. We don't express the mass of a grain of rice in tons, the length of a sofa in miles, or the volume of the sun in milliliters. **An atom is so small that its weight in grams or milligrams or even nanograms is an extremely small, inconvenient number to work with. The whole purpose of the unit amu is to easily work with the masses of individual atoms and subatomic particles.**

2.75 *The atomic weight of selenium is approximately 79 (the actual value is 78.96), and the atomic number of selenium is 34. Based on this, is it reasonable to conclude that a typical selenium atom has 34 protons and 45 neutrons? Justify your answer.*

It is likely that most stable atoms of selenium have a number of neutrons *close* to 45, but there don't actually have to be any atoms of selenium with 45 neutrons (^{79}Se). For example, you could get the same atomic weight with approximately 50% ^{78}Se and 50% ^{80}Se (it's easy to imagine many other combinations that would also be reasonable.) However, we wouldn't expect, for example, an atom of Se with 30 neutrons (^{64}Se) or 60 neutrons (^{94}Se) to be stable, because these values are far from the observed average atomic mass.

2.77 *Using electron arrangements, explain why nitrogen and phosphorus show similar chemical behavior.*

An atom of nitrogen has five valence electrons (five electrons in the outer shell, in this case shell 2). So does phosphorus, but in that case, the valence shell is shell 3. **Having the same valence-shell electron configuration gives them similar chemical behavior.**

(Since both are representative elements, chemical behavior depends entirely on the number of valence electrons.)

2.79 *The drawing below represents the arrangement of the atoms in acetaminophen. Write the chemical formula for acetaminophen.*

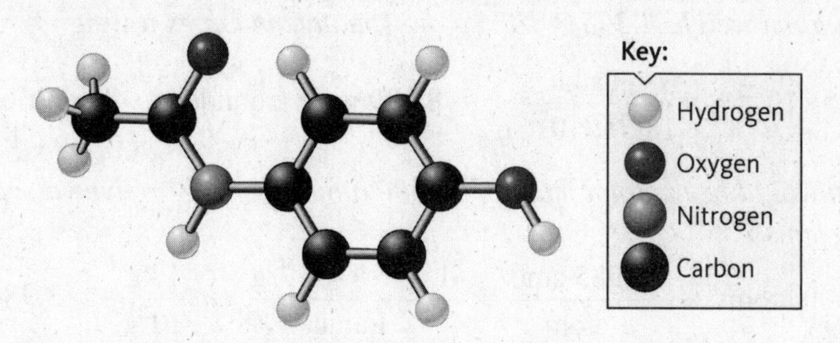

The structure contains eight carbon atoms, nine hydrogen atoms, two oxygen atoms and one nitrogen atom. The chemical formula is $\mathbf{C_8H_9O_2N}$.

Solutions to Summary and Challenge Problems

* indicates more challenging problems.

2.81 *Which of the following are true statements?*

a) *Elements that are in the same group generally show similar chemical behavior.*

True: the periodic table was originally arranged to put elements with similar behavior into columns (groups) together.

b) *Elements that are in the same period generally show similar chemical behavior.*

False: Periods are horizontal rows. Elements in the same period don't necessarily have behavior in common. Behavior generally changes from left to right across a period from that a metal to metalloid to nonmetal

c) *The nonmetals are on the left side of the periodic table.*

False: Nonmetals are to the right of the stair-step line.

d) *The elements in Group 7A are called halogens.*

True: This is one of the special group names you must memorize.

e) *The elements in Group 6A are representative elements.*

True: All elements in A groups (1A-8A) are, by definition, representative elements.

*2.83 *One amu is the same as 1.661×10^{-24} g.*

a) *A Mg-24 atom weighs 23.985 amu. How many grams does it weigh? How many micrograms?*

$$23.985 \text{ amu} \times \frac{1.661 \times 10^{-24}\text{ g}}{1 \text{ amu}} = \mathbf{3.984 \times 10^{-23}\ g}$$

$\text{g} = 10^6\ \mu\text{g}$, so

$$3.984 \times 10^{-23}\ \text{g} \times \frac{1 \times 10^{6}\ \mu\text{g}}{1\ \text{g}} = \mathbf{3.984 \times 10^{-17}\ \mu g}$$

Because the conversion factor given only has 4 significant figures, the answers are rounded to 4 significant figures.

b) *If an atom weighs 1.395×10^{-22} g, how many amu does it weigh?*

$$1.395 \times 10^{-22}\ \text{g} \times \frac{1\ \text{amu}}{1.661 \times 10^{-24}\ \text{g}} = \mathbf{83.99\ amu}$$ (rounded to 4 significant figures)

c) *What would be the total mass of 10,000 atoms of ^{24}Mg? Give your answer in kilograms.*

$$10{,}000\ \text{atoms} \times \frac{23.985\ \text{amu}}{1\ \text{atom}} \times \frac{1.661 \times 10^{-24}\ \text{g}}{1\ \text{amu}} \times \frac{1\ \text{kg}}{10^{3}\ \text{g}} = \mathbf{3.984 \times 10^{-22}\ kg}$$

2.85 *An ion has a charge of +2. This ion has a mass number of 42 and contains 22 neutrons.*

a) *What element is this?*

atomic number = mass number – neutrons = 42-22 = 20. The ion is **calcium.** (The charge or number of electrons doesn't determine the element identity of an atom or ion; only the number of protons specifies atom identity.)

b) *How many electrons are there in this atom?*

A neutral atom of calcium has electrons = protons = 20. Since this ion has a charge of +2, it's missing two electrons and must have **18 electrons**. (20 positive protons and 18 negative electrons give a total charge of +2.)

total charge of the protons	=	+20
total charge of the electrons	**=**	**–18**
Total charge of this ion	=	+2

2.87 *One formula unit of morphine contains 17 carbon atoms, 19 hydrogen atoms, 1 nitrogen atom, and 3 oxygen atoms. Write the chemical formula of morphine.*

The number of atoms of each element in the chemical formula is given as a subscript on its chemical symbol: $\mathbf{C_{17}H_{19}NO_3}$. (The "1" on nitrogen is understood. We haven't yet learned the convention for what order elements appear in chemical formulas, so we write them in the order given in the problem.)

2.89 a) *Explain why it is not possible to have a sample of morphine that contains five carbon atoms. (See Problem 2.87.)*

The formula of morphine is $C_{17}H_{19}NO_3$. **By definition, a formula unit is the smallest possible sample of a compound, just as an atom is the smallest possible unit of an element. The smallest possible unit of morphine has seventeen carbon atoms. If you break up the formula unit, it's not morphine anymore.**

b) *Is it possible to have a sample of morphine that contains five nitrogen atoms? Explain why or why not.*

It is possible. Since each formula unit of morphine contains one nitrogen atom, a sample of morphine that contained five formula units would contain five nitrogen atoms (along with $17 \times 5 = 85$ atoms of carbon, $19 \times 5 = 95$ atoms of hydrogen, and $3 \times 5 = 15$ atoms of oxygen.)

2.91 *Give one example of each of the following.*

a) *An element that has five valence electrons.*

This would apply to any Group 5A element, so take your pick: **N, P, As, Sb, Bi.**

b) *An element that has two electrons in shell 4. (Hint: It can have electrons in other shells.)*

The easy one here is **Ca**, with its two valence electrons. But generally **elements 21 (Sc) through 30 (Zn) also have two electrons in shell 4**, because the electrons that are being added to those elements are going into shell 3. (There are a couple of exceptions for reasons that are outside the scope of this text.)

c) *An element that has eight electrons in shell 2. (See the hint to part b.)*

Ne is the first element that has eight electrons in shell 2. But **every element after Ne also has eight electrons in shell 2, since that's the limit**—every later element in the periodic table is adding electrons into higher shells.

d) *An element that has a total of eight electrons.*

There's only one answer for this one: only **oxygen**, element 8, has exactly 8 electrons in its neutral atoms.

*2.93 *Lead has the following electron arrangement:*

shell 1: 2 electrons	*shell 2: 8 electrons*	*shell 3: 18 electrons*
shell 4: 32 electrons	*shell 5: ?? electrons*	*shell 6: ?? electrons*

Use the atomic number of lead and its position on the periodic table to figure out the number of electrons in shells 5 and 6.

See Sections 2.8 and 2.9. For the representative (A Group) elements, the group number (Group 4A for lead, Pb) equals the number of valence electrons (4 valence electrons in a lead atom), and for all elements, the period number (Period 6 for lead) equals the number of the outermost shell (shell 6 for lead). Therefore lead must have **four electrons in shell 6**. From there we can work backward to figure out how many electrons are in shell 5. In shells 1-4, there are a total of 60 electrons ($2 + 8 + 18 + 32 = 60$), and Shell 6 accounts for 4 more, a total of 64 electrons in shells 1-4 and 6. Lead is element number 82, and therefore each atom has 82 protons and 82 electrons; $82 - 64 =$ **18 electrons in shell 5**.

3

Chemical Bonds

Solutions to Section 3.1 Core Problems

3.1 *Which of the following elements satisfies the octet rule?*

"Satisfies the octet rule" means "has eight valence electrons," except in the case of the first few elements (hydrogen through beryllium, for which the "octet" is the helium electron arrangement, two valence electrons).

a) Br Bromine has seven valence electrons and does not, as a neutral atom, satisfy the octet rule.

b) Kr **Krypton has eight electrons in its valence shell.**

c) Ni Nickel is a transition metal; it has two valence electrons, and 16 electrons in the shell below the valence shell. Ni does not, as a neutral atom, satisfy the octet rule.

3.3 *How many covalent bonds does a phosphorus atom normally form, based on the number of valence electrons in the atom?*

Phosphorus is in Group 5A, so it **has 5 valence electrons**, meaning it has 3 spaces in the valence shell and **typically forms 3 covalent bonds**.

3.5 *Give an example of an element that normally forms two covalent bonds.*

An element that needs two electrons to complete its octet would normally form two covalent bonds. **The nonmetal elements in Group 6A have six valence electrons and would normally form two covalent bonds: O, S, Se.**

3.7 *Draw Lewis structures for each of the following molecules:*

In each case, we follow the steps: put the atom that forms the most bonds in the center of the molecule, arrange the other atoms so that they make the proper number of covalent bonds and lone pairs, and check to make sure all atoms have their octet.

NOTE:
In Section 3.2, we will see another way of drawing Lewis structures, in which lines are used to represent bonding electrons. Because the second convention is more common, and will be presented in the next section, both representations will be shown for these molecules.

a) HBr

H:Br: H—Br:

b) $SiCl_4$

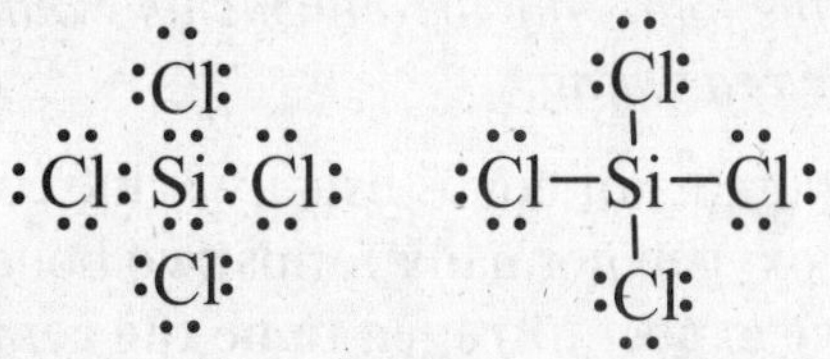

```
     ..              ..
    :Cl:            :Cl:
 ..  ..  ..      ..   |   ..
:Cl: Si :Cl:    :Cl—Si—Cl:
 ..  ..  ..      ..   |   ..
    :Cl:            :Cl:
     ..              ..
```

c) PH_3

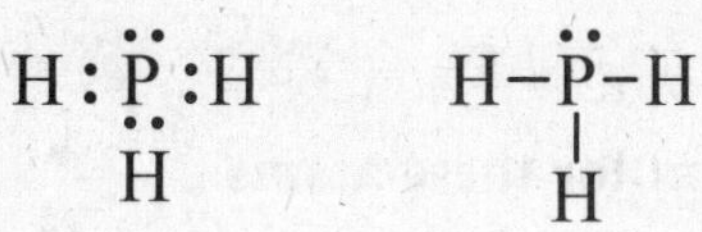

```
    ..          ..
H : P : H    H—P—H
    ..          |
    H           H
```

d) H_2S

```
    ..          ..
H : S : H    H—S—H
    ..          ..
```

3.9 *In the compound* SiH_2BrI*, which element is at the center of the molecule? Explain your answer, and draw the Lewis structure of this molecule.*

In general, the atom that typically makes the largest number of bonds is the central atom in the molecule—in this case, **Si**, which makes four bonds, while each of the other atoms only normally makes one.

NOTE:
In the structure below, it doesn't matter which atom you put in each of the four positions.

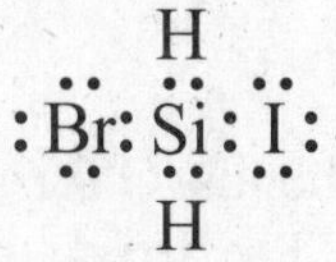

```
          H
  ..     ..    ..
: Br :  Si  : I :
  ..     ..    ..
          H
```

Solutions to Section 3.2 Core Problems

3.11 *Which of the following elements cannot form double bonds? Explain your reasoning.*

See Table 3.2. In principle any atom that can form more than one bond can form a double bond.

a) C Carbon typically makes four bonds, so it can form a double bond (one double + two single bonds, or even two double bonds).

b) N Nitrogen typically forms three bonds, so it can form a double bond (one double + one single).

c) O Oxygen typically forms two bonds, so it can form one double bond.

d) F **Fluorine normally only makes one bond, so it won't form double bonds.**

Alternatively, you can memorize the bonding arrangements in Table 3.2 and arrive at the same result (though this is probably more trouble than understanding the pattern).

3.13 *Nitrosyl bromide contains one bromine atom, one oxygen atom, and one nitrogen atom, linked by covalent bonds. Which of the following arrangements would you expect for these three atoms? Explain your reasoning.*

The atom that normally forms the largest number of bonds usually is the central atom. In this case, Br normally forms one bond, oxygen normally forms two bonds, and nitrogen normally forms three bonds, so **we expect nitrogen to be the central atom.**

a) Br—O—N

b) N—Br—O

c) O—N—Br **This is the best arrangement for these atoms.**

The Lewis structure that can give all atoms their normal bonding pattern and octet is:

:Ö=N̈—B̈r:

3.15 *Draw Lewis structures for each of the following molecules. Each molecule contains one double or triple bond.*

In each case, first look for the atom that normally makes the largest number of bonds, and put it in the center. Second, add bonds to fill the normal bonding patterns of all atoms. Then finish up by adding lone pairs to complete the octets of all atoms.

NOTE:
It doesn't matter (at this point) which directions the atoms are pointing from the central atom, the order of connection is what matters. We'll use the convention of representing bonding electron pairs as lines.

a) ClNO

N typically forms three bonds, O forms two bonds, and Cl only forms one bond, so N is the central atom. The arrangement that allows all atoms to have their normal bonding patterns is: :C̈l—N̈=Ö

b) CF_2O

C normally forms four bonds, F only one, and O two, so C is the central atom. The arrangement that allows all atoms to have their normal bonding patterns is:

Ö
‖
:F̈—C—F̈:

c) C_2H_2 (the two carbon atoms are bonded to each other, and each carbon atom is bonded to one hydrogen atom)

Carbon normally forms four bonds, and hydrogen can only form one. The only way to satisfy the bonding requirements of all atoms is to put a triple bond between the two carbon atoms, and a single bond between each carbon atom and a hydrogen atom:

H—C≡C—H

3.17 *Redraw each of the following structures using lines to represent bonding electron pairs.*

```
    :Cl:          H H
:Cl:P:Cl:      H:C:C::O
                  H
```

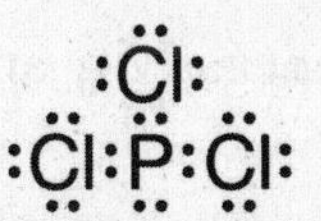

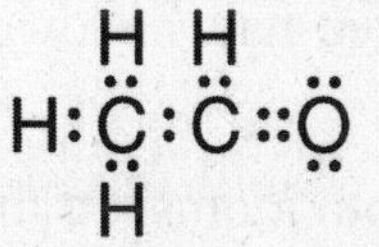

In each case, we simply find the electron pairs that are shown between two atoms, representing bonds, and turn each pair into a line connecting the two atoms. Remember one line = two electrons.

a)
```
      :Cl:
       |
:Cl—P—Cl:
```

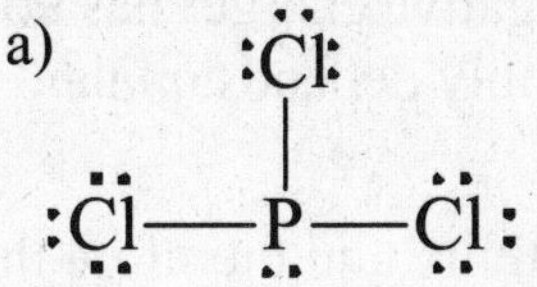

b)
```
   H  H
   |  |
H—C—C=O
   |
   H
```

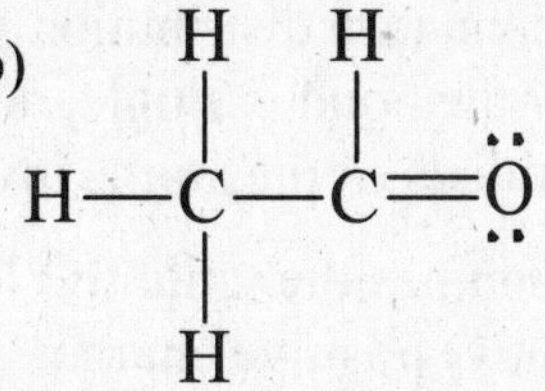

3.19 *Draw the structures of each of the following molecules, using lines to represent bonding electron pairs.*

Again, it doesn't matter (at this point) which directions the atoms are pointing from the central atom, the order of connection is what matters. Use all the same rules for drawing these Lewis structures as when using dots: put the atom that forms the most bonds in the center of the molecule, arrange the other atoms so that they make the proper number of covalent bonds and lone pairs, and check to make sure all atoms have their octet. The only difference between these structures and the dot structures you drew before is that bonding electron pairs are represented as lines.

a) NH_3
```
H—N—H
  |
  H
```

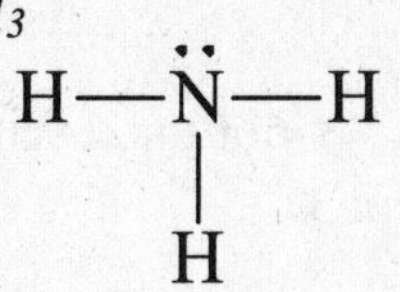

b) SiF_4
```
     :F:
      |
:F—Si—F:
      |
     :F:
```

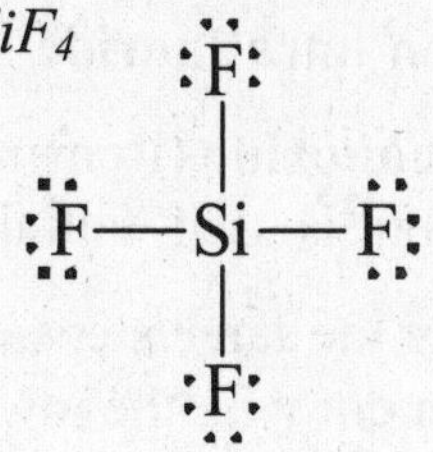

c) CH_2O
```
   :O:
    ||
H—C—H
```

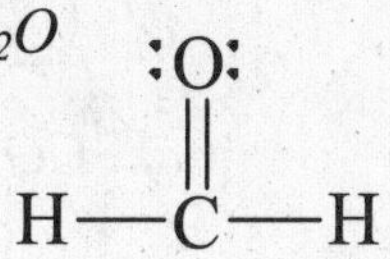

Solutions to Section 3.3 Core Problems

3.21 *Which of the Group 5A elements has the strongest attraction for electrons? Use the electronegativities in Figure 3.7 to answer this question.*

The electronegativity of an element is a measure of its attraction for electrons; the greater the electronegativity value, the stronger the attraction. **Nitrogen** has the greatest electronegativity of the Group 5A elements and therefore the strongest attraction for electrons.

3.23 *Which atom (if any) in each of the following molecules has a positive charge, based on the electronegativities of the elements?*

In a covalent bond between two atoms, the atom with the lower electronegativity value ends up with a smaller share of the electrons, and therefore a positive charge.

a) NO **Nitrogen** has the lower electronegativity and therefore the positive charge.

b) HCl **Hydrogen** has the lower electronegativity and therefore the positive charge.

c) N_2 Since both atoms are the same element and therefore have the same electronegativity, **neither atom is charged in this molecule.**

3.25 *In each of the following chemical bonds below, identify the positively charged atom.*

The polarity of a bond (the distribution of positive and negative charge) has nothing to do with whether the bond is single, double or triple; it only depends on the electronegativity values of the atoms involved.

a) Br–C **Carbon** has the (slightly) lower electronegativity and therefore the (slightly) positive charge.

b) N=O **Nitrogen** has the lower electronegativity and therefore the positive charge.

c) H–H Since both atoms are the same element and therefore have the same electronegativity, **neither atom is charged in this molecule.**

Solutions to Section 3.4 Core Problems

3.27 *Name the following compounds, using the IUPAC rules:*

a) ClF_3 **chlorine trifluoride** (Remember that "mono" is not used with the first element.)

b) N_2F_4 **dinitrogen tetrafluoride**

c) CO **carbon monoxide** (Remember that "mono" is not used with the first element, but is used with the second element.)

In each case, we follow the same steps: (1) Write the name of the first element (this will usually be the element with the lower electronegativity; since we have the formulas, we don't have to figure this out). (2) Write the name of the second element with the –ide ending from Table 3.4. (3) Add prefixes as appropriate from Table 3.5. Watch the spelling on each of the element names!

3.29 *Write chemical formulas for the following compounds:*

a) carbon tetrafluoride **CF_4** *b) sulfur dioxide* **SO_2**

The name tells us the two elements in the order they should appear, and the prefixes tell us how many atoms of each.

3.31 *Give the common names for the following compounds:*

a) H_2O **water** *b) NO* **nitric oxide**

These are in Table 3.6 and must be memorized.

Solutions to Section 3.5 Core Problems

3.33 *Use the electron arrangement of aluminum to show that aluminum satisfies the octet rule if it loses three electrons.*

The electron arrangement of aluminum is:

shell 1: 2 electrons   shell 2: 8 electrons   shell 3: 3 electrons

If the atom loses three electrons, its arrangement becomes:

shell 1: 2 electrons   shell 2: 8 electrons

The outermost occupied shell (shell 2) contains 8 electrons, so the ion satisfies the octet rule.

3.35 *Use Lewis structures to show how each of the following atoms can form an ion that satisfies the octet rule.*

a) chlorine An electrically neutral chlorine atom has seven valence electrons. If the atom gains one electron to form a –1 ion, it will have eight valence electrons, satisfying the octet rule.

$:\ddot{\underset{..}{Cl}}\cdot \longrightarrow :\ddot{\underset{..}{Cl}}:^{-}$

b) magnesium An electrically neutral magnesium atom has two valence electrons in shell 3. If the atom loses two electrons to form a +2 ion, the original valence shell will be empty, leaving the ion with eight electrons in the outermost occupied shell (shell 2).

$\cdot Mg\cdot \longrightarrow Mg^{+2}$

3.37 *Each of the following elements can form a stable ion. What is the charge on each ion?*

In each case, metals will tend to lose electrons to satisfy the octet rule, while nonmetals will tend to gain electrons. Also remember that electrons are **negative**, so a loss of electrons gives a positive charge to the ion, while a gain of electrons gives a negative charge to the ion.

a) Mg **+2.** A neutral magnesium atom has two valence electrons, so it forms a stable ion by losing them both to form an ion with a charge of +2.

b) Br **–1.** A neutral bromine atom has seven valence electrons and needs only one more to satisfy the octet rule, so it forms a stable ion with a charge of –1 by gaining one electron.

c) Se **–2.** A neutral selenium atom has six valence electrons and needs two more to satisfy the octet rule, so it forms a stable ion with a charge of –2 by gaining two electrons.

d) Cs **+1.** A neutral cesium atom has one valence electron, so it forms a stable ion with a charge of +1 by losing one electron.

3.39 *The Group 7A elements can all form ions. What is the charge on these ions?*

Each of the Group 7A elements has seven valence electrons in its neutral atoms, so by gaining one electron, each atom can satisfy the octet rule. **Each of the Group 7A elements forms a -1 ion.**

3.41 *Give an example of a representative element that can form an ion with a +2 charge.*

An atom (other than He) with two valence electrons would be expected to form an ion with a +2 charge—any Group 2A element: **Be, Mg, Ca, Sr, Ba, Ra.**

NOTE:

There are also transition elements that can form +2 ions, but the question specifies a representative element.

3.43 *Potassium and sulfur can combine to form an ionic compound.*

a) *Which element loses electrons, and how many electrons does each atom lose?*

Metals tend to lose electrons to form positive ions. **Potassium is the metal in this case, and would lose its one valence electron to form a +1 ion.**

b) *Which element gains electrons, and how many electrons does each atom gain?*

Nonmetals tend to gain electrons to form negative ions. **Sulfur is the nonmetal in this compound, and would gain two valence electrons to satisfy the octet rule and form a –2 ion.**

c) *What is the charge on each atom in the ionic compound?*

+1 on K, –2 on S.

d) *Write the formula of the ionic compound.*

To achieve charge balance (an overall charge of 0 for the compound), we need two potassium ions for every one ion for sulfur. We write the formula by listing the two elements in order, with the number of atoms of each as a subscript. The formula of the compound formed is **K_2S.**

Solutions to Section 3.6 Core Problems

3.45 *Using the rule of charge balance, explain why sodium and oxygen cannot combine to make a compound that has the formula NaO.*

Sodium forms a +1 ion, while oxygen forms a –2 ion. A compound with the formula NaO would have a total charge of (+1) + (–2) = –1, not the required 0 charge of a neutral compound. The formula of the compound of sodium and oxygen would be Na_2O, not NaO.

3.47 *Write the chemical formulas of the ionic compounds that are formed by each of the following pairs of ions:*

a) *Fe^{2+} and O^{2-}* Since the charges are equal and opposite, the ratio is 1:1 and the formula is **FeO.**

b) *Na^{+} and Se^{2-}* A ratio of 2 Na^{+}:1 Se^{2-} is required for charge balance: **Na_2Se.**

c) F^- and Sr^{2+} Two F^- ions are required to balance one Sr^{2+}, and we make sure to write the cation first, according to convention: $\mathbf{SrF_2}$.

d) Cr^{3+} and I^- The ratio is 1 Cr^{3+}:3 I^- for charge balance: $\mathbf{CrI_3}$.

e) N^{3-} and Mg^{2+} Again, we have to make sure to put the cation first. Three Mg^{2+} ions (for a total charge of +6) will balance with two N^{3-} ions (total charge –6): $\mathbf{Mg_3N_2}$.

In each case, we're looking for the combination of ions that would give an overall charge of 0 in the compound.

3.49 *Using the normal ion charges for the elements, give the chemical formula of the ionic compound that is formed by each of the following pairs of elements:*

a) potassium and bromine

Potassium (Group 1A) forms a +1 ion, and bromine (Group 7A) forms a –1 ion, so the ratio is 1:1, giving **KBr.**

b) zinc and chlorine

Zinc forms a +2 ion (see Table 3.8), while chlorine (Group 7A) forms a –1 ion. Charge balance requires two ions of chlorine for every one zinc ion: $\mathbf{ZnCl_2}$.

c) aluminum and sulfur

Aluminum (Group 3A) forms a +3 ion, while sulfur (Group 6A) forms a –2 ion. To achieve charge balance, two aluminum ions (total charge of +6) will combine with three ions of sulfur (total charge of –6): $\mathbf{Al_2S_3}$.

d) cobalt and chlorine (There are two possible compounds; give both of them.)

There are two possible compounds because cobalt has two common charges (Table 3.8): +2 and +3. To achieve charge balance, one Co^{2+} ion will combine with two Cl^- ions to give $\mathbf{CoCl_2}$, and one Co^{3+} ion will combine with three Cl^- ions to give $\mathbf{CoCl_3}$.

NOTE:

Make sure always to list the cation (the metal) first in the formula. In these examples the metal always happened to be listed first anyway, but it may not always be so.

Solutions to Section 3.7 Core Problems

3.51 *Name the following ionic compounds:*

a) K_2O **potassium oxide**

b) MgS **magnesium sulfide**

c) $AlCl_3$ **aluminum chloride**

d) $CuCl_2$ **copper(II) chloride**

Here we have to notice that copper is a transition metal with more than one possible charge. We figure out that it has a +2 charge in this compound, by seeing that the total charge on 2 Cl^- ions is $2 \times -1 = -2$, so the metal must have a +2 charge to achieve charge balance. Then we give the charge designation using a Roman numeral (II). We have to do this for almost any transition metal in an ionic compound (exceptions, as listed in Table 3.8, include Ni and Zn).

e) Cr_2O_3 **chromium(III) oxide**

Again, we have to notice that chromium is a transition metal with more than one possible charge. The total charge on 3 O^{2-} ions is $3 \times -2 = -6$, so the total charge on the two chromium ions must be +6 to achieve charge balance. Therefore each chromium ion must have a +3 charge in this compound. We give the charge designation using a Roman numeral (III).

f) MnS **manganese(II) sulfide**

Manganese is a transition metal with more than one possible charge. In this compound manganese is found in a 1:1 ratio with S^{2-}, so to achieve charge balance the cation must be Mn^{2+}. We use a Roman numeral (II) to designate the charge in the compound name.

In each case, the cation is named first, then the anion. Remember: (1) cations of representative elements are named the same as the element;(2) anions of representative elements get the –ide ending from Table 3.9; (3) transition metal cations usually need a Roman numeral designation to show the charge (Table 3.8); and (4) the prefixes from Table 3.5 are never used in ionic compounds!
Also, if you've gotten a little rusty on the names and symbols of the elements from Table 2.5 and the common charges for transition elements in Table 3.8, now is a good time to make sure you know those. Spelling counts!

3.53 *Write the chemical formulas of the following ionic compounds:*

a) sodium fluoride Na^+ and F^- combine to form **NaF.**

b) calcium iodide Ca^{2+} and I^- combine to form **CaI_2.**

c) chromium(II) sulfide Cr^{2+} and S^{2-} combine to form **CrS.**

d) ferric chloride Fe^{3+} and Cl^- combine to form **$FeCl_3$.**

e) zinc oxide Zn^{2+} and O^{2-} combine to form **ZnO.**

In each case, determine the charges on the two ions, figure out the ratio required for charge balance as in Section 3.6, and write the formula.

3.55 *The correct name for $CaCl_2$ is calcium chloride, but $CuCl_2$ is not called "copper chloride." Explain this difference, and give the correct name for $CuCl_2$.*

Because copper can form more than one stable ion, there is more than one compound with the name "copper chloride." In $CuCl_2$, to achieve charge balance with the two Cl^- ions, the charge on the copper ion is +2. The correct name of the compound is **copper(II) chloride**, with the Roman numeral (II) specifying the +2 charge on the copper ion.

Solutions to Section 3.8 Core Problems

3.57 *Complete each of the following statements by writing either a name or a formula.*

These are all from Table 3.11. Your instructor may expect you to memorize the formulas, with charges, and names of these polyatomic ions.

a) *The chemical formula of nitrate ion is* $\mathbf{NO_3^-}$.

b) *The chemical formula of* **ammonium** *ion is* NH_4^+.

Be careful to keep straight the ammonium ion, NH_4^+, and the molecular compound ammonia, NH_3.

c) *The chemical formula of hydrogen carbonate ion is* $\mathbf{HCO_3^-}$.

d) *The chemical formula of* **phosphate** *ion is* PO_4^{3-}.

3.59 *Write the chemical formulas of the ionic compounds that are formed by each of the following pairs of ions:*

a) *Zn^{2+} and OH^-* Charge balance requires two OH^- for each Zn^{2+}, and anytime we have more than one of a polyatomic ion in a formula we use parentheses to make it clear: $\mathbf{Zn(OH)_2}$.

b) *SO_4^{2-} and Ag^+* Remember that the cation is always listed first in the formula. Charge balance requires two Ag^+ for each SO_4^{2-}: $\mathbf{Ag_2SO_4}$. (The Ag^+ ion is monatomic and does not need parentheses.)

c) *K^+ and PO_4^{3-}* Charge balance requires three K^+ for every PO_4^{3-}: $\mathbf{K_3PO_4}$. (The K^+ ion is monatomic and does not need parentheses.)

d) *Br^- and NH_4^+* Charge balance requires a 1:1 ratio; just remember the cation is first: $\mathbf{NH_4Br}$. (The formula contains only one ammonium ion, so parentheses are not needed.)

3.61 *Write the chemical formulas of the following compounds:*

The process, using the rule of charge balance to determine the ratios of ions in compounds, is exactly the same for polyatomic ions as for monatomic ions (Section 3.6). In each case, we write the formulas for the ions, with their correct charges, and decide the ratio that will give charge balance. Then, if the formula includes more than one of a polyatomic ion, we write that ion in parentheses and put its subscript outside the parentheses.

a) *calcium carbonate* Ca^{2+} and CO_3^{2-} combine to form $\mathbf{CaCO_3}$.

b) *magnesium phosphate* Mg^{2+} and PO_4^{3-} combine to form $\mathbf{Mg_3(PO_4)_2}$. (We use parentheses because the formula includes two phosphate ions.)

c) *chromium(III) hydroxide* Cr^{3+} and OH^- combine to form $\mathbf{Cr(OH)_3}$. (We use parentheses because the formula includes two hydroxide ions.)

d) cobalt(II) nitrate — Co^{2+} and NO_3^- combine to form **$Co(NO_3)_2$**. (We use parentheses because the formula includes two nitrate ions.)

3.63 *Name the following compounds:*

a) $KHCO_3$ is made of K^+ ions and HCO_3^- ions: **potassium hydrogen carbonate**

b) $Cu_3(PO_4)_2$ is made of Cu^{2+} ions and PO_4^{3-} ions: **copper(II) phosphate**

c) $(NH_4)_2SO_4$ is made of NH_4^+ ions and SO_4^{2-} ions: **ammonium sulfate**

As with other ionic compounds, we give the name of the cation, remembering to include a Roman numeral charge designation if necessary, then the name of the anion. The trick here is knowing the polyatomic ions well enough to recognize them in the formulas. Compounds with more than two elements in their formulas often contain polyatomic ions.

Solutions to Section 3.9 Core Problems

3.65 *Draw Lewis structures that clearly show the difference between the bonds in IBr (a molecular compound) and NaBr (an ionic compound).*

:Ï—B̈r: Na^+ :B̈r:$^-$

In I–Br, a pair of electrons is shared between the two atoms. This compound contains only covalent bonds and is therefore molecular. In NaBr, the sodium atom gives up its electron to become Na^+, the bromine atom gains the electron to become Br^-, and the bond between the two ions is an electrostatic attraction between the + and – charges. This compound contains ions and is classified as ionic.

3.67 *Tell whether each of the following compounds is ionic or molecular:*

a) KCl metal + nonmetal: **ionic**

b) HCl two nonmetals: **molecular**

c) $CuSO_4$ metal + polyatomic ion: **ionic**

d) $SOCl_2$ three nonmetals, and no recognizable polyatomic ions: **molecular**

e) $Al(OH)_3$ metal + polyatomic ion: **ionic**

f) $C_6H_4(OH)_2$ three nonmetals, and no recognizable polyatomic ions: **molecular**

Remember that **compounds of two or more nonmetals are molecular** (unless they contain the NH_4^+ ion), while **compounds of a metal and a nonmetal, or those that contain polyatomic ions, are ionic.**

3.69 *Name the following compounds, using the correct naming system for each:*

a) SCl_2 two nonmetals, so we use molecular naming system: **sulfur dichloride**

b) $MgCl_2$ metal + nonmetal, so we use ionic naming system: **magnesium chloride**

Solutions to Concept Questions

* indicates more challenging problems.

3.71 *What is the octet rule?*

Atoms have a tendency to gain, lose, or share electrons such that they will have the same valence electron arrangement as a noble gas—usually eight valence electrons.

3.73 *Use Lewis structures to show that potassium and oxygen cannot both satisfy the octet rule if they form a covalently bonded K_2O molecule.*

$$\text{K}\cdot \;\; \cdot\ddot{\underset{\cdot\cdot}{\text{O}}}\cdot \;\; \cdot\text{K} \longrightarrow \text{K}\text{—}\ddot{\underset{\cdot\cdot}{\text{O}}}\text{—K}$$

In the covalently bonded molecule, O can easily have an octet, but the two K atoms cannot satisfy the octet rule (they end up with only 2 electrons each.) You could also draw a molecule in which the O atom forms a double bond to each K atom (though it's not the usual bonding arrangement for O); again, O would have an octet, but the K atoms would only have 4 electrons each. There just are not enough electrons to go around for covalent bonding in this molecule.

3.75 *The following are possible Lewis structures for H_2 and Li_2. One of these two molecules is very stable while the other is not. Which one is which? Explain your answer.*

$$\text{H:H} \qquad \text{Li:Li}$$

In the H_2 molecule, each H atom has a filled valence electron shell (remember that the "octet" for H and He is 2, because the first shell only holds 2 electrons), so this molecule is stable. In the Li_2 molecule, each Li atom has only 2 electrons; the Li atoms cannot satisfy the octet rule. Their valence electrons are in shell 2, which holds 8 electrons. The Li_2 molecule is not stable.

3.77 *Using electron arrangements, explain why removing one electron from a sodium atom makes an ion that satisfies the octet rule.*

Sodium (Na) has the following electron arrangement:

shell 1: 2 electrons shell 2: 8 electrons shell 3: 1 electron

An electrically neutral sodium atom has one valence electron. If the atom loses one electron to form a +1 ion, the original valence shell will be empty, leaving the ion with eight electrons in shell 2.

shell 1: 2 electrons shell 2: 8 electrons

3.79 *A student says "N_2O_3 is an ionic compound, made from N^{3-} and O^{2-} ions." Explain why this cannot be true.*

Compounds must have neutral (zero) charge overall, so it is never possible to combine only anions to form an ionic compound. N_2O_3, made from the ions listed, would have an overall charge of –12 per formula unit. In general, compounds formed of two or more nonmetals are nearly always molecular (an exception being the ionic compounds of NH_4^+).

3.81 a) *Which of the following drawings represent molecules, and which represent individual atoms?*

Only (b) is an individual atom (one circle by itself, not attached to any other circles.) Each circle represents an atom (the smallest particle of an element). Molecules are groups of atoms attached by covalent bonds (represented by lines connecting the circles.)

b) *Which of the following drawings represent compounds, and which represent elements?*

Compounds are composed of two or more different elements. **Only (e) represents a compound**. While (a), (c) and (d) are all molecules, each made up of more than one atom, only E contains a combination of two different elements. This is an important distinction to make, and one students often miss: molecules are made up of two or more *atoms*, but the atoms may be the same element or different elements; compounds are made up of two or more different *elements.*

c) *Write a chemical formula for each of these substances.*

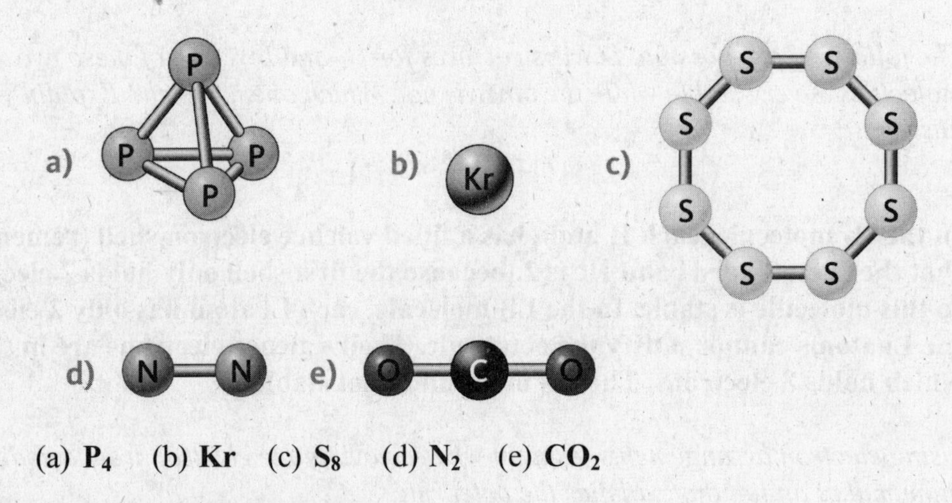

(a) $\mathbf{P_4}$ (b) **Kr** (c) $\mathbf{S_8}$ (d) $\mathbf{N_2}$ (e) $\mathbf{CO_2}$

Solutions to Summary and Challenge Problems

* indicates more challenging problems.

*3.83 *Aspartic acid is a vital component of most proteins. It has the basic structure shown here. Complete this structure by adding bonds, nonbonding electron pairs, or both so all atoms obey the octet rule and satisfy their normal bonding requirements.*

This looks harder than it is, and it's a really important problem for understanding a lot of things in Chapter 3.

Our approach to this problem is to first consider the normal bonding patterns for the atoms involved: H always has one bond and no lone pairs, C typically has four bonds and no lone pairs, N typically has three bonds and one lone pair, and O typically has two bonds and two lone pairs. Then we go through the molecule an atom at a time, considering how we can add bonds and lone pairs to complete everything. The molecule is shown here with the atoms (other than H) numbered for convenience.

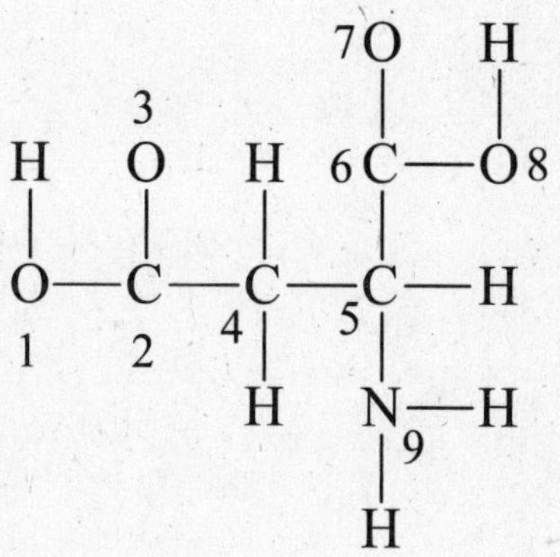

```
                      7O     H
          3            |     |
 H        O     H    6 C ---O 8
 |        |     |      |
 O ------ C ---- C ----C --- H
 1        2    4 |    5|
                H      N --- H
                       |9
                       H
```

First, all the hydrogen atoms in the molecule already have one bond, so we won't do anything to the hydrogen atoms. Oxygen atoms #1 and #8 have two bonds each, so we add two lone pairs to each to finish their octets. Carbon atoms #4 and #5 each have their four bonds, so they're finished. The nitrogen atom #9 has three bonds, so we add one lone pair to finish its octet. Here's what we have so far:

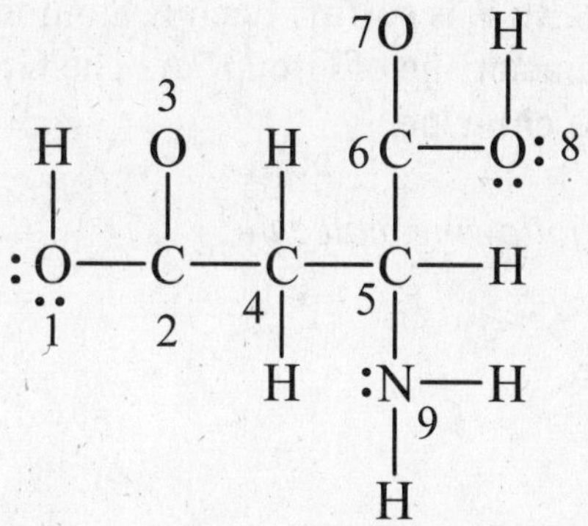

```
                      7O     H
          3            |     |
 H        O     H    6 C ---O: 8
 |        |     |      |    ..
:O ------ C ---- C ----C --- H
 ..
 1        2    4 |    5|
                H     :N --- H
                       |9
                       H
```

The more difficult cases are atoms #2, #3, #6 and #7. In each case we have a carbon with three bonds (needing one more bond) attached to an oxygen atom with one bond (needing one more bond). We can satisfy the bonding requirements of both atoms by adding a line to form a double bond between the carbon atom and the oxygen atom; this gives the carbon atoms, #2 and #6, 4 bonds each and the oxygen atoms, #3 and #7, 2 bonds each:

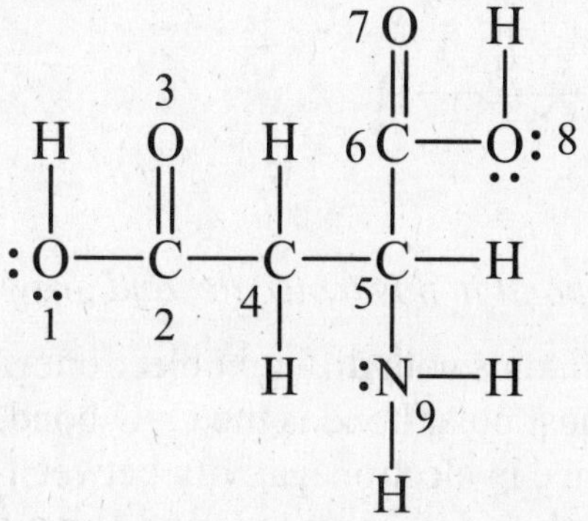

```
                      7O     H
          3            ||    |
 H        O     H    6 C ---O: 8
 |        ||    |      |    ..
:O ------ C ---- C ----C --- H
 ..
 1        2    4 |    5|
                H     :N --- H
                       |9
                       H
```

Now all carbons are satisfied, and the only thing remaining is to add two lone pairs to oxygen atoms #3 and #7. As a final step, we check back over the whole molecule to make sure each atom has its preferred bonding pattern.

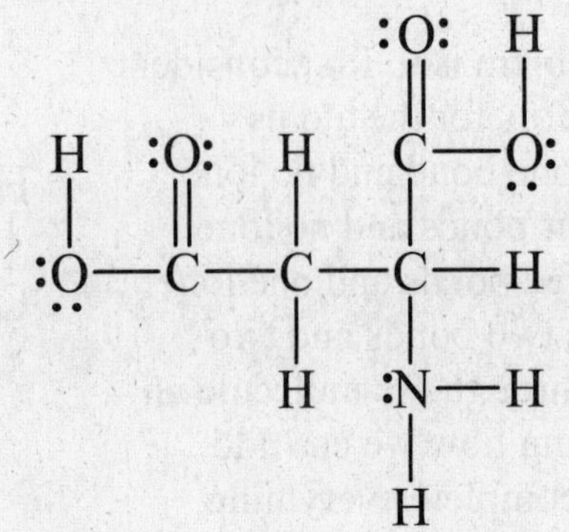

```
            :O:  H
            ||   |
  H  :O:  H  C—O:
  |   ||  |  |
:O—C——C——C—H
          |  |
          H :N—H
             |
             H
```

*3.85 *X and Z are elements in Period 3. They form the compound ZX_2, which has the Lewis structure shown here. Identify elements X and Z.*

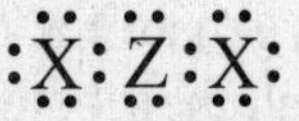

:X:Z:X:

Our most obvious clue here is in the bonding pattern of each atom. Z forms two bonds and has two lone pairs, characteristic of Group 6A. The Group 6A element in Period 3 is **sulfur, S, so Z is sulfur.** Each X atom forms one covalent bond and has three lone pairs, characteristic of Group 7A. The Group 7A element in Period 3 is **chlorine, Cl, so X is chlorine.**

Question 3.87 refers to the following molecule:

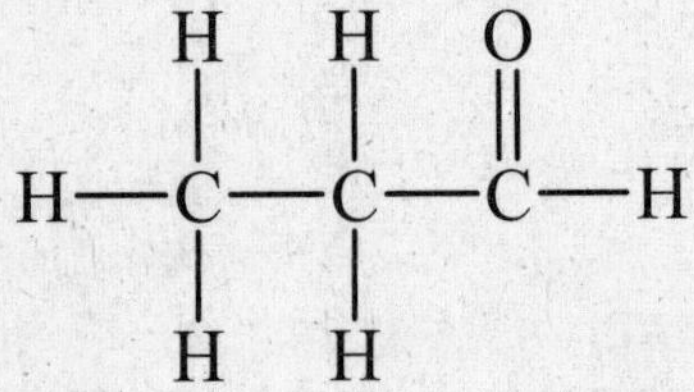

```
   H  H  O
   |  |  ||
H—C—C—C—H
   |  |
   H  H
```

3.87 *a) Find one nonpolar bond in this molecule, and draw an "X" through it.*

Bonds between atoms of the same element are nonpolar. In this molecule, there are two C-C bonds that are marked as nonpolar.

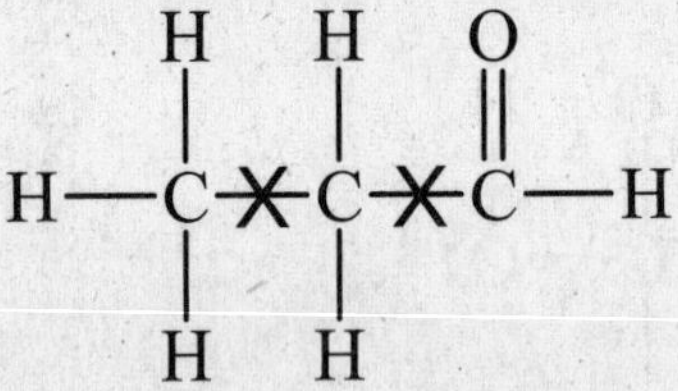

```
   H  H  O
   |  |  ||
H—C✕C✕C—H
   |  |
   H  H
```

b) Find one polar bond in this molecule, and draw a circle around it.

Bonds between atoms with different electronegativity values are polar. In this molecule, the most polar bond is the C=O bond, because that's the bond with the greatest difference in electronegativity between the two atoms, but the C-H bonds are all also slightly polar, because C and H have slightly different electronegativity values. Therefore any of the C-H bonds could also have been circled. (Organic chemists often approximate C-H bonds as nonpolar because the difference is small enough that it can often be ignored in practice, so make sure that you are consistent with your instructor; this should have been made clear in lecture.)

```
    H   H   O
    |   |   ‖
H — C — C — C — H
    |   |
    H   H
```

c) *One atom in this molecule is strongly negatively charged. Draw an arrow that points toward this atom.*

"Strongly" charged clues us in that we're looking for a significantly polar bond—in this case, the C=O bond. Because O has a much greater electronegativity than C, the O atom has a strong negative charge in this bond.

```
              ↙
    H   H   O
    |   |   ‖
H — C — C — C — H
    |   |
    H   H
```

d) *Are there any positively charged atoms in this molecule? If so, which ones are they?*

The biggest positive charge belongs to the C in the C=O, because that is the most polar bond (biggest difference in electronegativity between the two atoms); the O atom has a strong negative charge, and the C atom has a strong positive charge. But because each of the C-H bonds is also slightly polar, with C being more electronegative than H, the H atoms in this molecule are all also slightly positive. (Again, some instructors may treat the C-H bond as nonpolar, so make sure you're consistent with your instructor on this point.)

*3.89 *The element titanium can combine with oxygen to form the ionic compounds listed below. What is the charge on each titanium atom in each compound?*

In each case, remember that the oxide ion has a –2 charge, and use that to figure out the charge on Ti.

a) *TiO* **+2**: Ti^{2+} combines in a 1:1 ratio with O^{2-}.

b) *TiO_2* **+4**: Ti^{4+} combines in a 1:2 ratio with O^{2-}.

c) *Ti_2O_3* **+3**: Ti^{3+} combines in a 2:3 ratio with O^{2-}.

3.91 *Tell whether each of the compounds in Problem 3.90 is ionic or molecular. (You should not need to draw structures of the compounds.)*

In each case, if the compound is made of two or more nonmetals and does not include ammonium ions, it is molecular. If it is a combination of a metal (or ammonium ion, NH_4^+) and a nonmetal or a polyatomic ion, it is ionic.

NOTE:
Before you can even approach these problems, you need to have memorized the polyatomic ions in Table 3.11, and you need to be able to identify elements as metals or nonmetals based on their positions in the periodic table (Figure 2.3).

a) zinc bromide	**ionic**: metal + nonmetal
b) sulfur tetrafluoride	**molecular**: nonmetal + nonmetal
c) sodium nitride	**ionic**: metal + nonmetal
d) dinitrogen pentoxide	**molecular**: nonmetal + nonmetal
e) silver phosphate	**ionic**: metal + polyatomic anion
f) potassium sulfide	**ionic**: metal + nonmetal
g) nickel hydroxide	**ionic**: metal + polyatomic anion
h) nitrogen triiodide	**molecular**: nonmetal + nonmetal
i) chromium(III) nitrate	**ionic**: metal + polyatomic anion
j) cuprous oxide	**ionic**: metal + nonmetal
k) ammonium bromide	**ionic**: ammonium ion + nonmetal
l) carbon disulfide	**molecular**: nonmetal + nonmetal
m) nitrous oxide	**molecular**: nonmetal + nonmetal
n) potassium hydrogen carbonate	**ionic**: metal + polyatomic anion

3.93 *Each of the following names is incorrect. Explain why, and give the correct name for each compound.*

a) $CaCl_2$ *"calcium dichloride"* — **This is an ionic compound; numeric prefixes aren't used.** Correct name is **calcium chloride.**

b) FeO *"iron oxide"* — **Iron is a transition metal that can have more than one possible charge in compounds, so the charge must be specified with a Roman numeral.** Correct name is **iron(II) oxide.**

c) $NaNO_3$ *"sodium nitrogen trioxide"* — **The** NO_3 **group is actually a nitrate ion,** NO_3^-. Correct name is **sodium nitrate.**

d) ICl *"iodine chloride"* — **This is a molecular compound, and numeric prefixes must be used to specify how many atoms of each element.** Correct name is **iodine monochloride** ("mono" is not used with the first element).

Chapter 4

4

Energy and the Physical Behavior of Chemical Substances

Solutions to Section 4.1 Core Problems

4.1 *A baseball player hits a ball straight up. As the ball rises, does its potential energy increase, decrease, or remain the same? What about its kinetic energy?*

As the ball rises into the air, it gets farther and farther from its most stable position (on the ground), so **its potential energy is increasing as long as it keeps going up** (reaching a maximum value at the top of its rise, just before it starts to fall again.) Kinetic energy is energy of motion; **as the ball rises, its speed decreases, and the slower it moves, the less kinetic energy it has.** As the ball rises, kinetic energy is converted into potential energy.

4.3 *Which of the following are examples of kinetic energy, and which are examples of potential energy?*

a) *The energy in a gallon of gasoline* — While the molecules in a gallon of gasoline are moving and do have kinetic energy, when we talk about the "energy" in a fuel we're usually talking about **potential energy**—the energy we could get out of the fuel by burning it and converting it into other substances.

b) *The energy in a cup of boiling water* — We can't burn water and turn it into something else to get energy from it; the energy in a cup of boiling water is **kinetic energy** (the molecules of water are moving fast, compared to the same molecules at room temperature.)

c) *The energy in a spinning top* — **Kinetic energy**—the top is moving quickly.

4.5 *For each of the following pairs of objects, tell which one has more kinetic energy:*

Kinetic energy is energy of motion. Kinetic energy depends on the mass and speed of a moving object, and the kinetic energy of atoms and molecules is related to the temperature of an object or sample of matter.

a) *A car moving at 30 mph (miles per hour) or the same car moving at 40 mph.*

The faster an object is moving, the greater its kinetic energy, so **the car moving at 40 miles/hour has more kinetic energy.**

b) *A car moving at 15 mph or a bicycle moving at 15 mph.*

For two objects moving at the same speed, the one with more mass has more energy, so **the car has more kinetic energy.**

c) *The atoms in a cup of 80°C water or the atoms in a cup of 70°C water.*

For the same substance, the higher the temperature, the greater the thermal energy (kinetic energy of the molecules.) **The 80°C cup of water has more kinetic energy.**

d) *A new battery or a used battery.*

If the batteries are both still and at the same temperature, **they both have the same kinetic energy**.

4.7 *For each of the following pairs of objects, tell which one has more potential energy:*

Potential energy is stored energy; it depends on how far something is from its most stable position or condition.

a) *An airplane at 30,000 feet or an airplane at 20,000 feet.*

The airplane at 30,000 feet has more potential energy because it has farther to fall.

b) *A new battery or a used battery.*

The "energy" of a battery is potential energy—chemical energy that can be converted into kinetic energy through a chemical reaction. **A new or fully-charged battery has more potential energy than a battery that has been partially or fully discharged.**

c) *A warm piece of bread or a cool piece of bread.*

Temperature is a measure of kinetic energy, not potential energy. **Both pieces of bread have the same potential energy**, assuming they are in the same location (though the warm piece has more kinetic energy).

d) *A small stone on top of a building or a large stone on top of the same building.*

For two objects at the same height, the heavier one has more potential energy. The two stones are the same distance from the ground, but the large stone took more work to get there and will do more work upon hitting the ground. **The large stone has more potential energy.**

4.9 *Misti puts a bottle of water in the refrigerator.*

a) *As the water cools, do the water molecules speed up, slow down, or continue to move at the same speeds?*

Since the molecules don't change their mass as they slow down, the temperature of the sample is a measure of molecular motion. **As the water gets cooler, the molecules slow down.**

b) *Does the amount of thermal energy in the water increase, decrease, or remain the same?*

As something cools, it loses thermal energy to its surroundings. **The amount of thermal energy in the water decreases as the water cools.** (It's relatively simple when we are comparing equal amounts of the same substance; it would be a lot more complicated if we were comparing different sample sizes or different substances.)

4.11 a) *How many calories of heat do you need if you want to raise the temperature of 350 g of gasoline from 18.0°C to 22.0°C?*

heat = m × ΔT × c_p

We get c_p from table 4.2 (0.40 cal/g· °C for gasoline), and ΔT = 22 – 18 = 4°C.

$$\text{heat} = 350\text{ g} \times 4^\circ\text{C} \times \frac{0.40\text{ cal}}{\text{g} \bullet {}^\circ\text{C}} = 560\text{ cal}$$

NOTE:
The calculator answer already has 2 significant figures, the same as in the original specific heat measurement, so we don't need to round.

b) *Express your answer to part "a" in joules and in kilocalories.*

Joules: 4.184 J = 1 cal, so

560 cal × 4.184 J/cal = 2343.04 J (calculator answer)
= **2300 J** (rounded to 2 significant figures)

kilocalories: 1 kcal = 1000 cal, so

560 cal × 1 kcal/1000 cal = **0.56 kcal**

Again, our calculator answer already has 2 significant figures, so we don't need to round (remember that conversions within the metric system are exact, so the 1 kcal = 1000 cal conversion factor does not affect significant figures.)

Solutions to Section 4.2 Core Problems

4.13 *For each of the following statements, tell whether it describes a solid, a liquid, a gas, or more than one of these:*

a) *Sulfur dioxide has an extremely low density.*

"Extremely low" is a relative term, but in the context of comparing densities of solids, liquids, and gases, we know that gases have much lower densities than the other two states. **Sulfur dioxide is a gas.**

b) *Carbon disulfide can be poured from one container to another, but does not change its volume.*

Both liquids and gases can be poured, but liquids have constant volume, while gases expand or contract to fill their entire container. **Carbon disulfide is a liquid.**

c) *A piece of calcium chloride does not change its shape when it is put into a different container.*

Calcium chloride is a solid; only solids retain their shapes. (Actually, even the fact that you could talk about "a piece" of calcium chloride is a clue that it's a solid.)

4.15 *If you remove thermal energy from a glass of water, which of the following will happen to the water? (List all of the correct answers.)*

Thermal energy and temperature are closely related: the greater the thermal energy in a given sample, the higher its temperature. (Remember this only applies when you're talking about the same amount of the same substance—change amounts or substances and different reasoning applies.)

a) *The temperature of the water will go down.*

This is exactly what will happen—"lower temperature" actually means "less thermal energy."

b) *The temperature of the water will go up.*

No—you'd have to *add* thermal energy to increase the temperature of the sample.

c) *The water will freeze.*

If enough thermal energy is removed, the sample will eventually freeze.

d) *The water will boil.*

No—again, you'd have to *add* thermal energy to get this to happen.

4.17 *Describe the behavior of a helium atom inside a balloon that is filled with helium.*

The helium in a balloon is in the gas phase, so an atom of helium in a balloon would be moving randomly and fast. The molecule would travel in a straight line until it bounced off the wall of the container or another helium atom.

4.19 *Carbon dioxide is a gas at room temperature. At which of the following temperatures is it most likely to be a solid?*

–100°C *0°C* *100°C*

Solids have less thermal energy than gases or liquids, so the lower the temperature of a substance, the greater the probability that it will be a solid. **Carbon dioxide is most likely to be a solid at –100°C.**

4.21 *Explain why 1 g of gaseous nitrogen takes up far more space than 1 g of liquid nitrogen.*

The molecules in a sample of a liquid are essentially touching, while the molecules in a sample of gas are far apart. Gases are mostly empty space.

Solutions to Section 4.3 Core Problems

4.23 *Oxygen for medical use is sold in steel cylinders, because the container must be strong enough to withstand the high pressure when the cylinder is filled with a large amount of oxygen.*

a) *As the oxygen is drained from the cylinder, what happens to the pressure inside the cylinder?*

Fewer molecules of gas in the cylinder will have fewer collisions with the walls of the cylinder, so **the pressure will decrease.**

b) *If the temperature of the cylinder is lowered to 0°C, what happens to the pressure inside the cylinder?*

As the temperature of a sample of gas drops, the molecules move more slowly. They collide less frequently and less forcefully, and **the pressure decreases.**

c) *If the cylinder is dented, the volume of the oxygen inside the cylinder decreases. What happens to the pressure inside the cylinder?*

A decrease in the volume of the cylinder means that the molecules of gas have less space to move around in, resulting in an increase in the number of collisions between molecules and collisions with the walls of the container. **The pressure will increase.**

4.25 *Many household products are packaged in aerosol cans, which contain a high-pressure propellant gas. These cans always carry a warning against heating the can. Why is this?*

As the temperature of a sample of gas increases, the molecules collide with each other and the walls of the container more frequently and with more force, increasing the pressure. **Heating the gas could cause the pressure to increase to the point that the can (or its valve) could rupture.**

4.27 *A hiker drinks half of the water in her water bottle at the top of a mountain. When she gets to the bottom of the mountain, she notices that the water bottle has partially collapsed. Why is this?*

At the top of the mountain, air pressure is lower. **The amount of air that fills the space in the bottle when it's open on the mountaintop isn't enough to fill the same space at the higher atmospheric pressure in the valley, so the bottle collapses—the volume changes to equalize the pressure inside & outside the bottle.**

4.29 *The air pressure inside a bicycle tire is 80 psi. Convert this into the following units:*

a) *torrs* From Table 4.6, 1 psi = 51.7 torr.

$$\mathbf{80\ psi \times \frac{51.7\ torr}{1\ psi} = 4136\ torr\ (calculator\ answer)}$$

$$\mathbf{= 4100\ torr\ (rounded\ to\ 2\ sf)}$$

b) *bars* From Table 4.6, 1 bar = 750 torr and 1 psi = 51.7 torr.

$$\mathbf{80\ psi \times \frac{51.7\ torr}{1\ psi} \times \frac{1\ bar}{750\ torr} = 5.5146666667\ bar\ (calculator\ answer)}$$

$$\mathbf{= 5.5\ bar\ (rounded\ to\ 2\ sf)}$$

4.31 *A scuba-diving tank contains a mixture of oxygen and helium at a total pressure of 6.25 atm. If the partial pressure of the oxygen is 0.61 atm, what is the partial pressure of the helium?*

The total pressure is the sum of the pressures of the component gases. This mix only contains oxygen and helium, so whatever isn't oxygen is helium:
6.25 atm total – 0.61 atm oxygen = **5.64 atm helium**

Solutions to Section 4.4 Core Problems

4.33 *What is the correct term for each of the following changes of state?*

a) *Alcohol rapidly vanishes after it is put on the skin.* liquid alcohol to gas phase: **evaporation**

b) *Water turns to ice on a cold winter day.* liquid water to solid: **freezing**

c) *If air that contains water vapor is cooled, fog forms.* water vapor (gas) to liquid droplets: **condensation**

4.35 *Naphthalene has the chemical formula $C_{10}H_8$. It is the active ingredient in mothballs. The melting point of naphthalene is 80°C, and its boiling point is 218°C.*

a) *Is naphthalene a solid, a liquid, or a gas at room temperature?*

Room temperature is 20–25°C, so you would have to heat the naphthalene to melt it—**it is a solid at room temperature**.

b) *What is the normal state of naphthalene at 100°C?*

100°C is between the melting and boiling temperatures of naphthalene, so **it is a liquid at 100°C**. Think of it this way: if you took a sample at room temperature of 20°C and started heating it, the sample would melt to a liquid when it reached 80°C. If you kept heating it, the sample wouldn't boil until you reached 218°C; it would stay liquid through the 100°C mark.

4.37 *Sodium chloride (NaCl) melts at 801°C. If you have some liquid NaCl at 1000°C and you cool it down, at what temperature will it freeze?*

The melting point and freezing point of a substance are the same temperature (approached from opposite directions. If solid NaCl melts at 801°C, **then liquid NaCl will freeze at 801°C.**

4.39 *Methane (CH_4) is a gas at 25°C. Is the boiling point of methane higher than 25°C, or is it lower than 25°C? How can you tell?*

If methane is a gas at 25°C, then you would have to cool methane to a lower temperature to condense it. **The boiling point must be lower than 25°C**.

Another way to look at this: imagine liquid methane at some very low temperature. If you start warming the methane, it's going to boil at some temperature—we don't know what temperature, but it's going to boil before we reach room temperature (20–25°C), because we know it's a gas by the time it reaches room temperature.

4.41 *The sentence below is not entirely correct. How could you make it into a true statement? "At 20°C, water is a liquid."*

We know that there can be a significant amount of water in the gas phase at 20°C, because atmospheric humidity is gas-phase water in the air. One way to make this a true statement is to add the water vapor: **"At 20°C, water is a liquid, along with some water vapor."**

4.43 *Isopropyl alcohol can be used to cool the skin, because it evaporates easily and absorbs a large amount of heat as it does so. The heat of vaporization of isopropyl alcohol is 159 cal/g. How much heat is needed to evaporate 25 g of isopropyl alcohol?*

This is a problem we could have worked back in Chapter 1, without knowing anything about heat, evaporation, or isopropyl alcohol, by looking at the units. The heat of vaporization given in the problem has compound units (calories/gram), and values with compound units often end up being used as conversion factors (Section 1.6).

$$\textbf{25 grams} \times \frac{\textbf{159 calories}}{\textbf{1 gram}} = \textbf{3975 calories (calculator answer)}$$

$$= \textbf{4.0} \times \textbf{10}^3 \textbf{ calories (rounded to 2 sf)}$$

4.45 *Isopropyl alcohol is a liquid at room temperature. If you cool isopropyl alcohol, its temperature will drop to –89°C. However, the temperature then remains at –89°C for a few minutes before dropping further. Explain.*

–89°C must be the freezing point of isopropyl alcohol. As heat is removed from a sample of a liquid, the temperature drops smoothly until the freezing point is reached and the substance begins to freeze; then the heat that is removed all comes from turning the liquid into solid, until the entire sample has solidified. At that point, the temperature will begin to drop again.

4.47 *Use the following heating to answer the following questions.*

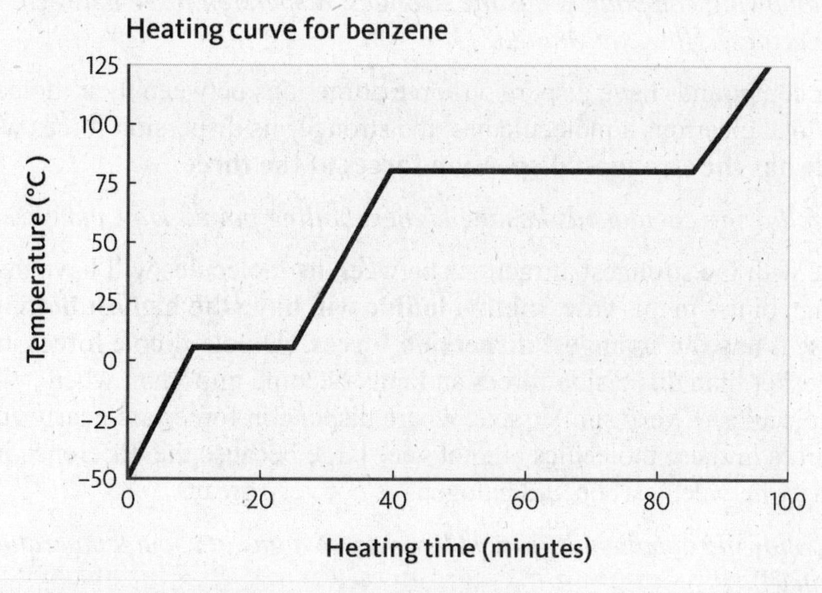

a) *What is the melting point of benzene?*

The melting point is the temperature at which the solid turns into liquid. As the solid is heated, its temperature increases smoothly until the melting point is reached; at that point, all the heat added goes into melting the sample instead of making it warmer, and the temperature remains constant even as heat is added. Once the entire sample is liquid, the temperature starts to increase again. **The flat portion of the graph at about 5°C is the melting point.**

b) *Is benzene a solid, a liquid, or a gas at 100°C?*

The two horizontal lines in the graph are the melting point and boiling point of benzene. At 100°C, the substance is past both melting and boiling, and the temperature is climbing again, so **benzene must be a gas at 100°C.**

c) *If you cool benzene from 25°C to –25°C, what will you observe?*

Benzene is a liquid at 25°C and a solid at –25°C, so **we will see benzene freeze as it is cooled over this range of temperatures.**

Solutions to Section 4.5 Core Problems

4.49 *Explain why NF_3 is a gas at room temperature, while CrF_3 is a solid that must be heated to 1100°C to melt it.*

NF_3 is a compound of two nonmetals; therefore it is a covalent compound and consists of individual NF_3 molecules, with no strong attractions between the molecules. The molecules are therefore free to float around in the gas phase at room temperature. Molecular compounds generally have low melting and boiling points compared to ionic compounds. CrF_3, a compound of a metal and a nonmetal, is ionic; the attractive forces (ion-ion attractions) in a crystal of CrF_3 are distributed evenly throughout the entire crystal and are relatively strong, so CrF_3 melts at a high temperature. Ionic compounds have high melting and boiling points compared to molecular compounds, and are always solid at room temp.

4.51 a) *Which of the following compounds has the strongest dispersion force between individual molecules? How can you tell?*

All molecular compounds have dispersion force attractions between their molecules, and the more total electrons a molecule has, the stronger its dispersion forces will be. **Methyl iodide has the strongest dispersion forces of the three.**

b) *Which of the following compounds has the highest boiling point? How can you tell?*

The substance with the strongest attractions between its molecules will have the highest boiling point—in this case, **methyl iodide will have the highest boiling point, because it has the strongest dispersion forces.** (Dipole-dipole forces in general are weaker than dispersion forces and only become important when comparing molecules of very similar size, where dispersion forces are nearly equal. The dipole forces in these molecules are not very large because the difference in electronegativity between carbon and halogens is not very great.)

c) *Which of the following compounds is most likely to be a gas at room temperature? How can you tell?*

The substance with the weakest attractions between its molecules will have the lowest boiling point, and is therefore the most likely to be a gas at room temperature. **Methyl chloride has the weakest dispersion forces, and therefore the lowest boiling point, of the three.** (Even though methyl chloride is the most polar, and therefore has the strongest dipole-dipole forces, of the three, it's not enough to overcome the strong dispersion forces between the methyl iodide molecules.)

Methyl chloride **Methyl bromide** **Methyl iodide**

4.53 *Each of the compounds below contains a covalent bond that is strongly polar.*

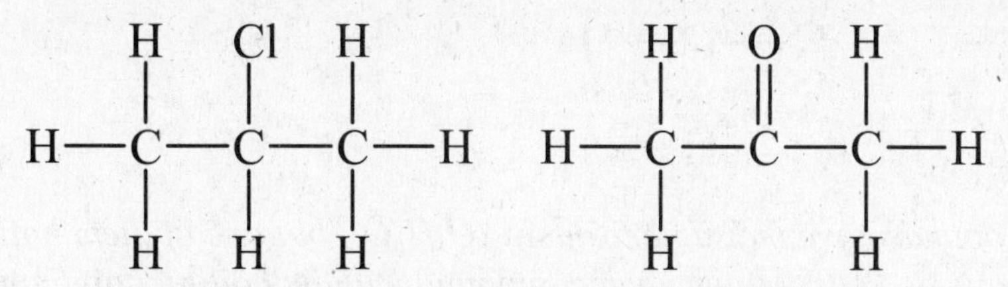

Isopropyl chloride Acetone

a) *Identify the polar bond in each molecule.*

Polar bonds, by definition, are bonds between atoms with very different electronegativity values. **The polar bonds are the C–Cl bond in isopropyl chloride and the C=O bond in acetone.**

b) *For which of these compounds should dipole–dipole attraction have a significant effect on the boiling point?*

Dipole-dipole attractions are described in Section 4.5. Both of these molecules are polar and therefore have dipole-dipole attractions, but these attractions are much more significant for compounds with N or O atoms than for compounds that contain halogens, so **the properties of the second compound, including its boiling point, should be more affected by dipole-dipole attractions.**

c) *Which compound should have the higher boiling point?*

The second compound has stronger dipole-dipole forces and is expected to have the higher boiling point.

4.55 *Draw structures to show how two molecules of the following compound below can form a hydrogen bond.*

A hydrogen bond is an attraction between an H atom covalently bonded to N, O or F on one molecule, and a lone pair on N, O or F on another molecule.

NOTE:

Carbon atoms and hydrogen atoms attached to carbon atoms, can **never** participate in hydrogen bonding. ALSO, we want to show the hydrogen bond as an attraction, not a full covalent bond, so we use a dashed line or arrows, never a solid line.

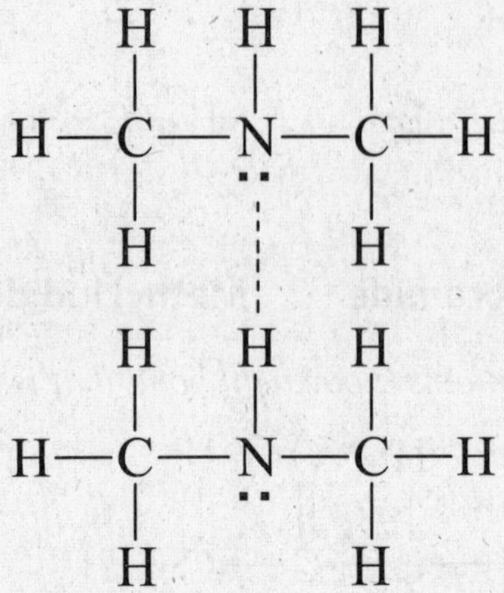

4.57 *The two following compounds have the same formula (C_3H_6O), but one of them boils at 95°C while the other boils at 49°C. Match each compound with its boiling point, and explain your answer.*

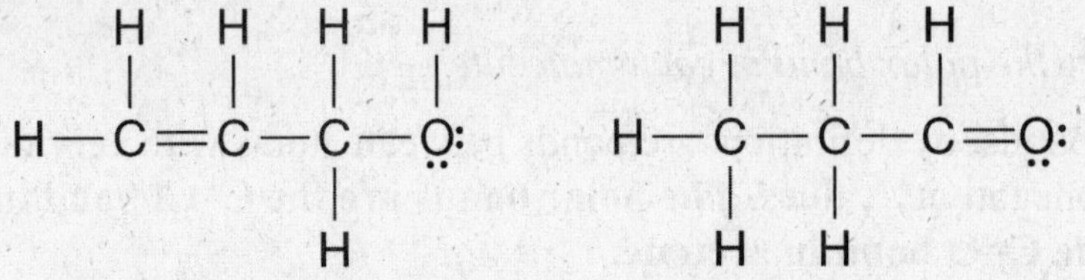

Compound 1 **Compound 2**

Both molecules are the same size, and therefore have about the same dispersion forces between molecules. Compound 1 is capable of forming hydrogen bonding attractions between its molecules, while compound 2 is not (all of the H atoms are attached to C, and for hydrogen bonding, we need H atoms attached to N, O or F.) Therefore **compound 1 will have stronger total attractions between its molecules and will have a higher boiling point. Compound 1 boils at 95°C, and compound 2 boils at 49°C.**

4.59 *The molecules that follow are roughly the same size, but they have very different boiling points. Explain the differences in their boiling points.*

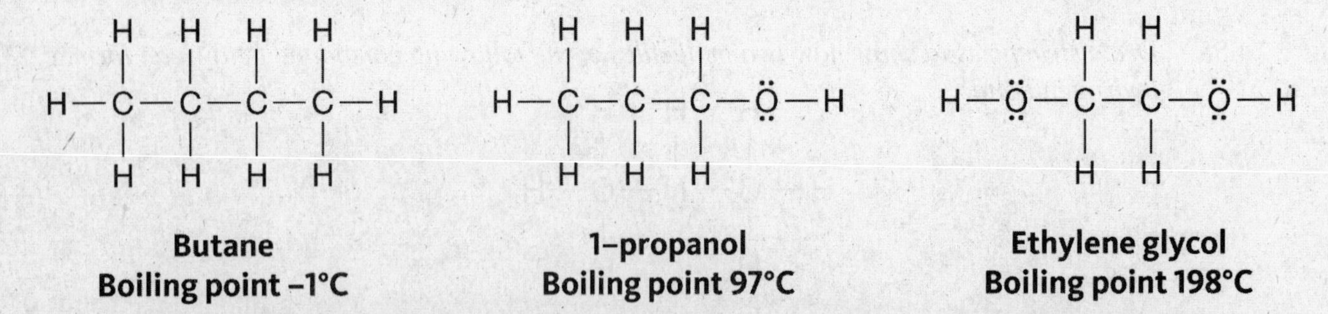

Butane
Boiling point −1°C

1–propanol
Boiling point 97°C

Ethylene glycol
Boiling point 198°C

Since all three molecules are about the same size, the dispersion forces between the molecules in each are about equal. Therefore, we compare the molecules' abilities to form hydrogen bonds. **Butane cannot form hydrogen bonds at all (all H atoms are attached to C, and for hydrogen bonding interactions to occur, there must be H atoms covalently bonded to N, O or F). Butane therefore has only dispersion forces between its molecules and will have the lowest boiling point of the three. Both 1-**

propanol and ethylene glycol can form hydrogen bonds, but ethylene glycol can form twice as many hydrogen bonding interactions as 1-propanol, giving it significantly stronger attractions between its molecules and a much higher boiling point.

4.61 *Asparagine is one of the building blocks of proteins. Identify all of the hydrogen atoms that can participate in hydrogen bonds in this compound by drawing a circle around each atom.*

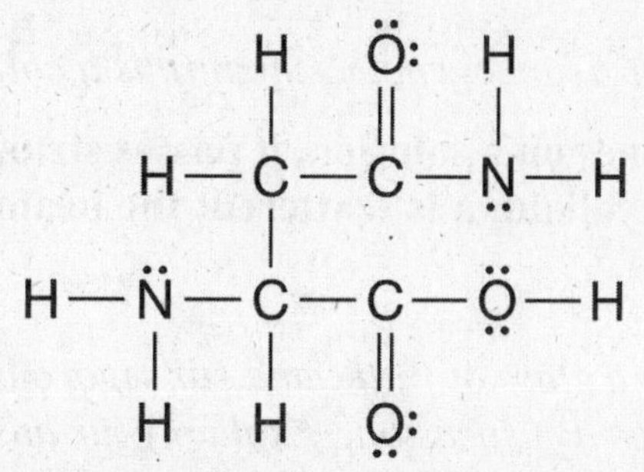

```
      H   Ö:  H
      |   ||  |
  H — C — C — N — H
      |       ..
H — N — C — C — Ö — H
    ..
    |   |   ||  ..
    H   H   O:
            ..
```

Hydrogen atoms that are covalently bonded to N, O or F can participate in hydrogen bonding. Hydrogen atoms attached to atoms of any other element (including, as in this molecule, C) cannot.

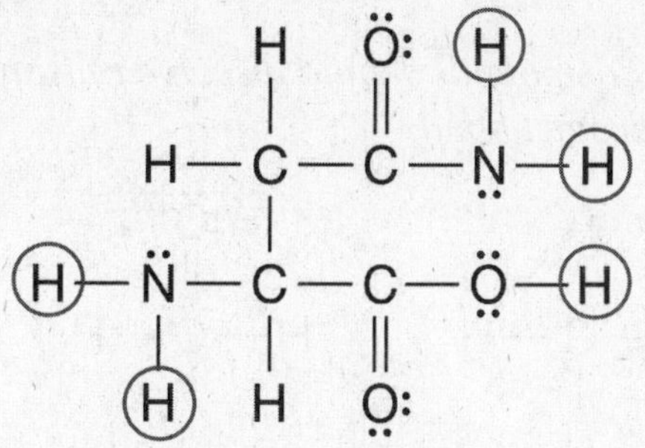

```
        H   Ö:  (H)
        |   ||   |
    H — C — C — N —(H)
        |       ..
(H)— N — C — C — Ö —(H)
     ..
     |   |   ||  ..
    (H)  H   O:
             ..
```

4.63 *Select from the following types of attractive forces: covalent bonds, ionic bonds, hydrogen bonds, and dispersion forces.*

a) *What types of attractive forces keep the hydrogen atoms attached to the oxygen atom in a water molecule?*

Covalent bonds attach atoms to each other within a molecule.

b) *What types of attractive forces exist between two separate water molecules?*

Dispersion forces exist between molecules of all molecular substances; in addition water molecules can form **hydrogen bonds** to other water molecules.

c) *Which of these forces (if any) must be overcome in order to boil water?*

When water or any other molecular substance is boiled, the molecules remain intact (the covalent bonds are *not* broken); only the attractions between molecules (in this case, **dispersion forces and hydrogen bonds**) are disrupted in boiling a molecular substance.

Solutions to Section 4.6 Core Problems

4.65 *When 10 grams of NaOH (a white solid) and 10 grams of water are mixed, the resulting mixture is a clear, colorless liquid. Is this mixture a solution? Which substance is the solute and which is the solvent in this mixture?*

The mixture is a single phase (all liquid), and is therefore a solution (a type of homogeneous mixture). The phase of the mixture (liquid) matches the phase of the water, and therefore water is the solvent.

4.67 *How could you tell whether a homogeneous mixture is a colloid or a solution?*

When a beam of light shines on a solution, it passes straight through, but when a beam of light shines on a colloid, it is scattered; the beam is visible from the side as it shines through a colloid.

4.69 *If you put a little flour into a glass of water and stir vigorously, will you produce a solution or will you produce a suspension? Explain your answer briefly.*

Suspension—the flour will not dissolve in the water. In order to be classified as a solution, the solute has to dissolve completely to the point that there are no solid particles left. A mixture of flour and water will settle over time.

4.71 *Which of the following compounds should dissolve reasonably well in water, based on their ability to form hydrogen bonds?*

```
     H  H  H             :Ö:              H   H
     |  |  |              ||              |   |
  H—C—C—C—H          H—C—H          H—C—N—H
     |  |  |                              |   ..
     H  H  H                              H

      a)                  b)                c)
```

Compound C has H covalently bonded to N, and lone pair electrons on N, both of which can participate in hydrogen bonding interactions with water. **Compound C is therefore likely to dissolve well in water**. Compound B has no H atoms capable of forming hydrogen bonds, but the lone pair electrons on O can act as hydrogen bond acceptors, so **Compound B is also likely to dissolve well in water. Compound A has no ability to participate in hydrogen bonding, and is not likely to dissolve in water.**

4.73 *The following compound can form two different types of hydrogen bonds with a water molecule.*

```
     H   H
     |   |
  H—N—N—H
     ..  ..
```

a) *Draw structures that show these two types of hydrogen bonds.*

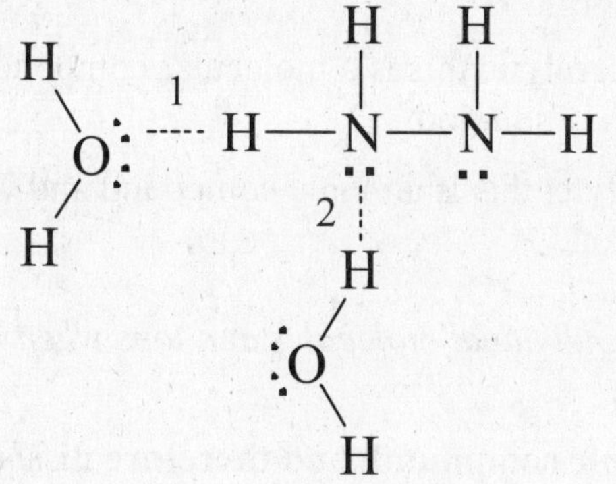

b) *For each type of hydrogen bond, tell which molecule is the donor and which is the acceptor.*

In attraction (1), N_2H_4 is the hydrogen bond donor and water is the hydrogen bond acceptor; since all of the H atoms in N_2H_4 are covalently bonded to N, any of them can form a hydrogen bond with a lone pair of electrons on the water molecule. (We could draw this same interaction between any H in this molecule and a water molecule.)

In attraction (2), N_2H_4 is the hydrogen bond acceptor and water is the hydrogen bond donor. The lone pair of electrons on either N atom can form a hydrogen bond with the H in the water molecule. (We could draw this same interaction from the lone pair on the other N atom.)

4.75 *Which of the following molecules can function as both a hydrogen bond donor and a hydrogen bond acceptor, and which can only function as a hydrogen bond acceptor?*

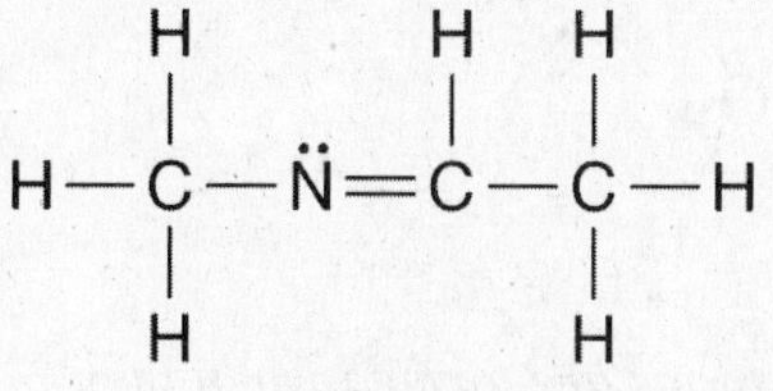

Compound 1 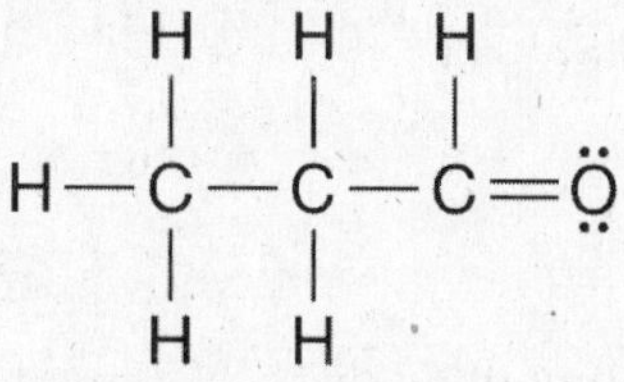 **Compound 2** 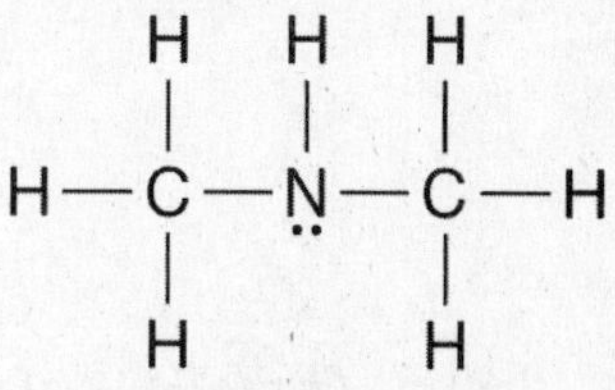 **Compound 3**

Hydrogen bond donors have H atoms covalently bonded to N or O. Hydrogen bond acceptors have lone pair electrons on N or O with δ- charges. Carbon atoms, and hydrogen atoms bonded to carbon atoms, cannot participate in hydrogen bonds at all. **Compounds 1 and 2 are hydrogen bond acceptors, while compound 3 can function as both a donor and acceptor.**

Solutions to Section 4.7 Core Problems

4.77 *All of the following compounds dissolve in water. Which of them are electrolytes? (You should not need to draw Lewis structures.)*

Soluble ionic compounds are electrolytes; most molecular compounds are not electrolytes. So we need to identify each substance as ionic or molecular.

a) $CaCl_2$ — **Electrolyte**: this is an ionic compound and will dissociate into ions in solution.

b) $HCONH_2$ dissociate — **Nonelectrolyte**: this is a molecular compound and will not into ions in solution.

c) $KC_2H_3O_2$ — **Electrolyte**: this is an ionic compound and will dissociate into ions in solution.

4.79 *Describe what happens to the individual molecules and ions when magnesium chloride dissolves in water.*

Magnesium chloride is an ionic compound, and therefore dissociates into Mg^{2+} ions and Cl^- ions when it dissolves in water. When the magnesium ions and chloride ions detach from the crystal lattice and disperse into the solution, each ion is surrounded with a cluster of water molecules. Each ion attracts the oppositely-charged part of the water molecule: a positive Mg^{2+} ion attracts the O in water, with its δ– charge; a negative Cl^- ion attracts the H in water, with its δ+ charge. (See Figure 4.31; all positive ions are going to behave like Na^+ in solution, and all negative ions are going to behave like Cl^- in solution.) This is one of those questions where a picture is worth, well, a lot of words:

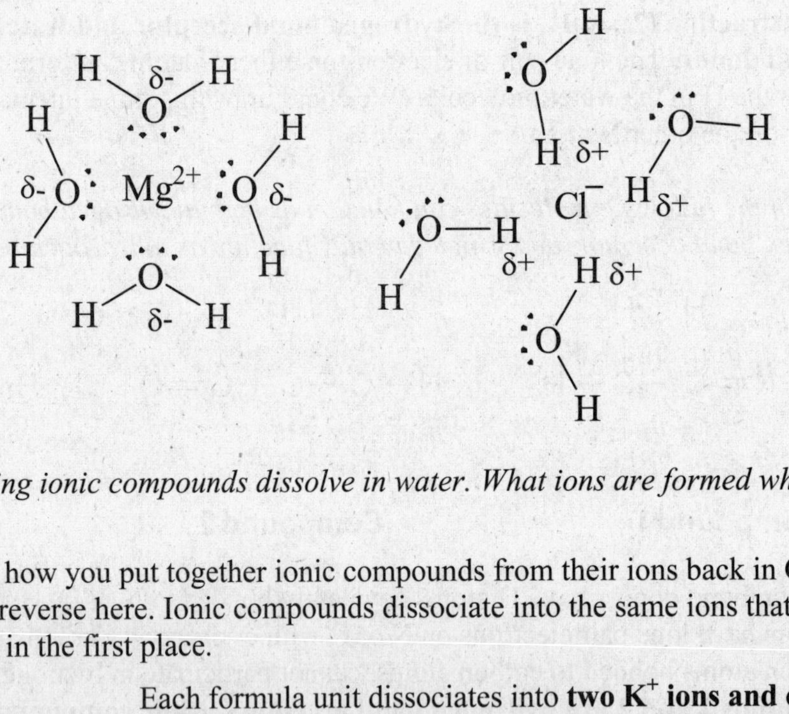

4.81 *The following ionic compounds dissolve in water. What ions are formed when they dissolve?*

Remember how you put together ionic compounds from their ions back in Chapter 3, and just do the reverse here. Ionic compounds dissociate into the same ions that were used to make them in the first place.

a) K_2S — Each formula unit dissociates into **two K^+ ions and one S^{2-} ion.**

b) $FeSO_4$ the — Each formula unit dissociates into **one Fe^{2+} ion and one SO_4^{2-} ion.** (In this case, we can identify the iron ion as Fe^{2+} because we know charge on the sulfate ion, and the two ions are in a 1:1 ratio. If this were Fe^{3+}, the formula of the compound would be $Fe_2(SO_4)_3$.)

c) $(NH_4)_2CO_3$ — Each formula unit dissociates into **two NH_4^+ ions and one CO_3^{2-} ion.**

4.83 *Draw a picture to show how calcium ions are solvated by water molecules.*

Since calcium ions are positive (Ca^{2+}), the water molecules surrounding them in solution will be oriented with the **δ– O atoms pointing in toward the positive calcium ion.**

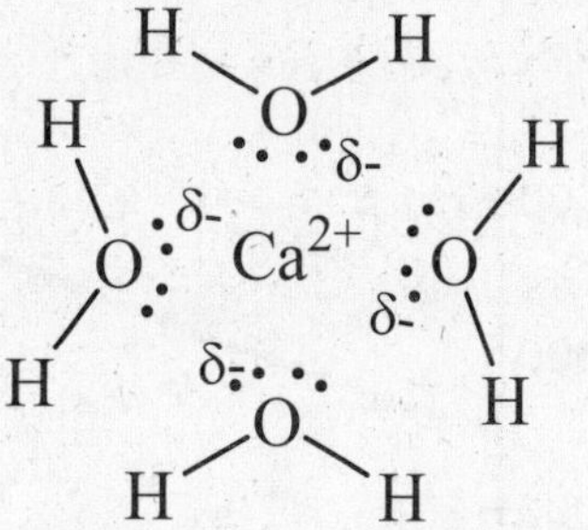

H H
H O δ- H
δ- O Ca^{2+} O δ-
H δ- O H
H H

4.85 *Molybdenum is a required nutrient. It occurs primarily in the molybdate ion, a polyatomic ion with the formula MoO_4^{2-}. Suggest a water-soluble compound that contains this ion and could be used in foods.*

While we have never seen the molybdate ion before and we don't know anything about its solubility, we do know that (1) we'll need to pair it with a positive ion to form a compound, and (2) all ionic compounds that contain Na^+ or K^+ are soluble in water. In addition, for use in foods, the ion we use must be non-toxic, and we know that Na^+ and K^+ are not toxic in small amounts. So (while there might also be other water-soluble compounds of this ion) we can be confident that **Na_2MoO_4 and K_2MoO_4 will be soluble in water**.

Solutions to Section 4.8 Core Problems

4.87 *Vitamin A is always described as insoluble in water. Does this really mean that you cannot dissolve vitamin A in water? If not, what does it mean?*

Nothing is really *completely* insoluble; when we call something "insoluble" we just mean it doesn't dissolve to a very great extent. In fact, very small amounts of vitamin A will dissolve in water, but the amount that will dissolve is much, much less than for water-soluble vitamins such as vitamin C. On the other hand, vitamin A is very soluble in (hydrophobic) fats, which is where it's stored in the body.

NOTE:
The solubility of most compounds in water changes with temperature, but predicting exactly *how* the solubility of a particular solid or liquid will change—whether increasing or decreasing—is far from simple (unlike the case of a gas dissolving in water, where solubility always decreases as temperature increases).

4.89 *The solubility of aspirin in water is around 3 g/L at 25°C. If you mix 2.5 grams of aspirin with a liter of water at 25°C, will all of the aspirin dissolve? Will you make a saturated solution or an unsaturated solution?*

Since the concentration of this solution (2.5 g/L) will be less than the solubility of aspirin at this temperature, all the aspirin will dissolve. The solution will be unsaturated, since more aspirin (another 0.5 g) could be dissolved in the same amount of water.

4.91 a) *Use Figure 4.34 to determine the temperature at which the solubility of sugar is 3000 g/L.*

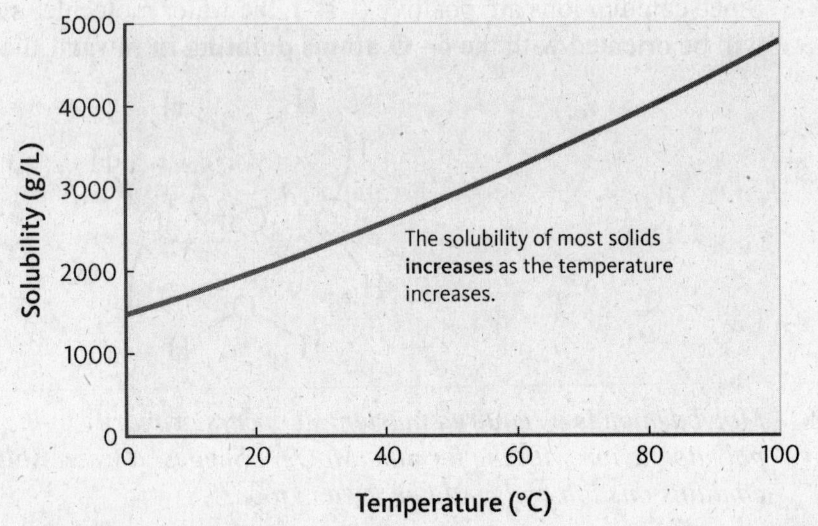

Figure 4.34

Go up the y axis to 3000 g/L, then follow straight across until you reach the solubility curve. From this point on the curve, follow another straight line down to the x axis. This line intersects the x axis at **about 53°C**.

b) *Use Figure 4.35 to determine the solubility of oxygen when the temperature is 20°C.*

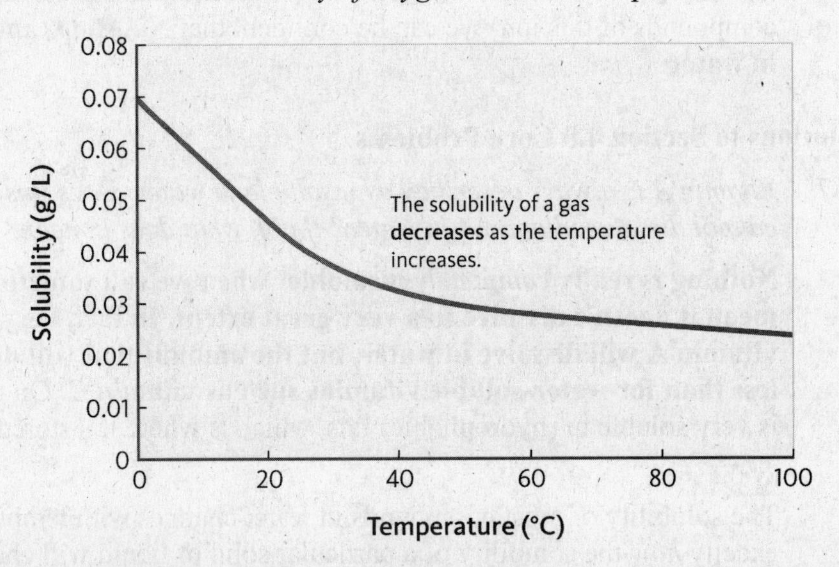

Figure 4.35

Go across the x axis to 20°C, then follow straight upward until you reach the solubility curve. From this point on the curve, follow another straight line to the left to the y axis. The line intersects the y axis at **about 0.045g/L.**

4.93 *Calcium sulfate is a solid at room temperature.*

a) *Will the solubility of calcium sulfate in water increase, remain the same, or decrease as you increase the temperature?*

For most solids, the solubility in water **increases** as the temperature increases.

b) Will the solubility of calcium sulfate in water increase, remain the same, or decrease as you increase the pressure?

For solids, pressure has no effect on solubility, so the solubility will **remain the same**.

Solutions to Section 4.9 Core Problems

4.95 *Identify the hydrophilic regions and hydrophobic regions in niacinamide.*

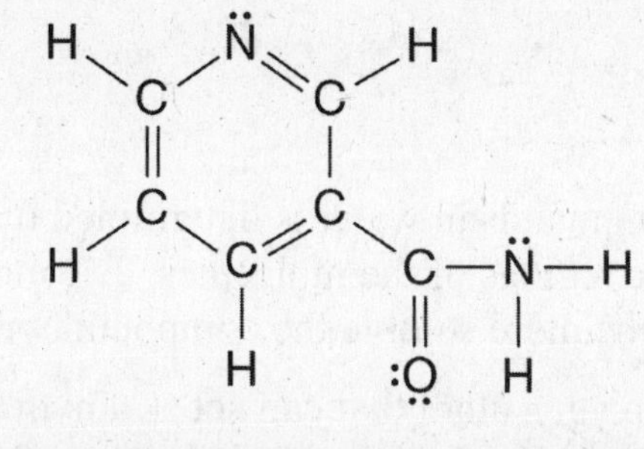

Niacinamide (one of the B vitamins)

Hydrophilic regions of a molecule are atoms or groups of atoms that can participate in hydrogen bonding as donors, acceptors, or both. Hydrogen bond donors have H atoms covalently bonded to N or O. Hydrogen bond acceptors have lone pair electrons on N or O with δ– charges. Carbon atoms, and hydrogen atoms bonded to carbon atoms, cannot participate in hydrogen bonds at all. **The hydrophilic regions of the niacin molecule are circled; all other areas are hydrophobic.**

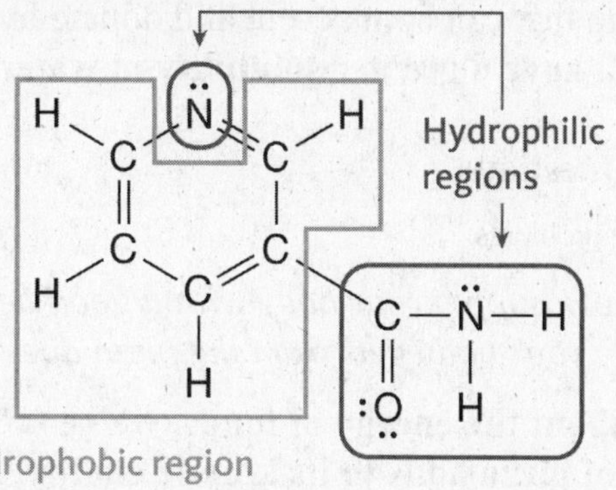

Hydrophilic regions

Hydrophobic region

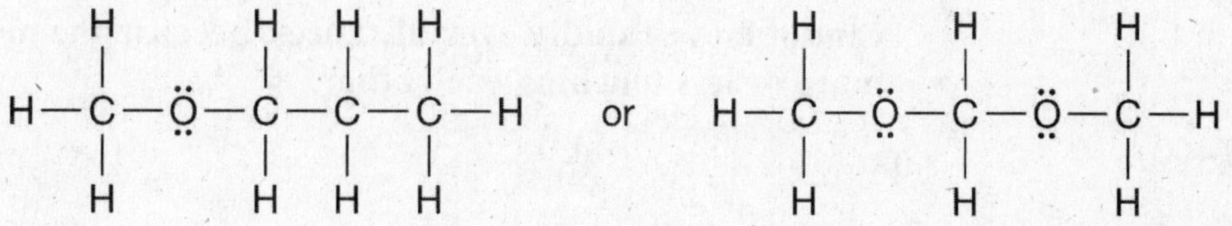

4.97 *From each of the following pairs of compounds, select the compound that will have the higher solubility in water.*

a) 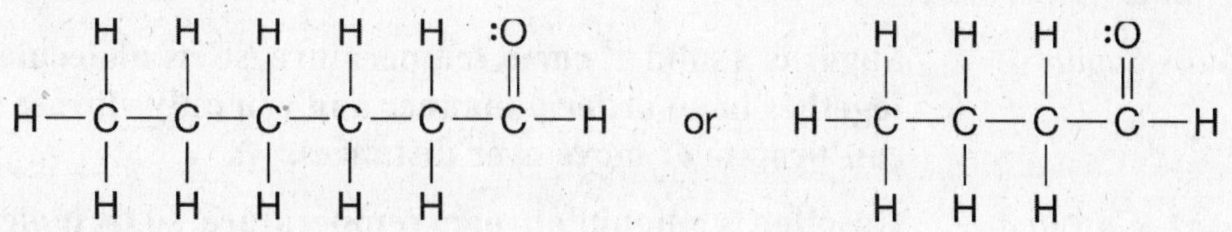 or

b) or

The solubility of organic compounds in water is determined by the balance between hydrophobic and hydrophilic regions of the molecules. The more hydrophilic or less hydrophobic a molecule is, the more soluble the compound will be in water.

(a) Each molecule has one oxygen atom that can act as a hydrogen bond acceptor. The molecule on the left has a much larger hydrophobic (non-hydrogen-bonding) area than the molecule on the right; **the compound on the right will have a greater solubility in water.**

(b) The molecule on the right has two oxygen atoms that can accept hydrogen bonds, while the molecule on the left has only one. The molecule on the right is more hydrophilic, so **the compound on the right will have a greater solubility in water.**

(c) The molecule on the left is entirely hydrophobic, while the molecule on the right has an –OH group that can both accept and donate hydrogen bonds. **The compound on the right will have a greater solubility in water.**

Solutions to Concept Questions

* indicates more challenging problems.

4.99 *A piece of pizza can supply your body with a good deal of energy. Explain why the amount of energy your body gets from the pizza does not change if you heat the pizza.*

When we talk about the energy of foods, we're talking about potential energy. Heating a slice of pizza adds to its kinetic energy, not its potential energy.

4.101 *Describe the motions and behavior of molecules of each of the following substances at room temperature.*

a) Sugar **Sugar is a solid at room temperature, so its molecules are stacked together in an orderly manner and can only vibrate in place. They can't rotate or move over distances.**

b) Gasoline **Gasoline is a liquid at room temperature, so its molecules can rotate and tumble as well as vibrating in place. However they cannot move rapidly over distances, because the molecules are more or less touching each other.**

4.103 *A student concludes that sand is a liquid, because he can pour the sand from one container to another, and the sand takes the shape of its container. Is this a reasonable conclusion? Why or why not?*

In fact the sand consists of tiny solid crystals that do retain their shape. When the sand is poured from one container into another, the relative positions of the crystals change, and the air space between the crystals changes shape, but the crystals don't change their shape.

4.105 *You have a bicycle tire that has been filled with air. List three ways to make the pressure inside the tire increase.*

The four variables for a gas are pressure, volume, temperature, and number of particles in the sample. To increase the pressure, you can:

- **decrease the volume** (squeezing the sample of air will increase its pressure, because the molecules will have less space to move in and will hit the sides of the container more often—this is obvious, if you think about it, in the word "compressing" which means both decreasing the volume and increasing the pressure);
- **increase the temperature** (heating the gas sample will cause the molecules to move faster, hitting the sides of the container more often and with greater force); or
- **add more gas to the bicycle tire** (if there are more molecules of air in the tire, there will be more collisions with the walls of the container—this is what a bicycle pump does.)

4.107 *Why do ionic compounds have much higher melting points and boiling points than molecular compounds do?*

Melting and boiling require adding enough energy to overcome the attractions between particles in a substance. In an ionic compound, each ion is surrounded by, and attracted to, several ions of the opposite charge. These attractions (ion-ion attractions or ionic bonds) are strong and are distributed evenly throughout an entire crystal of the compound. In a molecular compound, the only strong attractions are covalent bonds between atoms within a molecule; the attractions between molecules are very weak in comparison to ionic or covalent bonds. Therefore not as much energy is required to separate molecules from each other in a molecular compound as to separate ions from each other in an ionic compound. Make sure that you are making the distinction here: to melt or boil an ionic compound, the individual ions must actually be separated from each other, but when a molecular substance melts or boils, the molecules remain intact—the atoms are not separated from each other.

4.109 *When you boil a molecular substance, you must add enough energy to overcome some attractive forces. Which of the following must be overcome? (Select all of the correct answers.)*

a) Covalent bonds — **No**: the covalent bonds in a molecular substance are **not** disrupted in boiling, only the attractions between molecules. Covalent bonds hold atoms together within a molecule, they don't hold one molecule to another.

b) Dispersion forces — **Yes**: all molecular substances have dispersion forces attracting molecules to each other.

c) Hydrogen bonds — **Yes, in some cases**: if a substance is capable of forming hydrogen bonds between its molecules, then these attractive forces must be overcome to boil the substance. (If the substance cannot form hydrogen bonds between its molecules, then dispersion forces and dipole–dipole forces are all you have to worry about.)

d) Ion-ion attractions — **No**: there are no ions in molecular substances. Ion-ion attractions are found in ionic substances, not molecular substances.

4.111 Which *of the following is the most accurate description of the structure of water?*

a) Water is an ionic compound, containing H^+ *and* O^{2-} *ions.*

No: water is a molecular substance. Hydrogen and oxygen are both nonmetals and therefore form covalent, not ionic, bonds.

b) Water is an ionic compound, containing H^+ *and* OH^- *ions.*

No: water is a molecular substance. Hydrogen and oxygen are both nonmetals and therefore form covalent, not ionic, bonds.

c) Water is a covalent compound, with a small positive charge on each H and a small negative charge on the O.

Water is a covalent compound. Oxygen has a greater electronegativity than hydrogen, and the O atoms will draw the shared electrons away from the H atoms to a significant extent, causing a partial positive charge on H and a partial negative charge on O. **This is the most correct statement.**

d) Water is a covalent compound, with a small negative charge on each H and a small positive charge on the O.

No: while water is a covalent compound with polar bonds, the charges as described here are reversed.

e) Water is a covalent compound, and all three atoms in a water molecule have no charge.

No: This is not quite correct, because the electronegativity difference between O and H causes partial charges to exist. Oxygen has a greater electronegativity than hydrogen, and the O atoms will draw the shared electrons away from the H atoms to a significant extent, causing a partial positive charge on H and a partial negative charge on O.

4.113 *Carbon tetrachloride (CCl_4) does not dissolve in water. Which of the following statements is a reasonable explanation?*

First, both CCl_4 and H_2O are molecular compounds, so they will both have dispersion forces attracting molecules to each other, and neither will have ionic bonds. But water molecules can form hydrogen bonds, while CCl_4 molecules cannot.

a) *CCl_4 molecules and H_2O molecules repel each other.*

No: the molecules will attract each other (weakly) through dispersion forces.

b) *CCl_4 molecules are strongly attracted to one another and only weakly attracted to H_2O molecules.*

This is a reasonable statement: the dispersion forces between CCl_4 molecules are stronger than those that would exist between a CCl_4 molecule and a water molecule.

c) *H_2O molecules are strongly attracted to one another and only weakly attracted to CCl_4 molecules.*

This is a reasonable statement: the hydrogen bonds between water molecules are much stronger than the dispersion forces that would exist between a CCl_4 molecule and a water molecule.

d) *CCl_4 molecules are heavier than H_2O molecules.*

True but irrelevant in this situation: the mass of molecules has nothing to do with solubility.

4.115 *What is the difference between dissolving and dissociating?*

When molecular substances dissolve, the attractions between molecules of the solute are replaced by attractions between solute molecules and solvent molecules, but the connections between atoms within the molecule—the covalent bonds—remain intact. For example, when ethanol (C_2H_6OH) dissolves in water, the ethanol molecules separate from each other and form new associations (hydrogen bonds) with water, but the C, H and O atoms stay attached to each other. The molecules don't break—they have the same structure before and after dissolving.

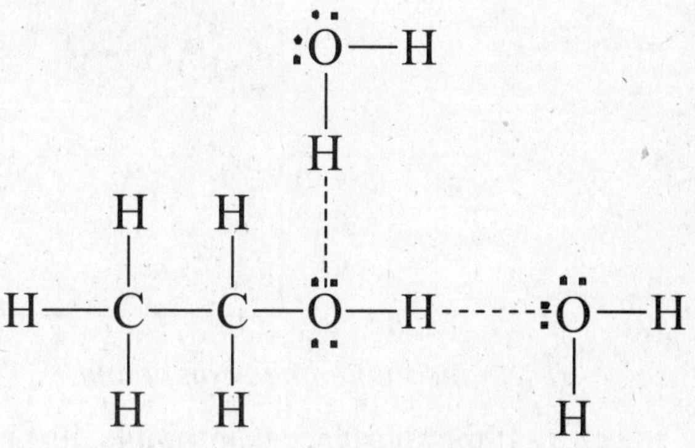

When ionic substances dissolve in water or other polar solvents, something more radical happens: the ionic bonds between the cations and anions actually break, and the ions form new associations with the solvent molecules. For example, when NaCl dissolves and dissociates, there are no longer any Na–Cl ionic bonds present, only

Na^+ ions surrounded by water molecules and Cl^- ions surrounded by water molecules.

4.117 *What is a hydrophobic region, and how is the size of a hydrophobic region related to solubility?*

"Hydrophobic" is a term applied to a molecule or a region of a molecule that cannot participate in hydrogen bonding with water. Any portion of a molecule is hydrophobic if it does not contain either (1) an H atom covalently bonded to an N, O or F atom, or (b) an N, O or F atom with lone pair electrons and a partial negative charge. (Areas that do meet one or both of these criteria and can participate in hydrogen bonding are called "hydrophilic.) If a molecule has large hydrophobic regions and small (or no) hydrophilic regions, then the substance will be insoluble in water.

4.119 *Each of the following pictures represents the results of mixing a compound with water. Match each of the following descriptions with the correct picture.*

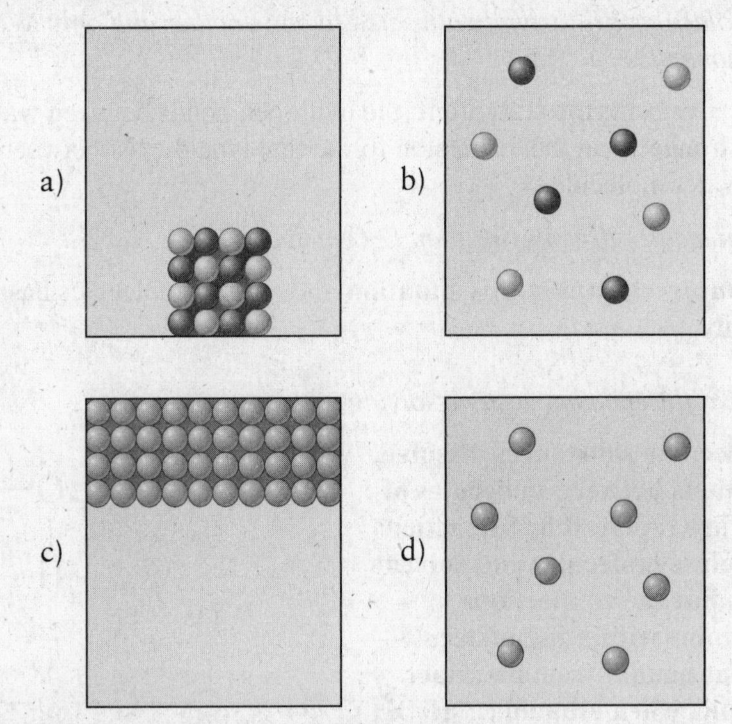

a) *An insoluble ionic compound*

If the substance is insoluble, its particles will be clustered together rather than dispersed through the solution. Pictures A and C both show this, but **picture A** makes clear that there are two different kinds of particles present (cations and anions, in this case) by showing two colors of dots. In picture C, while the solid is represented as being insoluble, using only one color of dots indicates that the atoms or molecules of the solid are all identical (therefore not an ionic compound).

NOTE:
Most ionic compounds are more dense than water and will sink to the bottom if they don't dissolve.

b) *An insoluble molecular compound*

In an insoluble compound, particles of the compound will be clustered together instead of dispersed. **Picture C** shows only one kind of particle (the molecules are all the same), clustered together at the top of the mixture. Many molecular compounds have lower densities than water and, therefore, float on water.

c) *A soluble ionic compound*

A soluble compound will have its particles dispersed through the solution (pictures B and D.) An ionic compound should be represented by different colors of dots for the cations and anions, so **picture B** is the soluble ionic compound (since all the dots are the same in D, it must represent a molecular compound).

d) *A soluble molecular compound*

A soluble compound will have its particles dispersed through the solution (pictures B and D.) A molecular compound should be represented by a single color of dots, so **picture D** is the soluble molecular compound (since B is represented with two different colors for cations and anions, it must represent an ionic compound).

Solutions to Summary and Challenge Problems

* indicates more challenging problems.

*4.121 *The specific heat of ethanol is 0.58 cal/g·°C, and the density of ethanol is 0.79 g/mL.*

a) *How much heat is needed to raise the temperature of 2.5 kg of ethanol from 20°C to 30°C?*. The equation is:

heat = m × ΔT × c_p

Our temperature change from 20°C to 30°C gives us ΔT = 10°C. We can either make the mass of the sample (2.5 kg) match the mass unit in the specific heat (g):

$$2.5\ \text{kg} \times \frac{1000\ \text{g}}{1\ \text{kg}} = 2.5 \times 10^3\ \text{g}$$

heat = 2.5×10^3 g × 10°C × 0.58 cal/g·°C = **1.5×10^4 cal** (rounded to 2 significant figures)

Or we can carry the "k" through the problem, giving us an answer in kcal instead of cal (this works because "k" just means "$\times10^3$" and can be carried through the multiplication steps like any other number or unit.) Since the units of energy are not specified in the problem, we can use either approach.

heat = 2.5 kg × 10°C × 0.58 cal/g·°C = **15 kcal** (rounded to 2 significant figures)

b) *How much heat must you remove from 25 pounds of ethanol to cool it from 70°F to 40°F?*

This time we really do have to change the units on our given measurements.

$$25\ \text{pounds} \times \frac{454\ \text{g}}{1\ \text{pound}} = 1.1 \times 10^4\ \text{g (rounded to 2 sf)}$$

to convert the temperatures to °C (both temperatures rounded to the nearest degree),
°C = (°F – 32) ÷ 1.8 = (70 – 32) ÷ 1.8 = 21°C
°C = (°F – 32) ÷ 1.8 = (40 – 32) ÷ 1.8 = 4°C
ΔT = 4°C – 21°C = –17°C
heat = 1.1×10^4 g × –17°C × 0.58 cal/g·°C = **-1.1×10^5 cal** (rounded to 2 significant figures)

The fact that this is a negative number means that the heat is leaving the sample, which is also accounted for in the question ("How much heat must you remove…") The amount of heat removed is **1.1×10^5 cal or 110 kcal.**

c) *How much heat must you add to 150 mL of ethanol to increase its temperature from 0°C to 50°C?*

This time the amount of ethanol is given in terms of volume; we need to get it into mass (grams) to use the heat equation. The relationship between mass and volume is density:
150 mL × 0.79 g/mL = 120 g (rounded to 2 significant figures)
The temperature change is 50°C – 0°C = 50°C.
heat = 120 g × 50°C × 0.58 cal/g·°C = **3.4×10^3 cal** (rounded to 2 significant figures)
= **3.4 kcal**

4.123 *At room temperature, you can have 100 mL of water in a 1000-mL container, but you can't have 100 mL of oxygen in a 1000-mL container. Why not?*

Water, a liquid, doesn't expand to fill its container; it will take the shape of the bottom of the container but retain its own volume. 100 mL of water will leave 900 mL of empty space in a 1000-mL container. Oxygen, a gas, expands to fill the entire volume of any container it is put into, so 100 mL of oxygen in a 1000-mL container will expand to 1000 mL.

4.125 *An unusually high number of car tires blow out on desert highways in the summer. Why is this? (Hint: What happens to the air in the tire?)*

The tire and the air inflating it get very hot in contact with the hot pavement, and as the air heats up, its pressure increases. In some cases the pressure gets so high that the tire can rupture, especially if it already has a weak spot or excessive wear.

4.127 *How much heat would you need to do the following?*

a) *Melt 30 g of ice.*

The heat of fusion of ice—the amount of energy required to convert one gram of ice at 0°C to one gram of water at 0°C—is 80 cal/g. This value has a compound unit, which means it is likely to be used as a conversion factor. So to melt 30 g of ice,
30 g × 80 cal/g = **2400 cal = 2.4 kcal**

b) *Boil 30 g of water.*

The heat of vaporization of water—the amount of energy required to convert one gram of water at 100°C to one gram of steam at 100°C—is 540 cal/g. To boil 30 g of ice,

30 g × 540 cal/g = **1.6×10^4 cal = 16 kcal**

*c) *Warm 30 g of water from 20°C to 100°C and boil the water.*

We'll have to calculate the amount of energy needed for the warming process (an 80°C temperature change), then add the energy required to boil (which we calculated in step b).
heat = 30 g × 80°C × 1.0 cal/g·°C = **2400 cal** = **2.4 kcal** for the warming
2.4 kcal (to heat the water) + 16 kcal (to boil it) = **18.4 kcal total**

4.129 *What types of bonds or attractive forces are responsible for each of the following?*

a) *Calcium and chlorine atoms remain essentially in fixed positions in a crystal of $CaCl_2$.*

This is an ionic compound (metal + nonmetal) made up of Ca^{2+} ions and Cl^- ions, held together in a crystal lattice by **ion-ion attractions.**

b) *Hydrogen atoms are attached to oxygen atoms in a water molecule.*

The H and O atoms in water share electrons, forming **covalent bonds.**

c) *Water molecules tend to stay close to one another at room temperature.*

As a molecular substance, water has **dispersion forces** between its molecules. But in water, a much greater factor is **hydrogen bonding.**

4.131 *A student is asked to compare the melting points of H_2O and NaOH. He draws the following Lewis structures:*

$$\mathrm{H{-}\ddot{\underset{\cdot\cdot}{O}}{-}H} \qquad \mathrm{Na{-}\ddot{\underset{\cdot\cdot}{O}}{-}H}$$

He then states that H_2O should have the higher melting point, because it has two hydrogen atoms that can participate in hydrogen bonds, while NaOH has only one. Would you agree with this reasoning? Explain.

While the reasoning for H_2O is fine, the melting point of NaOH has nothing to do with hydrogen bonding, because it's an ionic compound. The Lewis structure of NaOH is incorrect because it shows a covalent bond between Na and O rather than an ionic bond. In fact NaOH, an ionic compound, has a much higher melting point than H_2O, a molecular compound. Ionic compounds have ion-ion attractions (also called ionic bonds) throughout an entire crystal. This gives ionic compounds in general much higher melting and boiling points than molecular compounds, which have relatively weak attractions between molecules.

4.133 *Methanol (wood alcohol) is a liquid that can dissolve some ionic compounds, including NaCl. Show how sodium ions and chloride ions can be solvated by methanol molecules.*

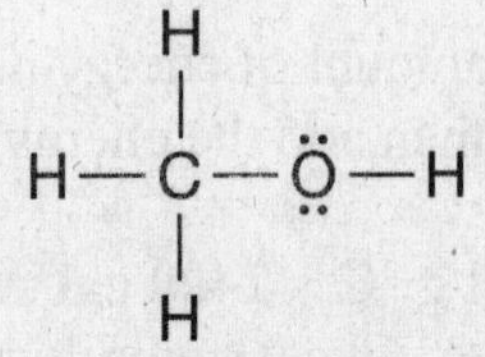

H
|
H—C—Ö—H
|
H

Methanol

The solvation of ions by methanol is very similar to the solvation of ions in water: the δ– O atoms will be attracted to the cations, while the δ+ H atoms (only those attached to O) will be attracted to the anions.

NOTE:

The H atoms bonded to C will **not** participate in this interaction, because they don't have enough of a partial positive charge. (There's nothing special about the number of molecules of methanol shown for each ion, it's just how many conveniently fit in the picture.)

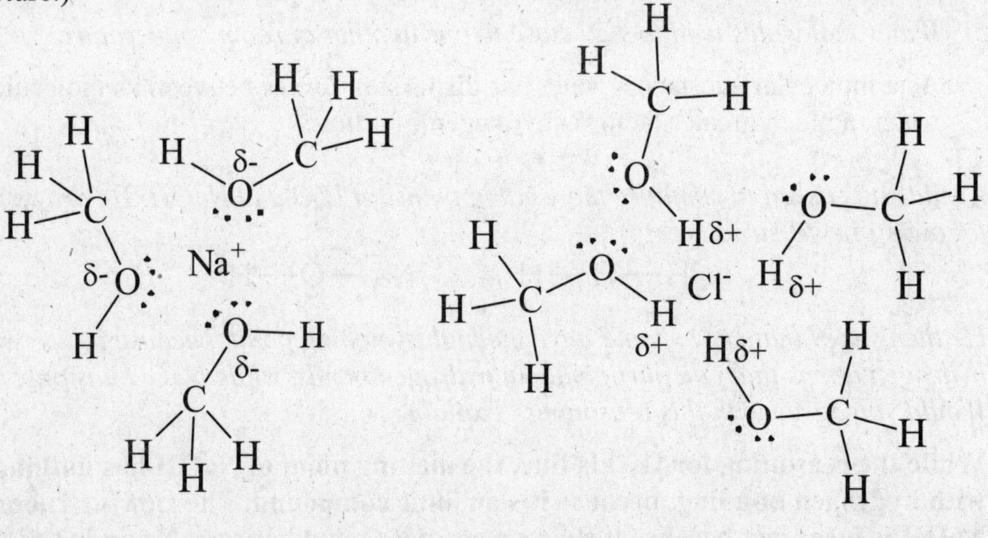

4.135 A *solution that contains 3 g of aspirin in 1 L of water is a saturated solution at 25°C, but it is an unsaturated solution at 35°C. Explain how this is possible.*

This is possible if the solubility of aspirin (like the solubility of most solids) increases as the solution temperature increases. While only 3 g will dissolve in a liter of water at 25°C, more than 3 g can dissolve at 35°C.

*4.137 *a) Some intravenous solutions contain magnesium ions. Magnesium chloride can be used to supply the magnesium ions in these solutions, but magnesium carbonate cannot. What is the most likely reason for this?*

Magnesium carbonate has a very low solubility in water, so it can't be dissolved in intravenous solutions. Ionic compounds containing sodium, potassium, and nitrate ions are always soluble, and chloride compounds (like magnesium chloride) are usually soluble. Other combinations are generally assumed to be insoluble (unless we have specific information about their solubility).

b) *Many intravenous solutions contain chloride ions. $HgCl_2$ dissolves in water, but it is never used as a source for the chloride ions in intravenous solutions. Can you think of a possible reason?*

The mercury ions are toxic, so even though $HgCl_2$ is soluble, it couldn't be used in medical applications.

Solution Concentration

Solutions to Section 5.1 Core Problems

5.1 *Write conversion factors that correspond to the following concentrations:*

For each of these, look at how the concentration unit is defined.

a) 0.15% (v/v) acetic acid The definition of %v/v is mL solute per 100 mL of solution:

$$\frac{\textbf{0.15 mL acetic acid}}{\textbf{100 mL soln}} \textbf{ or } \frac{\textbf{100 mL soln}}{\textbf{0.15 mL acetic acid}}$$

b) 50 ppm Br^- One definition of ppm is μg solute per mL solution:

$$\frac{\textbf{50 }\mu\textbf{g Br}}{\textbf{1 mL soln}} \textbf{ or } \frac{\textbf{1 mL soln}}{\textbf{50 }\mu\textbf{g Br}}$$

c) 3 mg/dL fructose This concentration unit already contains both units needed to make the conversion factors:

$$\frac{\textbf{3 mg fructose}}{\textbf{1 dL soln}} \textbf{ or } \frac{\textbf{1 dL soln}}{\textbf{3 mg fructose}}$$

d) 6.5% (w/v) $MgSO_4$ The definition of %w/v is g solute per 100 mL of solution:

$$\frac{\textbf{6.5 g MgSO}_4}{\textbf{100 mL soln}} \textbf{ or } \frac{\textbf{100 mL soln}}{\textbf{6.5 g MgSO}_4}$$

e) 20 ppb Pb^{2+} One definition of ppb is nanograms (ng) solute per milliliter (mL) solution:

$$\frac{\textbf{20 ng Pb}^{2+}}{\textbf{1 mL soln}} \textbf{ or } \frac{\textbf{1 mL soln}}{\textbf{20 ng Pb}^{2+}}$$

5.3 *Calculate the percent concentration of each of the following solution:.*

a) 2.31 g of sucrose dissolved in enough water to make 25.0 mL of solution

Since we're given the mass of a solid solute, we'll be calculating %w/v:

$$\frac{\textbf{2.31 g sucrose}}{\textbf{25.0 mL soln}} \times \textbf{100\%} = \textbf{9.24\% w/v}$$

b) 177 mL of isopropyl alcohol dissolved in enough water to make 243 mL of solution

Here we are given mL of a liquid solute, so we calculate %v/v:

$$\frac{\textbf{177 mL isopropyl alcohol}}{\textbf{243 mL soln}} \times \textbf{100\%} = \textbf{72.8\% v/v}$$

c) 275 g of $MgSO_4$ dissolved in enough water to make 3.25 L of solution

Again, because the amount of solute is given in grams, we calculate %w/v, but this time we have to convert the volume of the solution to mL first:

$$3.25\ \text{L} \times \frac{1000\ \text{mL}}{1\ \text{L}} = 3250\ \text{mL}$$

$$\frac{\textbf{275 g MgSO}_\mathbf{4}}{\textbf{3250 mL soln}} \times \textbf{100\%} = \textbf{8.46\% w/v}$$

5.5 *Calculate the mass or volume of solute that would be needed to make each of the following solutions:*

This is where we put into practice the skill from 5.1. Each of these concentration units gives a set of conversion factors; we just have to decide which to use.

a) 500.0 mL of 0.75% (w/v) $CaCl_2$

Our conversion factors are:

$$\frac{0.75\ \text{g CaCl}_2}{100\ \text{mL soln}} \text{ or } \frac{100\ \text{mL soln}}{0.75\ \text{g CaCl}_2}$$

Then, $\textbf{500.0 mL soln} \times \frac{\textbf{0.75 g CaCl}_\mathbf{2}}{\textbf{100 mL soln}} = \textbf{3.8 g CaCl}_\mathbf{2}$

b) 625 mL of 15.0% (v/v) ethylene glycol

Conversion factors:

$$\frac{15\ \text{mL ethylene glycol}}{100\ \text{mL soln}} \text{ or } \frac{100\ \text{mL soln}}{15\ \text{mL ethylene glycol}}$$

Then, $\textbf{625 mL} \times \frac{\textbf{15.0 mL ethylene glycol}}{\textbf{100 mL soln}} = \textbf{93.8 mL ethylene glycol}$

c) 2.50 L of 0.125% (w/v) vitamin C

Conversion factors:

$$\frac{0.125\ \text{g Vitamin C}}{100\ \text{mL soln}} \text{ or } \frac{100\ \text{mL soln}}{0.125\ \text{g Vitamin C}}$$

We also need to convert the volume to mL:

$$2.50\ \text{L} \times \frac{1000\ \text{mL}}{1\ \text{L}} = 2500\ \text{mL}$$

$$2500 \text{ mL soln} \times \frac{0.125 \text{ g Vitamin C}}{100 \text{ mL soln}} = 3.13 \text{ g Vitamin C}$$

5.7 *A solution contains 57.3 mg of niacin in a total volume of 125 mL. Calculate the concentration of this solution in each of the following units:*

NOTE:
Throughout this chapter, it is assumed that you can do metric conversions. If you are having trouble with this skill, it is very important to go back to Chapter 1 and review!

In most problems of this kind, there are two general strategies, either of which is fine. All concentration units are expressed in terms of the amount of solute over the amount of solution, so we can convert the mass of solute and volume of solution separately into the units required by the problem, and set up the ratio at the end. Or, we can start by setting up the ratio, then carry out as many conversion steps as needed to get to the desired units. Both strategies will be shown.

a) mg/dL

The amount of solute (57.3 mg) is already in the needed units (mg). To convert the volume, we use the relationship 1000 mL = 1 L = 10 dL.

$$125 \text{ mL soln} \times \frac{10 \text{ dL}}{1000 \text{ mL}} = 1.25 \text{ dL soln}$$

Then, we set up the ratio and calculate the concentration, rounding to 3 significant figures:

$$\frac{57.3 \text{ mg niacin}}{1.25 \text{ dL soln}} = 45.8 \text{ mg/dL}$$

In the second strategy, we set up the entire problem in one string, starting from the ratio (amount of solute)/(amount of solvent) and then performing any necessary conversions:

$$\frac{57.3 \text{ mg niacin}}{125 \text{ mL soln}} \times \frac{1000 \text{ mL}}{10 \text{ dL}} = 45.8 \text{ mg/dL}$$

b) μg/dL

The amount of solute (57.3 mg) needs to be converted to μg, using the relationship

10^6 μg = 1 g = 10^3 mg:

$$57.3 \text{ mg niacin} \times \frac{10^6 \ \mu\text{g}}{10^3 \text{ mg}} = 5.73 \times 10^4 \ \mu\text{g niacin}$$

The amount of solution (125 mL) needs to be converted to μg, using the relationship 1000 mL = 1 L = 10 dL:

$$125 \text{ mL soln} \times \frac{10 \text{ dL}}{1000 \text{ mL}} = 1.25 \text{ dL soln}$$

Then, we set up the ratio and calculate the concentration, rounding to 3 significant figures:

$$\frac{\mathbf{5.73 \times 10^4\ \mu g\ niacin}}{\mathbf{1.25\ dL\ soln}} = \mathbf{4.58 \times 10^4\ \mu g/dL}$$

In the second strategy, we set up the entire problem in one string, starting from the ratio (amount of solute)/(amount of solvent) and then performing any necessary conversions:

$$\frac{\mathbf{57.3\ mg\ niacin}}{\mathbf{125\ mL\ soln}} \times \frac{\mathbf{10^6\ \mu g}}{\mathbf{10^3\ mg}} \times \frac{\mathbf{1000\ mL}}{\mathbf{10\ dL}} = \mathbf{4.58 \times 10^4\ \mu g/dL}$$

c) *ppm*

One definition of ppm is μg solute per mL solution:

$$\text{ppm} = \frac{\mu\text{g solute}}{\text{mL soln}}$$

We convert the amount of solute (57.3 mg) to μg, using the relationship 10^6 μg = 1 g = 10^3 mg:

$$57.3\text{ mg niacin} \times \frac{10^6\ \mu\text{g}}{10^3\text{ mg}} = 5.73 \times 10^4\ \mu\text{g niacin}$$

The amount of solution (125 mL) is already in the needed unit. We set up the ratio and calculate the concentration, rounding to 3 significant figures:

$$\frac{\mathbf{5.73 \times 10^4\ \mu g\ niacin}}{\mathbf{125\ mL\ soln}} = \mathbf{458\ \mu g/mL = 458\ ppm}$$

In the second strategy, we set up the entire problem in one string, starting from the ratio (amount of solute)/(amount of solvent) and then performing any necessary conversions:

$$\frac{\mathbf{57.3\ mg\ niacin}}{\mathbf{125\ mL\ soln}} \times \frac{\mathbf{10^6\ \mu g}}{\mathbf{10^3\ mg}} = \mathbf{458\ \mu g/mL = 458\ ppm}$$

d) *ppb*

One definition of ppb is ng solute per mL solution:

$$\text{ppb} = \frac{\text{ng solute}}{\text{mL soln}}$$

The amount of solute (57.3 mg) needs to be converted to ng, using the relationship 10^9 ng = 1 g = 10^3 mg:

$$57.3\text{ mg niacin} \times \frac{10^9\ ng}{10^3\text{ mg}} = 5.73 \times 10^7\ \mu\text{g niacin}$$

The amount of solution (125 mL) is already in the needed unit. We set up the ratio and calculate the concentration, rounding to 3 significant figures:

$$\frac{\mathbf{5.73\times10^7\ ng\ niacin}}{\mathbf{125\ mL\ soln}}=\mathbf{4.58\times10^5\ ng/mL=4.58\times10^5\ ppb}$$

In the second strategy, we set up the entire problem in one string, starting from the ratio (amount of solute)/(amount of solvent) and then performing any necessary conversions:

$$\frac{\mathbf{57.3\ mg\ niacin}}{\mathbf{125\ mL\ soln}}\times\frac{\mathbf{10^9\ ng}}{\mathbf{10^3\ mg}}=\mathbf{4.58\times10^5\ ng/mL=4.58\times10^5\ ppb}$$

5.9 *Calculate the mass of solute in 25.0 mL of each of the following solutions:*

NOTE:
Throughout this chapter, it is assumed that you can do metric conversions. If you are having trouble with this skill, it is very important to go back to Chapter 1 and review!

Concentrations are always given in the form

$$\text{concentration}=\frac{\text{amount of solute}}{\text{amount of solution}}$$

Since concentration values have compound units, we can usually use a given concentration, with units, as a conversion factor as we did in Chapter 1. (See Problem 5.1 for a review.)

a) *water that contains 2.5 ppm of fluoride ions*

Our setup is:

25.0 mL solution = ? g fluoride

One definition of ppm is μg solute per mL solution, so we can rewrite this concentration value:

$$2.5\text{ ppm fluoride}=\frac{2.5\ \mu\text{g fluoride}}{\text{mL soln}}$$

This gives us two conversion factors:

$$\frac{2.5\ \mu\text{g fluoride}}{\text{mL soln}}\text{ or }\frac{1\text{ mL soln}}{2.5\ \mu\text{g fluoride}}$$

We can now use the concentration as a conversion factor to get from the volume of the solution to the mass of the solute (rounded to 2 significant figures):

$$\mathbf{25\ mL\ solution}\times\frac{\mathbf{2.5\ \mu g\ fluoride}}{\mathbf{mL\ soln}}=\mathbf{63\ \mu g\ fluoride}$$

b) *blood plasma that contains 92 mg/dL of glucose*

In this problem, the units in the concentration are already spelled out for us; we just have to do a conversion to make the concentration unit match our given volume unit. We can use the relationship 1000 mL = 1 L = 10 dL for this conversion:

$$\mathbf{25\ mL\ solution} \times \frac{\mathbf{10\ dL}}{\mathbf{1000\ mL}} \times \frac{\mathbf{92\ mg\ glucose}}{\mathbf{dL\ soln}} = \mathbf{23\ mg\ glucose}$$

c) *water that contains 31 ppb of lead ions*

One definition of ppb is ng solute per mL solution:

$$31\ \text{ppb} = \frac{31\ \text{ng solute}}{\text{mL soln}}$$

This gives us two conversion factors:

$$\frac{31\ \text{ng lead}}{\text{mL soln}} \text{ or } \frac{1\ \text{mL soln}}{31\ \text{ng lead}}$$

We can now use the concentration as a conversion factor to get from the volume of the solution to the mass of the solute (rounded to 2 significant figures):

$$\mathbf{25\ mL\ solution} \times \frac{\mathbf{31\ ng\ lead}}{\mathbf{mL\ soln}} = \mathbf{780\ ng\ lead}$$

Solutions to Section 5.2 Core Problems

5.11 *How much does each of the following weigh? Be sure to use the appropriate unit for each answer.*

Remember from Section 2.5 that the average atomic weight of an element, in amu or atomic mass units, is given in the periodic table. What's new in this section is the definition of a mole of an element: *a mole is the atomic weight of an element expressed in grams, and contains 6.022 x 10^{23} atoms of that element.* In every case, **the average mass of an atom of an element, given in amu, is the same value as the mass of a mole of atoms of that element, given in grams.** The number is the same; we just use amu for individual atoms, molecules or formula units and g for moles of a substance.

Remember: amu for atoms, g for moles!

a) *One atom of sulfur* — The atomic mass of sulfur, as given in the periodic table, is **32.06 amu**. (Remember from Chapter 2 that this is the *average* mass of an atom of sulfur–there are no atoms of sulfur that actually have a mass of 32.06 amu.)

b) *One mole of sulfur* — If the average mass of an atom of sulfur is 32.06 amu, then the mass of a mole of sulfur is **32.06 g**.

c) *One molecule of N_2O* — One molecule of N_2O contains two nitrogen atoms at 14.01 amu each (on average), and one oxygen atom with a mass of 16.00 amu each (on average). The average mass of a molecule of N_2O is therefore

2 N atoms	=	2 × 14.01 amu	=	28.02 amu
+ 1 O atom	=	16.00 amu	=	16.00 amu
			=	**44.02 amu**

d) One mole of N_2O

If the average mass of one molecule of N_2O is 44.02 amu, then the mass of a mole of N_2O is **44.02 g.**

e) One formula unit of $(NH_4)_3PO_4$

Since this substance is ionic, we can't meaningfully talk about a "molecule" of it, which is why we have to use the term "formula unit." We add up the atomic masses of the atoms in the formula (rounding our answer to two decimal places):

3 N atoms	=	3 × 14.01 amu	=	42.03 amu
12 H atoms	=	12 × 1.008 amu	=	12.096 amu
1 P atom	=	30.97 amu	=	30.97 amu
+ 4 O atoms	=	4 × 16.00 amu	=	64.00 amu
			=	**149.10 amu**

f) One mole of $(NH_4)_3PO_4$

If one formula unit of $(NH_4)_3PO_4$ has a mass of 143.05 amu, then the mass of one mole of $(NH_4)_3PO_4$ is **149.10 g.**

5.13 *a) If you have 25 moles of iron, how many grams of iron do you have?*

We look in the periodic table to find the atomic mass of iron (Fe), 55.84 amu. From the definition of a mole, if one atom of Fe has a mass of 55.84 amu, then one mole of iron has a mass of 55.84 grams. We can use this relationship to write two conversion factors:

$$\frac{1 \text{ mole iron}}{55.84 \text{ grams iron}} \text{ or } \frac{55.84 \text{ grams iron}}{1 \text{ mole iron}}$$

Then we choose the correct form of the conversion factor to make the units cancel properly, and set up the calculation (rounding our answer to 2 significant figures):

$$\mathbf{25 \text{ moles iron} \times \frac{55.84 \text{ g iron}}{1 \text{ mole iron}} = 1400 \text{ g iron}}$$

b) If you have 0.0615 moles of sodium lactate ($NaC_3H_5O_3$), how many grams of sodium lactate do you have?

For this problem we need to add up the molar mass of sodium lactate, using the atomic mass values from the periodic table. We add up the atoms in the formula (rounding our answer to two decimal places):

1 Na atom			=	22.99 amu
3 C atoms	=	3 × 12.01 amu	=	36.03 amu
5 H atoms	=	5 × 1.008 amu	=	5.040 amu
+ 3 O atoms	=	3 × 16.00 amu	=	48.00 amu
			=	112.06 amu

If one formula unit of $NaC_3H_5O_3$ has a mass of 112.06 amu, then one mole of $NaC_3H_5O_3$ has a mass of 112.06 grams. We can use this relationship to write two conversion factors:

$$\frac{1\text{ mole NaC}_3\text{H}_5\text{O}_3}{112.06\text{ grams NaC}_3\text{H}_5\text{O}_3}\text{ or }\frac{112.06\text{ grams NaC}_3\text{H}_5\text{O}_3}{1\text{ mole NaC}_3\text{H}_5\text{O}_3}$$

Then we choose the correct form of the conversion factor to make the units cancel properly, and set up the calculation (rounding our answer to 3 significant figures):

$$\mathbf{0.0615\text{ moles NaC}_3\text{H}_5\text{O}_3 \times \frac{112.06\text{ g NaC}_3\text{H}_5\text{O}_3}{1\text{ mole NaC}_3\text{H}_5\text{O}_3} = 6.89\text{ g NaC}_3\text{H}_5\text{O}_3}$$

5.15 *Convert each of the following masses into moles:*

a) 6.75 g of aluminum

We look in the periodic table to find the atomic mass of aluminum (Al), 26.98 amu. From the definition of a mole, if one atom of Al has a mass of 26.98 amu, then one mole of Al has a mass of 26.98 grams. We can use this relationship to write two conversion factors:

$$\frac{1\text{ mole Al}}{26.98\text{ grams Al}}\text{ or }\frac{26.98\text{ grams Al}}{1\text{ mole Al}}$$

Then we choose the form of the conversion factor that makes the units cancel properly, and set up the calculation (rounding our answer to 3 significant figures):

$$\mathbf{6.75\text{ g Al} \times \frac{1\text{ mole Al}}{26.98\text{ g Al}} = 0.250\text{ mol Al}}$$

b) 12.99 g of leucine ($C_6H_{13}NO_2$)

For this problem we need to add up the molecular mass of leucine, using the atomic mass values from the periodic table:

6 C atoms	=	6 × 12.01 amu	=	72.06 amu
13 H atoms	=	13 × 1.008 amu	=	13.10 amu
1 N atom			=	14.01 amu
+ 2 O atoms	=	2 × 16.00 amu	=	32.00 amu
			=	131.17 amu

If one molecule of leucine has a mass of 131.17 amu, then one mole of leucine has a mass of 131.17 grams. We can use this relationship to write two conversion factors:

$$\frac{1 \text{ mole leucine}}{131.17 \text{ grams leucine}} \text{ or } \frac{131.17 \text{ grams leucine}}{1 \text{ mole leucine}}$$

Then we choose the form of the conversion factor that makes the units cancel properly, and set up the calculation (rounding our answer to 4 significant figures):

$$\mathbf{12.99 \text{ g leucine} \times \frac{1 \text{ mole leucine}}{131.17 \text{ g leucine}} = 0.09903 \text{ mol leucine}}$$

c) 1.35 kg of $Cu(NO_3)_2$

First, we add up the formula weight of $Cu(NO_3)_2$:

1 Cu			=	63.55 amu
2 N	=	2 × 14.01 amu	=	28.02 amu
+ 6 O	=	6 × 16.00 amu	=	96.00 amu
			=	187.57 amu

If one formula unit of $Cu(NO_3)_2$ has a mass of 187.57 amu, then one mole of $Cu(NO_3)_2$ has a mass of 187.57 grams. We can use this relationship to write two conversion factors:

$$\frac{1 \text{ mole } Cu(NO_3)_2}{187.57 \text{ grams } Cu(NO_3)_2} \text{ or } \frac{187.57 \text{ grams } Cu(NO_3)_2}{1 \text{ mole } Cu(NO_3)_2}$$

Then we set up the calculation, first converting the given mass in kg to grams, choosing the correct form of the concentration conversion factor to make the units cancel properly, and rounding our answer to 3 significant figures:

$$\mathbf{1.35 \text{ kg } Cu(NO_3)_2 \times \frac{1000 \text{ g}}{1 \text{ kg}} \times \frac{1 \text{ mole } Cu(NO_3)_2}{187.57 \text{ g } Cu(NO_3)_2} = 7.20 \text{ mole } Cu(NO_3)_2}$$

Solutions to Section 5.3 Core Problems

5.17 *Calculate the molarity of each of the following solutions:*

There are two general strategies here. Molarity is expressed in terms of the moles of solute per liter of solution, so we can calculate the moles of solute and liters of solution separately, then set up the ratio moles/L at the end. Or, we can start by setting up the ratio (amount of solute)/(amount of solution), then carry out as many conversion steps as needed to get to the desired final units of mol/L. Both strategies will be shown if applicable.

a) 0.350 moles of acetic acid ($HC_2H_3O_2$) in a total volume of 0.250 L.

The amount of solute (0.350 moles) and the amount of solution (0.250 L) are already both in the needed units to calculate moles/liter; we only have to set up the ratio (and make sure to report the answer to 3 significant figures):

$$\mathbf{\frac{0.350 \text{ moles acetic acid}}{0.250 \text{ L soln}} = 1.40 \text{ mol/L} = 1.40 \text{ M}}$$

b) *0.0624 moles of thiamine hydrochloride ($C_{12}H_{18}N_4OSCl_2$) in a total volume of 85.5 mL.*

The amount of solute (0.0624 moles) is already in the needed unit (mol). To convert the volume, we use the relationship 1000 mL = 1 L:

$$85.5 \text{ mL soln} \times \frac{1 \text{ L}}{1000 \text{ mL}} = 0.0855 \text{ L soln}$$

Then, we set up the ratio and calculate the concentration, rounding to 3 significant figures:

$$\frac{\mathbf{0.0624 \text{ mol thiamine hydrochloride}}}{\mathbf{0.0855 \text{ L soln}}} = \mathbf{0.730 \text{ mol thiamine hydrochloride/L}}$$

$$= \mathbf{0.730 \text{ M}}$$

In the second strategy, we set up the entire problem in one string, starting from the ratio (amount of solute)/(amount of solvent) and performing any necessary conversions:

$$\frac{\mathbf{0.0624 \text{ mol thiamine hydrochloride}}}{\mathbf{85.5 \text{ mL soln}}} \times \frac{\mathbf{1000 \text{ mL}}}{\mathbf{1 \text{ L}}}$$

$$= \mathbf{0.730 \text{ mol thiamine hydrochloride/L}}$$

$$= \mathbf{0.730 \text{ M}}$$

c) *6.1×10^{-4} moles of nitric acid (HNO_3) in a total volume of 7.5 mL.*

The amount of solute (6.1×10^{-4} moles) is already in the needed unit (mol). To convert the volume, we use the relationship 1000 mL = 1 L:

$$7.5 \text{ mL soln} \times \frac{1 \text{ L}}{1000 \text{ mL}} = 0.0075 \text{ L soln}$$

Then, we set up the ratio and calculate the concentration, rounding to 2 significant figures:

$$\frac{\mathbf{6.1 \times 10^{-4} \text{ mol HNO}_3}}{\mathbf{0.0075 \text{ L soln}}} = \mathbf{0.081 \text{ mol HNO}_3\text{/L} = 0.081 \text{ M}}$$

We could also set up the problem in one string, starting from the ratio (amount of solute)/(amount of solvent):

$$\frac{\mathbf{6.1 \times 10^{-4} \text{ mol HNO}_3}}{\mathbf{7.5 \text{ mL soln}}} \times \frac{\mathbf{1000 \text{ mL}}}{\mathbf{1 \text{ L}}} = \mathbf{0.081 \text{ mol HNO}_3\text{/L} = 0.081 \text{ M}}$$

5.19 *Calculate the molarity of each of the following solutions:*

Again, there are two general strategies. Molarity is expressed in terms of the moles of solute per liter of solution, so we can calculate the moles of solute and liters of solution separately, then set up the ratio moles/L at the end. Or, we can start by setting up the ratio (amount of solute)/(amount of solution), then carry out as many conversion steps as needed to get to the desired final units of mol/L. Both strategies will be shown.

a) *62.4 g of $MgSO_4$ in a total volume of 2.50 L.*

The given volume (2.50 L) is already in the needed unit (L). To convert 62.4 g of $MgSO_4$ into moles, we first need to add up its formula mass:

1 Mg			=	24.30 amu
1 S			=	32.06 amu
+ 4 O	=	4 × 16.00	=	64.00 amu
			=	120.36 amu

If one formula unit of $MgSO_4$ has a mass of 120.36 amu, then one mole of $MgSO_4$ has a mass of 120.36 grams. We can now convert the mass of $MgSO_4$ into moles (we don't round yet because there are more steps to carry out):

$$62.4\ \text{g MgSO}_4 \times \frac{1\ \text{mole MgSO}_4}{120.36\ \text{g MgSO}_4} = 0.518444666\ \text{mol MgSO}_4$$

Then, we set up the ratio and calculate the concentration, rounding to 3 significant figures:

$$\frac{\mathbf{0.518444666\ mol\ MgSO_4}}{\mathbf{2.50\ L\ soln}} = \mathbf{0.207\ mol\ MgSO_4/L = 0.207\ M\ MgSO_4}$$

We could also set up the problem in one string, starting from the ratio (amount of solute)/(amount of solvent):

$$\frac{\mathbf{62.4\ g\ MgSO_4}}{\mathbf{2.50\ L\ soln}} \times \frac{\mathbf{1\ mole\ MgSO_4}}{\mathbf{120.36\ g\ MgSO_4}} = \mathbf{0.207\ mol\ MgSO_4/L = 0.207\ M\ MgSO_4}$$

b) *3.37 g of tryptophan ($C_{11}H_{12}N_2O_2$) in a total volume of 453 mL.*

We need to convert the given volume (453 mL) to 0.453 L (you should be pretty good at this type of Chapter 1 conversion by now).

To convert 3.37 g of tryptophan into moles, we first need to add up its formula mass:

11 C	=	11 × 12.01	=	132.11 amu
12 H	=	12 × 1.008	=	12.10 amu
2 N	=	2 × 14.01	=	28.02 amu
+ 2 O	=	2 × 16.00	=	32.00 amu
			=	204.23 amu

If one molecule of tryptophan has a mass of 204.23 amu, then one mole of tryptophan has a mass of 204.23 grams. We convert the given mass of tryptophan into moles (we don't round yet because there are more steps to carry out):

$$3.37\ \text{g tryptophan} \times \frac{1\ \text{mole}}{204.23\ \text{g}} = 0.016501004\ \text{mol tryptophan}$$

Then, we set up the ratio and calculate the concentration, rounding to 3 significant figures:

$$\frac{\textbf{0.016501004 mol tryptophan}}{\textbf{0.453 L}} = \textbf{0.0364 mol tryptophan/L}$$

$$= \textbf{0.0364 M tryptophan}$$

We could also set up the problem in one string:

$$\frac{\textbf{03.37 g tryptophan}}{\textbf{453 mL}} \times \frac{\textbf{1 mole}}{\textbf{204.23 g}} \times \frac{\textbf{1000 mL}}{\textbf{1 L}} = \textbf{0.0364 mol tryptophan/L}$$

$$= \textbf{0.0364 M tryptophan}$$

c) *45.5 mg of vitamin C ($C_6H_8O_6$) in a total volume of 2.75 mL.*

We need to convert the given volume, 2.75 mL, to 0.00275 L.

To convert 45.5 mg of $C_6H_8O_6$ into moles, we first need to add up its formula mass:

6 C	=	6 × 12.01	=	72.06 amu
8 H	=	8 × 1.008	=	8.064 amu
+ 6 O	=	6 × 16.00	=	96.00 amu
			=	176.12 amu

If one molecule of $C_6H_8O_6$ has a mass of 176.12 amu, then one mole of $C_6H_8O_6$ has a mass of 176.12 grams. We convert the given mass (45.5 mg) of $C_6H_8O_6$ into grams, then moles (we don't round yet because there are more steps to carry out):

$$45.5 \text{ mg } C_6H_8O_6 \times \frac{1 \text{ g}}{1000 \text{ mg}} \times \frac{1 \text{ mole}}{176.12 \text{ g}} = 0.000258347 \text{ mol } C_6H_8O_6$$

Then, we set up the ratio and calculate the concentration, rounding to 3 significant figures:

$$\frac{\textbf{0.000258347 mol } \mathbf{C_6H_8O_6}}{\textbf{0.00275 L}} = \textbf{0.0939 mol } \mathbf{C_6H_8O_6}\textbf{/L} = \textbf{0.0939 M } \mathbf{C_6H_8O_6}$$

We could also set up the problem in one string:

$$\frac{\textbf{45.5 mg } \mathbf{C_6H_8O_6}}{\textbf{2.75 mL}} \times \frac{\mathbf{10^3}\textbf{ mL}}{\textbf{1 L}} \times \frac{\textbf{1 g}}{\mathbf{10^3}\textbf{ mg}} \times \frac{\textbf{1 mole}}{\textbf{176.12 g}}$$

$$= \textbf{0.0939 mol } \mathbf{C_6H_8O_6}\textbf{/L} = \textbf{0.0939 M } \mathbf{C_6H_8O_6}$$

5.21 a) *How many grams of NaOH would you need in order to prepare 2.50 L of 0.100 M NaOH?*

Let's go back to Chapter 1 principles and restate the problem mathematically:

2.50 L solution = ? g NaOH

The units of molarity, M, are mol/L, and it's usually helpful to rewrite the given molarity as mol/L (0.100 M = 0.100 mol/L). Concentration units are compound units and are often used as conversion factors. We can use the given values of volume and concentration to determine the number of moles of NaOH needed:

$$2.50 \text{ L solution} \times \frac{0.100 \text{ mol NaOH}}{\text{L solution}} = 0.250 \text{ mol NaOH}$$

To convert from moles of NaOH to grams of NaOH, we need to calculate the molar mass.

1 Na	=	22.99 amu
1 H	=	1.008 amu
+ 1 O	=	16.00 amu
	=	40.00 amu (rounded to 2 decimal places)

If one formula unit of NaOH has a mass of 40.00 amu, then one mole of NaOH has a mass of 40.00 g. We can use this to calculate the mass of NaOH in 0.250 mol:

$$\mathbf{0.250 \text{ mol NaOH} \times \frac{40.00 \text{ g NaOH}}{\text{mol NaOH}} = 10.0 \text{ g NaOH}}$$

We also could have set up the entire calculation in one string:

$$\mathbf{2.50 \text{ L solution} \times \frac{0.100 \text{ mol NaOH}}{\text{L solution}} \times \frac{40.00 \text{ g NaOH}}{\text{mol NaOH}} = 10.0 \text{ g NaOH}}$$

b) *How many grams of $Cu(NO_3)_2$ would you need in order to prepare 100.0 mL of 0.255 M $Cu(NO_3)_2$?*

The setup: 100.0 mL solution = ? g $Cu(NO_3)_2$

We rewrite the given molarity as mol/L (0.255 M = 0.255 mol/L) and use it as a conversion factor. We also need to convert the given volume, 100.0 mL, to 0.1000 L. Then we can calculate the moles of $Cu(NO_3)_2$ needed:

$$0.1000 \text{ L solution} \times \frac{0.255 \text{ mol Cu(NO}_3)_2}{\text{L solution}} = 0.0255 \text{ mol Cu(NO}_3)_2$$

To convert from moles of $Cu(NO_3)_2$ to grams of $Cu(NO_3)_2$, we need to calculate the molar mass.

First, we add up the formula weight of $Cu(NO_3)_2$:

1 Cu			=	63.55 amu
2 N	=	2 × 14.01 amu	=	28.02 amu
+ 6 O	=	6 × 16.00 amu	=	96.00 amu
			=	187.57 amu

If one formula unit of $Cu(NO_3)_2$ has a mass of 187.57 amu, then one mole of $Cu(NO_3)_2$ has a mass of 187.57 grams. We can use this to calculate the mass of $Cu(NO_3)_2$ in 0.0255 mol (rounding to 3 significant figures):

$$\mathbf{0.0255 \text{ mol Cu(NO}_3)_2 \times \frac{187.57 \text{ g}}{\text{mol}} = 4.78 \text{ g Cu(NO}_3)_2}$$

We also could have set up the entire calculation in one string:

$$\mathbf{100.0\ mL\ solution} \times \frac{\mathbf{1\ L}}{\mathbf{1000\ mL}} \times \frac{\mathbf{0.255\ mol\ Cu(NO_3)_2}}{\mathbf{L\ solution}} \times \frac{\mathbf{187.57\ g}}{\mathbf{mol}} = \mathbf{4.78\ g\ Cu(NO_3)_2}$$

c) *How many grams of glucosamine ($C_6H_{13}NO_5$) would you need in order to prepare 550 mL of 1.4 M glucosamine?*

The setup: 550 mL solution = ? g glucosamine

We rewrite the given molarity as mol/L (1.4 M = 1.4 mol/L) and use it as a conversion factor. We convert the given volume, 550 mL, to 0.550 L. (Remember [see Appendix A] that in whole-number measurements with one or two zeros at the end, the zeros are assumed to be significant; however in this problem it doesn't matter anyway, as we're ultimately going to round to 2 significant figures because of the molarity.) Then we can calculate the moles of glucosamine needed:

$$0.550\ \text{L solution} \times \frac{1.4\ \text{mol}\ C_6H_{13}NO_5}{\text{L solution}} = 0.77\ \text{mol}\ C_6H_{13}NO_5$$

NOTE:

The value 0.77 moles is the calculator answer, not rounded, because we still have another step to do; we'll round to 2 significant figures at the end.

To convert from moles of $C_6H_{13}NO_5$ to grams of $C_6H_{13}NO_5$, we need to calculate the molar mass.

First, we add up the formula weight of $C_6H_{13}NO_5$:

6 C	=	6 × 12.01	=	72.06 amu
13 H	=	13 × 1.008	=	13.10 amu
1 N			=	14.01 amu
+ 5 O	=	5 × 16.00	=	80.00 amu
			=	179.17 amu

If one formula unit of $C_6H_{13}NO_5$ has a mass of 179.17 amu, then one mole of $C_6H_{13}NO_5$ has a mass of 179.17 grams. We can use this to calculate the mass of $C_6H_{13}NO_5$ in 0.77 mol (rounding to 2 significant figures):

$$\mathbf{0.77\ mol\ C_6H_{13}NO_5} \times \frac{\mathbf{179.17\ g}}{\mathbf{mol}} = \mathbf{140\ g\ C_6H_{13}NO_5}$$

We also could have set up the entire calculation in one string:

$$\mathbf{550\ mL\ solution} \times \frac{\mathbf{1\ L}}{\mathbf{1000\ mL}} \times \frac{\mathbf{1.4\ mol\ C_6H_{13}NO_5}}{\mathbf{L\ solution}} \times \frac{\mathbf{179.17\ g}}{\mathbf{mol}} = \mathbf{140\ g\ C_6H_{13}NO_5}$$

The value 140 g has an ambiguous number of significant figures; it isn't completely clear whether the zero at the end is significant or not. Since our answer should have 2 significant figures, we could eliminate the ambiguity by reporting the final value as

1.4×10^2 g $C_6H_{13}NO_5$.

Solutions to Section 5.4 Core Problems

5.23 *Two solutions are separated by a membrane. Solution A contains 0.05 M sucrose, while solution B contains 0.07 M sucrose. If the membrane is permeable to water but not to sucrose, in which direction will osmosis occur?*

Solvent always travels from the solute with lower solute concentration to the solution with greater solute concentration. You can think of this as nature trying to make things equal: when the solvent leaves the more dilute solution (in this case A), it becomes more concentrated, and when solvent enters the more concentrated solution (in this case B), it becomes more dilute, until the two concentrations are equal. **Solvent will flow from Solution A to Solution B.**

5.25 *Calculate the total molarity of each of the following solutions:*

a) *0.125 M NaCl*

NaCl dissociates into Na^+ and Cl^- ions, giving two total moles of ions for every one mole of NaCl. Multiply the labeled concentration of NaCl by 2 to get total ion concentration:

$$\frac{\textbf{0.125 mol NaCl}}{\textbf{L}} \times \frac{\textbf{2 mol ions}}{\textbf{1 mol NaCl}} = \frac{\textbf{0.250 mol ions}}{\textbf{L}} = \textbf{0.250 M total}$$

b) *0.22 M Na_2CO_3*

1 mole of Na_2CO_3 dissociates into 2 moles of Na^+ and 1 mole of CO_3^{2-} ions, giving three total moles of ions for every one mole of Na_2CO_3. Multiply the labeled concentration of Na_2CO_3 by 3 to get total ion concentration:

$$\frac{\textbf{0.22 mol Na}_2\textbf{CO}_3}{\textbf{L}} \times \frac{\textbf{3 mol ions}}{\textbf{1 mol Na}_2\textbf{CO}_3} = \frac{\textbf{0.66 mol ions}}{\textbf{L}} = \textbf{0.66 M total}$$

c) *A solution that contains 0.12 M glucose (a nonelectrolyte) and 0.21 M KBr.*

This solution contains two solutes, whose concentrations must be added together. Since the problem says glucose is a nonelectrolyte, we know it doesn't dissociate, so its total contribution is its labeled concentration (0.12 M). KBr dissociates into K^+ and Br^-, giving two moles of ions for every one mole KBr, or a total ion concentration from KBr of 0.21M × 2 = 0.42 M total ion concentration:

$$\frac{0.21 \text{ mol KBr}}{\text{L}} \times \frac{2 \text{ mol ions}}{1 \text{ mol KBr}} = \frac{0.42 \text{ mol ions}}{\text{L}} = 0.42 \text{ M ions}$$

The overall total solute concentration is therefore the sum of the total contributions from the two solutes:
0.42 M (from KBr) + 0.12 M glucose = 0.54 M total solute concentration.

5.27 *Label each of the following solutions as isotonic, hypotonic, or hypertonic. Be sure to account for the dissociation of electrolytes.*

See Table 5.2 for a summary of the relationship between tonicity and solution concentration.

a) 0.14 M lactose ($C_{12}H_{22}O_{11}$)

Lactose is a nonelectrolyte (covalent/molecular compound), so its labeled concentration is the total concentration. **This solution has a total molarity of less than 0.28 M and is hypotonic.**

b) 0.14 M KCl

KCl, an ionic compound, dissociates into K^+ and Cl^-, giving two total moles of ions for every one mole KCl, or a total ion concentration of **0.14 M × 2 = 0.28 M total ion concentration**. This solution is therefore **isotonic.**

c) 0.14 M $MgCl_2$

1 mole of $MgCl_2$ dissociates into 1 mole of Mg^{2+} and 2 moles of Cl^- ions, giving three total moles of ions for every one mole of $MgCl_2$. Multiply the labeled concentration by 3 to get total ion concentration: **0.14 M × 3 = 0.42 M total ion concentration**, which is greater than the 0.28 M isotonic concentration. This solution is therefore **hypertonic.**

5.29 *What will happen to a red blood cell if it is placed into each of the solutions in Problem 5.27?*

Solution (a) is hypotonic—too dilute—and the cell will swell and burst (hemolysis). Solution (b) is isotonic—the same concentration as the solution inside the cell—so nothing will happen to the cell (as far as osmosis is concerned anyway). Solution (c) is hypertonic—too concentrated—and the cell will shrivel as water osmoses out (crenation).

5.31 *You need to make 500 mL of an isotonic solution of urea (N_2H_4CO). How many grams of urea do you need? Urea is a nonelectrolyte.*

Strategy: the setup of the problem, as in Chapter 1, is

500 mL of solution = ? g of urea

"Isotonic" means that the molarity is 0.28 M, and since urea is a nonelectrolyte we don't have to worry about dissociation—we really do need a 0.28 M solution. Usually when we see a concentration given in M, we're better off writing it in terms of the component units, mol/L (giving us 0.28 mol/L). We can now see that we'll need to convert the mL in the problem to L, and then our molarity (mol/L) will get us to moles.

500 mL → L → mol → ? = ? g

To get from moles to grams, we need to add up the molar mass of urea by looking at the atomic weights from the periodic table:

2 N	=	2 × 14.01	=	28.02 g/mol
4 H	=	4 × 1.008	=	4.032
1 C	=	1 × 12.01	=	12.01
+ 1 O	=	1 × 16.00	=	16.00
total			=	60.06 g/mol

So:

$$500\ \text{mL} \times \frac{1\ \text{L}}{1000\ \text{mL}} \times \frac{0.28\ \text{mol}}{\text{L}} \times \frac{60.06\ \text{g}}{\text{mol}} = 8.4\ \text{g urea}$$

Our answer is rounded to two significant figures, limited by the isotonic molarity (0.28 M). Remember (see Appendix A) that in whole-number measurements with one or two zeros at the end, the zeros are assumed to be significant; therefore the needed volume, 500 mL, has three significant figures.

5.33 *A solution contains 0.04 M $MgSO_4$, 0.05 M NaCl, and 0.15 M sucrose. (Sucrose is a nonelectrolyte.)*

a) *Calculate the total molarity of solute particles in this solution.*

First calculate the total ion concentrations for the electrolytes. $MgSO_4$ dissociates into Mg^{2+} and SO_4^{2-}, 2 mol ions for each 1 mol solute: 0.04 M × 2 = 0.08 M (total contribution from $MgSO_4$). NaCl dissociates into Na^+ and Cl^-, 2 mol ions for each 1 mol solute: 0.05 M × 2 = 0.10 M (total contribution from NaCl).

Since sucrose is a nonelectrolyte, we can use its molarity (0.15 M) as given. Then we add all solute molarities to get total solute concentration:
0.08 M + 0.10 M + 0.15 M = 0.33 M.

b) *Is this solution isotonic, hypotonic, or hypertonic?*

Isotonic concentration is 0.28 M; this solution is more concentrated than 0.28 M and is therefore **hypertonic.**

5.35 *In the diagram below, solutions A and B are separated by a membrane that is permeable to water, glucose, and ions.*

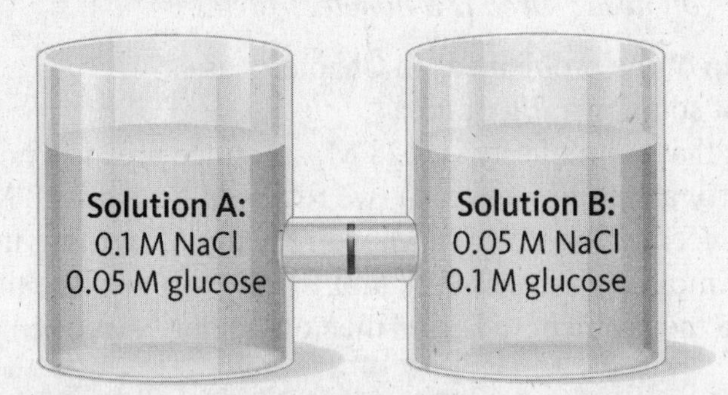

a) *In which direction does osmosis occur?*

To figure this out, we have to first add up the total solute concentration in each solution.

Solution A: 0.1 M NaCl gives a total ion concentration of 0.2 M; adding 0.05 M for the glucose gives a total concentration for all solutes of **0.25 M.**

Solution B: 0.05 M NaCl gives a total ion concentration of 0.1 M; adding 0.1 M for the glucose gives a total concentration for all solutes of **0.2 M.**

Solution A is more concentrated than Solution B, so the net flow of water (osmosis) goes from Solution B to Solution A.

b) *In which direction does Na^+ dialyze?*

Dialysis carries solutes from the more concentrated solution to the less concentrated solution. **Since the concentration of Na^+ is greater in Solution A, Na^+ will dialyze from Solution A to Solution B.**

c) *In which direction does glucose dialyze?*

Since the concentration of glucose is greater in Solution B, glucose will dialyze from Solution B to Solution A.

Solutions to Section 5.5 Core Problems

5.37 *If a solution contains 0.2 moles of CO_3^{2-}, how many equivalents does it contain?*

The number of equivalents is the number of moles times the charge (ignoring the negative sign on anions and just looking at the size of the charge), so for carbonate, **0.2 moles × 2 = 0.4 Eq.**
To look at this another way, one mole of CO_3^{2-} would have a total charge of 2 mol:

$$1 \text{ mol } CO_3^{2-} = 2 \text{ mol charge} = 2 \text{ Eq } CO_3^{2-}$$

$$\mathbf{0.2\,mol\,CO_3^{2-} \times \frac{2\,Eq}{mol} = 0.4\,Eq}$$

5.39 *A solution contains 0.15 Eq of Mg^{2+} ions.*

a) *How many moles of Mg^{2+} does this solution contain?*

moles × charge = equivalents

$$\mathbf{moles = \frac{Eq}{charge} = \frac{0.15\,Eq}{2} = 0.075\,moles}$$

b) *How many grams of Mg^{2+} does this solution contain?*

From the periodic table, the molar mass of Mg is 24.31 g/mol, so

$$\mathbf{0.075\,mol\,Mg^{2+} \times \frac{24.31\,g}{mol} = 1.8\,g}$$

5.41 a) *Convert 6.25 g of Ca^{2+} into equivalents and into milliequivalents.*

Strategy: g → mol → Eq. Molar mass of Ca (40.08 g/mol) comes from the periodic table.

$$\mathbf{6.25\,g\,Ca^{2+} \times \frac{mol}{40.08\,g} \times \frac{2\,Eq}{mol} = 0.312\,Eq \times \frac{1000\,mEq}{Eq} = 312\,mEq}$$

b) Convert 27.3 g of $C_4H_4O_4^{2-}$ (succinate ions) into equivalents and into milliequivalents.

Add up atomic weights from the periodic table:

4 C	=	4 × 12.01	=	48.04 g/mol
4 H	=	4 × 1.008	=	4.032
+ 4 O	=	4 × 16.00	=	64.00
total			=	116.07 g/mol

$$\mathbf{27.3\ g\ succinate \times \frac{mol}{116.07\ g} \times \frac{2\ Eq}{mol} = 0.470\ Eq \times \frac{1000\ mEq}{Eq} = 470\ mEq}$$

5.43 *a) A solution contains 127 mEq of Cu^{2+}. How many grams is this? How many milligrams?*

For the Cu^{2+} ion, 1 mole ions = 2 eq. To get from moles to grams, we need the molar mass of copper from the periodic table, 63.55 g/mol. From there, the question is a unit conversion problem:

$$\mathbf{127\ mEq\ Cu^{2+} \times \frac{1\ Eq}{1000\ mEq} \times \frac{mol}{2\ Eq} \times \frac{63.55\ g}{mol} = 4.04\ g \times \frac{1000\ mg}{g} = 4040\ mg}$$

b) A solution contains 34 mEq of CO_3^{2-}. How many grams is this? How many milligrams?

For CO_3^{2-}, 1 mol = 2 Eq. To convert from moles to grams, we'll need to add up the molar mass of CO_3^{2-}:

1 C	=	1 × 12.01	=	12.01 g/mol
+ 3 O	=	3 × 16.00	=	48.00
total			=	60.01 g/mol

From there, it's a unit conversion problem:

$$\mathbf{34\ mEq\ CO_3^{2-} \times \frac{1\ Eq}{1000\ mEq} \times \frac{mol}{2\ Eq} \times \frac{60.01\ g}{mol} = 1.02\ g \times \frac{1000\ mg}{g} = 1020\ mg}$$

5.45 *You have a 0.25 M solution of $CaCl_2$. Calculate the concentrations of Ca^{2+} and Cl^- in this solution, in milliequivalents per liter.*

The concentration of Ca^{2+} = the stated concentration of $CaCl_2$, since 1 mol $CaCl_2$ contains 1 mol Ca^{2+}:

$$\mathbf{\frac{0.25\ mol\ CaCl_2}{L} \times \frac{1\ mol\ Ca^{2+}}{mol\ CaCl_2} \times \frac{2\ Eq}{mol} \times \frac{1000\ mEq}{Eq} = 500\ mEq\ Ca^{2+}/L}$$

This answer should have two significant figures; to clarify, we can report it as **5.0×10^2 mEq/L.**

The concentration of Cl^- = 2 × the stated concentration of $CaCl_2$ or since 1 mol $CaCl_2$ contains 2 mol Cl^-:

$$\frac{\textbf{0.25 mol CaCl}_2}{\textbf{L}} \times \frac{\textbf{2 mol Cl}^-}{\textbf{mol CaCl}_2} \times \frac{\textbf{1 Eq}}{\textbf{mol}} \times \frac{\textbf{1000 mEq}}{\textbf{Eq}} = \textbf{500 mEq Cl}^-\textbf{/L}$$

Again, our answer should have two significant figures, so we report it as **5.0×10^2 mEq/L.**

NOTE:

The total equivalents of cations and anions must always be equal in a solution, because the overall charge must be zero.

5.47 *A solution contains 2.88 g of sulfate ions in a total volume of 3.38 L. Calculate the concentration of sulfate ions in each of the following units:*

a) Eq/L

Setup: $\frac{\text{2.88 g sulfate}}{\text{3.38 L soln}} = \text{? Eq/L}$

We're going to have to get from grams to moles, so add up molar mass of sulfate:

1 S	=	1 × 32.06	=	32.06 g/mol
+ 4 O	=	4 × 16.00	=	64.00
total			=	96.06 g/mol

$$\frac{\textbf{2.88 g sulfate}}{\textbf{3.38 L}} \times \frac{\textbf{1 mol}}{\textbf{96.06 g}} \times \frac{\textbf{2 Eq}}{\textbf{mol}} = \frac{\textbf{0.0177 Eq sulfate}}{\textbf{L}}$$

b) mEq/L

$$\frac{\textbf{0.0177 Eq sulfate}}{\textbf{L}} \times \frac{\textbf{1000 mEq}}{\textbf{Eq}} = \textbf{17.7 mEq sulfate/L}$$

5.49 *You have 750 mL of a solution that contains 4.2 mEq/L of SO_4^{2-} ions. Calculate the mass of the sulfate ions in this solution.*

Setup: 750 mL = ? g sulfate ions

Our strategy: the compound unit in the concentration, 4.2 mEq/L, indicates that it will be used as a conversion factor. We'll need to convert the volume 750 mL to L, then use the concentration to get mEq of sulfate, then convert that to Eq using the relationship 1000 mEq = 1 Eq.

NOTE:

We do not round the answer to any of these intermediate steps, though we will round the final answer to 2 significant figures.

$$750\text{ mL soln} \times \frac{1\text{ L}}{1000\text{ mL}} \times \frac{4.2\text{ mEq SO}_4^{2-}}{\text{L}} = 3.15\text{ mEq SO}_4^{2-} \times \frac{1\text{ Eq}}{1000\text{ mEq}}$$

$$= 0.00315\text{ mEq SO}_4^{2-}$$

Since moles × charge = equivalents,

$$\text{moles} = \frac{\text{Eq}}{\text{charge}} = \frac{0.00315\ \text{Eq}}{2} = 0.001575\ \text{moles}$$

Lastly, we use the molar mass of SO_4^{2-} (96.06 g/mol, see Problem 5.47) to convert to grams of sulfate, and round our final answer to 2 significant figures:

$$\mathbf{0.001575\ mol\ SO_4^{2-} \times \frac{96.06\ g}{mol} = 0.15\ g\ SO_4^{2-}}$$

NOTE:
We could also have set up the entire problem in one string, treating it as a unit conversion problem:

$$\mathbf{750\ mL\ soln \times \frac{1\ L}{1000\ mL} \times \frac{4.2\ mEq\ SO_4^{2-}}{L} \times \frac{1\ Eq}{1000\ mEq} \times \frac{1\ mol\ SO_4^{2-}}{2\ Eq\ SO_4^{2-}} \times \frac{96.06\ g}{mol} = 0.15\ g\ SO_4^{2-}}$$

Solutions to Section 5.6 Core Problems

5.51 a) *You have 100 mL of 0.50 M sodium lactate solution. If you add water until the total volume reaches 750 mL, what will be the molarity of sodium lactate in the resulting solution?*

Assume the given volume 100 mL has three significant figures. This is a dilution problem, with C_1= 0.50 M, V_1 = 100 mL, and V_2 = 750 mL; C_2 is the unknown quantity.

NOTE:
In dilution problems, it usually doesn't matter what units your C and V values have, as long as the units of C_1 and C_2 match and the units of V_1 and V_2 match (the C and V don't have to match each other).

$$C_1 \times V_1 = C_2 \times V_2$$

Solve for C_2 to get:

$$\frac{C_1 \times V_1}{V_2} = C_2$$

$$\frac{\mathbf{(0.50\ M)(100\ mL)}}{\mathbf{750\ mL}} = \mathbf{0.067\ M}$$

b) *You have 33.5 mL of 2.50% (w/v) NaCl solution. If you add water until the total volume reaches 150.0 mL, what will be the percent concentration of NaCl in the resulting solution?*

C_1= 2.50%, V_1 = 33.5 mL, and V_2 = 150.0 mL. C_2 is the unknown quantity.

$$C_1 \times V_1 = C_2 \times V_2$$

Solve for C_2 to get:

$$\frac{C_1 \times V_1}{V_2} = C_2$$

$$\frac{(2.50\,\%)(33.5\,\text{mL})}{150.0\,\text{mL}} = 0.558\,\%\ \text{w/v}$$

5.53 a) *If you mix 50 mL of water with 10 mL of 1.14% (w/v) H_2SO_4, what will be the percent concentration of H_2SO_4 in the resulting solution?*

This is a dilution problem, with C_1= 1.14%, V_1 = 10 mL, and V_2 = 60 mL—this is the only tricky bit here, realizing that the total volume will be 50 mL + 10 mL (we are assuming that the volumes are additive, which is usually true for dilute solutions in the same solvent). Always read the wording of the problem carefully; sometimes it will be written in terms of the total volume (as in Problem 5.51), sometimes it will be written in terms of the amount added and you'll have to figure out the total volume. C_2 is the unknown quantity.

$$C_1 \times V_1 = C_2 \times V_2$$

Solve for C_2 to get:

$$\frac{C_1 \times V_1}{V_2} = C_2$$

$$\frac{(1.14\,\%)(10\,\text{mL})}{60\,\text{mL}} = 0.19\,\%\ \text{w/v}$$

b) *If you mix 1.50 L of water with 250 mL of 0.60 M KI, what will be the molarity of KI in the resulting solution?*

Again, assume volumes are additive. C_1= 0.60 M, V_1 = 250 mL or 0.25 L, and V_2 = 1.75 L (1.50 L + 0.25 L). C_2 is the unknown quantity. The two volume measurements must match, so let's use both in liters.

$$C_1 \times V_1 = C_2 \times V_2$$

Solve for C_2 to get:

$$\frac{C_1 \times V_1}{V_2} = C_2$$

$$\frac{(0.60\,\text{M})(0.25\,\text{L})}{1.75\,\text{L}} = 0.086\,\text{M}$$

5.55 *Intravenous sodium lactate solutions contain 1.72% (w/v) sodium lactate in water. If you have 100 mL of 5.00% (w/v) sodium lactate, and you need to dilute it to 1.72%, what must the final volume be? How much water will you add?*

Dilution problem with C_1= 5.00%, V_1 = 100 mL, and C_2 = 1.72 %. V_2 is the unknown quantity.

$$C_1 \times V_1 = C_2 \times V_2$$

Solve for V_2 to get:

$$\frac{C_1 \times V_1}{C_2} = V_2$$

$$\frac{(5.00\%)(100\,\text{mL})}{1.72\,\%} = 291\,\text{mL (total volume)}$$

To get to 300 mL total volume, you'll have to add approximately **191 mL of water** (291 mL total – 100 mL initial), assuming that the volumes of the mixed solutions are additive (this is usually true for dilute solutions).

5.57 *You need to make 100 mL of 0.90% (w/v) sodium chloride solution. You have a bottle of 5.0% (w/v) sodium chloride solution, which you must dilute to the correct concentration. What volume of the 5.0% solution should you use, and how much water must you add to it?*

Dilution problem with $C_1 = 5.0\%$, $C_2 = 0.90\%$, and $V_2 = 100$ mL.

$$C_1 \times V_1 = C_2 \times V_2$$

Solve for V_1 to get:

$$\frac{C_2 \times V_2}{C_1} = V_1$$

$$\frac{\mathbf{(0.90\%)(100\,mL)}}{\mathbf{5.0\,\%}} = \mathbf{18\,mL}$$

To get to 100 mL total volume, you'll have to add **82 mL of water** (100 mL total – 18 mL initial), assuming that the volumes of the mixed solutions are additive.

Solutions to Concept Questions

* indicates more challenging problems.

5.59 *Define each of the following concentration units.*

These definitions come from the chapter.

a) percentage (w/v) **is defined as grams solute per 100 mL solution.**

$$\frac{\textbf{grams solute}}{\textbf{100 mL solution}} = \textbf{\% (w/v)}$$

b) percentage (v/v) **is defined as mL solute per 100 mL solution.**

$$\frac{\textbf{mL solute}}{\textbf{100 mL solution}} = \textbf{\% (v/v)}$$

c) molarity **is defined as mol solute per L solution.**

$$\frac{\textbf{mol solute}}{\textbf{L solution}} = \textbf{M} = \textbf{molarity}$$

d) ppm **is parts per million, defined as either mg solute per L solution, or µg solute per mL solution.**

$$\frac{\textbf{mg solute}}{\textbf{L solution}} = \frac{\boldsymbol{\mu}\textbf{g solute}}{\textbf{mL solution}} = \textbf{ppm}$$

(These two definitions are mathematically equivalent; memorize one but not both.)

e) *mg/dL* **is defined as mg solute per deciliter of solution** (the definition of this one is, conveniently, contained in the units.)

$$\frac{\textbf{mg solute}}{\textbf{dL solution}} = \textbf{mg/dL}$$

f) *ppb* **is defined as either µg solute per L solution, or ng solute per mL solution.**

$$\frac{\mu\textbf{g solute}}{\textbf{L solution}} = \frac{\textbf{ng solute}}{\textbf{mL solution}} = \textbf{ppb}$$

(These two definitions are mathematically equivalent; memorize one but not both.)

g) *mEq/L* **is defined as milliequivalents of solute per liter of solution** (again, the definition is contained in the units.)

$$\frac{\textbf{mEq solute}}{\textbf{L solution}} = \textbf{mEq/L}$$

5.61 *Charlene needs to prepare a 30% (v/v) solution of ethanol in water. She mixes 30 mL of ethanol with 100 mL of water.*

a) *Why doesn't this produce a 30% (v/v) solution?*

A 30% (v/v) solution will contain 30 mL of ethanol per 100 mL of *solution*. Mixing 30 mL of ethanol with 100 mL of water will give much more than 100 mL of solution, and the solution will be too dilute. She'll need closer to 70 mL of water (though not exactly, because when two different liquids are mixed the volumes are often not exactly additive.)

b) *How could Charlene have made the solution correctly?*

She should take 30 mL of ethanol, put it in a container that is marked for a 100-mL volume, and add water until the total volume of the solution is 100 mL. This will take much less than 100 mL of water.

5.63 *The chemical formula of calcium oxide is CaO. Does this mean that calcium oxide contains equal weights of calcium and oxygen? Explain your answer.*

Chemical formulas give the ratio of atoms or ions (or of moles of atoms or ions) in the formula, not the ratio of masses. CaO contains equal numbers of calcium and oxide ions, but since calcium and oxide ions do not have the same mass, CaO does not contain equal weights of calcium and oxygen. (In fact, it contains 40 g of calcium for every 16 g of oxygen.)

5.65 *Samson sees a bottle labeled "2 M HCl" in his laboratory. He tells Pierre that this bottle contains two moles of HCl. Is this correct? If not, what can you say about the number of moles of HCl in the bottle?*

Capital M, as a concentration unit, stands for moles solute per liter of solution. (In fact, it's usually useful, in calculation problems, to cross out the M and write in mol/L, since you almost always need the mol/L units to work your problem out anyway.) **If the container happens to hold exactly one liter of solution, then it contains two moles of HCl, but that would be a coincidence; if it contains any other volume of solution, then the amount of HCl won't be exactly equal to two moles. The number of moles of HCl in the bottle is actually equal to the volume of the bottle (in L) times 2 mol/L.**

5.67 *You can make a pickle by putting a cucumber into an aqueous solution that contains salt, vinegar, and some flavoring ingredients.*

a) *Why does the cucumber shrink as it sits in this solution?*

If the solution outside the cucumber has a higher total solute concentration than the solution inside the cucumber's cells, then osmosis will cause water to flow out of the cells, and the cucumber will lose volume (shrink).

b) *Why does the cucumber taste like vinegar after a while?*

The molecules responsible for the flavor of the vinegar (acetic acid, in fact, though we don't have to know this yet) dialyze into the cells of the cucumber.

c) *Why doesn't the cucumber become salty?*

If the cucumber doesn't become salty, then the salt in the solution doesn't dialyze into the cells; we can infer that the cell membranes must be impermeable to ions (or at least to sodium ions and chloride ions).

5.69 *A red blood cell is placed into a 1.0% (w/v) solution of sodium lactate. The cell rapidly swells and bursts.*

a) *Why did the cell do this?*

Osmosis of water into the cell caused it to swell. If the solution outside the membrane has a lower total solute concentration than the solution inside, then water will osmose into the cell, causing it to swell and burst.

b) *What does this observation tell you about the percent concentration of isotonic sodium lactate solutions?*

The solution must have been hypotonic; isotonic solutions of sodium lactate must be more concentrated than 1.0% (w/v).

5.71 *Javier says "isotonic solutions always have a concentration around 0.28 M." LaShawndra then asks why the concentration of isotonic NaCl is only 0.14 M. How should Javier respond?*

NaCl is an ionic compound, so it dissociates when it dissolves in aqueous solution, forming Na^+ and Cl^- ions. For purposes of osmosis, the total

concentration of solute particles in the solution is therefore actually twice the labeled value of the solution, or 0.28 M (0.14 M Na^+ plus 0.14 M Cl^-).

5.73 *What is the difference between osmosis and dialysis?*

Osmosis is the flow of *solvent* across a semipermeable membrane, while dialysis is the flow of *solute* across a semipermeable membrane. (In both cases, though, the direction of flow is the direction that would tend to make both sides equal for the species in question; solvent will flow from the less concentrated solution to the more concentrated solution, while each solute will flow from its more concentrated solution to its more dilute solution.)

5.75 *Why do electrolyte solutions always contain the same numbers of equivalents of cations and anions?*

Equivalents are moles of charge, and the total charge of an electrolyte solution must always be 0 because the solution must be electrically neutral (that is, the total + charge and the total – charge must be equal).

5.77 *If you add water to 1% (w/v) solution of sugar, will the concentration of the solution increase, decrease, or remain the same? Explain your answer.*

Adding solvent to a solution always makes it less concentrated, so adding water to a 1% (w/v) solution of sugar will cause the concentration to decrease.

Solutions to Summary and Challenge Problems

* indicates more challenging problems.

5.79 *You have a solution that contains 175 mEq/L of citrate ($C_6H_5O_7^{3-}$) ions. Calculate the concentration of citrate ions in this solution in each of the following units:*

a) Eq/L Conversion factor: 1000 mEq = 1 Eq.

$$\frac{\mathbf{175\ mEq}}{\mathbf{L}} \times \frac{\mathbf{1\ Eq}}{\mathbf{1000\ mEq}} = \mathbf{0.175\ Eq/L}$$

b) M Conversion factor: 1 mol citrate = 3 Eq citrate. M stands for mol/L, so that's the units we're trying to reach:

$$\frac{\mathbf{175\ mEq}}{\mathbf{L}} \times \frac{\mathbf{1\ Eq}}{\mathbf{1000\ mEq}} \times \frac{\mathbf{1\ mol\ citrate}}{\mathbf{3\ Eq\ citrate}} = \mathbf{0.0583\ mol/L = 0.0583\ M}$$

*5.81 *Folic acid is an essential vitamin and is linked to the prevention of neural tube defects in developing embryos. One cup (236 mL) of fresh orange juice contains around 75 μg of folic acid. Calculate the concentration of folic acid in orange juice in each of the following units:*

a) mg/dL There are several ways to convert μg to mg, but one is to set both equal to the root unit g: 10^3 mg = 1 g and 10^6 μg = 1 g, so 10^3 mg = 10^6 μg. A similar strategy gets us the relationship 10^3 mL = 1 L = 10 dL.

$$\frac{75\ \mu g}{236\ mL}\times\frac{10^3\ mg}{10^6\ \mu g}\times\frac{10^3\ mL}{10\ dL}=0.032\ mg/dL$$

b) μg/dL

$$\frac{75\ \mu g}{236\ mL}\times\frac{10^3\ mL}{10\ dL}=32\ \mu g/dL$$

c) ppm Using the μg/mL definition of ppm:

$$\frac{75\ \mu g}{236\ mL}=0.32\ \mu g/mL=0.32\ ppm$$

d) ppb To use the ng/mL definition of ppb, we need the conversion factor between μg and ng. We use the same strategy as in (a) to get 10^6 μg = 1 g = 10^9 ng:

$$\frac{75\ \mu g}{236\ mL}\times\frac{10^9\ ng}{10^6\ \mu g}=320\ ng/mL=320\ ppb$$

Another strategy here is to realize that the concentration in ppb is always 1000 (10^3) times larger than the concentration in ppm, and move the decimal three places to the right. (If you deal with ppm and ppb a lot, this is a useful thing to know!)

*5.83 *Concentrations of nitrate above 45 ppm are considered hazardous to infants, because they interfere with the ability of the blood to carry oxygen. If you drink 8.0 fluid ounces of water that contains 45 ppm of nitrate, what mass of nitrate ions are you consuming?*

Setup: 8.0 fl oz (fluid ounces) = ? g nitrate (or possibly mg or μg, as appropriate, depending on answer)
Concentration of 45 ppm can be rewritten as 45 μg/mL. To use this unit we'll have to convert the volume in fl oz to mL (from Table 1.4, 1 fl oz = 29.57 mL). We may also have to be ready convert the mass of nitrate ions from μg to mg or g.

$$8.0\ fl\ oz\times\frac{29.57\ mL}{1\ fl\ oz}\times\frac{45\ \mu g}{mL}=1.1\times10^4\ \mu g\times\frac{1\ g}{10^6\ \mu g}=0.011\ g\ nitrate$$

Since the problem in this case doesn't specify the mass unit, μg is fine, but often on a multiple-choice test the unit may be different; always be prepared to do the Chapter 1-type conversion to get whatever unit is needed.

5.85 *A millimole (mmol) equals $^1/_{1000}$ of a mole. Convert 38 mmol of NaCl into the following units:*

a) moles The equivalence given in the problem, 1 mmol = 1/1000 mol, could also be written as 1000 mmol = 1 mol. This is a straight-up Chapter 1 conversion; you don't even have to understand the unit you're using, you just have to apply the rules for the metric prefixes.

$$38 \text{ mmol} \times \frac{1 \text{ mol}}{1000 \text{ mmol}} = 0.038 \text{ mol}$$

b) grams

We can save a step by starting with the answer from (a). To get molar mass of NaCl, add up atomic weights from the periodic table:

1 Na	=	22.99 g/mol
+ 1 Cl	=	35.45
total		58.44 g/mol

$$0.038 \text{ mol} \times \frac{58.44 \text{ g}}{\text{mol}} = 2.2 \text{ g}$$

c) milligrams

If we use our answer from (b) this becomes a simple Chapter 1 unit conversion:

$$2.2 \text{ g} \times \frac{1000 \text{ mg}}{\text{g}} = 2200 \text{ mg}$$

*5.87 *A solution is prepared by dissolving 2.5 ounces of $AgNO_3$ in enough water to make 32 fluid ounces of solution. Calculate the molarity of this solution. You may use the information in Tables 1.4 and 1.5.*

Setup: $\frac{2.5 \text{ oz}}{32 \text{ fl oz}} = ? \frac{\text{mol}}{\text{L}}$

Since concentration units are always in terms of (amount solute)/(amount solution), we can start our setup with 2.5 oz/32 fl oz and go from there. Molarity is defined by the units mol/L, so we'll need to get the amount of $AgNO_3$ from ounces to moles, going through g: ounces→g→mol. We'll also have to convert our volume from fluid ounces to liters; Table 1.4 doesn't have a direct conversion, so we'll go fluid ounces→mL→L (there are other paths).

To get molar mass of $AgNO_3$, add up atomic weights from the periodic table:

1 Ag	=	107.9 g/mol
1 N	=	14.01
+ 3 O	=	48.00
total		169.91 g/mol

$$\frac{2.5 \text{ oz}}{32 \text{ fl oz}} \times \frac{28.35 \text{ g}}{1 \text{ oz}} \times \frac{\text{mol}}{169.91 \text{ g}} \times \frac{1 \text{ fl oz}}{29.57 \text{ mL}} \times \frac{1000 \text{ mL}}{\text{L}} = \frac{0.44 \text{ mol}}{\text{L}} = 0.44 \text{ M}$$

*5.89 *Boric acid (H_3BO_3) has been used to treat eye infections. Calculate the percent concentration of a 0.28 M solution of boric acid.*

Boric acid is a solid, so we'll be calculating a concentration in % (w/v), or g solute/100 mL.

Setup: 0.28 mol/L = ? g/100 mL

Strategy: convert mol→g and L→mL to find g/mL, then multiply by 100 to get % (w/v).

To get molar mass of boric acid, add up atomic weights from the periodic table:

1 B	=	1 × 10.81	=	10.81 g/mol
3 H	=	3 × 1.008	=	3.024
+ 3 O	=	3 × 16.00	=	48.00
total				61.83 g/mol

$$\mathbf{\frac{0.28\ mol}{L} \times \frac{61.83\ g}{1\ mol} \times \frac{1\ L}{1000\ mL} \times 100\% = 1.7\%\ (w/v)}$$

*5.91 *An isotonic solution contains glucose and KCl. The concentration of KCl is 0.08 M. What is the approximate molarity of glucose in this solution?*

If the solution is isotonic, the total concentration of all solutes is 0.28 M. Since KCl is an electrolyte, dissociating into two moles of ions for every mole of KCl, the total ion concentration of the solution is 2 × 0.08 M = 0.16 M. (We can justify giving our answer to two decimal places, because what we're really doing is adding 0.08 M K^+ + 0.08 M Cl^-; this means we gain a significant figure in this calculation.) The rest of the concentration has to be made up by glucose, and since glucose is a nonelectrolyte, we don't have to do anything to its concentration:

0.28 M (total) – 0.16 M (from KCl) = 0.12 M glucose.

5.93 *A solution contains 480 mg of SO_4^{2-}. Convert this into the following units:*

a) g This part, at least, is an easy Chapter 1 review problem, no concentrations to deal with, just mg→g. (One or two zeroes at the end of a measured quantity are assumed to be significant, so the mass has three significant figures.)

$$\mathbf{480\ mg \times \frac{1\ g}{1000\ mg} = 0.480\ g}$$

b) mol Save a step by starting with the answer from (a). We need the molar mass of sulfate ion as well:

1 S	=	1 × 32.06	=	32.06 g/mol
+ 4 O	=	4 × 16.00	=	64.00
total				96.06 g/mol

$$\mathbf{0.480\ g \times \frac{1\ mol}{96.06\ g} = 0.00500\ mol}$$

c) Eq The charge on sulfate ion is -2, so 1 mol sulfate = 2 Eq sulfate. Save steps by starting with the answer from (b):

$$\mathbf{0.00500\ mol \times \frac{2\ Eq}{1\ mol} = 0.0100\ Eq}$$

d) mEq Convert the answer from (c) into mEq:

$$\mathbf{0.0100\ Eq \times \frac{1000\ mEq}{1\ Eq} = 10.0\ mEq}$$

If you'd like to see the entire calculation, mg to mEq, in one string:

$$\mathbf{480\ mg \times \frac{1\ g}{1000\ mg} \times \frac{1\ mol}{96.06\ g} \times \frac{2\ Eq}{1\ mol} \times \frac{1000\ mEq}{1\ Eq} = 10.0\ mEq}$$

5.95 *A solution contains 2.0 mEq/L of $C_4H_4O_5^{2-}$ (malate ions).*

a) *How many milligrams of $C_4H_4O_5^{2-}$ are there in 350 mL of this solution?*

Setup: 350 mL = ? mg
Strategy: mL→L, then use given concentration to convert L→mEq, then to Eq, then to mol (using 1 mol malate = 2 Eq, because malate ion has a charge of -2), then to grams (using molar mass of malate).

To get molar mass of malate ion, add up atomic weights from the periodic table:

$$\begin{aligned} 4\ C &= 4 \times 12.01 = 48.04\ g/mol \\ 4\ H &= 4 \times 1.008 = 4.032 \\ 5\ O &= 5 \times 16.00 = 80.00 \\ \text{total} &\quad 132.07\ g/mol \end{aligned}$$

$$\mathbf{350\ mL \times \frac{1\ L}{1000\ mL} \times \frac{2.0\ mEq}{L} \times \frac{1\ Eq}{1000\ mEq} \times \frac{1\ mol}{2\ Eq} \times \frac{132.07\ g}{mol} \times \frac{1000\ mg}{g} = 46\ mg\ malate\ ions}$$

b) *The entire bottle of solution contains 0.52 g of $C_4H_4O_5^{2-}$ ions. What is the total volume of the solution in the bottle?*

Setup: 0.52 g = ? L
Strategy: this is essentially the problem in part (a) in reverse. Convert g→mol using the molar mass (132.07 g/mol, see above), mol→Eq (using 1 mol malate = 2 Eq), then Eq→mEq, then use given concentration to convert mEq→L.

$$\mathbf{0.52\ g \times \frac{1\ mol}{132.07\ g} \times \frac{2\ Eq}{mol} \times \frac{1000\ mEq}{1\ Eq} \times \frac{1\ L}{2.0\ mEq} = 3.9\ L\ solution}$$

5.97 *A solution contains 250 mg of Ca^{2+} in a total volume of 500 mL. Calculate the concentration of calcium in this solution in each of the following units:*

Since concentration units are always in the form of (amount solute)/(amount solution), we can start all of our conversions with 250 mg/500 mL. (Another option is to do two separate calculations, one to convert the amount of solute from given into desired units, one to convert the amount of solution from given into desired units, and find the ratio at the end—this is a perfectly fine strategy as well and is mathematically equivalent.) Also, since we're always looking at the same solute (calcium ion), we don't have to explicitly mention the solute in the calculation.

In some of the conversions below, there may be a shorter way to do the calculation using a previous answer, but the solution shown starts from the original given information each time.

NOTE:
About significant figures here: since one or two zeroes at the end of a measured quantity are assumed to be significant, the given volumes, and all of the concentration values calculated from them, have three significant figures.

a) M It's usually helpful to replace the capital M of molarity with the component units, mol/L. Then our setup involves using the molar mass of calcium, 40.08 g/mol, to convert 250 mg of calcium ions into moles of calcium ions. We also need to convert mL of solution into L of solution:

$$\frac{\mathbf{250\ mg}}{\mathbf{500\ mL}} \times \frac{\mathbf{1\ g}}{\mathbf{1000\ mg}} \times \frac{\mathbf{1\ mol}}{\mathbf{40.08\ g}} \times \frac{\mathbf{1000\ mL}}{\mathbf{L}} = \mathbf{0.0125\ mol\ Ca^{2+}/L = 0.0125\ M\ Ca^{2+}}$$

b) mEq/L

$$\frac{\mathbf{250\ mg}}{\mathbf{500\ mL}} \times \frac{\mathbf{1\ g}}{\mathbf{1000\ mg}} \times \frac{\mathbf{1\ mol}}{\mathbf{40.08\ g}} \times \frac{\mathbf{2\ Eq}}{\mathbf{1\ mol}} \times \frac{\mathbf{1000\ mEq}}{\mathbf{1\ Eq}} \times \frac{\mathbf{1000\ mL}}{\mathbf{L}} = \mathbf{25.0\ mEq\ Ca^{2+}/L}$$

c) mg/dL This one's a lot easier—all we have to do is convert the solution volume from mL to dL, using the relationship 1000 mL = 1 L = 10 dL:

$$\frac{\mathbf{250\ mg}}{\mathbf{500\ mL}} \times \frac{\mathbf{1000\ mL}}{\mathbf{10\ dL}} = \mathbf{50.0\ mg/dL}$$

d) ppm We use the definition ppm = μg/mL, and convert mg to μg using the relationship 10^3 mg = 1 g = 10^6 μg:

$$\frac{\mathbf{250\ mg}}{\mathbf{500\ mL}} \times \frac{\mathbf{10^6\ \mu g}}{\mathbf{10^3\ mg}} = \mathbf{500\ ppm}$$

5.99 *Frozen orange juice concentrate contains 166 mg/dL of vitamin C. If you mix one can of the concentrate with three cans of water, what will be the concentration of vitamin C in the resulting juice?*

This one may look tricky because we don't know the volume of a can—but remember, the volume units don't have to match the concentration units, they just have to match each other. It's enough to call the initial volume "1 can," and to recognize that adding three cans of water (by filling the same can to the same level with water) will give a final approximate total volume of four cans (since all solutions are aqueous, volumes should be additive). C_1= 166 mg/dL, V_1 = 1 can, and V_2 = 4 cans; C_2 is the unknown quantity.

$$C_1 \times V_1 = C_2 \times V_2$$

Solve for C_2 to get:

$$\frac{C_1 \times V_1}{V_2} = C_2$$

$$\frac{\mathbf{(166\ mg/dL)(1\ can)}}{\mathbf{4\ cans}} = \mathbf{42\ mg/dL}$$

This problem requires a judgment call on the number of significant figures. Realistically, we should be able to fill the can with water accurately enough to get two significant figures, but probably not more than two.

*5.101 *You have 100 mL of 1.0 M NaCl. You need to add enough water to make this into an isotonic solution. How much water must you add?*

Dilution problem with C_1= 1.0 M, V_1 = 100 mL, and C_2 = 0.28 M (isotonic concentration). V_2 is the unknown quantity.

$$C_1 \times V_1 = C_2 \times V_2$$

Solve for V_2 to get:

$$\frac{C_1 \times V_1}{C_2} = V_2$$

$$\frac{\mathbf{(1.0\ M)(100\ mL)}}{\mathbf{0.28\ M}} = \mathbf{360\ mL\ (rounded\ to\ 2\ sf)}$$

To get to 360 mL total volume, you'll have to add **260 mL of water** (360 mL total – 100 mL initial). This assumes that the volumes are additive, which is generally a reasonable assumption for dilute aqueous solutions.

6

Chapter 6

Chemical Reactions

Solutions to Section 6.1 Core Problems

6.1 *Which of the following are physical changes, and which are chemical reactions?*

a) *You bend a piece of steel.*

Since the steel doesn't turn into any other substance when you bend it—it's still steel—this is a **physical change.**

b) *A piece of steel rusts.*

When the steel rusts, it isn't steel anymore; it turns into a difference substance (different color, different brittleness and structure, different density, etc.) This is a **chemical reaction**.

c) *Your body burns fat.*

When the fat is burned it isn't fat anymore. This is a **chemical reaction**.

d) *You compress the air in a bicycle pump.*

The air doesn't change into a different substance; you're just packing more of it into the same space. This is a **physical change.**

6.3 *Which of the following are physical properties, and which are chemical properties?*

a) *Hydrogen burns if it is mixed with air.*

Burning is a chemical reaction, so this statement describes a **chemical property.** While we don't know the chemical formula of hydrogen, once the hydrogen is burned, we don't have hydrogen anymore but other substances. The hydrogen reacts with oxygen, producing products that have different chemical formulas from the original materials.

b) *Hydrogen is a gas at room temperature.*

This observation doesn't mention or depend on a chemical reaction; this is a **physical property.**

c) *Hydrogen combines with nitrogen to make ammonia.*

Since the statement describes a chemical reaction, this is a **chemical property.**

d) *Hydrogen condenses at –253ºC.*

This observation doesn't mention or depend on a chemical reaction (condensation is a physical change); this is a **physical property.** The hydrogen does not change its chemical formula, and the process is reversible.

6.5 *When you boil water, the water turns to steam. If you boil 5 g of water, how much does the resulting steam weigh?*

This is a physical change; the water is still water, so the resulting steam has **the same mass as the liquid, 5 g.** The law of mass conservation states that in any change, the mass of the final products equals the mass of the starting materials. Since water is the only reactant and water vapor is the only product, the masses of both must be equal.

6.7 *Does the following sentence describe a typical physical change or a typical chemical change?*

"The starting materials and the products have different chemical formulas."

Different chemical formulas means the starting materials and products are different substances, so this is a chemical change.

Solutions to Section 6.2 Core Problems

6.9 *A student is asked to balance the following equation: $Fe + Cl_2 \rightarrow FeCl_3$. The student gives the following answer: $Fe + Cl_2 \rightarrow FeCl_2$. Is this a reasonable answer? Explain why or why not.*

This is not a reasonable answer. In balancing chemical equations, changing the formula of any substance is not allowed. By changing the formula of the product from $FeCl_3$ to $FeCl_2$, the student has written a completely different chemical reaction from the one given. The only change that can be made in balancing a reaction is the addition of coefficients.

6.11 *Write a chemical equation that represents the following reaction: "One molecule of solid PCl_3 reacts with three molecules of liquid water, forming one molecule of aqueous H_3PO_3 and three molecules of aqueous HCl."*

The number of molecules becomes the coefficient in the equation; the descriptions of phases (solid, liquid, aqueous) become phase labels in parentheses; "reacts with" and "and" translate as + signs in the equation; and the transformation word "to form" is represented by a reaction arrow (→). Coefficients of one are understood and are never written.

NOTE:
We're also expected to know the formula for water (H_2O).

PCl_3 (s) + 3 H_2O (l) → H_3PO_3 (aq) + 3 HCl (aq)

6.13 *Tell whether each of the following is a balanced equation. If it is not balanced, explain why not.*

a) $2\ Ca + Cl_2 \rightarrow 2\ CaCl_2$

Not balanced: there are two Cl atoms on the reactant (left) side and four on the products (right) side. The correctly balanced equation is:

$$Ca + Cl_2 \rightarrow CaCl_2$$

b) $Mg(OH)_2 + 2\ HF \rightarrow MgF_2 + 2\ H_2O$

Balanced. Left (reactants) side: one Mg, two O, four H, two F. Right (products) side: one Mg, two F, four H, two O.

c) $C_5H_{12}O + 8\ O_2 \rightarrow 5\ CO_2 + 6\ H_2O$

Not balanced: while C and H are balanced, there are 17 atoms of O on the left (reactants) side and only 16 on the right (products) side. The correctly balanced equation is:

$$2\ C_5H_{12}O + 15\ O_2 \rightarrow 10\ CO_2 + 12\ H_2O$$

It often happens that, when a reaction is almost balanced, you end up with an odd number of atoms of an element on one side (in this case, O on the left side of the equation) and even numbers on the other side (the O atoms on the right side of the equation). This even/odd problem is almost always easily solved by doubling everything—start from 2 $C_5H_{12}O$ and everything works out.

Problems 6.15 through 6.22 are equation-balancing questions. They are ordered from easiest (6.15 and 6.16) to hardest (6.21 and 6.22).

NOTE:
When balancing reaction equations, **always** take a moment to check the balance of all atoms after you finish. It's easy to make a small mistake somewhere, and this can cost you big points. But the good news is, you can always tell if you did it right!

A couple of strategy hints: (1) if the equation contains mostly compounds but there's an element that appears by itself (for example, O_2 in 6.17c and 6.19a), it may be helpful to balance that element last (when you change its coefficient, it doesn't throw other elements in the reactants & products out of balance). (2) If you get hopelessly stuck and can't get an equation to balance, get a clean sheet of paper, write the equation down again, and start from scratch. Often the problem is a copying error in one of the formulas or an arithmetic error in the balancing, and it may be much more difficult (and take longer!) to find the error than just to solve the problem from the beginning—chances are you won't make the same error again.

6.15 *Balance the following equations.*

a) $S + \underline{2}\ Cl_2 \rightarrow SCl_4$

b) $\underline{2}\ Ag + S \rightarrow Ag_2S$

c) $\underline{2}\ K + Cl_2 \rightarrow \underline{2}\ KCl$

6.17 *Balance the following equations.*

a) $CaO + \underline{2}\ HCl \rightarrow CaCl_2 + H_2O$

b) $\underline{4}\ Fe + \underline{3}\ O_2 \rightarrow \underline{2}\ Fe_2O_3$

c) $CH_4 + \underline{2}\ O_2 \rightarrow CO_2 + \underline{2}\ H_2O$

6.19 *Balance the following equations.*

a) **2** C_4H_{10} + **13** $O_2 \rightarrow$ **8** CO_2 + **10** H_2O

The carbon and hydrogen are easy to deal with, and everything is tidy until you try to finish up with O_2, at which point you have an odd number of O on the right and can only have even numbers on the left (the O atoms come in pairs on the left). This even/odd problem is almost always easily solved by doubling everything—start from 2 C_4H_{10} and everything works out.

b) **2** $AlCl_3$ + **3** $H_2O \rightarrow Al_2O_3$ + **6** HCl

c) **2** $AgNO_3 + MgI_2 \rightarrow$ **2** $AgI + Mg(NO_3)_2$

6.21 *Balance the following equations.*

a) **2** $Al(OH)_3$ + **3** $H_2SO_4 \rightarrow Al_2(SO_4)_3$ + **6** H_2O

b) **4** $C_5H_{11}NO_2$ + **27** $O_2 \rightarrow$ **20** CO_2 + **22** H_2O + **2** N_2

Look carefully at this one for a moment first. The N_2 in the product is going to require 2 $C_5H_{11}NO_2$; going back and forth to balance C and H is simple until you try to finish up with O_2, at which point you have an odd number of O on the right and can only have even numbers on the left (the O atoms come in pairs on the left). This even/odd problem is easily solved by doubling everything; once you have 4 $C_5H_{11}NO_2$, everything else comes out pretty neatly.

Solutions to Section 6.3 Core Problems

6.23 *Methane has the chemical formula CH_4, and is the principal constituent of natural gas. When methane burns, it reacts with oxygen according to the following balanced equation:*

$$CH_4 + 2\ O_2 \rightarrow CO_2 + 2\ H_2O$$

What is the mass relationship between methane and water in this reaction?

One mole of methane reacts to form two moles of water. One mole of methane has a mass of 12.01 + 4(1.008) = 16.04 g; one mole of water has a mass of 16.00 + 2(1.008) = 18.02 g; therefore **16.04 g of methane reacts to form 2(18.016) = 36.03 g water.** Since these numbers constitute our final answer in this problem, we round them appropriately to 2 decimal places; however, if we were continuing on with further calculations (as in the next problem 6.25), we wouldn't round until the end of the problem.

6.25 *Aluminum hydroxide is an ingredient in some antacids. It reacts with hydrochloric acid (the acid in your stomach) according to the following equation:*

$$Al(OH)_3(s) + 3\ HCl(aq) \rightarrow AlCl_3(aq) + 3\ H_2O(l)$$

a) How many grams of HCl will react with 2.50 g of $Al(OH)_3$?

Add up the molar masses of HCl (35.45 + 1.008 = 36.458 g/mol) and $Al(OH)_3$ (26.98 + 3×16.00 + 3×1.008 = 78.00 g/mol). From the balanced equation, we see that 3 moles of HCl (3×36.458 = 109.374 g) reacts with 1 mole of $Al(OH)_3$ (78.004 g). Using this mass ratio,

$$2.50 \text{ g Al(OH)}_3 \times \frac{109.374 \text{ g HCl}}{78.004 \text{ g Al(OH)}_3} = 3.51 \text{ g HCl}$$

b) *When 2.50 g of $Al(OH)_3$ reacts with HCl, what mass of water is formed?*

The molar mass of H_2O is 18.016 g/mol. From the balanced equation, we see that 1 mole of $Al(OH)_3$ (78.004 g) produces 3 moles of H_2O (3×18.016 = 54.048 g). Using this mass ratio,

$$2.50 \text{ g Al(OH)}_3 \times \frac{54.048 \text{ g H}_2\text{O}}{78.004 \text{ g Al(OH)}_3} = 1.73 \text{ g H}_2\text{O}$$

c) *If you want to make 2.50 g of $AlCl_3$ using this reaction, how many grams of HCl do you need?*

Add up the molar mass of $AlCl_3$:

1 Al	=	1 × 26.98	=	26.98 g/mol
+ 3 Cl	=	3 × 35.45	=	106.35
total				133.33 g/mole

One mole of $AlCl_3$ (133.33 g) requires three moles of HCl:
3 mol HCl × 36.458 g/mol = 109.374 g

$$2.50 \text{ g AlCl}_3 \times \frac{109.374 \text{ g HCl}}{133.33 \text{ g AlCl}_3} = 2.05 \text{ g HCl}$$

Solutions to Section 6.4 Core Problems

6.27 *Barium hydroxide reacts with ammonium chloride to produce barium chloride, ammonia and water. The balanced equation for this reaction is:*

$Ba(OH)_2(s) + 2\ NH_4Cl(s) + 5.5\ kcal \rightarrow BaCl_2(s) + 2\ NH_3(g) + 2\ H_2O(l)$

a) *When this reaction occurs, do the surroundings become cooler, or do they become warmer?*

The reaction is taking in heat from its surroundings as it goes, as shown by the heat term on the reactant side of the equation; as reactants are used up, heat is also used up. Therefore **the surroundings become cooler as they give heat to the reaction.**

b) *Is this reaction exothermic, or is it endothermic?*

Reactions that take in heat as they go are **endothermic.**

c) *What is the correct sign (positive or negative) of ΔH for this reaction?*

ΔH is positive for endothermic reactions.

d) *What is the relationship between the mass of NH_4Cl and the heat for this reaction?*

As 5.5 kcal of heat is consumed, 2 moles of NH_4Cl react. The molar mass of NH_4Cl is 53.49 g, so 2 moles of NH_4Cl is 106.98 g NH_4Cl. **The reaction of 106.98 g NH_4Cl consumes 5.5 kcal of heat.**

6.29 *For the reaction that follows, ΔH is –287 kcal.*

$$2\ Mg(s) + O_2(g) \rightarrow 2\ MgO(s)$$

a) *During this reaction, do the surroundings become hotter, or do they become cooler?*

Since ΔH is negative, the reaction is exothermic, giving off heat to the surroundings, and **the surroundings become warmer as the reaction progresses.**

b) *Is heat a product in this reaction, or is it a reactant?*

Negative ΔH means the reaction produces heat.

c) *Rewrite the chemical equation so it includes the heat.*

$$\mathbf{2\ Mg(s) + O_2(g) \rightarrow 2\ MgO(s) + 287\ kcal}$$

6.31 *Benzene (C_6H_6) is a chemical compound in gasoline. When gasoline burns, the benzene reacts with oxygen:*

$$2\ C_6H_6 + 15\ O_2 \rightarrow 12\ CO_2 + 6\ H_2O + 1562\ kcal$$

How much heat will be given off when 4.32 g (one teaspoon) of benzene reacts with oxygen?

One mole of benzene has a mass of 78.11 g. For every 1562 kcal of heat produced, 2 mol or 156.22 g of benzene is consumed:

$$\mathbf{4.32\ g\ benzene \times \frac{1562\ kcal}{156.22\ g\ benzene} = 43.2\ kcal}$$

6.33 *Your body breaks down glucose (blood sugar) into lactic acid when it needs an immediate source of energy. The chemical equation for this reaction (called a fermentation reaction) is:*

$$C_6H_{12}O_6 \rightarrow 2\ C_3H_6O_3 + 16.3\ kcal$$

a) *Calculate the amount of heat that is formed when your body makes 1.00 g of lactic acid ($C_3H_6O_3$) using this reaction.*

First, add up the molar mass of lactic acid (90.08 g/mol). When two moles (180.16 g) of lactic acid are formed, 16.3 kcal of energy is produced; this gives us the relationship we need for the calculation:

$$\mathbf{1.00\ g\ lactic\ acid \times \frac{16.3\ kcal}{180.16\ g\ lactic\ acid} = 0.0905\ kcal = 90.5\ cal}$$

b) *How many grams of glucose must your body break down in this fashion in order to produce 5.00 kcal of heat?*

Molar mass of glucose: 180.16 g/mol, and 1 mol glucose yields 16.3 kcal:

$$\mathbf{5.00\ kcal \times \frac{180.16\ g\ glucose}{16.3\ kcal} = 55.3\ g\ glucose}$$

6.35 *One cup of cottage cheese contains 6 g of carbohydrate, 28 g of protein, and 10 g of fat. How many Calories are there in a cup of cottage cheese?*

Carbohydrates and proteins supply about Cal/g; fats supply about 9 Cal/g. So we can calculate the energy value of each component, using the nutritive values as conversion factors:

carbohydrates:	6 g × 4 Cal/g	=	24 Cal
protein:	28 g × 4 Cal/g	=	112 Cal
fat:	10 g × 9 Cal/g	=	90 Cal
total:			**226 Cal**

Because the nutritive values we use only have one significant figure, we should appropriately round our answer to one significant figure as well: **230 Cal**. (Nutritional energies, by convention, are rounded to the nearest 10 Calories.)

6.37 *A Mega-Sweet snack cake contains 160 Calories. The manufacturer wants to remove enough sugar to lower the Calorie value to 140 Calories. How many grams of sugar must the manufacturer remove? (Sugar is a carbohydrate.)*

160 – 140 = 20 Calories to remove. We use the nutritive value of sugar (a carbohydrate) as a conversion factor:

20 Calories × 1 g/4 cal = 5 grams of sugar.

Solutions to Section 6.5 Core Problems

6.39 *Write a chemical equation that represents the combustion reaction of each of the following:*

In each case, remember: (1) combustion is *always* reaction with O_2; (2) if there are carbon atoms in the reactants, CO_2 will be one of the products; (3) if there are H atoms in the reactants, H_2O will be one of the products.

NOTE:

Any oxygen in the reactant molecules is indistinguishable from the oxygen supplied by O_2, once it gets to the products, so we don't try to account for it separately but treat it as any other atom in balancing (that is, O in the reactants could end up in CO_2 or H_2O).

a) *C_7H_8 (toluene, a compound that increases the octane rating of gasoline)*

$$C_7H_8 + 9\ O_2 \rightarrow 7\ CO_2 + 4\ H_2O$$

b) *C_4H_{10} (butane, the fuel in cigarette lighters)*

$$2\ C_4H_{10} + 13\ O_2 \rightarrow 8\ CO_2 + 10\ H_2O$$

Remember, when you get to the point of balancing O, that doubling the reaction solves the even/odd problem.

c) *$C_4H_{10}O$ (t-butyl alcohol, a compound that increases the octane rating of gasoline)*

$$C_4H_{10}O + 6\ O_2 \rightarrow 4\ CO_2 + 5\ H_2O$$

6.41 *Which of the following equations represent combustion reactions, which represent precipitation reactions, and which represent neither of these types?*

For this one you have to know the definitions of these reaction types. Combustion reactions always have O_2 as a reactant. Precipitation reactions always have two soluble (aqueous) ionic compounds combining to form at least one insoluble (solid) ionic compound. Phase labels in parentheses are used to show the states or phases of the reactants and products.

a) $C_2H_6O + 3\ O_2 \rightarrow 2\ CO_2 + 3\ H_2O$

Combustion reaction.

b) $NaOH(aq) + HBr(aq) \rightarrow NaBr(aq) + H_2O(l)$

Neither. O_2 isn't a reactant and there is no insoluble, solid ionic product.

c) $2\ PbBr_2 + O_2 \rightarrow 2\ PbO + 2\ Br_2$

Combustion reaction. O_2 is a reactant.

d) $AlCl_3(aq) + 3\ NaOH(aq) \rightarrow Al(OH)_3(s) + 3\ NaCl(aq)$

Precipitation reaction. Two aqueous ionic reactants producing at least one insoluble, solid ionic product defines a precipitation reaction.

6.43 *Write a net ionic equation for each of the following precipitation reactions:*

In each case, first show all **aqueous ionic compounds** as the dissociated ions (do NOT dissociate molecular compounds, or ionic compounds labeled as solid). Then cross off any ions that are completely unchanged between reactants and products.

a) $AgNO_3(aq) + KBr(aq) \rightarrow AgBr(s) + KNO_3(aq)$

Dissociate aqueous ionic compounds:
$Ag^+(aq) + NO_3^-(aq) + K^+(aq) + Br^-(aq) \rightarrow AgBr(s) + NO_3^-(aq) + K^+(aq)$

NOTE:
Polyatomic ions are **not** broken up when ionic compounds are dissociated; the polyatomic ions we learned in Chapter 3 remain intact.

Once we've determined which substances dissociate, the next step is to cross off any ions that are identical on both sides:
$Ag^+(aq) + \cancel{NO_3^-(aq)} + \cancel{K^+(aq)} + Br^-(aq) \rightarrow AgBr(s) + \cancel{NO_3^-(aq)} + \cancel{K^+(aq)}$
Rewrite without crossed-off species to get **net ionic equation**:
$\mathbf{Ag^+(aq) + Br^-(aq) \rightarrow AgBr(s)}$

b) $3\ MgCl_2(aq) + 2\ Na_3PO_4(aq) \rightarrow Mg_3(PO_4)_2(s) + 6\ NaCl(aq)$

Dissociate aqueous ionic compounds:
$3\ Mg^{2+}(aq) + 6\ Cl^-(aq) + 6\ Na^+(aq) + 2\ PO_4^{3-}(aq) \rightarrow$
$Mg_3(PO_4)_2(s) + 6\ Na^+(aq) + 6\ Cl^-(aq)$

We use the subscripts in the formulas to keep the reactions balanced: the dissociation of 3 moles $MgCl_2$ gives 3 moles of Mg^{2+} and 6 moles of Cl^-.

NOTE:
Polyatomic ions are **not** broken up when ionic compounds are dissociated; the polyatomic ions we learned in Chapter 3 remain intact.

Once we've determined which substances dissociate, the next step is to cross off any ions that are identical on both sides:
$3\ Mg^{2+}(aq)$ + ~~$6\ Cl^-(aq)$~~ + ~~$6\ Na^+(aq)$~~ + $2\ PO_4^{3-}(aq) \rightarrow$
$Mg_3(PO_4)_2(s)$ + ~~$6\ Na^+(aq)$~~ + ~~$6\ Cl^-(aq)$~~
Rewrite without crossed-off species to get **net ionic equation**:
$\mathbf{3\ Mg^{2+}(aq) + 2\ PO_4^{3-}(aq) \rightarrow Mg_3(PO_4)_2(s)}$

c) $Pb(NO_3)_2(aq) + CaI_2(aq) \rightarrow Ca(NO_3)_2(aq) + PbI_2(s)$

Dissociate aqueous ionic compounds:
$Pb^{2+}(aq) + 2\ NO_3^-(aq) + Ca^{2+}(aq) + 2I^-(aq) \rightarrow Ca^{2+}(aq) + 2\ NO_3^-(aq) + PbI_2(s)$
Cross off any ions that are identical on both sides:
$Pb^{2+}(aq)$ + ~~$2\ NO_3^-(aq)$~~ + ~~$Ca^{2+}(aq)$~~ + $2I^-(aq) \rightarrow$ ~~$Ca^{2+}(aq)$~~ + ~~$2\ NO_3^-(aq)$~~ + $PbI_2(s)$
Rewrite without crossed-off species to get **net ionic equation:**
$\mathbf{Pb^{2+}(aq) + 2I^-(aq) \rightarrow PbI_2(s)}$

Solutions to Section 6.6 Core Problems

6.45 *Iron reacts with sulfuric acid (H_2SO_4) according to the following equation:*

$$Fe(s) + H_2SO_4(aq) \rightarrow FeSO_4(aq) + H_2(g)$$

Will the rate of this reaction increase, decrease, or remain the same if...

a) *the concentration of sulfuric acid is increased?*

Increase: most reactions go faster if the concentration of reactants is increased.

b) *the temperature is decreased?*

Decrease: almost all reactions go slower at lower temperatures.

c) *the iron is ground up into powder?*

Increase: an increase in the surface area of a solid reactant will allow the reaction to go faster, because it makes it easier for the reactants to reach each other.

d) *a catalyst is added?*

Increase: a catalyst increases the rate of a reaction by lowering the activation energy.

6.47 *Food spoils when it is left at room temperature, because bacteria convert some of the nutrients in the food into unpleasant (and occasionally toxic) products. Explain why refrigerating the food slows down the spoiling process.*

Spoilage is a chemical process, and colder temperatures slow down the rates of chemical reactions.

6.49 *Both carbon and sodium can react with oxygen, but sodium reacts much more rapidly than carbon does.*

$$C(s) + O_2(g) \rightarrow CO_2(g)$$
$$4\ Na(s) + O_2(g) \rightarrow 2\ Na_2O(s)$$

Which of these reactions has the larger activation energy?

All other things being equal, reactions that have greater activation energy are slower than those with low activation energy. These reactions seem pretty similar, so **if the reaction of carbon with oxygen is much slower, it must have a greater activation energy.**

6.51 *Carbonic acid breaks down into water and carbon dioxide according to the following equation:*

$$H_2CO_3(aq) + 1\ kcal \rightarrow CO_2(aq) + H_2O(l)$$

The activation energy of this reaction is 21 kcal, but a catalyst in human blood reduces the activation energy to 12 kcal. Draw an energy diagram for this reaction, showing both the uncatalyzed and the catalyzed reaction.

The inclusion of heat as a reactant tells us that the reaction is endothermic, so the products will be 1 kcal higher in energy than the reactants. The activation energy curve will therefore go up 21 kcal and come down 20 kcal (for the uncatalyzed reaction). For the catalyzed reaction, the reactants and products stay at the same energy levels; the activation energy curve will go up 12 kcal and come down 11 kcal to the products.

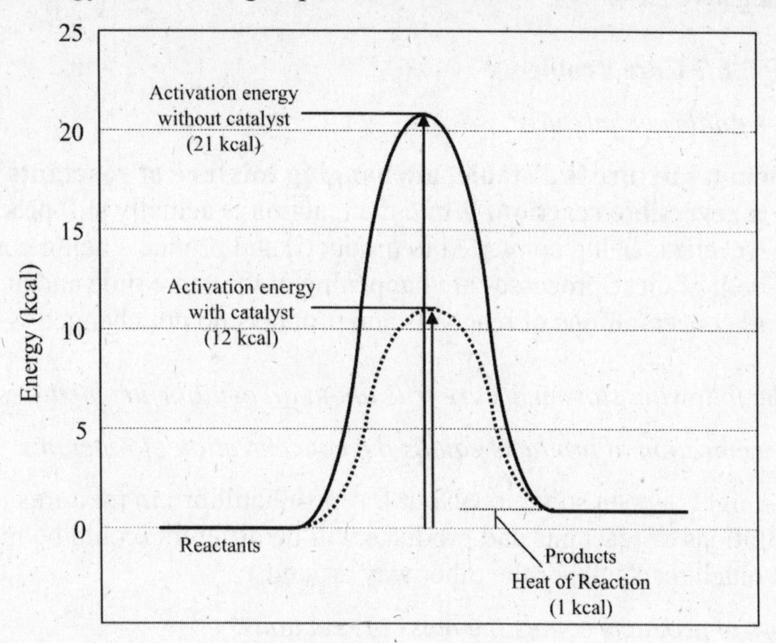

6.53 *Using the energy diagram that follows, estimate the activation energy and the heat of reaction. Also, tell whether this is an exothermic or an endothermic reaction.*

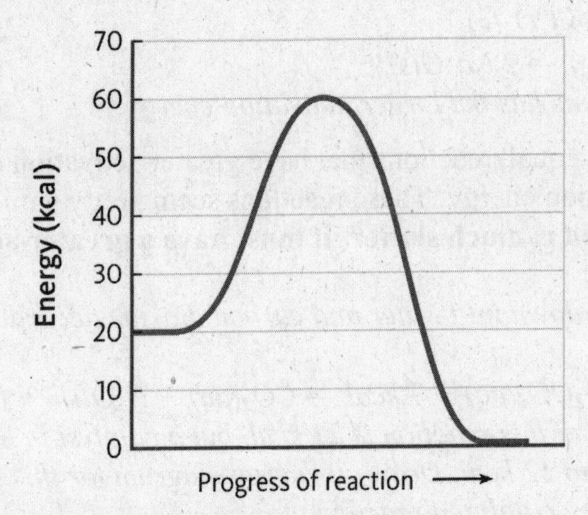

The **activation energy is the difference in energy** between the reactants (here, about 20 kcal) and the top of the curve (here, about 60 kcal), or **40 kcal for this reaction.** The **heat of the reaction** is the difference between the reactants (20 kcal) and the products (about 2 kcal), a difference of **18 kcal for this reaction.** Because the products are lower in energy than the reactants, **this is an exothermic reaction** (negative ΔH).

Solutions to Section 6.7 Core Problems

6.55 *What is an equilibrium mixture?*

An equilibrium mixture is a stable, unchanging mixture of reactants and products formed by a reversible reaction. While the reaction is actually still proceeding in both directions—reactants being converted to products, and products being converted back to reactants—both of these processes are happening at the same time and at the same rate, so the overall concentrations of reactants and products do not change.

6.57 *Which of the following statements are true about <u>all</u> equilibrium mixtures?*

a) *The concentration of products equals the concentration of reactants.*

False—might happen sometimes, but for most equilibrium mixtures the concentrations of reactants and products will be different (could be mostly products and not much reactants, or the other way around.)

b) *The mass of products equals the mass of reactants.*

False, for the same reasons given in (a).

c) *The rate of the forward reaction equals the rate of the reverse reaction.*

True—this is part of the definition of equilibrium and is true for all equilibrium mixtures.

d) The forward and reverse reactions stop.

False—the forward and reverse reactions are both still going, they're just going at the same rate so that the overall composition of the mixture doesn't change.

6.59 *Grain alcohol (ethanol) that is not intended for human consumption is often made by the following reaction:*

$$C_2H_4(g) + H_2O(g) \rightarrow C_2H_6O(g)$$

If you mix some C_2H_4 and some water and heat them to 100°C, you end up with a mixture of C_2H_4, H_2O, and C_2H_6O.

a) Explain why neither of the reactants runs out.

If the reaction appears to stop with some reactant molecules still remaining, then it must be an equilibrium process; it doesn't really stop, but the rate of the reverse reaction catches up to the rate of the forward reaction and some of the products are continually being converted back to reactants. The forward and reverse reactions have the same rate and the reaction appears to have stopped—an equilibrium mixture.

b) If you heat some C_2H_6O to 100°C, what will happen?

The reverse reaction will occur and produce some $C_2H_4(g)$ and $H_2O(g)$, until this process also reaches equilibrium.

6.61 *One way to make hydrogen gas is to heat a mixture of steam and carbon monoxide. (This reaction also makes carbon dioxide, which can easily be separated from the hydrogen.)*

$$H_2O(g) + CO(g) \rightleftarrows H_2(g) + CO_2(g)$$

NOTE:
All of the changes that follow will be demonstrations of LeChatelier's principle: any disturbance we make to a reaction mixture will cause the reaction to go in the direction that partially counteracts the disturbance. If we add more of a reactant or product, the reaction will go in the direction that uses it up; if we remove a reactant or product, the reaction will go in the direction that forms more of it.

Suppose that you have an equilibrium mixture of these four substances. Which way will the reaction proceed (forward or backward) if you...

a) add some carbon monoxide to the mixture?

Carbon monoxide, CO, is a reactant; adding more will cause the reaction to go **forward** (using up reactants and producing more products).

b) add some carbon dioxide to the mixture?

Carbon dioxide, CO_2, is a product; adding more will cause the reaction to go **backward** (converting products back into reactants).

c) remove some hydrogen from the mixture?

Hydrogen, H_2, is a product. If it is removed from the mixture, the system will try to replace it, causing the reaction to go **forward** (using up reactants and producing more products).

6.63 *In the equilibrium in Problem 6.61, which of the following would increase the amount of hydrogen gas in the equilibrium mixture?*

Again, this is an application of LeChatelier's principle, that any disturbance to an equilibrium mixture will be partially counteracted by the equilibrium reaction. To increase the amount of H_2, we would need the reaction to go forward.

a) *Adding some carbon monoxide to the mixture.*

Adding more of a reactant makes the reaction go forward and **would increase the amount of H_2 in the equilibrium mixture.**

b) *Removing some carbon monoxide from the mixture.*

Removing some of a reactant makes the reaction go backward and would actually **decrease** the amount of H_2 in the equilibrium mixture.

c) *Adding some carbon dioxide to the mixture.*

Adding more of a product makes the reaction go backward and would actually **decrease** the amount of H_2 in the equilibrium mixture.

d) *Removing some carbon dioxide from the mixture.*

Removing some of a product makes the reaction go forward and **would increase the amount of H_2 in the equilibrium mixture.**

Solutions to Concept Questions

* indicates more challenging problems

6.65 a) *What is the primary difference between a physical change and a chemical reaction?*

A physical change does not break or form chemical bonds; the same substances (elements or compounds) are present before and after the change. In a chemical reaction, chemical bonds are broken or formed (or, usually, both). The substances in the reactants are changed to different substances in the products.

b) *Which is more likely to be reversible?*

Physical changes are usually easily reversible; chemical changes are often difficult to reverse and may be considered irreversible. (Equilibrium reactions, chemical reactions that are reversible, are an exception to this.)

6.67 *What is the difference between writing "2 N" and writing "N_2" in a chemical equation?*

"2 N" says that there are two separate atoms of N. "N_2" says that there are two atoms of N that are chemically bonded to each other to make one molecule.

6.69 *Jackie is asked to balance the equation $Na + Cl_2 \rightarrow NaCl$. She writes the following answer: $Na + Cl_2 \rightarrow NaCl_2$.*

a) *Is this answer a balanced equation? If not, explain why not.*

It is a balanced equation, because each element has the same number of atoms on both sides of the equation.

b) *Is this answer correct? If not, explain why not.*

It is not correct, because she changed the chemical species present in the reaction. The compound formed by Na and Cl is NaCl, not $NaCl_2$. Chemical formulas cannot be altered for the convenience of balancing an equation.

6.71 *What is the nutritive value of a food?*

The nutritive value of a food is the amount of energy it produces when it burns–the amount of energy that an organism can gain by metabolizing the nutrients it contains.

6.73 *What is a precipitation reaction?*

Precipitation reactions always have at least two soluble (aqueous) ionic compounds combining to form at least one insoluble (solid) ionic compound.

6.75 *Explain why each of the following increases the rate of a reaction.*

a) *Increasing the temperature*

At higher temperatures, molecules collide more frequently and with greater energy, increasing the probability that a collision will result in a successful reaction.

b) *Stirring the mixture (if it is heterogeneous)*

Stirring a reaction mixture brings the different reactant substances into better contact with each other. The reactants have to be able to reach each other in order to react.

c) *Breaking up a solid reactant into smaller pieces*

This increases the contact between the reactant substances. The reactants have to be able to reach each other in order to react.

d) *Increasing the concentration of an aqueous reactant*

This increases the likelihood of contact between the reactant substances, making the reaction faster.

6.77 *Stefan says "when you increase the temperature, the activation energy increases."*

a) *Explain why this is not correct.*

Activation energy is part of the nature of the reaction. It can only be changed by adding a catalyst (which, technically, actually makes the reaction happen by a different pathway anyway—same reactants and products, but strictly speaking a different reaction.) **Changing the temperature does not change the activation energy,** but it does change the number of collisions that have the required activation energy.

b) *Explain why increasing the temperature speeds up reactions, using the concept of activation energy in your answer.*

At higher temperatures, molecules collide more frequently and more energetically, increasing the probability that a given collision will have the required activation energy to result in the formation of products.

6.79 *When you eat table sugar (sucrose), your body breaks down the sucrose into two other sugars, glucose and fructose, using a reversible reaction:*

$$\text{sucrose} + H_2O \rightleftarrows \text{glucose} + \text{fructose}$$

Suppose you have an equilibrium mixture of these four substances. If you add some glucose to this mixture and then wait a few minutes, what will happen to the concentration of fructose in the mixture?

This is an application of LeChatelier's principle, that any disturbance to an equilibrium mixture will be partially counteracted by the equilibrium reaction. Addition of more of a product to an equilibrium mixture causes the reaction to go backward, converting products back into reactants. Therefore, addition of glucose to this reaction mixture will use up some fructose to form more sucrose and water, **decreasing the concentration of fructose in the mixture.**

Solutions to Summary and Challenge Problems

* indicates more challenging problems

6.81 *Balance each of the following equations.*

a) $Mg + Cl_2 \rightarrow MgCl_2$ **already balanced**

b) $\underline{2}\ Al + \underline{3}\ S \rightarrow Al_2S_3$

c) $P_4 + \underline{6}\ Cl_2 \rightarrow \underline{4}\ PCl_3$

d) $\underline{2}\ MgI_2 + O_2 \rightarrow \underline{2}\ MgO + \underline{2}\ I_2$

e) $Na_2S + \underline{2}\ AgF \rightarrow \underline{2}\ NaF + Ag_2S$

6.83 *Aluminum reacts with HCl to form aluminum chloride and hydrogen gas (H_2).*

a) *Write a balanced chemical equation for this reaction.*

Aluminum, as a metal, is just written as Al (don't put a charge on it as it is not in a compound, nor is it an ion in solution!). We go back to Chapter 3 principles to figure out the formula of aluminum chloride (you remember how to do that, don't you?): aluminum, a Group 3A metal, forms an ion with a +3 charge, and chloride, a halogen, has a –1 charge, giving us the formula $AlCl_3$. (If your Chapter 3 skills are a bit rusty, now is a good time to go back and review; many problems will assume that you know how to get from the name of a compound to a formula.)

$$\mathbf{2\ Al + 6\ HCl \rightarrow 2\ AlCl_3 + 3\ H_2}$$

b) *If 10.0 g of aluminum reacts with HCl, how many grams of hydrogen are formed?*

First we need to use the periodic table (Chapter 5 review!) to add up the molar masses of aluminum (26.98 g/mol) and H_2 (2.016 g/mol). Then we can get the relationship between Al and H_2: 2 moles of Al (53.96 g) react to form 3 moles of H_2 (6.048 g).

$$\mathbf{10.0\ g\ Al \times \frac{6.048\ g\ H_2}{53.96\ g\ Al} = 1.12\ g\ H_2}$$

*6.85 *Vitamin C ($C_6H_8O_6$) reacts with oxygen as follows:*

$$2\ C_6H_8O_6 + O_2 \rightarrow 2\ C_6H_6O_6 + 2\ H_2O$$

a) *How many grams of oxygen will react with 500 mg of vitamin C (the mass in a typical vitamin C supplement)?*

We'll need to convert the mass from mg to g, which gives us 0.500 g of vitamin C (we will assume that the 500-mg number is good to 3 significant figures.) Then we add up the molar mass of vitamin C:

6 C	=	6 × 12.01	=	72.06 g/mol
8 H	=	8 × 1.008	=	8.064
+ 6 O	=	6 × 16.00	=	96.00
total				176.12 g/mol, and the molar mass of O_2 is 32.00 g/mol.

The balanced reaction equation tells us that two moles (352.24 g) of vitamin C will react with one mole (32.00 g) of O_2.

$$\mathbf{0.500\ g\ vitamin\ C \times \frac{32.00\ g\ O_2}{352.24\ g\ vitamin\ C} = 0.0454\ g\ O_2}$$

b) *The density of oxygen is roughly 1.3 g/L under typical atmospheric conditions. Using this fact and your answer to part a, calculate the number of milliliters of oxygen that will react with 500 mg of vitamin C.*

Chapter 1 review! The density is essentially a conversion factor between the volume and mass of a substance; then we just have to watch the volume units and convert L into mL.

$$\mathbf{0.0454\ g\ O_2 \times \frac{1\ L}{1.3\ g} \times \frac{1000\ mL}{L} = 34.9\ mL\ O_2}$$

*6.87 *Silver nitrate solution reacts with sodium hydroxide solution as follows:*

$$2\ AgNO_3(aq) + 2\ NaOH(aq) \rightarrow Ag_2O(s) + H_2O(l) + 2\ NaNO_3(aq)$$

a) *If you mix 50.0 mL of 0.150 M $AgNO_3$ with a large amount of 0.100 M NaOH, how many grams of Ag_2O will you form?*

This problem mixes molarities (Chapter 5) with molar masses and the relationship from the reaction. There are several ways to approach this problem, but the main point of all of them is to realize that the reaction gives the relationship between **moles** of all the different substances. So we first calculate the moles of $AgNO_3$ reacting (and since M stands for mol/L, we'll need to convert mL to L):

$$50.0\ \mathbf{mL\ AgNO_3} \times \frac{1\ \mathbf{L}}{1000\ \mathbf{mL}} \times \frac{0.150\ \mathbf{mol}}{\mathbf{L}} = 0.00750\ \mathbf{mol\ AgNO_3}$$

Using the fact that (according to the reaction) we get 1 mole of Ag_2O for every 2 moles of $AgNO_3$, and the molar mass of Ag_2O is 231.8 g/mol,

$$0.00750\ \mathbf{mol\ AgNO_3} \times \frac{1\ \mathbf{mol\ Ag_2O}}{2\ \mathbf{mol\ AgNO_3}} \times \frac{231.8\ \mathbf{g\ Ag_2O}}{\mathbf{mol\ Ag_2O}} = 0.869\ \mathbf{g\ Ag_2O}$$

b) *How many mL of the 0.100 M NaOH do you need in order to consume all of the $AgNO_3$?*

The strategy here is similar: instead of looking at the *mass* relationship between the species, we look at the *mole* relationship, and handle the concentration problem however we need to, to get the correct units (mL) in the answer. In this case we know (from the balanced equation) that it takes 2 moles of NaOH for every 2 moles of $AgNO_3$, and the concentration of the NaOH solution is 0.100 mol/L:

$$0.00750\ \mathbf{mol\ AgNO_3} \times \frac{2\ \mathbf{mol\ NaOH}}{2\ \mathbf{mol\ AgNO_3}} \times \frac{1\ \mathbf{L\ NaOH\ soln}}{0.100\ \mathbf{mol\ NaOH}} \times \frac{1000\ \mathbf{mL}}{\mathbf{L}}$$

$$= 75.0\ \mathbf{mL\ NaOH\ soln}$$

*6.89 *Acetylene (C_2H_2) is used in welding torches, because it produces an extremely hot flame when it burns. How many grams of CO_2 will be formed in the combustion of 5.00 g of acetylene?*

First write the balanced equation for the combustion:

$2\ C_2H_2 + 5\ O_2 \rightarrow 4\ CO_2 + 2\ H_2O$

Then calculate the molar masses of the two species mentioned, acetylene (26.04 g/mol) and CO_2 (44.01 g/mol). From the reaction equation, we see that two moles of acetylene (2 mol × 26.04 g/mol = 52.08 g) produce four moles of CO_2 (4 mol × 44.01 g/mol = 176.04 g):

$$5.00\ \mathbf{g\ C_2H_2} \times \frac{176.04\ \mathbf{g\ CO_2}}{52.08\ \mathbf{g\ C_2H_2}} = 16.9\ \mathbf{g\ CO_2}$$

6.91 *Mammals can "burn" amino acids (the building blocks of proteins) to obtain energy. An example is the burning of the amino acid alanine ($C_3H_7NO_2$):*

$2\ C_3H_7NO_2 + 6\ O_2 \rightarrow N_2H_4CO + 5\ CO_2 + 5\ H_2O + 624\ kcal$

a) *If a mammal burns 1.00 g of alanine, how much heat will be produced?*

Add up the molar mass of alanine: 89.10 g/mol. According to the reaction equation, two moles of alanine (178.20 g) react to produce 624 kcal of heat:

$$1.00\ \mathbf{g\ alanine} \times \frac{624\ \mathbf{kcal}}{178.20\ \mathbf{g\ alanine}} = 3.50\ \mathbf{kcal}$$

b) *How does your answer to part a compare with the Calorie value for proteins (4 Cal/g)? Remember that a "Calorie" (a nutritional calorie) is actually a kilocalorie.*

3.50 kcal = 3.50 Calorie, which looks like a little less than the given Calorie value for proteins of 4 Calories per gram; however, remember that that was given as

"*about* 4 calories per gram" and is actually an approximate, average value for all proteins, to only one significant figure. Proteins are composed of combinations of many amino acids, not just alanine. In fact, if we could only have reported our calculated answer to one sf, it would also have been 4 kcal.

6.93 *Fred wants to know how much of each nutrient there is in the Twinkie® he is about to eat. He sees that the Twinkie contains 1 g of protein, 27 g of carbohydrate, and 160 Calories. However, the nutritional label has been damaged, so he cannot read the amount of fat. How many grams of fat are there in Fred's Twinkie?*

First, calculate the energy value of the components we actually know about:

carbohydrates:	$27\text{ g} \times 4\text{ Cal/g} =$	108 Cal
protein:	$1\text{ g} \times 4\text{ Cal/g} =$	4 Cal
total from carbohydrates and protein:		112 Cal accounted for.

If there are 160 total Calories, then the fat must be supplying 160 Cal - 112 Cal= 48 Cal:

$$\mathbf{48\,Cal \times \frac{1\,g\,fat}{9\,Cal} = about\ 5\,g\,fat}$$

*6.95 *When your body produces excess heat, you perspire. The perspiration (which is mostly water) evaporates, removing the excess heat.*

a) *If 10.0 g of water evaporates from your skin, how much heat does it remove from your body? (The heat of vaporization of water is 540 cal/g.)*

The heat of vaporization is a conversion factor between g of water and heat:

$$\mathbf{10.0\,g\,water \times \frac{540\,cal}{1\,g\,water} = 5400\,cal}$$

b) *How many grams of fat must your body burn to supply the heat that was removed from your body in part a?*

Now we have to remember that nutritional Calories are actually kilocalories; our number from above, 5400 cal, is really 5.4 kcal or 5.4 Cal.

$$\mathbf{5.4\,Cal \times \frac{1\,g\,fat}{9\,Cal} = 0.6\,g\,fat}$$ or less than one gram of fat burned to produce enough energy to evaporate 10 g of water from the skin.

6.97 *Sucrose (table sugar) breaks down in your digestive tract, producing two other types of sugar, glucose (blood sugar) and fructose (fruit sugar). In your stomach, this reaction is catalyzed by HCl and the activation energy is 26 kcal. In your intestine, the reaction is catalyzed by an enzyme called sucrase, and the activation energy is 9 kcal. In which part of your digestive system is sucrose broken down more rapidly?*

Sucrase is a much more effective catalyst for this reaction, bringing the activation energy to a much lower value than with HCl as a catalyst. The lower the activation energy, the faster the reaction, so **this reaction is much more rapid in the intestine than in the**

stomach. (While changes in temperature also affect reaction rates, the temperature in different parts of the body is nearly the same.)

6.99 *A chemist puts 2.00 grams of N_2O_4 into a container. Some of the N_2O_4 breaks down into NO_2, forming an equilibrium mixture.*

$$N_2O_4(g) \rightarrow 2\ NO_2(g)$$

If the chemist then puts 2.00 grams of NO_2 into an identical container, which of the following statements will be true?

If this reaction is an equilibrium as we are told, then the reaction can't be complete in either direction. A sample of N_2O_4 will turn into an equilibrium mixture of N_2O_4 and NO_2, and a sample of NO_2 will also turn into an equilibrium mixture of N_2O_4 and NO_2.

a) *Some of the NO_2 will turn into N_2O_4.* **This is the only true statement.**

b) *All of the NO_2 will turn into N_2O_4.* **This implies the reverse reaction going to completion, not an equilibrium process.**

c) *None of the NO_2 will turn into N_2O_4.* **This implies the forward reaction going to completion, not an equilibrium process.**

Chapter 7

7

Acids and Bases

Solutions to Section 7.1 Core Problems

7.1 *Write the chemical equation for the self-ionization of water.*

See Section 7.1:

$\mathbf{2\ H_2O \rightleftarrows H_3O^+ + OH^-}$

7.3 *Why is the chemical equation that follows not an accurate representation of the self-ionization of water?*

$$H_2O(l) \rightleftarrows H^+(aq) + OH^-(aq)$$

The reaction shown makes it look as though there is actually an H^+ ion formed in the self-ionization, when in fact the H^+ ion is covalently attached to a water molecule, forming H_3O^+, hydronium ion.

7.5 *Calculate the molar concentrations of the following ions. Give your answer to each of the following questions as a power of ten.*

For each of these, we will use the relationship

$$[H_3O^+] \times [OH^-] = 1.0 \times 10^{-14}$$

a) *OH^- in a solution that contains 10^{-11} M H_3O^+.*

Solve the equation above for $[OH^-]$:

$$[OH^-] = \frac{1.0 \times 10^{-14}}{[H_3O^+]} = \frac{1.0 \times 10^{-14}}{10^{-11}} = 10^{-3}\ M$$

b) *H_3O^+ in a solution that contains 10^{-5} M OH^-.*

Solve the equation above for $[H_3O^-]$:

$$[H_3O^+] = \frac{1.0 \times 10^{-14}}{[OH^-]} = \frac{1.0 \times 10^{-14}}{10^{-5}} = 10^{-9}\ M$$

c) *OH^- in a solution that contains 0.001 M H_3O^+.*

Solve the equation above for $[OH^-]$:

$$[OH^-] = \frac{1.0 \times 10^{-14}}{[H_3O^+]} = \frac{1.0 \times 10^{-14}}{0.001} = 1 \times 10^{-11}\ M$$

7.7 *Calculate the molar concentration of H_3O^+ in each of the following solutions:*

For each of these, we will use the relationship

$[H_3O^+] \times [OH^-] = 1.0 \times 10^{-14}$

a) *A solution that contains 4.1×10^{-13} M OH^-.*

Solve the equation above for $[H_3O^+]$:

$$[H_3O^+] = \frac{1.0\times10^{-14}}{[OH^-]} = \frac{1.0\times10^{-14}}{4.1\times10^{-13}} = 0.024\text{ M or }2.4\times10^{-2}\text{ M}$$

b) *A solution that contains 0.0075 M OH^-.*

Solve the equation above for $[OH^-]$:

$$[H_3O^+] = \frac{1.0\times10^{-14}}{[OH^-]} = \frac{1.0\times10^{-14}}{0.0075} = 1.3\times10^{-12}\text{ M}$$

7.9 *Explain why the concentration of H_3O^+ increases when you add HCl to water.*

HCl ionizes in water to form H^+ ions and Cl^- ions; the H^+ ions combine with water molecules to form H_3O^+ ions.

Solutions to Section 7.1 Core Problems

7.11 *Which of the following solutions are acidic, which are neutral, and which are basic?*

Regardless of the chemical identities of solutes in a solution, solutions with pH values below 7 are acidic, solutions with pH values greater than 7 are basic, and solutions with pH = 7 are neutral.

a) *A solution of NH_4Cl in water (pH = 3.86).* **Acidic**—pH is less than 7.

b) *A solution of Na_2CO_3 in water (pH = 10.95).* **Basic**—pH is greater than 7.

c) *A solution of NaCl in water (pH = 7.00).* **Neutral.**

7.13 *Which of the following solutions is the most acidic, and which is the least acidic?*

The more acidic a solution, the lower its pH value.

0.1 M NH_4Cl (pH = 5.12) **This solution has the highest pH value of the three and is therefore the least acidic.**

0.1 M $NaHSO_3$ (pH = 4.00)

0.1 M $NaHSO_4$ (pH = 1.54) **This solution has the lowest pH value of the three and is therefore the most acidic.**

7.15 *Calculate the pH of each of the following solutions. Give your answers as whole numbers.*

We'll use the relationship $pH = -\log[H_3O^+]$.

a) A solution that contains 10^{-5} M H_3O^+

$pH = -\log[H_3O^+] = -\log(10^{-5}) = \mathbf{5}$

b) A solution that contains 10^{-3} M OH^-

First we'll need to use the ion product of water: $[H_3O^+] \times [OH^-] = 1.0 \times 10^{-14}$

and solve for $[H_3O^+]$: $[H_3O^+] = \dfrac{1.0\times10^{-14}}{[OH^-]} = \dfrac{1.0\times10^{-14}}{10^{-3}} = 10^{-11}$ M

$pH = -\log[H_3O^+] = -\log(10^{-11}) = \mathbf{11}$

c) A solution that contains 0.0001 M H_3O^+

This concentration, written in decimal form, is 1×10^{-4}:
$pH = -\log[H_3O^+] = -\log(1 \times 10^{-4}) = \mathbf{4}$

d) A solution that contains 0.00001 M OH^-

First we'll need to use the ion product of water: $[H_3O^+] \times [OH^-] = 1.0 \times 10^{-14}$

to find $[H_3O^+]$: $[H_3O^+] = \dfrac{1.0\times10^{-14}}{[OH^-]} = \dfrac{1.0\times10^{-14}}{0.00001} = 1\times10^{-9}$ M

$pH = -\log[H_3O^+] = -\log(1 \times 10^{-9}) = \mathbf{9}$

7.17 *The concentration of H_3O^+ in a solution of soapy water is 3.1×10^{-9} M.*

a) What is the concentration of OH^- in the soapy water?

$$\mathbf{[OH^-] = \frac{1.0\times10^{-14}}{[H_3O^+]} = \frac{1.0\times10^{-14}}{3.1\times10^{-9}} = 3.2\times10^{-6}\ M}$$

b) What is the pH of the soapy water?

$\mathbf{pH = -\log[H_3O^+] = -\log(3.1 \times 10^{-9}) = 8.51}$

7.19 *The pH of a cup of coffee is 5.13. Calculate the concentration of H_3O^+ and OH^- in the coffee.*

$\mathbf{[H_3O^+] = 10^{-pH} = 10^{-5.13} = 7.4 \times 10^{-6}\ M}$

$$\mathbf{[OH^-] = \frac{1.0\times10^{-14}}{[H_3O^+]} = \frac{1.0\times10^{-14}}{7.4\times10^{-6}} = 1.3\times10^{-9}\ M}$$

Solutions to Section 7.3 Core Problems

7.21 *Write the chemical equations for the ionization of the following acids in water:*

When writing the equation for the ionization of an acid in water, the reactants and products are always the same: the acid plus water yields a hydronium ion plus the anion of the acid. Remember to use a normal one-way reaction arrow for strong acids and an equilibrium arrow for weak acids.

a) *HNO_3 (nitric acid, a strong acid that is used in the manufacture of fertilizers)*

Strong acids ionize completely in water, so we use a one-way arrow:

$\mathbf{HNO_3 + H_2O \rightarrow H_3O^+ + NO_3^-}$

b) *$HC_3H_5O_3$ (lactic acid, a weak acid that is found in sour milk)*

Weak acids ionize only slightly, as indicated by equilibrium arrows:

$\mathbf{HC_3H_5O_3 + H_2O \rightleftarrows H_3O^+ + C_3H_5O_3^-}$

c) *$H_2C_4H_4O_4$ (succinic acid, a weak acid that is formed when sugars are burned by your body)*

Weak acids ionize only slightly, so we use equilibrium arrows:

$\mathbf{H_2C_4H_4O_4 + H_2O \rightleftarrows H_3O^+ + HC_4H_4O_4^-}$

7.23 *Which solution has the higher pH: 0.1 M formic acid or 0.01 M formic acid?*

The more concentrated a solution of an acid, the lower the pH (lower pH = more acidic). Therefore **0.01 M formic acid, the more dilute solution, will have the higher pH.**

7.25 *If you dissolve 0.1 moles of formic acid in 1 L of water, the resulting solution contains 0.004 mol of H_3O^+. Based on this information, is formic acid a strong acid, or is it a weak acid?*

A weak acid, because 0.1 mol of a strong acid would have ionized completely to give 0.1 mol H_3O^+.

7.27 *The pH of 0.01 M benzoic acid is 3.12, and the pH of 0.01 M cyanoacetic acid is 2.33. Based on this information, which is the stronger acid, benzoic acid or cyanoacetic acid?*

At the same concentration, cyanoacetic acid has the lower pH and therefore forms a more acidic solution; **cyanoacetic acid is a stronger acid than benzoic acid.**

7.29 *Nitric acid (HNO_3) is stronger than nitrous acid (HNO_2). Based on this information, answer the following questions:*

a) *Which contains a higher concentration of H_3O^+ ions, 0.01 M nitric acid or 0.01 M nitrous acid?*

0.01 M nitric acid: at the same concentration, stronger acids will produce more H_3O^+ ions.

b) *Which solution has the higher pH?*

0.01 M nitrous acid will have a lower concentration of H_3O^+ ions and therefore a higher pH (higher pH = less acidic solution.)

c) *Which solution contains a higher concentration of OH^- ions?*

0.01 M nitrous acid has a lower $[H_3O^+]$ and therefore higher $[OH^-]$.

7.31 *Both formic acid and carbonic acid contain two hydrogen atoms. Why is the chemical formula of formic acid written $HCHO_2$ (with the two hydrogen atoms listed separately), while the chemical formula of carbonic acid is written H_2CO_3 (with the two hydrogen atoms written together)?*

Acid formulas are written with the acidic H atom(s) first. Formic acid, $HCHO_2$, only has one H that it can donate, while carbonic acid, H_2CO_3, has two H atoms to donate.

7.33 *Methylphosphoric acid has the chemical formula CH_5PO_4.*

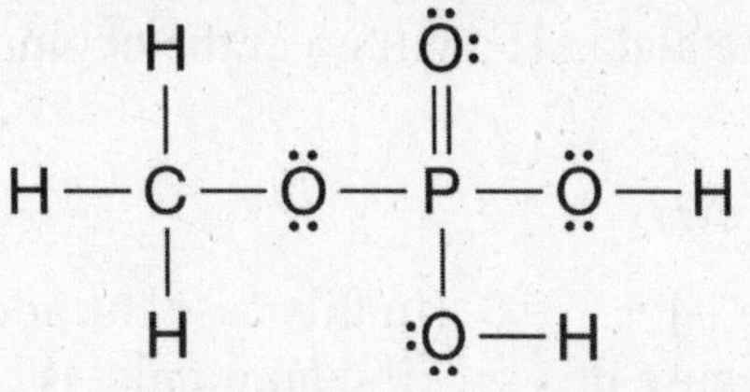

```
     H          Ö:
     |          ||
 H — C — Ö — P — Ö — H
     |          |
     H         :Ö — H
```

a) *Based on the structure of this compound, how many hydrogen ions can be removed from a molecule of methylphosphoric acid?*

Only two—the H atoms on O are acidic, but H covalently attached to C are never acidic.

b) *Write the chemical formula of methylphosphoric acid in a way that clearly shows the number of ionizable hydrogen atoms.*

This means writing the formula with the two ionizable H atoms at the beginning, leaving three H atoms to appear in their normal position following C in the formula: **$H_2CH_3PO_4$**

7.35 *Carbonic acid (H_2CO_3) is a polyprotic acid. When carbonic acid dissolves in water, which is higher, the concentration of HCO_3^- ions or the concentration of CO_3^{2-} ions?*

The first ionization of a polyprotic acid is always the strongest, while the subsequent ionizations take place to a smaller extent. HCO_3^- is the product of the first ionization and CO_3^{2-} is the product of the second ionization; therefore **the concentration of HCO_3^- ions is much higher than the concentration of CO_3^{2-} ions.**

Solutions to Section 7.4 Core Problems

7.37 *Methylamine is a base because it can form a bond to H^+. Draw Lewis structures to show how methylamine reacts with water to form a hydroxide ion (see Figure 7.6).*

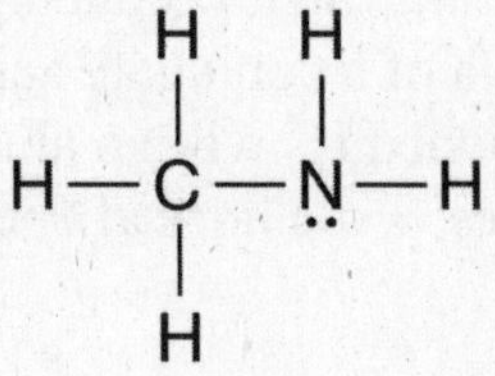

```
     H   H
     |   |
 H — C — N — H
     |   ··
     H
```

$$
\begin{array}{ccc}
\mathrm{H\ \ H} & & \\
\mathrm{H{-}C{-}N{:}} & + \ \mathrm{H{-}\ddot{\underset{\cdot\cdot}{O}}{-}H} & \longrightarrow \\
\mathrm{H\ \ H} & &
\end{array}
\quad
\begin{array}{c}
\mathrm{H\ \ H} \\
\mathrm{H{-}C{-}\overset{+}{N}{-}H} \\
\mathrm{H\ \ H}
\end{array}
\ + \ \mathrm{{:}\overset{-}{\ddot{\underset{\cdot\cdot}{O}}}{-}H}
$$

7.39 *Write the chemical formula for each of the following:*

To write the formula of a conjugate base, always remove an H^+ (and change the charge on what's left by –1). To write the formula of a conjugate acid, always add an H^+ (and change the charge by +1). Usually the added H^+ will go at the beginning of the formula, because it's an acidic H.

a) *The conjugate acid of amide ion, NH_2^-*

Add H^+: $\mathbf{NH_3}$. The final charge is $-1 + 1 = 0$. (In this case the added H^+ does not go at the beginning of the formula, because this species, ammonia, is one that's already familiar to us as NH_3.)

b) *The conjugate base of nitric acid, HNO_3*

Remove H^+: $\mathbf{NO_3^-}$. The final charge is $0 - 1 = -1$.

c) *The conjugate acid of nicotine, $C_{10}H_{14}N_2$*

Add H^+: $\mathbf{HC_{10}H_{14}N_2^+}$. The final charge is $0 + 1 = +1$.

d) *The conjugate base of sulfurous acid, H_2SO_3*

Remove H^+: $\mathbf{HSO_3^-}$. The final charge is $0 - 1 = -1$.

e) *The conjugate acid of dihydrogen citrate ion, $H_2C_6H_5O_7^-$*

Add H^+: $\mathbf{H_3C_6H_5O_7}$. The final charge is $-1 + 1 = 0$.

f) *The conjugate base of dihydrogen citrate ion, $H_2C_6H_5O_7^-$*

Remove H^+: $\mathbf{HC_6H_5O_7^{2-}}$. The final charge is $-1 - 1 = -2$.

NOTE:
The H is removed from the beginning of the formula (the answer $H_2C_6H_4O_7^{2-}$ would be incorrect).

7.41 *Write chemical equations that show how the following bases react with water to produce hydroxide ions:*

In each case, the base will take an H^+ from water, producing OH^- and the conjugate acid of the original base. To write the formula of a conjugate acid, always add an H^+ (and change the charge by +1). Usually the added H^+ will go at the beginning of the formula, because it's an acidic H. For strong bases, write normal one-way reaction arrows; for weak bases, use equilibrium arrows.

a) *methoxide ion (OCH_3^-), a strong base*

$\mathbf{OCH_3^- + H_2O \rightarrow HOCH_3 + OH^-}$

b) hypochlorite ion (ClO^-), a weak base

$$ClO^- + H_2O \rightleftarrows HClO + OH^-$$

c) imidazole ($C_3H_4N_2$), a weak base

$$C_3H_4N_2 + H_2O \rightleftarrows HC_3H_4N_2^+ + OH^-$$

d) sulfite ion (SO_3^{2-}), a weak base

$$SO_3^{2-} + H_2O \rightleftarrows HSO_3^- + OH^-$$

7.43 *Explain why you get a basic solution when you dissolve NaF in water.*

Fluoride ion, F^-, reacts with water to produce hydroxide ions:

$$F^- + H_2O \rightleftarrows HF + OH^-$$

7.45 *The pH of 0.01 M NaF is 7.57 and the pH of 0.01 M NaCN (sodium cyanide) is 10.60. Based on this information, which is the stronger base, F^- or CN^-? Explain your answer.*

The higher the pH of a solution, the more basic it is. At the same concentration, NaCN produces a higher pH than NaF, so **CN^- must be a stronger base than F^-.** (Remember from Chapter 5 that both of these ionic compounds will dissociate in solution.)

7.47 *Piperidine ($C_5H_{11}N$) is a stronger base than piperazine ($C_4H_{10}N_2$). Use this information to answer the following questions:*

a) Which solution has a higher pH, 0.01 M piperidine or 0.01 M piperazine?

At the same concentration, the stronger base will produce a higher pH: **0.01 M piperidine will have a higher pH than 0.01 M piperazine.**

b) Which solution contains a higher concentration of OH^- ions?

By definition, the stronger base is the one that produces more OH^- ions. **0.01 M piperidine will have a higher concentration of OH^- ions than 0.01 M piperazine.**

c) Which solution contains a higher concentration of H_3O^+ ions?

The solution of the *weaker* base will have lower concentration of OH^- ions and a higher concentration of H_3O^+. **0.01 M piperazine will have a higher concentration of H_3O^+ ions, because it is the weaker base.**

Solutions to Section 7.5 Core Problems

7.49 *Draw Lewis structures to show how H^+ is transferred when HNO_2 and NH_3 react with one another. The Lewis structure of HNO_2 is:*

$$H—\ddot{\underset{\cdot\cdot}{O}}—\ddot{N}=\ddot{\underset{\cdot\cdot}{O}}$$

In any acid-base reaction, the acid gives up an H^+ ion, while the base accepts an H^+ ion.

$$\mathrm{H{-}\underset{\underset{H}{|}}{\overset{\overset{H}{|}}{N}}\!:} \;+\; \mathrm{H{-}\ddot{\underset{..}{O}}{-}\ddot{N}{=}\ddot{\underset{..}{O}}} \;\longrightarrow\; \mathrm{H{-}\underset{\underset{H}{|}}{\overset{\overset{H}{|}}{N^{+}}}{-}H} \;+\; \mathrm{{}^{-}\!:\ddot{\underset{..}{O}}{-}\ddot{N}{=}\ddot{\underset{..}{O}}}$$

Things to notice in this reaction: as the acid (HNO_2) donates H^+, the pair of electrons that forms the O–H bond in HNO_2 becomes an unshared pair of electrons in the conjugate base, NO_2^-, while the unshared pair of electrons in NH_3 is converted into a bonding pair in NH_4^+.

7.51 *Write chemical equations for the acid–base reactions that occur when solutions of the following substances are mixed:*

In acid-base reactions, we always do the same thing: take H^+ off the acid and put it on the base. If both substances are weak, use equilibrium arrows; if either the acid or the base is strong, use a one-way reaction arrow.

a) *HNO_2 (nitrous acid) and C_2H_7NO (ethanolamine, a base)*

Take the H^+ off the acid (HNO_2) and put it on the base (C_2H_7NO). Both of these species are weak, so we use equilibrium arrows in the reaction:

$\mathbf{HNO_2 + C_2H_7NO \rightleftarrows NO_2^- + HC_2H_7NO}$

b) *H_3O^+ and F^-*

Take the H^+ off the acid (H_3O^+) and put it on the base (F^-). H_3O^+ is a strong acid, so we use a one-way arrow:

$\mathbf{H_3O^+ + F^- \rightarrow H_2O + HF}$

b) *OH^- and $H_2PO_4^-$*

Take the H^+ off the acid ($H_2PO_4^-$) and put it on the base (OH^-). OH^- is a strong base, so we use a one-way arrow:

$\mathbf{OH^- + H_2PO_4^- \rightarrow H_2O + HPO_4^{2-}}$

d) *C_5H_5N (pyridine, a base) and $HC_2H_3O_2$ (acetic acid)*

Take the H^+ off the acid ($HC_2H_3O_2$) and put it on the base (C_5H_5N). Both species are weak, so we use equilibrium arrows:

$\mathbf{C_5H_5N + HC_2H_3O_2 \rightleftarrows HC_5H_5N^+ + C_2H_3O_2^-}$

7.53 *Write chemical equations for the acid–base reactions that occur when:*

a) *solutions of $HC_2H_3O_2$ (acetic acid) and KOH are mixed.*

We can begin by recognizing that KOH is an ionic compound and will dissociate into K^+ ions and OH^- ions. Hydroxide ion is a strong base and will react with acetic acid. Potassium has no acid-base properties, so it is a spectator ion and is omitted from the reaction equation. The acid in this reaction is acetic acid; we can see from the formula that acetic acid contains only one acidic hydrogen atom (the formula begins with one H). As in all acid-base reactions, a hydrogen ion moves from the acid to the base:

$\mathbf{HC_2H_3O_2\ (aq) + OH^-\ (aq) \rightarrow C_2H_3O_2^-\ (aq) + H_2O\ (l)}$

b) *solutions of HCN (hydrocyanic acid) and Na_2CO_3 are mixed.*

We can begin by recognizing that Na_2CO_3 is an ionic compound. All ionic compounds dissociate when they dissolve in water, so the sodium carbonate solution really contains Na^+ ions and CO_3^{2-} ions. The next step is to identify the acid and the base; hydrocyanic acid has the word "acid" in the name and its formula starts with H, so we know that HCN is the acid and that it acid has only one acidic H atom. CO_3^{2-} acts as the base in this reaction. Na^+ has no acid-base properties; it's a spectator ion and is left out of the reaction equation. The overall reaction is:

$HCN\ (aq) + CO_3^{2-}\ (aq) \rightarrow HCO_3^{2-}\ (aq) + CN^-\ (aq)$

7.55 *Identify the conjugate pairs in the following acid–base reaction.*

$$H_2CO_3(aq) + C_5H_5N(aq) \rightarrow HCO_3^-(aq) + HC_5H_5N^+(aq)$$

H_2CO_3 (an acid) and HCO_3^- (a base) are conjugates.
C_5H_5N (a base) and $HC_5H_5N^+$ (an acid) are conjugates.

7.57 *If you add a solution of NaOH to a solution of H_2CO_3, two reactions occur, one after the other. Write the chemical equations for these two reactions. (Remember that NaOH dissociates into Na^+ and OH^-, and the hydroxide ion is the actual base.)*

We can see from the formula that H_2CO_3 contains two ionizable H atoms, so the reaction with OH^- will take place in two steps, with one hydrogen ion moving from the acid to the base in each step. First,

$H_2CO_3\ (aq) + OH^-\ (aq) \rightarrow HCO_3^-\ (aq) + H_2O\ (l)$

HCO_3^-, the conjugate base of H_2CO_3, still contains an ionizable H atom and reacts with a second hydroxide ion:

$HCO_3^-\ (aq) + OH^-\ (aq) \rightarrow CO_3^{2-}\ (aq) + H_2O\ (l)$

Since Na^+ has no acid-base properties, it is left out of the reaction equations

Solutions to Section 7.6 Core Problems

7.59 *Sodium bisulfite is an ionic compound that is used as a mild bleaching agent and a food preservative. It contains the bisulfite ion, an amphiprotic ion with the chemical formula HSO_3^-.*

a) *Write the chemical equation for the reaction of HSO_3^- with OH^-. Is the bisulfite ion functioning as an acid in this reaction, or is it functioning as a base?*

"Amphiprotic" means the species (usually an anion with an initial H in its formula) can act as both an acid (if it's put together with a base) and a base (if it's put together with an acid). **OH^- is a strong base, so bisulfite will act as an acid:**

$HSO_3^- + OH^- \rightarrow H_2O + SO_3^{2-}$

b) *Write the chemical equation for the reaction of HSO_3^- with H_3O^+. Is the bisulfite ion functioning as an acid in this reaction, or is it functioning as a base?*

H_3O^+ is a strong acid, so bisulfite will act as a base:

$HSO_3^- + H_3O^+ \rightarrow H_2SO_3 + H_2O$

7.61 *Alanine ($HC_3H_6NO_2$) is an amino acid, one of the building blocks of proteins. Like all amino acids, it is an amphiprotic molecule.*

An amphiprotic species will act as an acid when combined with a stronger base, but will act as a base when combined with a stronger acid.

a) *Write the chemical equation for the reaction of alanine with OH^-.*

Since OH^- is a base, alanine will act as the acid and donate H^+:

$$\mathbf{HC_3H_6NO_2 + OH^- \rightarrow C_3H_6NO_2^- + H_2O}$$

b) *Write the chemical equation for the reaction of alanine with H_3O^+.*

Since H_3O^+ is an acid, alanine will act as the base and accept H^+:

$$\mathbf{HC_3H_6NO_2 + H_3O^+ \rightarrow H_2C_3H_6NO_2^+ + H_2O}$$

7.63 *Taurine is an amphiprotic compound that your body uses to make bile salts.*

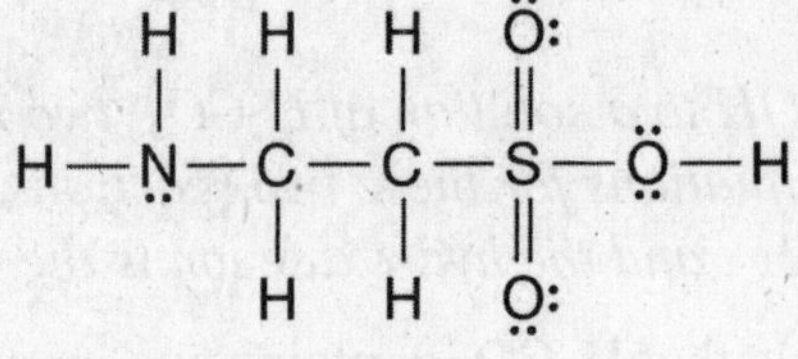

```
     H   H   H   Ö:
     |   |   |   ||
H—N—C—C—S—Ö—H
     ¨   |   |   ||   ¨
         H   H   O:
                 ¨
```

Taurine

a) *Which hydrogen atom in this compound is acidic (i.e. which one can be removed by a base)?*

This question refers back to Figure 7.5. **Only the H on O is likely to be acidic**; it shares the common structure of the acids in Figure 7.5. H atoms covalently bonded to nitrogen are not usually acidic (except when an H^+ has been added to a nitrogen base).

b) *When taurine functions as a base, where does H^+ bond?*

Nitrogen atoms with lone pairs are usually basic, so the **H^+ will add to the nitrogen atom** (the lone pair on N will become a covalent bond to H).

Solutions to Section 7.7 Core Problems

7.65 *The following table lists five buffer solutions. In each case, one of the two ingredients in the buffer is missing. Supply the missing chemical.*

A buffer solution always contains a weak acid and its conjugate base (or a weak base and its conjugate acid—which is another way of saying the same thing, since really the two species are a conjugate acid/base pair.) So if we have an acid, we just write its conjugate base, and vice versa, by adding or removing H^+ and changing the charge as necessary.

	Acidic component	*Basic component*
Buffer #1	$HCHO_2$	$\mathbf{CHO_2^-}$
Buffer #2	$\mathbf{HC_3H_5O_3}$	$C_3H_5O_3^-$
Buffer #3	$HC_2O_4^-$	$\mathbf{C_2O_4^{2-}}$
Buffer #4	$\mathbf{H_2C_2O_4}$	$HC_2O_4^-$
Buffer #5	H_2SO_3	$\mathbf{HSO_3^-}$

7.67 *You need to make a buffer that contains the $C_4H_4O_4^{2-}$ ion. What chemical compound could you use to supply this ion?*

This is similar to Problem 4.85. We need to find a soluble ionic compound of the $C_4H_4O_4^{2-}$ ion, without knowing anything about the ion (in this case, even its name!) But we know it's an anion, and therefore needs to be paired with a cation; and we know that ionic compounds containing Na^+ and K^+ are soluble in water. So two options are **$Na_2C_4H_4O_4$ and $K_2C_4H_4O_4$**. (Review: Section 4.7)

7.69 *Which of the following pairs of chemicals produce a buffer solution when dissolved in water?*

a) *HCN and NaCN*

This pairing contains both partners in a weak conjugate acid-base pair, HCN and CN^-, and **is therefore a buffer.**

b) *HCl and NaOH*

This is a combination of a strong acid and a strong base. Buffers have to be made of a *weak* conjugate acid-base pair, so **this will not produce a buffer solution.**

c) *$H_3C_6H_5O_7$ (citric acid) and $NaH_2C_6H_5O_7$*

This pairing contains both partners in a weak conjugate acid-base pair, $H_3C_6H_5O_7$ and $H_2C_6H_5O_7^-$, and **is therefore a buffer.**

d) *$H_2C_2O_4$ (oxalic acid) and $Na_2C_2O_4$*

The acid and base in this pairing, $H_2C_2O_4$ and $C_2O_4^{2-}$, are not conjugates of each other, but are the conjugate acid and base of $HC_2O_4^-$ (that is, the species in the middle was skipped.) This pairing **will not produce a buffer solution.**

7.71 *The pK_a of mandelic acid ($HC_8H_7O_3$) is 3.8.*

a) *What is the pH of a buffer that contains equal concentrations of $HC_8H_7O_3$ and $NaC_8H_7O_3$?*

A buffer that contains equal molar concentrations of an acid and its conjugate base has a pH equal to the pK_a of the acid, so **the pH of this buffer is 3.8.**

b) *If you make a buffer that contains 0.1 M $HC_8H_7O_3$ and 0.2 M $NaC_8H_7O_3$, what is the approximate pH of the solution? (You can give a pH range.)*

This buffer has a higher concentration of the conjugate base than of the conjugate acid, so we'd expect it to have a **slightly higher (more basic) pH than the equal-concentration buffer in (a)—maybe 4.0-4.2.**

c) *What substance in this buffer neutralizes acids? Write a balanced equation for the reaction between this substance and H_3O^+.*

The conjugate base, $C_8H_7O_3^-$, will neutralize acids: **$C_8H_7O_3^- + H_3O^+ \rightarrow HC_8H_7O_3 + H_2O$**

d) *What substance in this buffer neutralizes bases? Write a balanced equation for the reaction between this substance and OH^-.*

The conjugate acid, $HC_8H_7O_3$, will neutralize bases: $HC_8H_7O_3 + OH^- \rightarrow C_8H_7O_3^- + H_2O$

7.73 *You need to prepare a buffer that has a pH of 6.9, using $H_2PO_4^-$ ($pK_a = 7.21$) and HPO_4^{2-}.*

a) *Should you use equal concentrations of the two substances in the buffer? If not, which one should be present in higher concentration, and why?*

For a buffer that contains equal concentrations of a weak acid and its conjugate base, the pH of the solution is equal to the pKa of the buffer. **A buffer containing equal concentrations of these two substances would have a pH of 7.21. If we need a lower pH (a more acidic solution), we should use a greater concentration of the conjugate acid, $H_2PO_4^-$.**

b) *Which of the two substances in this buffer neutralizes acids?*

The conjugate base, HPO_4^{2-}, will neutralize acids.

c) *Write the chemical equation that shows how this buffer would react with OH^- ions.*

The conjugate acid, $H_2PO_4^-$, would react with OH^- ions:
$H_2PO_4^- + OH^- \rightarrow HPO_4^{2-} + H_2O$

7.75 *A buffer has a pH of 8.3. If a little NaOH is added to this buffer, which of the following is the most likely pH value for the mixture?*

a) 4.1 *b) 7.0* *c) 8.1* *d) 8.3* ***e) 8.5*** *f) 11.5*

NaOH is a strong base, so adding a little of it will cause the pH to **increase.** However, the whole purpose of a buffer is to limit changes in pH, so we would expect only a small increase in the pH of the solution. Options (a), (b) and (c) represent *decreases* in pH (more acidic solutions), which would require adding acid, not base. Option (d) has the same pH, so it's not right either. Options (e) and (f) are increases in pH, and (f) is unlikely because it's a large increase. **Option (e) 8.5 is the most likely outcome.**

Solutions to Section 7.8 Core Problems

7.77 *What are the two primary buffers in intracellular fluid?*

See section 7.8: **the protein buffer system and the phosphate buffer system, $H_2PO_4^-$ and HPO_4^{2-}.**

7.79 *One of the important buffer systems in living cells is the phosphate buffer.*

a) *What two substances make up this buffer?*

$H_2PO_4^-$ and HPO_4^{2-}. See Section 7.8.

b) *Which of these substances neutralizes acids?*

The conjugate base of the pair, HPO_4^{2-}, neutralizes acids.

c) Which of these substances neutralizes bases?

The conjugate acid of the pair, $H_2PO_4^-$, neutralizes bases.

7.81 *Explain why proteins that contain histidine can help to maintain the pH of body fluids around 7.*

To make a buffer with a pH near 7 requires an acid with a pK_a near 7. When histidine is incorporated into a protein, its conjugate acid can have a pK_a of around 7.

7.83 *Carbonated beverages contain CO_2 dissolved in water. As carbon dioxide dissolves in water, the pH of the water changes. Does the pH go up, or does it go down? Explain your answer.*

When carbon dioxide dissolves in water, it also reacts to form H_2CO_3, carbonic acid, which ionizes in two steps producing H^+ ions. Therefore carbon dioxide acts as an acid in water. When CO_2 dissolves in water, the solution becomes more acidic, and the pH goes down.

7.85 *Explain why the pH of your blood plasma goes up when you breathe too fast.*

The most direct way to think of this is to simply know that CO_2 behaves as an acid in aqueous solution. **Breathing too fast causes the body to expel more CO_2 from the blood. Since CO_2 behaves as an acid in aqueous solution, its removal from blood causes the blood to become more basic, and the pH rises.**

When CO_2 dissolves in aqueous solution, the molecules of CO_2 actually react with water molecules to form carbonic acid, H_2CO_3, which then partially ionizes to form hydrogen carbonate ions, HCO_3^-, and hydronium ions, H_3O^+:

$$CO_2 + H_2O \rightleftarrows H_2CO_3,$$

$$H_2CO_3 + H_2O \rightleftarrows H_3O^+ + HCO_3^-$$

The H_3O^+ ions that form are responsible for the acidity of CO_2 solutions. The greater the concentration of CO_2 in the solution, the more acidic the solution will be, and the lower its pH.

7.87 *If the concentration of HCO_3^- ions in your blood becomes too high, how does your body get rid of the excess bicarbonate ions?*

The kidneys are capable of excreting excess HCO_3^- ions to prevent the blood from becoming too basic.

7.89 *People with severe kidney failure excrete excessive amounts of HCO_3^- in their urine. The HCO_3^- comes from blood plasma. How does this affect the pH of the blood plasma?*

The buffer in blood plasma consists of H_2CO_3 and HCO_3^-. **Since HCO_3^- is the base in the carbonate buffer, excreting too much of it would leave the remaining buffer with too much acid, causing a decrease in the pH of the blood plasma.**

Solutions to Concept Questions

* indicates more challenging problems

7.91 *All aqueous solutions contain H_3O^+ ions and OH^- ions. Where do these ions come from?*

The self-ionization of water causes these ions to form, when one water molecule loses an H^+ ion to another water molecule:

$$2\ H_2O \rightleftarrows H_3O^+ + OH^-$$

7.93 *Why are acid–base reactions often called proton transfer reactions?*

Acid-base reactions involve the transfer of H^+ ions. Since an H^+ ion has one proton and no electrons, it is often called a "proton."

7.95 *What element must all acids contain, and why?*

Acids are defined as donors of H^+ ion, so all acids must contain H atoms.

7.97 a) *What is the difference between a strong acid and a weak acid?*

All acids contain ionizable H atoms. Strong acids ionize completely in water, while weak acids ionize only to a small extent, with most molecules in the solution maintaining the covalent bond to hydrogen.

b) *What is the difference between a strong base and a weak base?*

All bases form OH^- ions in water, and all bases can accept H^+ ions from acids. **Strong bases dissociate or react completely in water to form OH^- ions, while weak bases only partially react to form OH^- ions in aqueous solution.**

7.99 *Solutions of sodium acetate ($NaC_2H_3O_2$) conduct electricity much better than solutions of acetic acid ($HC_2H_3O_2$). Why is this?*

Sodium acetate, a soluble ionic compound, is a strong electrolyte: it dissociates completely in water, forming 2 moles of ions for every 1 mole of compound dissolved. These ions can move through the solution, carrying charge, so solutions of sodium acetate are good conductors of electricity. Acetic acid, a weak acid, actually ionizes relatively little in water; most of the acetic acid remains as $HC_2H_3O_2$ molecules, which don't conduct electricity, while only a few molecules ionize to form ions. With fewer ions in the solution to carry a charge, solutions of acetic acid are relatively poor conductors of electricity.

7.101 *What are the two common types of bases?*

First, anions are often basic, as they can combine with H^+ ions. The strong bases fall into this category; the strong bases are soluble ionic compounds containing hydroxide ion, OH^- (especially NaOH and KOH). Second, compounds containing a nitrogen atom with a lone pair of electrons and covalent bonds to carbon or hydrogen are also typically bases.

7.103 *What is an amphiprotic substance?*

A substance that is capable of acting both as an acid and as a base, depending on the other species present in a reaction, is called "amphiprotic." The most common type of amphiprotic species is an anion (therefore, a base) with an ionizable hydrogen (therefore, an acid).

7.105 *Explain why solutions of CO_2 in water do not have a pH of 7.00.*

When carbon dioxide dissolves in water, it also reacts to form H_2CO_3, carbonic acid, which ionizes in two steps, producing H^+. Therefore carbon dioxide acts as an acid in water. When CO_2 dissolves in water, the solution becomes more acidic, and the pH goes down.

Solutions to Summary and Challenge Problems

* indicates more challenging problems

*7.107 *How many grams of acetoacetic acid ($HC_4H_5O_3$) are needed to prepare 250 mL of a 0.075 M solution?*

First, we can use the volume and molarity to determine the number of moles of acetoacetic acid (don't round off yet, as there are more steps to do):

$$250\text{ mL} \times \frac{1\text{ L}}{1000\text{ mL}} \times \frac{0.075\text{ mol}}{\text{L}} = 0.01875\text{ mol}$$

To convert moles to grams, we'll need the molar mass of acetoacetic acid:

4 C	=	4 × 12.01	=	48.04 g/mol
6 H	=	6 × 1.008	=	6.048
+ 3 O	=	3 × 16.00	=	48.00
			=	102.09 g/mol

We then use the molar mass as a conversion factor and round to 2 significant figures:

$$\mathbf{0.01875\ mol \times \frac{102.09\ g}{1\ mol} = 1.9\ g}$$

*7.109 a) *HBr is a strong acid. What is the pH of a solution that is made by dissolving 450 mg of HBr in enough water to make 100 mL of solution?*

This problem is somewhat involved. Let's start with the pH and work backward to find a strategy: to calculate the pH, we need the concentration of H^+ in moles/L (usually written as $[H^+]$). Since HBr is a strong acid, it will dissociate completely in aqueous solution:

$HBr(aq) \rightarrow H^+ (aq) + Br^- (aq)$

Therefore the molarity of H^+ ion is equal to the calculated molarity of HBr. To calculate the molarity of the HBr solution (moles/L), we need to convert the 450 mg mass to moles using the molar mass of HBr, 80.91 g/mol (don't round yet, as there are more steps to go):

$$450\text{ mg} \times \frac{1\text{ g}}{1000\text{ mg}} \times \frac{1\text{ mol}}{80.91\text{ g}} = 5.561873 \times 10^{-3}\text{ mol}$$

The volume of the solution is 100 mL, or 0.1 L (one significant figure). The molarity of HBr, and therefore of H^+ in the solution, is then calculated and rounded to 1 significant figure:

$$\frac{5.561873 \times 10^{-3} \text{ mol}}{0.1 \text{ L}} = 5.56 \times 10^{-2} \text{ mol/L} = 5.56 \times 10^{-2} \text{ M} = [\text{HBr}] = [\text{H}^+]$$

Now we can calculate the pH of the solution:
$pH = -\log[H^+] = -\log(5.6 \times 10^{-2}) = 1.3$

b) *What is the pH of a solution that is made by dissolving 525 mg of $Ba(OH)_2$ in enough water to make 75 mL of solution?*

The pH of the solution is calculated from $[H^+]$, the molarity of H^+ ion in the solution, but $Ba(OH)_2$ is a base, so there will be several steps involved here.
$Ba(OH)_2$, an ionic compound containing the hydroxide ion, dissociates completely in water:
$Ba(OH)_2 \text{ (s)} \rightarrow Ba^{2+} \text{ (aq)} + 2\ OH^- \text{ (aq)}$

Since $Ba(OH)_2$ dissociates to form OH^-, it is a strong base. The molarity of OH^- is twice the concentration of $Ba(OH)_2$ in the solution.
To calculate the molarity of $Ba(OH)_2$, we need to calculate moles of $Ba(OH)_2$, using the molar mass, 171.3 g/mol (don't round yet, since there are more steps to do):

$$525 \text{ mg} \times \frac{1 \text{ g}}{1000 \text{ mg}} \times \frac{1 \text{ mol}}{171.3 \text{ g}} = 3.064799 \times 10^{-3} \text{ mol Ba(OH)}_2$$

The volume of the solution is 75 mL or 0.075 L; we can calculate the concentration of $Ba(OH)_2$ and of OH^- (again, don't round yet):

$$\frac{3.064799 \times 10^{-3} \text{ mol Ba(OH)}_2}{0.075 \text{ L}} = \frac{4.0863981 \times 10^{-2} \text{ mol Ba(OH)}_2}{\text{L}}$$

Since each mole of $Ba(OH)_2$ dissociates to produce 2 moles of OH^- in solution:

$$\frac{4.0863981 \times 10^{-2} \text{ mol Ba(OH)}_2}{\text{L}} \times \frac{2 \text{ mol OH}^-}{\text{mol Ba(OH)}_2} = \frac{8.1727963 \times 10^{-2} \text{ mol OH}^-}{\text{L}}$$

With the concentration of OH^-, we can calculate the concentration of H^+ by using the ion product of water:
$[H_3O^+] \times [OH^-] = 1.0 \times 10^{-14}$

Solve the equation above for $[H_3O^-]$, rounding to 2 significant figures:

$$[\text{H}_3\text{O}^+] = \frac{1.0 \times 10^{-14}}{[\text{OH}^-]} = \frac{1.0 \times 10^{-14}}{8.1727963 \times 10^{-2}} = 1.2 \times 10^{-13} \text{ M H}_3\text{O}^+$$

Now we can calculate the pH of the solution:
$pH = -\log[H^+] = -\log(1.2 \times 10^{-13}) = 12.91$
To recap, our overall strategy was:
$[Ba(OH)_2] \rightarrow [OH^-] \rightarrow [H_3O^+] \rightarrow pH$

7.111 *From each of the following pairs of solutions, tell which solution has the higher pH.*

Higher pH = less acidic or more basic.

a) *A solution of NaOH or a solution of HCl.*

Since NaOH is a strong base, the solution of NaOH will have the higher pH.

b) *A 1.0 M solution of HCl or a 0.1 M solution of HCl.*

The 0.1 M solution has a lower concentration of H_3O^+ and is therefore less acidic, and will have the higher pH.

c) *A 1.0 M solution of NaOH or a 0.1 M solution of NaOH*

The 1.0 M solution has the higher concentration of OH^- ions and is therefore more basic, and will have the higher pH.

d) *A solution that contains 6.0×10^{-4} M H_3O^+ or a solution that contains 5.0×10^{-3} M H_3O^+*

The 6.0×10^{-4} M solution has a lower H_3O^+ concentration, is less acidic, and will have a higher pH.

e) *A solution that contains 0.0025 M OH^- or a solution that contains 3.1×10^{-4} M OH^-*

The 0.0025 M solution has a higher OH^- concentration, is more basic, and has a higher pH.

f) *A 0.1 M solution of a strong acid or a 0.1 M solution of a weak acid*

At the same concentration, the solution of the weak acid is less acidic and will have a higher pH.

7.113 *A molecule of malonic acid contains three carbon atoms, four hydrogen atoms, and four oxygen atoms, and the chemical formula of malonic acid is normally written $H_2C_3H_2O_4$. Based on this formula, which of the following ionic compounds probably do not exist?*
$NaHC_3H_2O_4$ $Na_2C_3H_2O_4$ $Na_3C_3HO_4$ $Na_4C_3O_4$

The given formula for the acid, $H_2C_3H_2O_4$, has only two of its H atoms listed at the beginning of the formula; therefore it is diprotic (it has only two ionizable hydrogen atoms). The ions that will exist are $HC_3H_2O_4^-$ (the result of removing one H^+) and $C_3H_2O_4^{2-}$ (the result of removing both H^+), so we can write the formulas for the compounds of these two ions with sodium:

$NaHC_3H_2O_4$ exists **$Na_2C_3H_2O_4$ exists**

The other two compounds are based on ions from which more than two H^+ have been removed. The H atoms listed after C are not acidic and will not be removed from the acid to form anions:

$Na_3C_3HO_4$ does not exist **$Na_4C_3O_4$ does not exist**

7.115 *Only one of the six hydrogen atoms in lactic acid can be removed by a base. Which one is it?*

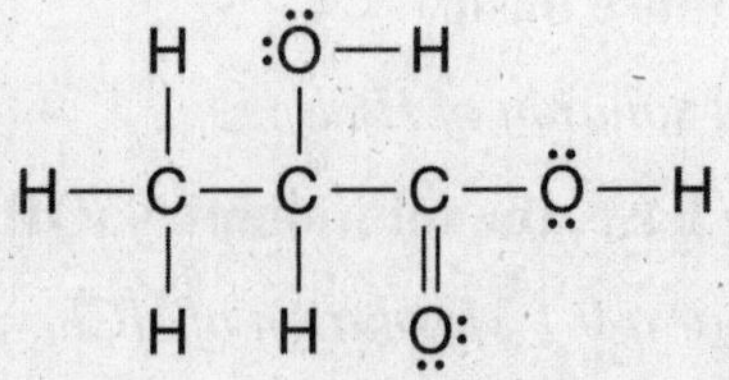

$$\mathrm{H{-}C(H)(H){-}C(H)(\ddot{O}{-}H){-}C({=}\ddot{O}){-}\ddot{O}{-}H}$$

Lactic acid

Only the circled H is likely to be acidic; it fits the general structure of common acids given in Figure 7.5 (central atom, in this case C, attached to an OH group and at least one other O atom.) The other H on an O atom (on the top of the molecule) doesn't have the extended arrangement of another O two bonds away. Hydrogen atoms bonded to C are **never** ionizable.

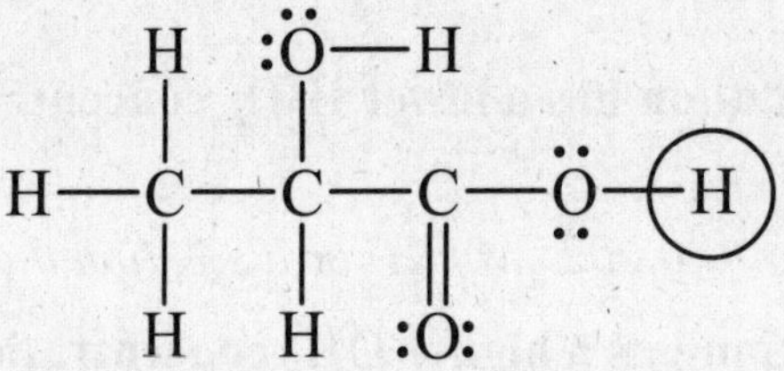

$$\mathrm{H{-}C(H)(H){-}C(H)(\ddot{O}{-}H){-}C({=}\ddot{O}){-}\ddot{O}{-}\boxed{H}}$$

*7.117 *When acetic acid dissolves in water, it reacts with the water as follows:*

$$HC_2H_3O_2(aq) + H_2O(l) \rightleftarrows C_2H_3O_2^-(aq) + H_3O^+(aq)$$

Use le Châtelier's Principle to explain why adding some $NaC_2H_3O_2$ to this solution makes the pH of the solution go up.

$NaC_2H_3O_2$ dissolves in water to form Na^+ ions and $C_2H_3O_2^-$ ions, so it is really just a source of more $C_2H_3O_2^-$. **Adding $C_2H_3O_2^-$ (a product in the reaction) to the equilibrium mixture will cause the reaction to go backward, using up some H_3O^+, making the solution less acidic, and causing an increase in pH.**

7.119 *When solutions of HCN and C_2H_7N are mixed, the following acid–base reaction occurs:*

$$HCN(aq) + C_2H_7N(aq) \rightarrow CN^-(aq) + HC_2H_7N^+(aq)$$

a) *Which reactant is the acid and which is the base?*

HCN is the species that donates H^+, so it is the acid (and CN^- is its conjugate base). C_2H_7N accepts H^+, so it is the base (and $HC_2H_7N^+$ is its conjugate acid).

b) *If you mix equal volumes of 0.1 M HCN and 0.1 M C_2H_7N, the product mixture will have a pH of 10.00. Based on this information, what can you say about the relative strengths of HCN and C_2H_7N?*

Equal amounts of the two substances combine to form a solution that is significantly basic; therefore the base, C_2H_7N, must be stronger than the acid, HCN. If they had equal strengths, we'd expect an equal mixture of the two to have a pH of 7; if the acid were stronger, we'd expect the mixture to be acidic.

7.121 *For each of the following solutions, give the formula of a chemical that you could add to the solution in order to make a buffer.*

A buffer solution always contains a weak acid and its conjugate base (or a weak base and its conjugate acid—which is another way of saying the same thing, since really the two species are a conjugate acid/base pair.) So if we have an acid, we need its conjugate base, and vice versa. If the species needed is an ion, we'll have to determine the formula of a soluble compound of that ion. See Problems 7.65 and 7.67.

a) 0.1 M $HC_7H_5O_2$ (benzoic acid)

We need a soluble compound of the benzoate ion, $C_7H_5O_2^-$; one option is **$NaC_7H_5O_2$.**

b) 0.1 M $NaCHO_2$ (sodium formate)

This time we have the anion, which will act as a base, and we need the conjugate acid, **$HCHO_2$** (formic acid).

c) 0.1 M C_3H_9N (trimethylamine, a weak base)

We need to supply the conjugate acid of trimethylamine: **$HC_3H_9N^+$**. Since it's a cation, we'll have to add an anion to make a soluble ionic compound. One option is **$(HC_3H_9N)Cl$.**

d) 0.1 M K_2HPO_4

The HPO_4^{2-} ion is amphiprotic, so there are two directions we could go here, either of which will make a buffer (though the two buffers would have very different pH values).
If we use HPO_4^{2-} ion as the *base* in the buffer, we need a compound of its conjugate acid, $H_2PO_4^-$ ion. One possible compound is **KH_2PO_4.** (Using Na instead of K will also work, and there are other reasonable answers.)

NOTE:
This is the same combination as the phosphate buffer in intracellular fluid (Section 7.8).

If we use HPO_4^{2-} ion as the *acid* in the buffer, we need a compound of its conjugate base, PO_4^{3-} ion. One possible compound is **K_3PO_4.**

*7.123 *A buffer that contains 0.10 M NaH_2PO_4 and 0.10 M Na_2HPO_4 has a pH of 7.21. If you need to prepare 3.0 liters of this buffer, how many grams of NaH_2PO_4 and how many grams of Na_2HPO_4 will you need?*

The easiest way to do this calculation is to ignore the fact that both of the solutes are in the same solution and that it's a buffer problem—really, this is a Chapter 5 calculation in which the two solutes just happen to be mixed in the same 3.0 L of solution. We can rewrite the question in two parts:

a) *How many grams of NaH_2PO_4 will you need to prepare 3.0 liters of 0.10 M NaH_2PO_4?*

First, add up the molar mass of NaH_2PO_4: 119.98 g/mol. Then replace the M with its component units, mol/L, and we have both of our conversion factors:

$$\mathbf{3.0\,L \times \frac{0.10\,mol}{L} \times \frac{119.98\,g}{mol} = 36\,g\,NaH_2PO_4}$$

b) *How many grams of Na_2HPO_4 will you need to prepare 3.0 liters of 0.10 M Na_2HPO_4?*

The setup is the same an in part (a). Molar mass of Na_2HPO_4: 141.96 g/mol.

$$\mathbf{3.0\,L \times \frac{0.10\,mol}{L} \times \frac{141.96\,g}{mol} = 43\,g\,Na_2HPO_4}$$

*7.125 *A patient who is suffering a bout of severe vomiting loses a substantial amount of acid, because the stomach contents are very acidic. His body moves H_3O^+ from his blood plasma into his stomach to replace the lost acid, but this makes the pH of his blood plasma go up. As a result, the patient's breathing rate changes, to return the plasma pH to the correct value. Does the patient's breathing rate increase, or does it decrease? Explain your answer.*

If the patient's blood plasma pH goes up (for whatever reason), that means it's becoming more basic, and it needs more acid to correct the pH. Since CO_2 acts as an acid in aqueous solutions, the body will retain more CO_2 to bring the pH down to the correct value. Breathing more slowly causes the body to retain more (or expel less) CO_2, making the blood more acidic and decreasing the pH, so the patient's breathing rate will decrease.

Chapter 8

8

Hydrocarbons:
An Introduction to Organic Molecules

Solutions to Section 8.1 Core Problems

8.1 *Why do carbon atoms always share four electron pairs in chemical compounds?*

Carbon atoms have four valence electrons. To achieve an octet of valence electrons, a carbon atom has to get four electrons from other atoms, which means forming four covalent bonds (each bond made of one electron from C and one electron from the other atom).

8.3 *What is a tetrahedral arrangement? Give an example of a compound that has this arrangement of atoms.*

"Tetrahedral arrangement" refers to the three-dimensional arrangement of bonds around a carbon atom that makes four single bonds. Any carbon atom that has covalent bonds to four other atoms will have a tetrahedral arrangement, so some examples are CH_4, CCl_4 and CF_4. See Figure 8.1. (Later in the chapter, you'll be able to provide a lot more examples!)

NOTE:
Strictly speaking, for some other atoms besides carbon, a tetrahedral arrangement doesn't have to be four single bonds but rather four "electron groups"–electron groups meaning single bonds, double bonds, triple bonds or lone pairs. But for our purposes, we're only using the term to describe the four-single-bond arrangement, mostly for carbon atoms. The possible electron arrangements around other atoms besides C, N and O are outside the scope of this course.

8.5 *Which of the following molecules contain at least one tetrahedral arrangement of atoms?*

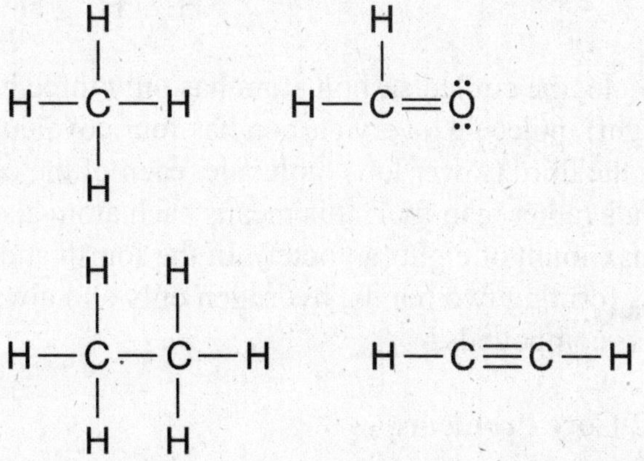

```
    H              H
    |              |
H—C—H         H—C=Ö:
    |
    H

    H   H
    |   |
H—C—C—H       H—C≡C—H
    |   |
    H   H
```

Carbon atoms with four single bonds to other atoms have a tetrahedral arrangement. Carbon atoms with any other bonding pattern are *not* tetrahedral. **The first and third structures have tetrahedral arrangements of atoms.**

8.7 *Using your knowledge of valence electrons and the octet rule, explain why the following compound is unlikely to exist.*

```
    H
    |
H — C — H
```

The carbon atom in this structure only has 6 valence electrons, and therefore does not satisfy the octet rule. Carbon's normal bonding arrangement is to make four covalent bonds.

8.9 *In each of the following molecules, find the atom or atoms (if any) that do not form the correct number of covalent bonds, and draw a circle around each atom.*

```
    H   H   H   H   H              H
    |   |   |   |   |              |
H — C = C — C — C = C — H      H — C — C ≡ C — H
                                   |
                                   H

    H   H   H                  H
    |   |   |                  |
H — C = C — C — H          H — C — H — H
    |   |   |                  |
    H   H   H                  H
```

Carbon atoms normally form a total of four covalent bonds; hydrogen atoms always form exactly one bond in molecular compounds. Look for carbon atoms with more or less than four bonds, and hydrogen atoms with more or less than one bond.

```
    H   H    H    H   H              H     H    H              H
    |   |    |    |   |              |     |    |              |
H — C = C — (C) — C = C — H     H — (C) = (C) — C — H     H — C — (H) — H
                                     |     |    |              |
                                     H     H    H              H
```

In the first molecule, the circled carbon atom has only three bonds rather than four. In the second (upper right) molecule, every carbon has four covalent bonds and every hydrogen atom has one. In the third (lower left) molecule, each of the circled carbon atoms is forming five bonds rather than four; this means each atom also has ten valence electrons rather than the maximum of eight (an octet). In the fourth and last molecule, the circled hydrogen atom is forming two bonds; hydrogen only and always forms one covalent bond in molecular compounds.

Solutions to Section 8.2 Core Problems

8.11 *Which of the following molecules (if any) are alkanes?*

```
    H   H              H   H              H
    |   |              |   |              |
H — C — C — H      H — C = C — H      H — C — Ö — H
    |   |                                 |   ··
    H   H                                 H
```


Alkanes are hydrocarbons (compounds containing *only* C and H) in which all C—C connections are single bonds. If the molecule contains other elements besides C and H, or if there are *any* double or triple bonds between C atoms, the molecule is not an alkane. **Only the first molecule listed is an alkane.** The second molecule contains a double bond (and is therefore an alkene), while the third contains an oxygen atom (alkanes by definition are hydrocarbons).

8.13 *a) Draw the condensed structural formula that corresponds to the following full structural formula:*

```
    H   H   H   H   H   H
    |   |   |   |   |   |
H — C — C — C — C — C — C — H
    |   |   |   |   |   |
    H   H   H   H   H   H
```

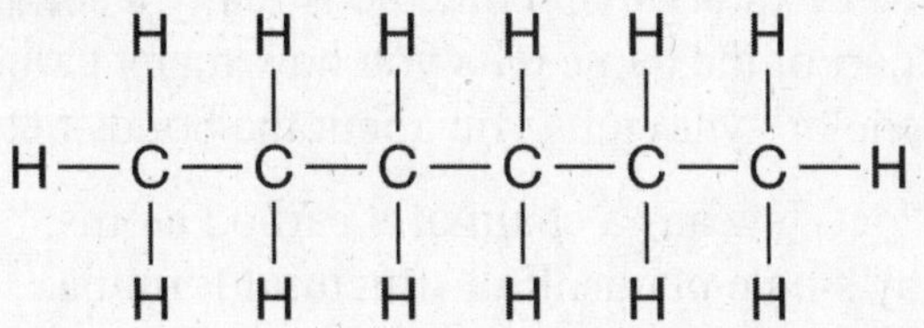

In a condensed structural formula, each carbon atom's H atoms are listed after it with a subscript. So a C atom with three H atoms attached is listed as CH_3, a C atom with two H atoms attached is listed as CH_2, etc. This molecule's condensed structural formula is **$CH_3–CH_2–CH_2–CH_2–CH_2–CH_3$**. Other options:

—Single bonds can be shown but are optional: **$CH_3CH_2CH_2CH_2CH_2CH_3$**

—The formula can be further condensed by taking the repeating CH_2 groups and putting them together in parentheses, with a subscript to show how many of these groups are in a row:

$CH_3(CH_2)_4CH_3$

b) Draw the full structural formula that corresponds to the following condensed structural formula:

$$CH_3-CH_2-CH_2-CH_3$$

A full structural formula shows every bond. To convert a condensed structural formula to a full structural formula, spread the H atoms on each carbon back out again:

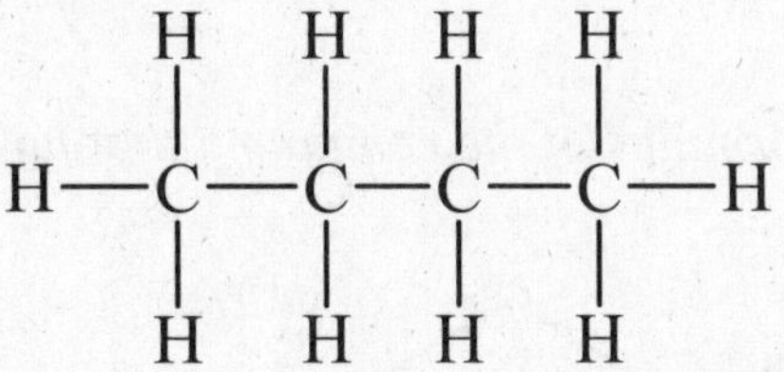

```
    H   H   H   H
    |   |   |   |
H — C — C — C — C — H
    |   |   |   |
    H   H   H   H
```

8.15 *Draw the line structures of the molecules in Problem 8.13.*

A line structure shows only the bonds between carbon atoms, leaving the H atoms to be inferred (the fact that carbon *always* forms four bonds in organic molecules makes this possible.) Remember that "ends & bends" are carbon atoms.

NOTE:

It doesn't matter whether the first bond you draw is sloping up or down.

The molecule in a) has six carbon atoms connected by five bonds:

The molecule in b) has four carbon atoms connected by three bonds:

8.17 *Draw the full and condensed structural formulas of each of the following molecules:*

Remember that the name of an organic molecule is really a set of instructions for drawing the structure. The root part of the name tells you how many carbons in a row to start with, and the ending (-ane, -ene or –yne) tells you about the bonds between them.

a) octane "oct-" means a chain of 8 carbon atoms, "-ane" means they're all by single bonds. Full structural formula:

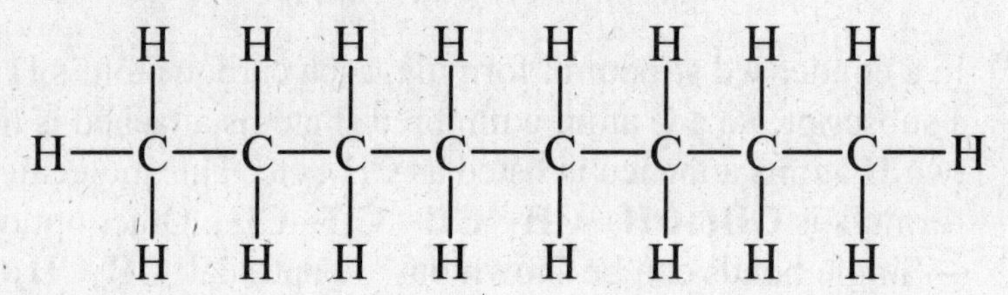

Condensed structural formula:

$\mathbf{CH_3CH_2CH_2CH_2CH_2CH_2CH_2CH_3}$ or $\mathbf{CH_3(CH_2)_6CH_3}$

b) propane "prop-" means a chain of 3 carbon atoms, "-ane" means they're all connected by single bonds. Full structural formula:

H H H
| | |
H—C—C—C—H
| | |
H H H

Condensed structural formula: $\mathbf{CH_3CH_2CH_3}$

8.19 *a) Draw the full and condensed structural formulas of the compound that has the following line structure.*

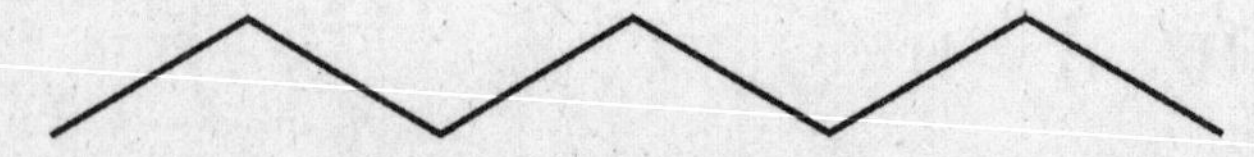

The molecule has seven carbon atoms connected by 6 single bonds. Full structural formula:

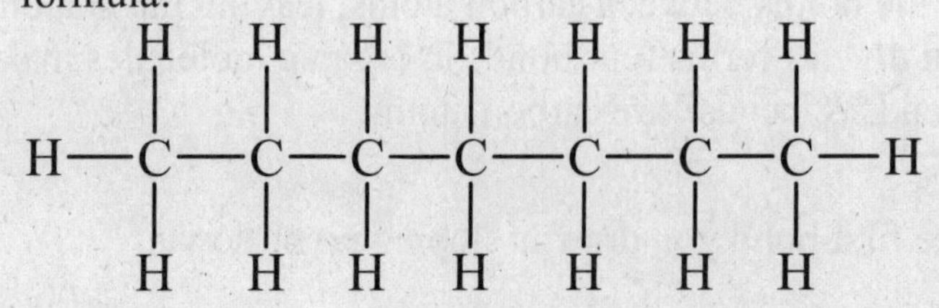

Condensed structural formula: $\mathbf{CH_3CH_2CH_2CH_2CH_2CH_2CH_3}$ or $\mathbf{CH_3(CH_2)_5CH_3}$

b) *Write the molecular formula of this molecule.*

The molecular formula shows atom counts, listing each element only once: $\mathbf{C_7H_{16}}$

c) *Name this molecule.*

Seven carbons = "hept-", all single bonds = "-ane": **heptane**

Solutions to Section 8.3 Core Problems

8.21 *Identify each of the following molecules as a linear alkane, a branched alkane, or a cycloalkane.*

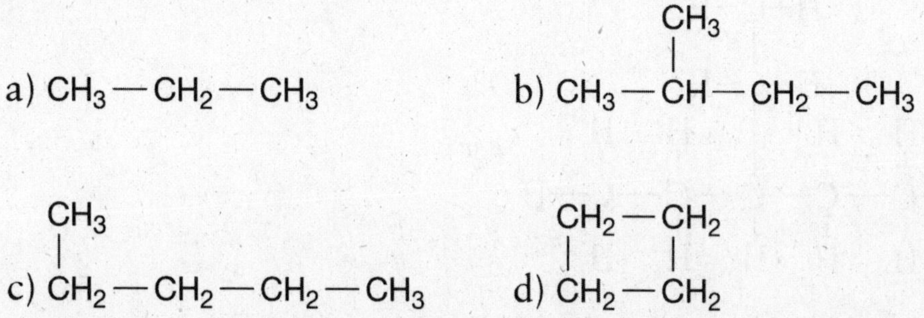

a) $CH_3-CH_2-CH_3$

b)
$$\begin{array}{l} CH_3 \\ | \\ CH_3-CH-CH_2-CH_3 \end{array}$$

c)
$$\begin{array}{l} CH_3 \\ \,| \\ CH_2-CH_2-CH_2-CH_3 \end{array}$$

d)
$$\begin{array}{ccc} CH_2 & - & CH_2 \\ | & & | \\ CH_2 & - & CH_2 \end{array}$$

"Linear" alkanes have all the carbon atoms in one long string, and each carbon atom has a maximum of two other C atoms attached; "branched alkanes" have branching points where one or more carbon atoms have attachments to three or four other carbons; and "cycloalkanes" have the ends of the carbon chain bonded together into a loop (ring).

a) linear alkane **b) branched alkane** **c) linear alkane** **d) cycloalkane**

8.23 *a) Draw the condensed structural formula that corresponds to the following full structural formula:*

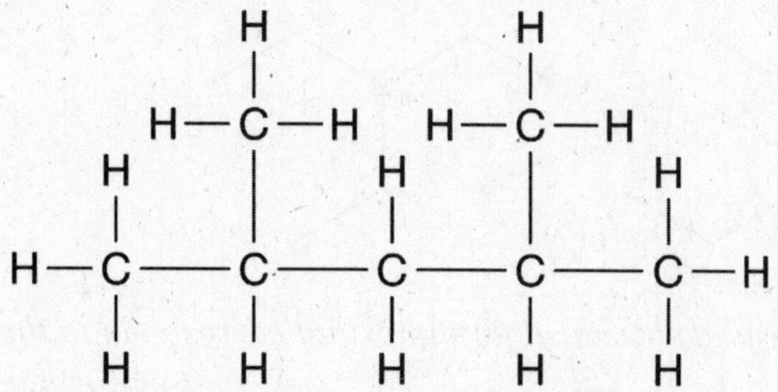

$$\begin{array}{ccccccccccc} & & & & H & & & & H & & \\ & & & & | & & & & | & & \\ & & & H- & C & -H & & H- & C & -H & \\ & & H & & | & & H & & | & & H \\ & & | & & | & & | & & | & & | \\ H & - & C & - & C & - & C & - & C & - & C-H \\ & & | & & | & & | & & | & & | \\ & & H & & H & & H & & H & & H \end{array}$$

In the condensed structure of a branched alkane, it may not be possible to write the condensed structural formula all on one line as with a linear alkane:

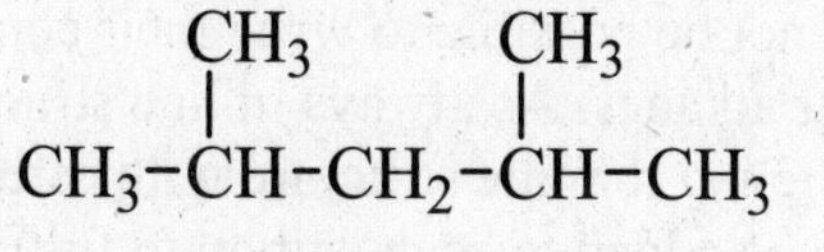

$$\begin{array}{l} CH_3 \qquad\quad CH_3 \\ | \qquad\qquad\quad\;\; | \\ CH_3-CH-CH_2-CH-CH_3 \end{array}$$

The branch alkyl groups can also be written as condensed formulas in parentheses, after the carbon atoms to which they are attached:

$\mathbf{CH_3CH(CH_3)CH_2CH(CH_3)CH_3}$ or $\mathbf{CH_3CH(CH_3)CH_2CH(CH_3)_2}$

b) Draw the full structural formula that corresponds to the following condensed structural formula:

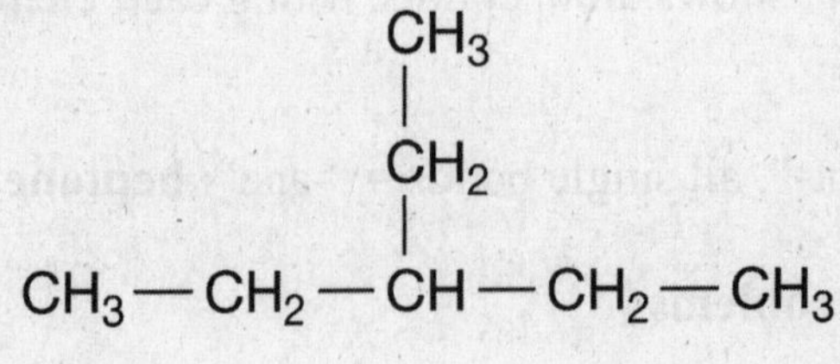

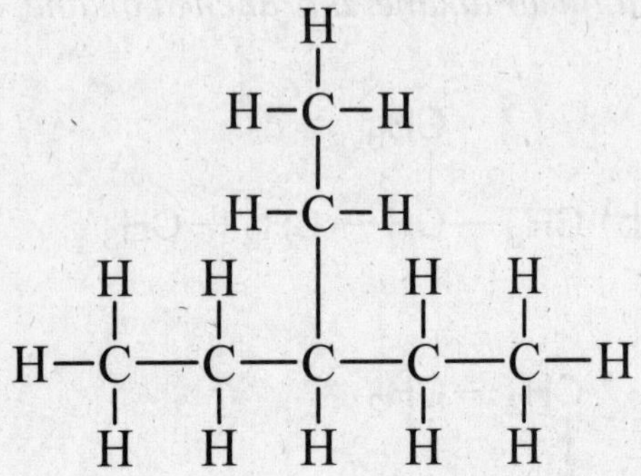

8.25 *Draw the line structures of the molecules in Problem 8.23.*

Line structures show the C–C bonds and leave the carbon and hydrogen atoms to be inferred. The structures can be oriented in any direction, but the lines are usually spread out from each other (for example, if there are three lines from one point, they're 120° apart.)

a) 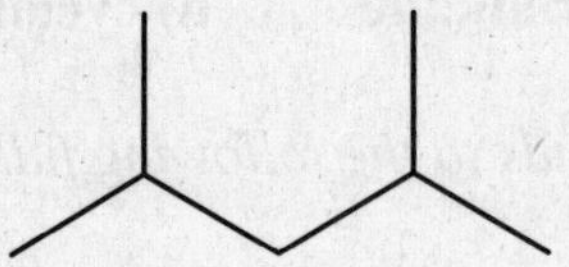 or 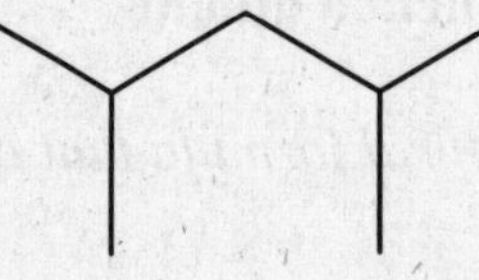

b) 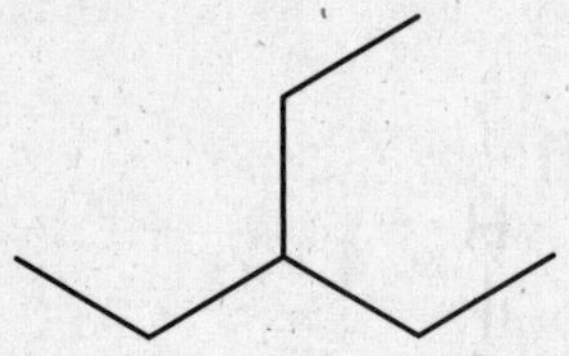 or 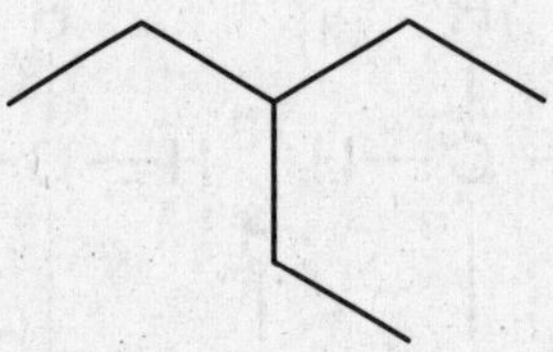

8.27 *Draw the condensed structural formulas that correspond to the following line structures:*

a) 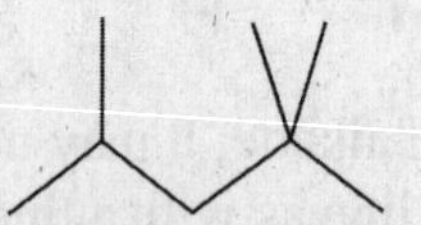 b) 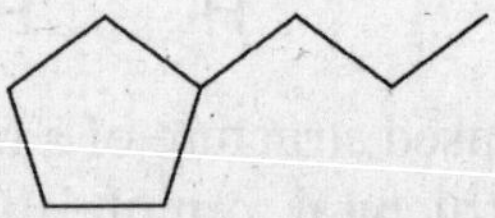

When structures are branched or cyclic, it may not be possible to write their condensed structural formulas all on one line as with linear alkanes. As always in line structures, "ends and bends" are carbon atoms; start by labeling the line structure with carbon atoms, then decide how many hydrogen atoms need to be added to each carbon to fulfill its four bonds.

a)

```
   CH3          CH3
   |            |
CH3-CH-CH2-C-CH3
                |
                CH3
```

b)

```
      CH2
     /   \
  CH2     CH-CH2-CH2-CH3
     \   /
   CH2-CH2
```

There's another option for the first structure: in a branched alkane, branch alkyl groups can also be written as condensed formulas in parentheses, after the carbon atoms to which they are attached:

a) **$CH_3CH(CH_3)CH_2C(CH_3)_3$**

These structures are harder to read and usually are only used in cases where there's a big advantage to having the entire structure in one line (for example, in a text document where the structure can be typed out like this but drawing a structure would be much more inconvenient.) This usually does not work well with cycloalkanes.

8.29 *What, if anything, is wrong with the following condensed structural formula?*

```
                CH3
                |
CH3—CH2—CH—CH2—CH2—CH3
```

The carbon atom where the methyl group is attached should only have one hydrogen atom; as it is, this carbon is forming 5 bonds instead of 4.

8.31 *Tell whether each of the following pairs of molecules are isomers. If they are not, explain why not.*

Isomers have different structures, but the same molecular formula. If two molecules have the same structure, they are not isomers but are in fact the same molecule. If two molecules have different molecular formulas, then they are not isomers (and may have no particular relationship to each other at all).

a) $CH_3—CH_2—CH_2—CH_2—CH_2—CH_3$ and

```
      CH3  CH3
      |    |
CH3—CH—CH—CH3
```

b)

```
CH3
|
CH2—CH2—CH2—CH3
```

and $CH_3—CH_2—CH_2—CH_2—CH_3$

c) $CH_3—CH_2—CH_2—CH_2—CH_2—CH_3$ and

```
    CH2-CH2
   /       \
CH2         CH2
   \       /
    CH2-CH2
```

a) **The molecules are isomers.** They have different structures but the same molecular formula: C_6H_{14}.

b) **The molecules are NOT isomers. They are the same molecule (longest chain is 5 carbons, no branches, and all C–C bonds are single).**

c) **The molecules are NOT isomers. They have different structures and different molecular formula: C_6H_{14} and C_6H_{12}.**

Solutions to Section 8.4 Core Problems

8.33 *What is the name of the following alkyl group?*

$$CH_3–CH_2–CH_2–$$

The group has three carbons, so its root is **prop-**, and all alkyl groups get the ending **–yl** when they are branches of a larger molecule, so the name of the alkyl group is **propyl**.

8.35 *Name the following compounds using the IUPAC system.*

The steps are always the same:

(1) Identify the principal chain. (For a branched alkane, this will be the longest continuous chain you can find; for a cycloalkane, the ring is the principle chain.) Identify and name the alkyl groups attached to the principle chain.

(2) Number the carbon atoms in the principal chain, starting from the end that is closest to an alkyl group. (For a cycloalkane with one alkyl group, no number is needed–the alkyl group is automatically at Carbon #1 and is therefore not assigned a number.)

(3) Assemble the complete name by writing the names of the branches in front of the name of the principal chain, and giving a number as the "address" of each alkyl group.

a) $CH_3—CH_2—CH(CH_3)—CH_2—CH_3$

b) $CH_3—CH_2—CH_2—CH_2—CH(CH(CH_3)—CH_3)—CH_2—CH_2—CH_3$

c) $CH_3—CH(CH_3)—CH_2—CH(CH_2—CH_3)—CH_2—CH_2—CH_3$

d) $CH_3—CH(CH_3)—CH(CH_3)—CH_2—CH_2—CH_3$

$$\begin{array}{c} \quad CH_3 \quad CH_3 \\ \text{e) } CH_3-\overset{|}{\underset{|}{C}}-\overset{|}{CH}-CH_2-CH_3 \\ \quad CH_3 \end{array}$$

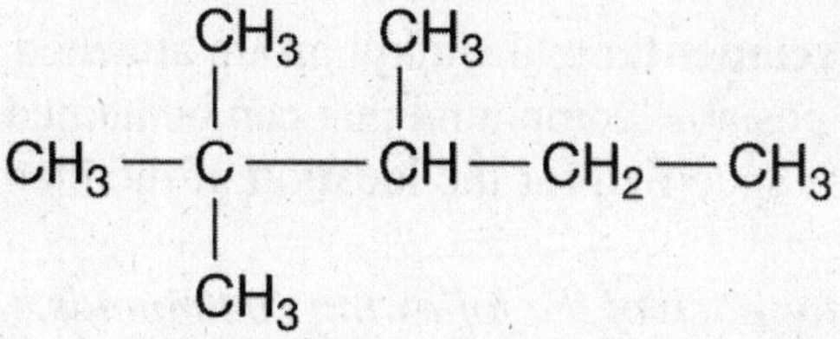

$$\begin{array}{c} \quad CH_2-CH_3 \qquad\qquad CH_3 \\ \text{f) } CH_3-CH_2-\overset{|}{\underset{|}{C}}-CH_2-CH_2-\overset{|}{CH}-CH_3 \\ \quad CH_2-CH_3 \end{array}$$

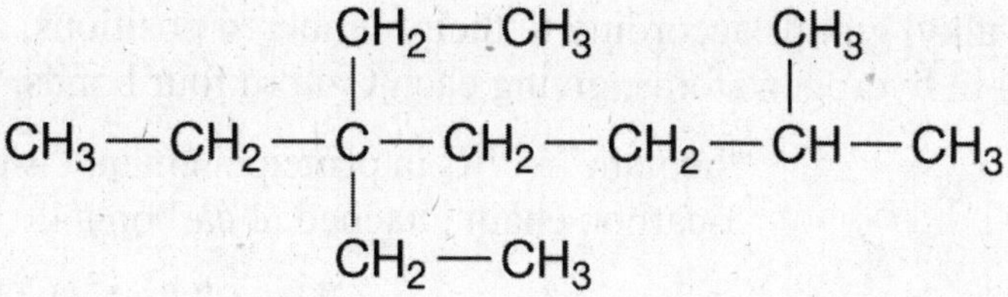

g)

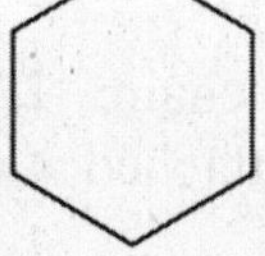

h) 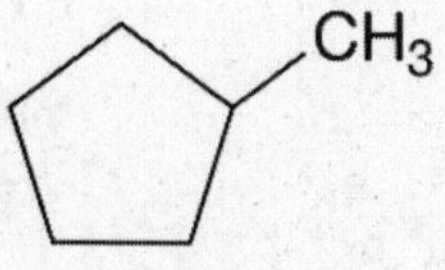 CH_3

a) The principle chain has 5 carbons (pent-), the alkyl group has one carbon (methyl-), and the methyl group is on carbon #3: **3-methylpentane.**

b) The longest chain has 8 carbons (**octane**), and the group attached to carbon #4 is a three-carbon group attached by its center carbon atom (**isopropyl**): **4-isopropyloctane.**

c) The longest chain has 7 carbons (**heptane**), with an ethyl group on carbon #4 and a methyl group on carbon #2. The two alkyl groups are listed in alphabetical order in the name of the compound: **4-ethyl-2-methylheptane.**

d) No tricks here! The longest chain is 6 carbons (**hexane**), numbering is from left to right (because the left end is nearest to the first alkyl group), and there are two methyl groups, on carbons 2 and 3: **2,3-dimethylhexane**.

e) The longest chain is 5 carbons (**pentane**), numbering is from left to right (because the left end is nearest to the first alkyl group), and there are three methyl groups, two on carbon 2 and one on carbon 3. Since there are a total of three methyl groups, we need to use the prefix "tri-". Just remember that you have to give an address for every alkyl group, even if two alkyl groups are attached at the same carbon: **2,2,3-trimethylpentane**.

f) The longest chain is 7 carbons (**heptane**), numbering is from right to left (because the right end is nearest to the first alkyl group), and there are two ethyl groups (requiring the addition of the prefix "di-") on carbon 5 and one methyl group on carbon 2: **5,5-diethyl-2-methylheptane**.

g) A 6-membered ring; the ring structure makes this a "cyclo-" compound, and the fact that it has 6 carbon atoms makes it "hex-". There are no alkyl groups attached: **cyclohexane**.

h) Start with the 5-membered ring, **cyclopentane**; the alkyl group attached is a methyl group, and since there is only one possible compound that can be named **methylcyclopentane**, no number is specified for the location of the methyl group.

8.37 *Draw condensed structural formulas for each of the following compounds:*

For each of these, remember to think of the name as a set of instructions for drawing the molecule. Go to the end of the name first, because that tells you the length of the principle chain; add the alkyl groups according to their numbered positions, and wait until the last thing to fill in hydrogen atoms, giving each C atom four bonds.

a) 4-propylheptane "heptane" = 7 C in principal chain; "4-propyl" = a 3-carbon chain attached at carbon #4:

```
                 CH2-CH2-CH3
                 |
CH3-CH2-CH2-CH-CH2-CH2-CH3
```

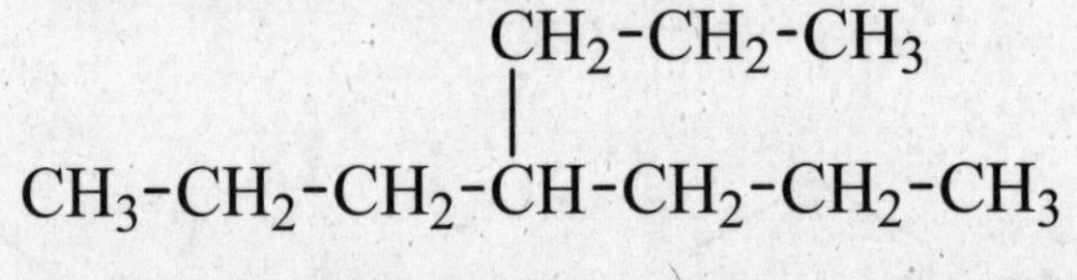

b) 2,2-dimethylpentane "pentane" = 5 C in principal chain; "2,2-dimethyl" = two methyl groups, both attached at carbon #2:

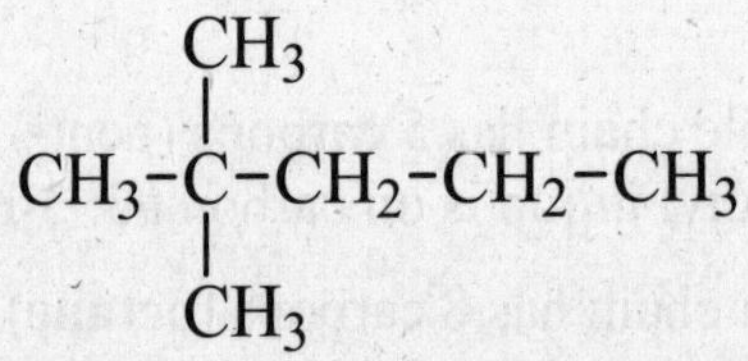

```
     CH3
     |
CH3-C-CH2-CH2-CH3
     |
     CH3
```

c) 3-ethyl-4-propylnonane "nonane" = 9 C in principal chain; "3-ethyl" = a two-carbon chain attached at carbon #3; "4-propyl" = a 3-carbon chain attached at carbon #4:

```
                CH2-CH2-CH3
                |
CH3-CH2-CH-CH-CH2-CH2-CH2-CH2-CH3
            |
            CH2-CH3
```

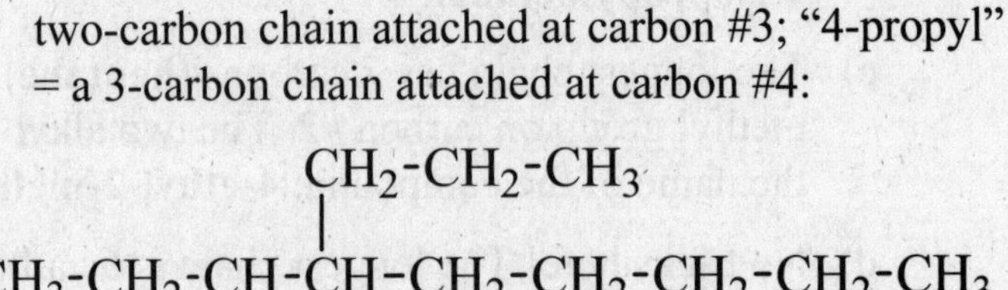

d) 5-butyl-2,3,4-trimethyldecane "decane" = 10 C in principal chain; "5-butyl" = a four-carbon chain attached at carbon #5; "2,3,4-trimethyl" = three methyl groups attached at carbons #2, 3 and 4:

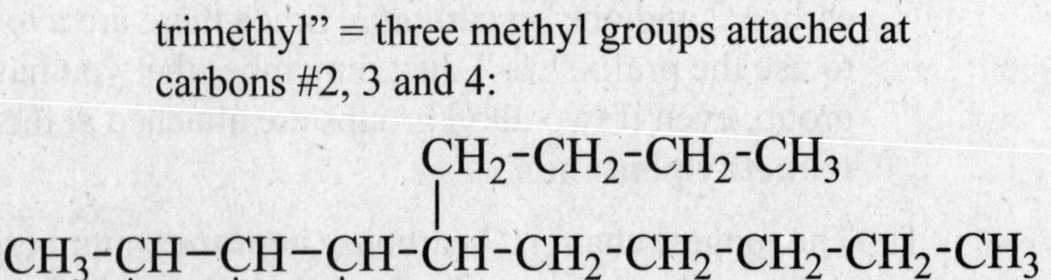

```
                    CH2-CH2-CH2-CH3
                    |
CH3-CH-CH-CH-CH-CH2-CH2-CH2-CH2-CH3
    |   |   |
    CH3 CH3 CH3
```


e) cyclopentane "pentane" = 5 C, "cyclo" = in a ring:

```
        CH2
      /     \
   CH2       CH2
     \       /
     CH2-CH2
```

f) isopropylcyclobutane "butane" = 4 C, "cyclo" = in a ring, "isopropyl" = a 3-carbon chain attached by the middle carbon:

```
CH2−CH−CH−CH3
 |    |   |
CH2−CH2  CH3
```

8.39 *Are 3-ethylpentane and 2-methylhexane isomers? Explain why or why not.*

Draw both structures:

The two molecules have different structures but the same molecular formula (C_7H_{16}), so they are isomers.

Solutions to Section 8.5 Core Problems

8.41 *Which of the following molecules contain a functional group? Circle the functional group (if any) in each molecule.*

```
       CH3                  OH
        |                   |
CH3 — CH — CH3       CH3 — CH — CH3

                            CH2
       CH2               /      \
       ||             CH2        CH2
CH3 — C — CH3           \        /
                         CH2−CH2
```

Functional groups are pretty much anything that's not alkane. C–C single-bonded parts of a molecule and C–H bonds are not functional groups, but any parts of a molecule that have double or triple bonds, or contain other elements, are functional groups.

```
       CH3               (OH)                 (CH2)                  CH2
        |                  |                    ||                  /    \
CH3 — CH — CH3     CH3 — (CH) — CH3    CH3 — (C) — CH3          CH2      CH2
                                                                  \      /
                                                                  CH2−CH2
alkane–no functional group    alcohol         alkene        alkane–no functional group
```

8.43 *The following molecule is called 1-butene.*

$CH_2{=}CH{-}CH_2{-}CH_3$

Which of the following compounds should show similar chemical behavior to 1-butene, based on their functional groups?

We would expect other compounds with C=C double bonds (alkene functional groups) to show similar chemical behavior.

a) $O{=}CH{-}CH_2{-}CH_3$ — This molecule has a C=O double bond, not a C=C double bond, and is **not** expected to have similar c hemistry to 1-butene.

b) $NH{=}CH{-}CH_2{-}CH_3$ — This molecule has a C=N double bond, not a C=C double bond, and is **not** expected to have similar chemistry to 1-butene.

c) $CH_3{-}CH{=}CH{-}CH_2{-}CH_3$ — **This molecule has a C=C double bond, and is expected to have similar chemistry to 1-butene.**

d) $CH_2{=}CH{-}CH_3$ — **This molecule has a C=C double bond, and is expected to have similar chemistry to 1-butene.**

Solutions to Section 8.6 Core Problems

8.45 *What structural feature is present in an alkene?*

The word "alkene" or the ending –ene in a molecule name indicates a C=C double bond in the molecule.

8.47 *Classify each of the following molecules as an alkane, an alkene, or an alkyne:*

a) 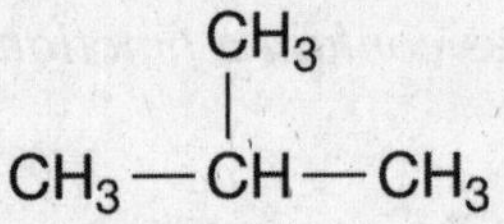 $CH_3{-}CH(CH_3){-}CH_3$

b) 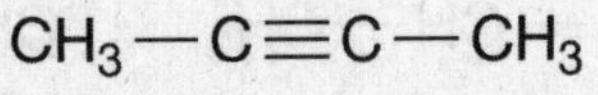 $CH_3{-}C{\equiv}C{-}CH_3$

c) $CH_3{-}CH(CH_3){-}CH{=}CH_2$

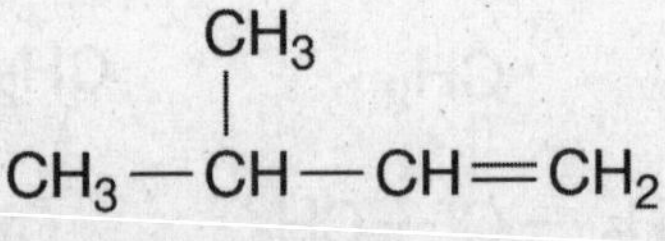

If all C–C bonds are single, the molecule is an alkane; if there is one or more C=C, the molecule is an alkene; and if there is one or more C≡C, the molecule is an alkyne.

a) **alkane** (all C–C bonds are single)

b) **alkyne** (contains C≡C)

c) **alkene** (contains C=C)

8.49 *Draw a full structure of perchloroethylene that shows the actual arrangement of the atoms.*

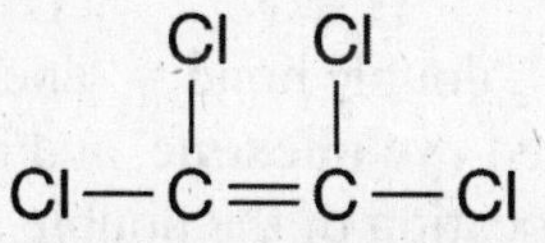

$$Cl-C(Cl)=C(Cl)-Cl$$

Perchloroethylene

Carbon atoms whose bonding arrangement is 1 double bond and 2 single bonds have the three bonded atoms in the same plane at 120° from each other.

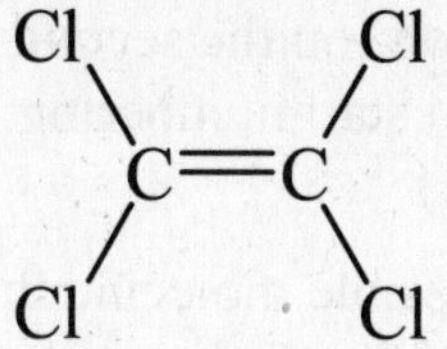

$$(Cl)_2C=C(Cl)_2$$

8.51 *Name the following compounds, using the IUPAC system.*

Start the same way as for alkanes; then, if a double or triple bond is present, change the –ane ending to –ene or –yne as appropriate, and decide whether you need to give the location of the alkene or alkyne by numbering the bonds. (Remember that while we number the carbon atoms in the principal chain to give the location of alkyl groups, we number the bonds between carbon atoms in the principal chain to give the location of alkene or alkyne functional groups.)

a) $CH_2=CH-CH_2-CH_2-CH_2-CH_3$

b) 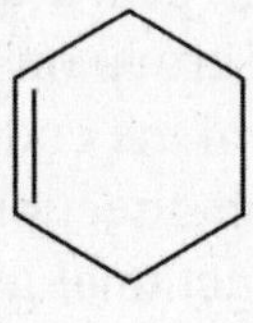 (cyclohexene)

c) $CH_3-CH_2-CH_2-CH_2-C\equiv C-CH_3$

d) $HC\equiv C-CH_3$

e) $CH_3-CH(CH_3)-CH_2-CH=CH-CH_2-CH_3$

f) 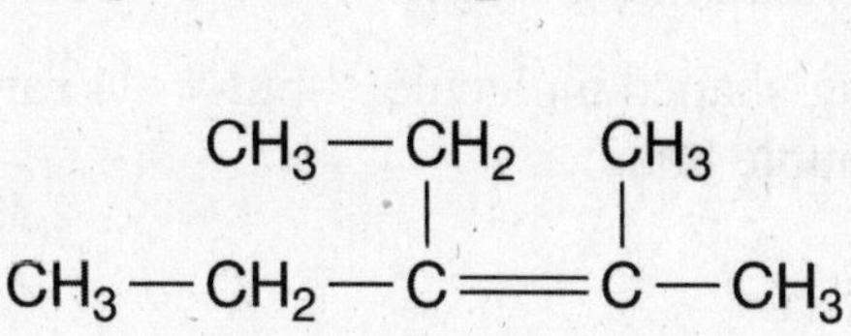 $CH_3-CH_2-C(CH_2-CH_3)=C(CH_3)-CH_3$

a) 6 C = "hex"; double bond = "ene." Since there is more than one possible molecule that could be named "hexene," we have to specify the location of the double bond; since the double bond is the first bond in the molecule, it gets the number 1: **1-**

hexene. (Remember to start numbering from the end nearest the functional group, which in this case is an alkene–this molecule is numbered left to right.)

b) 6 C = "hex"; bent into ring = "cyclo"; double bond = "ene". Since there is only one possible molecule that could be named **cyclohexene**, and because there are no substituents, we **do not** specify the location of the double bond. (Regardless of where you put the double bond, you get the same molecule–try it.)

c) 7 C = "hept"; triple bond = "yne"; and since there is more than one possible molecule that could be named "heptyne," we have to specify the location (on the second bond of the molecule) of the triple bond: **2-heptyne**. (Remember to start numbering from the end nearest the functional group–in this case, right to left.)

d) 3 C = "prop"; triple bond = "yne". Since there is only one possible molecule that could be named "**propyne**," we **do not** specify the location of the triple bond. (Either place you put the triple bond becomes bond #1.)

e) 7 C = "hept"; double bond = "ene"; and since there is more than one possible molecule that could be named "heptene," we have to specify the location of the double bond on the third bond of the molecule: "3-heptene." Then there's a methyl group on carbon #6: **6-methyl-3-heptene.**

NOTE:
Remember to start numbering from the end nearest the functional group–in this case, right to left. Once we decide which direction to number the molecule, we keep the same numbering for both functional groups and substituents.

f) Longest carbon chain that includes the double bond is 5 carbons, "pent"; double bond on bond #2 = "2-pentene"; methyl group on carbon #2 and ethyl group on carbon #3, **3-ethyl-2-methyl-2-pentene**. (Remember to start numbering from the end nearest the functional group–in this case, right to left. Once we choose which direction to number the molecule, we keep the same numbering system in naming functional groups and substituent groups. Count the bonds for the location of the double bond, count the atoms for the locations of the alkyl groups, and list the alkyl groups in alphabetical order when finally assembling the name.)

8.53 *Draw condensed structural formulas for the following molecules:*

Remember to think of the name as a set of instructions for drawing the molecule. Go to the end of the name first, because that tells you the length and bonding (–ane, –ene or –yne) of the principle chain; add the alkyl groups according to their numbered positions, and wait until the last thing to fill in hydrogen atoms, giving each C atom four bonds.

a) cyclobutene "cyclo-" = ring-shaped molecule; "-but-" = 4 carbons; "-ene" = one double bond:

$$\begin{array}{ccc} CH_2 & - & CH_2 \\ | & & \| \\ CH_2 & - & CH \end{array}$$

b) 3-octene — "-oct-" = 8 carbons; "3- -ene" = one double bond, on the third (#3) bond of the molecule:

$$CH_3\text{-}CH_2\text{-}CH{=}CH\text{-}CH_2\text{-}CH_2\text{-}CH_2\text{-}CH_3$$

c) 1-hexyne — "-hex-" = 6 carbons; "1- -yne" = one triple bond, on the f irst (#1) bond of the molecule:

$$CH{\equiv}C\text{-}CH_2\text{-}CH_2\text{-}CH_2\text{-}CH_3$$

d) acetylene — This is a common name for ethyne ("eth-" = 2 carbons; "-yne" = one triple bond):

$$CH{\equiv}CH$$

e) 3-ethyl-1-hexene — "-hex-" = 6 carbons; "1- -ene" = one double bond, on the first (#1) bond of the molecule; "3-ethyl-" = a 2-carbon ethyl group on the third (#3) carbon of the molecule:

$$\begin{array}{l} \qquad\qquad\quad CH_2\text{-}CH_3 \\ \qquad\qquad\quad | \\ CH_2{=}CH\text{-}CH\text{-}CH_2\text{-}CH_2\text{-}CH_3 \end{array}$$

f) 2,2-dimethyl-3-octyne — "-oct-" = 8 carbons; "3- -yne" = one triple bond, on the third (#3) bond of the molecule; "2,2-dimethyl-" = two methyl ($-CH_3$) groups, both on carbon #2:

$$\begin{array}{l} \qquad\; CH_3 \\ \qquad\;\; | \\ CH_3\text{-}C\text{-}C{\equiv}C\text{-}CH_2\text{-}CH_2\text{-}CH_2\text{-}CH_3 \\ \qquad\;\; | \\ \qquad\; CH_3 \end{array}$$

8.55 *Draw the line structure of 4-methyl-1-pentene.*

Look at the parts of the name: "-pent-" means the molecule has a 5-carbon chain; "1- -ene" means that the first bond (bond #1) is a double bond; "4-methyl-" means that there is a methyl group on bond #4.

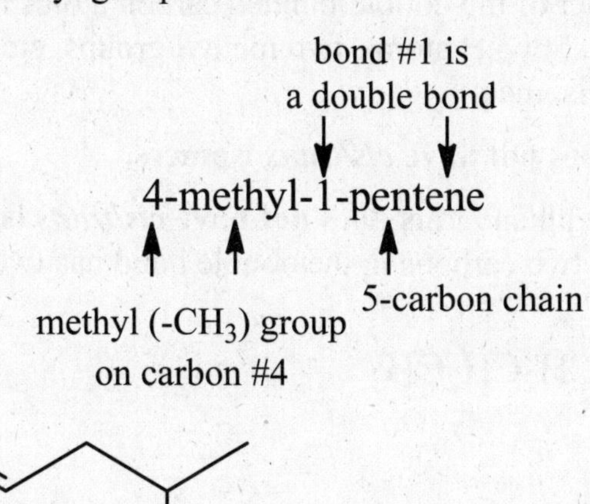

Remember that in deciding which way to number the chain, alkene & alkyne groups take precedence over alkyl groups, and once you decide which end to number from, you have to stick with it.

8.57 *Draw the condensed structural formula that corresponds to the following line structure:*

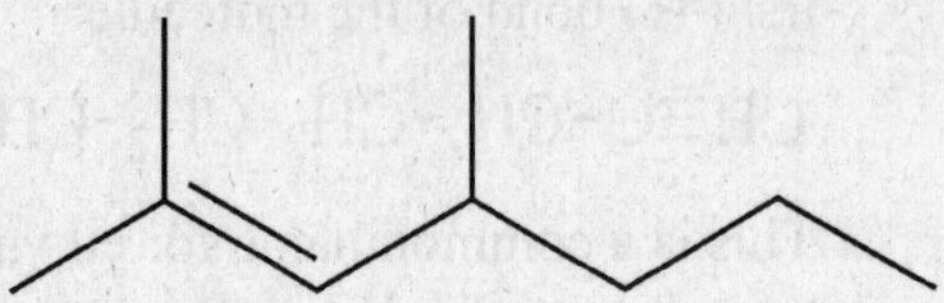

Remember that ends and bends are carbon atoms. Once you have identified all the carbon atoms, fill in with H to complete the 4 bonds on each carbon.

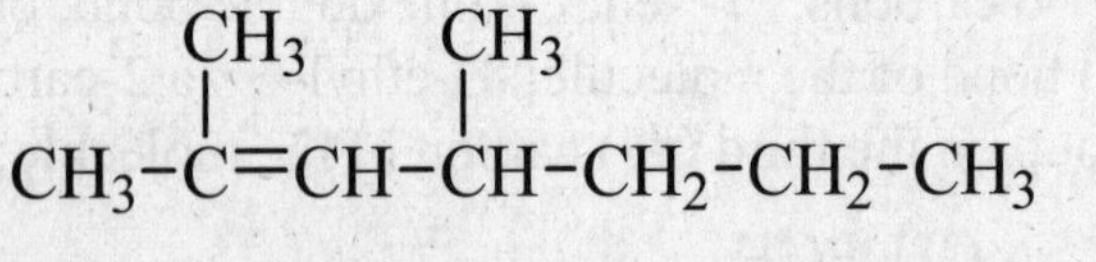

$$CH_3-\overset{\overset{CH_3}{|}}{C}=CH-\overset{\overset{CH_3}{|}}{CH}-CH_2-CH_2-CH_3$$

8.59 *What (if anything) is wrong with the following condensed structural formula?*

$$CH_3-\overset{\overset{CH_3}{|}}{CH}=CH-CH_3$$

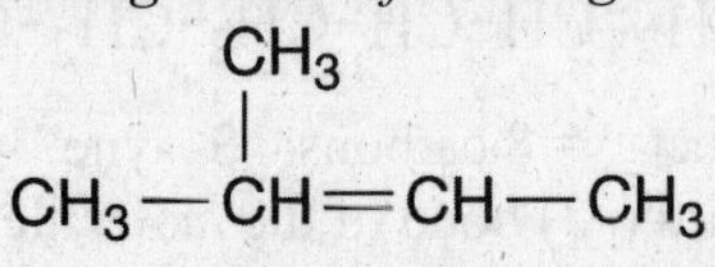

The second carbon atom in the chain has 5 bonds instead of only 4. It shouldn't have an H atom.

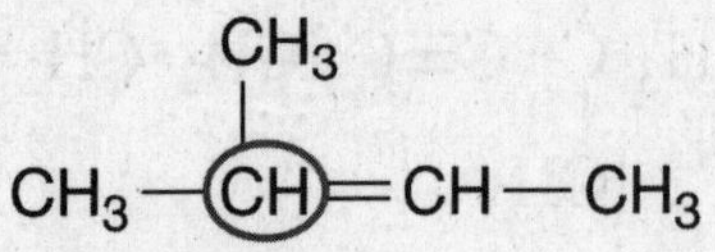

$$CH_3-\overset{\overset{CH_3}{|}}{\boxed{CH}}=CH-CH_3$$

Solutions to Section 8.7 Core Problems

8.61 *Which of the following have* cis *and* trans *isomeric forms?*

Cis/trans isomerism is only a consideration in alkenes, not in alkynes or alkanes. Not all alkenes have *cis/trans* isomers; if either of the double-bonded carbon atoms has two of the same groups attached (for example, two H atoms, two methyl groups, etc.), then the molecule does not have *cis* and *trans* isomers.

a) hexane is an alkane and **does not have *cis/trans* isomers.**

b) 1-hexene even though it's an alkene, **this does not have *cis/trans* isomers,** because one of the two carbons in the double bond has two H atoms attached:

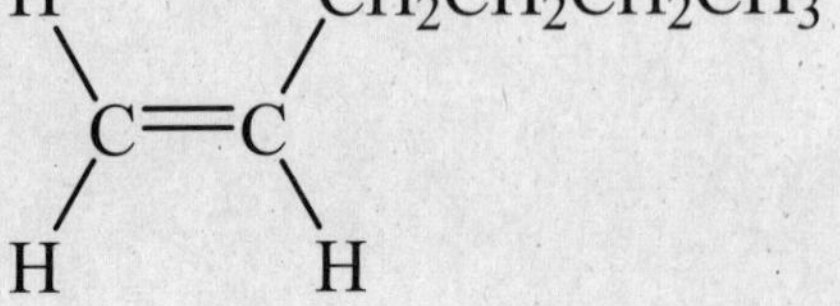

$$(H)(H)C=C(H)(CH_2CH_2CH_2CH_3)$$

c) 3-hexene **does have *cis* and *trans* forms, shown below:**

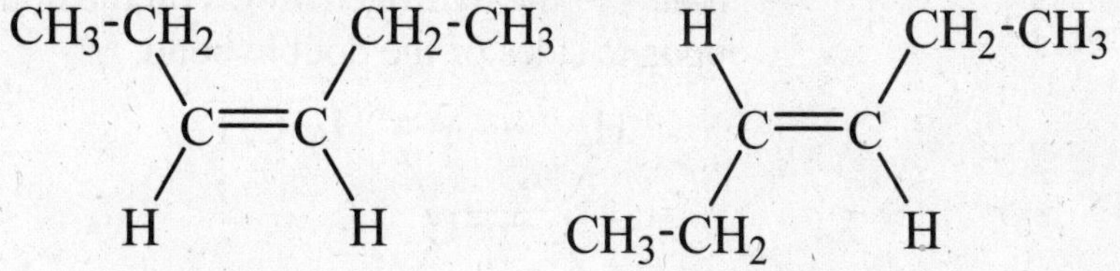

```
CH3-CH2        CH2-CH3      H              CH2-CH3
      \        /              \            /
       C====C                  C====C
      /        \              /            \
     H          H        CH3-CH2            H
           cis                      trans
```

d) 1-hexyne is an alkyne and **does not have *cis/trans* isomers.**

e) 3-hexyne is an alkyne and **does not have *cis/trans* isomers.**

8.63 *Name the following compounds, using the IUPAC system. Be sure to specify whether each molecule is the cis or the trans form.*

a) 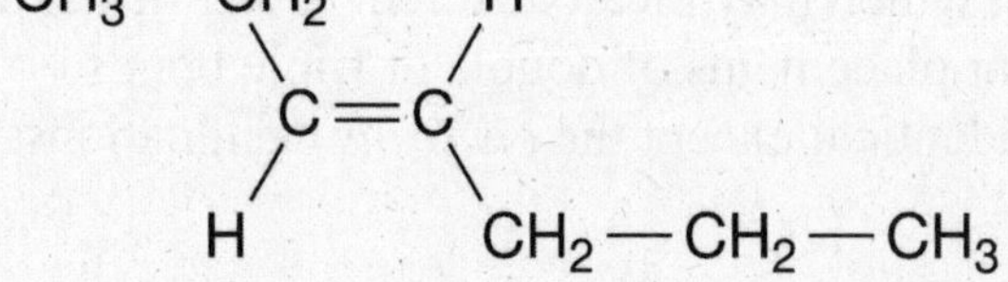
```
CH3—CH2     H
       \   /
        C=C
       /   \
      H     CH2—CH2—CH3
```

b) 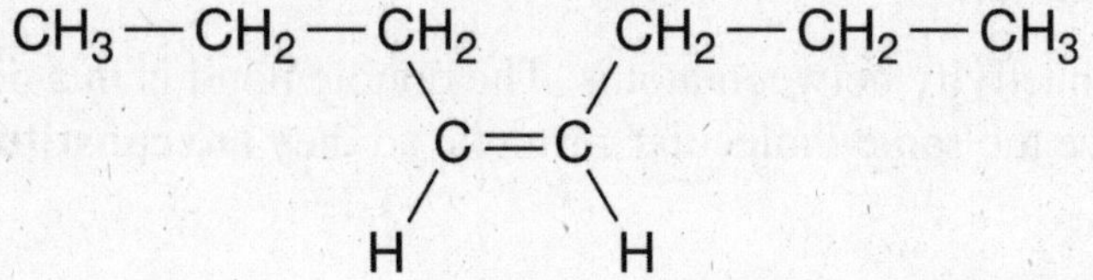
```
CH3—CH2—CH2     CH2—CH2—CH3
            \   /
             C=C
            /   \
           H     H
```

a) The longest chain is 7 carbons, "hept-"; the third bond (numbering from the left, the end closest to the double bond) is double, "3- -ene"; the double bond is a *trans* alkene; and there are no alkyl groups. The name of the molecule is ***trans*-3-heptene.**

b) The longest chain is 8 carbons, "oct-"; the fourth bond (numbering from either end, because it doesn't matter in this molecule) is a double bond, "4- -ene"; the double bond is a *cis* alkene; and there are no alkyl groups. The name of the molecule is ***cis*-4-octene.**

8.65 *Draw condensed structural formulas for the following molecules. Be sure to show the* cis–trans *geometry clearly in your structure.*

a) cis-*2-heptene* "-hept-" = 7-carbon chain; "2- -ene" = double bond on bond #2; "cis-" = the hydrogen atoms on the double-bond carbons are on the same side of the double bond.

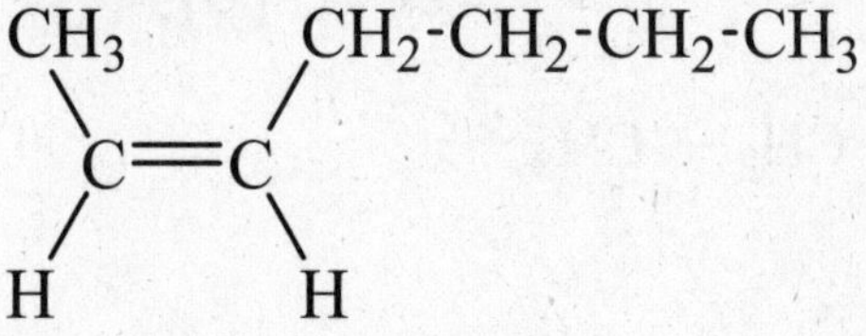

```
CH3        CH2-CH2-CH2-CH3
   \       /
    C====C
   /       \
  H         H
```

b) trans-*3-hexene* "-hex-" = 6-carbon chain; "3- -ene" = double bond on bond #3; "trans-" = the hydrogen atoms on the double-bond carbons are on opposite sides of the double bond.

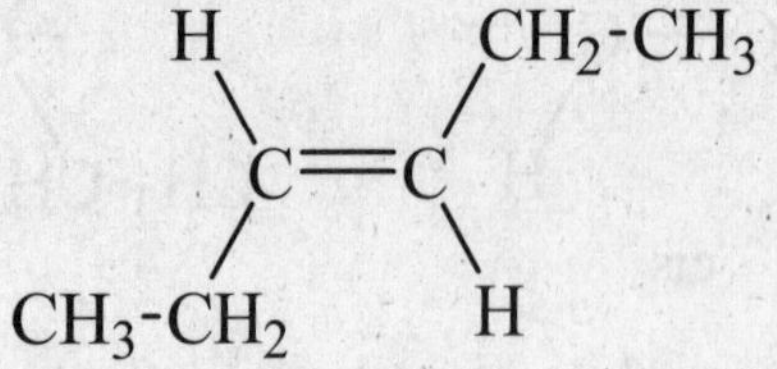

H, CH_2-CH_3 / C=C / CH_3-CH_2, H

8.67 *Which of the following pairs of molecules are constitutional isomers, which are stereoisomers, and which are not isomers?*

You always have the option of drawing the two molecules to decide whether they are isomers or not, but often we can figure this out from the names. Constitutional isomers have different arrangements of atoms, which may mean different substituents, different placements of substituents, or different placements of double or triple bonds. Stereoisomers must have everything identical except the *cis/trans* designations.

a) cis-*3-hexene and* trans-*3-hexene*

Everything is the same except the *cis* and *trans* designations; these are **stereoisomers.**

b) cis-*3-hexene and* trans-*2-hexene*

These have different connectivity between atoms. The double bond is in a different place. However, they have the same molecular formula, so they are **constitutional isomers.**

c) cis-*3-hexene and 1-hexene*

These have different connectivity between atoms. The double bond is in a different place. However, they have the same molecular formula, so they are **constitutional isomers.**

d) cis-*3-hexene and cyclohexene*

These have different molecular formulas (C_6H_{12} and C_6H_{10}, so they are **not isomers.**

Solutions to Section 8.8 Core Problems

8.69 *Which of the following compounds (if any) contain an aromatic ring?*

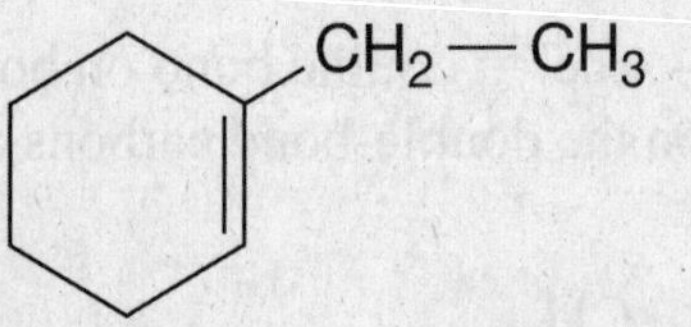

CH_2-CH_3

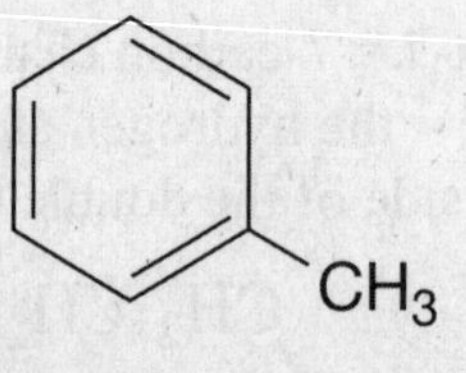

CH_3

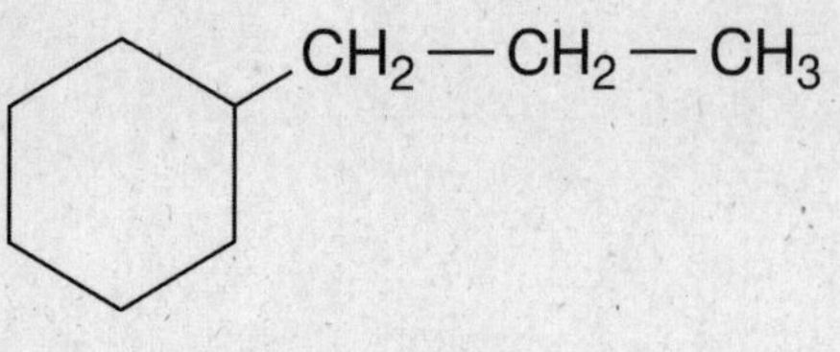

$CH_2-CH_2-CH_3$

For our purposes, only 6-membered rings with an alternating pattern of three double bonds are considered aromatic. Any other arrangement is not aromatic.

The first molecule has only one double bond and is **not aromatic.** The second molecule has the correct arrangement and **does contain an aromatic ring.** The third molecule just has a cyclohexane ring with no double bonds–**not aromatic.**

8.71 *Draw the structures of the following molecules:*

a) *toluene* — There is no way around memorizing this if your instructor wants you to know it, because the name does not give instructions for the structure. Toluene is the common name for methylbenzene, a benzene ring with a methyl group attached:

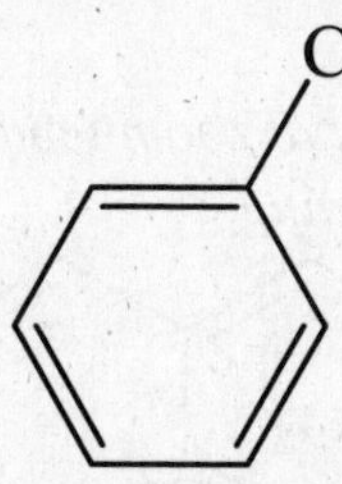

CH_3

b) *propylbenzene* — is a benzene ring with a propyl group attached:

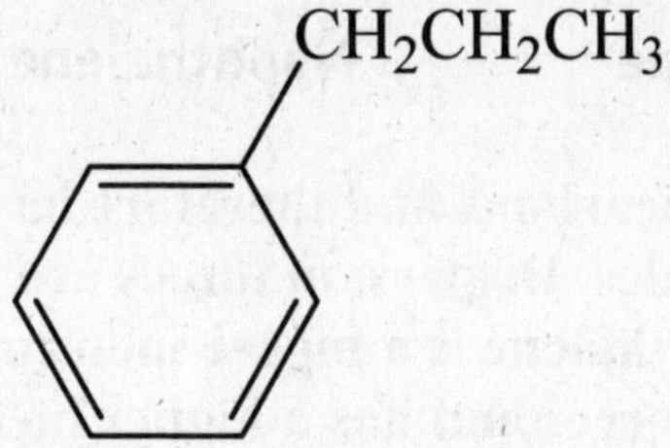

$CH_2CH_2CH_3$

8.73 *Name the following molecule, using the IUPAC system.*

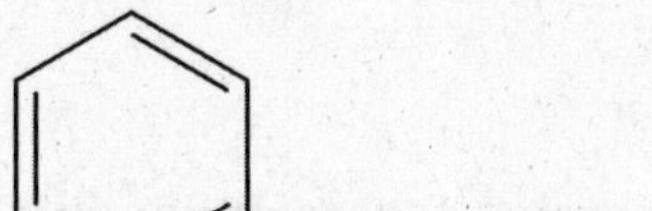

$CH_2—CH_3$

The molecule is a **benzene** ring with a two-carbon alkyl group attached, **ethyl-**: **ethylbenzene**. No number is needed, as long as there is only one alkyl group, because all positions on the ring are the same.

8.75 *The following molecule contains several of the functional groups you have studied in this chapter. Identify each functional group in this molecule, and tell whether it is an alkene, alkyne, or aromatic group.*

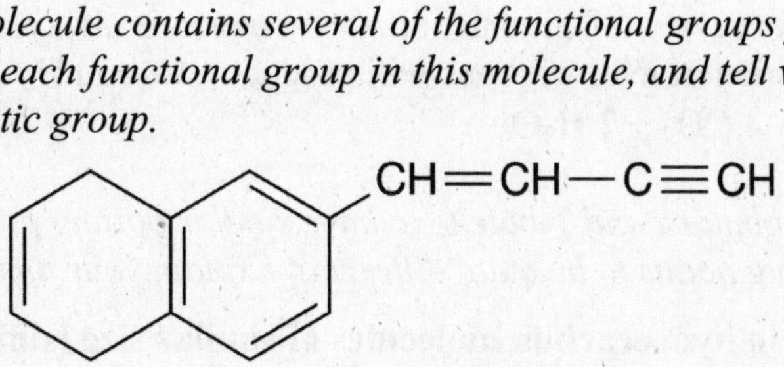

$CH{=}CH{-}C{\equiv}CH$

Alkenes have double bonds between carbon atoms; alkynes have triple bonds between carbon atoms; and aromatic functional groups have the special arrangement of a 6-membered ring with alternating double and single bonds.

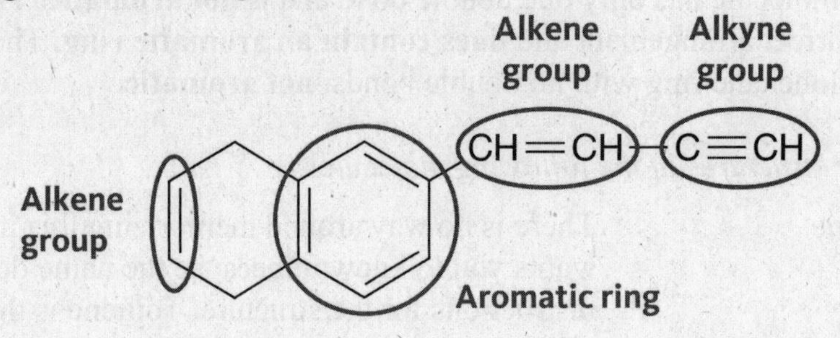

Solutions to Section 8.9 Core Problems

8.77 *One of the following compounds is a liquid at room temperature, while the other is a solid. Which is which? Explain your reasoning.*

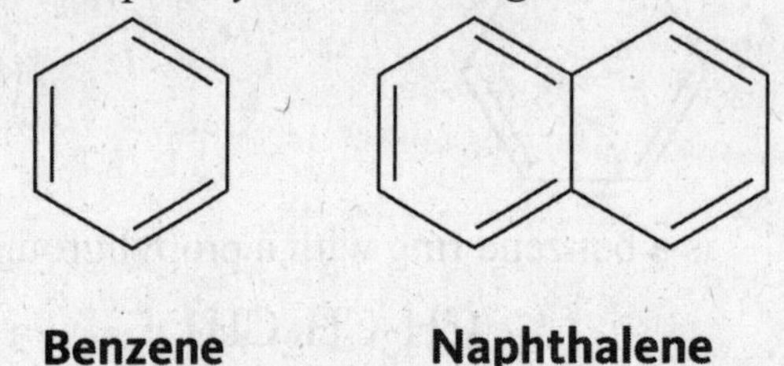

Benzene **Naphthalene**

Both compounds are hydrocarbons and therefore have dispersion forces as the only attractions between molecules. Dispersion forces are stronger in molecules that have more electrons. Since naphthalene is a bigger molecule and will have more electrons, it has stronger dispersion forces and has a higher melting and boiling point than benzene. Naphthalene is a solid at room temperature, benzene is a liquid at room temperature.

8.79 *Explain why pentane is insoluble in water.*

Pentane, a hydrocarbon, has only dispersion forces attracting its molecules to each other; pentane molecules are not capable of forming hydrogen bonds to water molecules. Water molecules form relatively weak dispersion force attractions, and generally only mix well with molecules that can participate in hydrogen bonding.

8.81 *Write the balanced chemical equation for the combustion of acetylene, C_2H_2.*

See Section 6.5 for review of combustion. Combustion is always a reaction with O_2, and complete combustion of hydrocarbons will produce CO_2 and H_2O:
$\mathbf{2\ C_2H_2 + 5\ O_2 \rightarrow 4\ CO_2 + 2\ H_2O}$

8.83 *Would you expect butane and 1-butene to have similar boiling points, or would you expect their boiling points to be quite different? Explain your answer.*

Multiple bonds, in hydrocarbon molecules of similar size (similar number of electrons, similar dispersion forces) have relatively little effect on the boiling points of the substances. Butane and 1-butene should have similar boiling points.

8.85 *Arrange the following compounds in order from lowest boiling point to highest boiling point. If two compounds should have roughly the same boiling point, list them together.*
butane *2-methylbutane* *2,2-dimethylbutane* *pentane*

Boiling points of hydrocarbons are determined by the strength of the dispersion force attractions between the molecules, which in turn are determined by the number of electrons in the molecules: larger molecules have more electrons, which gives them stronger dispersion forces between the molecules. Stronger dispersion forces in a compound make the boiling point higher.

When we're comparing several hydrocarbons (and therefore don't have to contend with hydrogen bonding effects), we can shortcut this by just looking at the molecular formula or even just the number of carbon atoms: **more carbon atoms = higher boiling point.**

So the ranking, from lowest to highest boiling point, is:
butane (C_4H_{10}) < 2-methylbutane and pentane (C_5H_{12}) < 2,2-dimethylbutane (C_6H_{14})

8.87 *What is the significance of combustion reactions in modern society?*

Basically our entire economy is driven by turning carbon and hydrocarbons into carbon dioxide and water–most of our electricity comes from combustion of coal, almost all of our vehicles (cars, trucks, planes, trains, boats) are powered by combustion of petroleum products. Even our bodies are powered by combustion of carbohydrates, fats and proteins–which is not itself unique to modern society, but the scale of it (over 7 billion people), and the extent to which our food production, transportation and energy sources use fossil fuel, are unprecedented in the history of the world.

Solutions to Concept Questions

* indicates more challenging problems.

8.89 *Sulfur atoms are similar to carbon atoms in that they can form long chains, linked by covalent bonds. Give a reason why sulfur would not be likely to be the fundamental element for living organisms on some other planet.*

Since sulfur (Group 6A) atoms only typically form 2 bonds, the wide array of branched structures and double- or triple-bonded structures that are formed by carbon are simply not available from sulfur. If we assume that life requires a lot of different possible chemicals, sulfur is not a great candidate.

8.91 *Chloroform has the molecular formula $CHCl_3$. Which of the following could be the correct structural formula for chloroform, based on the bonding properties of carbon, hydrogen, and chlorine?*

```
     Cl              Cl              H
     |               |               |
  C—H—Cl          H—C—Cl          C—Cl—Cl
     |               |               |
     Cl              Cl              Cl
```

The normal bonding patterns are one bond for H, four bonds for C, and one bond for Cl. Only the **second structure** matches all these bonding patterns.

8.93 *Define each of the following terms:*

a) *alkane* — **a hydrocarbon with no double or triple bonds between carbon atoms; a saturated hydrocarbon.**

b) *alkene* — **a hydrocarbon with at least one double bond between carbon atoms.**

c) *alkyne* — **a hydrocarbon with at least one triple bond between carbon atoms**

d) *branched alkane* — **an alkane in which not all carbons are in one continuous chain; an alkane in which at least one carbon atom is bonded to three or four other carbon atoms.**

e) *aromatic compound* — **a hydrocarbon containing a 6-membered ring of carbon atoms with alternating double and single bonds.**

8.95 *The IUPAC rules allow us to write "methylpropane" instead of 2-methylpropane, but they do not allow us to write "methylhexane" instead of 2-methylhexane. Why is this?*

IUPAC names have to contain enough information to specify a unique compound. Any other number on 2-methylpropane gives a compound that actually has a different name (1-methylpropane and 3-methylpropane are both actually butane; draw them and check it out!) **So there is only one possible compound with the name "methylpropane." But 2-methylhexane and 3-methylhexane are different compounds, so just saying "methylhexane" is not enough to specify a single compound.** (1- and 6-methylhexane are both properly named heptane; 4-methylhexane should be named 3-methylhexane, and 5-methylhexane is actually 2-methylhexane.)

8.97 *Beside each of the following molecules is an incorrect name. Explain why each name is wrong, and give the correct name.*

a) $CH_3-CH_2-CH_2-CH(CH_3)-CH_3$ **4-Methylpentane**

b) $CH_3-C(CH_3)_2-CH_2-CH_3$ **2-Dimethylbutane**

d) (cyclohexane ring with CH_3) **3-Methylcyclohexane**

a) **Numbering always starts from the end nearest a branch. This molecule should be numbered from right to left: 2-methylpentane.**

b) **When there are multiple alkyl groups, the location of each one has to be specified, by a number designating which carbon of the principle chain is attached to the group. Since both methyl groups are on carbon #2 of the principle chain, this molecule is 2,2-dimethylbutane.**

d) **For cycloalkanes with only one alkyl group, numbers are not used. This molecule is methylcyclohexane.** (Check it out: any number you put in front of methylcyclohexane is actually the same molecule.)

8.99 *Why isn't it necessary to use a number to show the position of the double bond in cyclopentene (i.e. why don't we write "1-cyclopentene", "2-cyclopentene", etc.)?*

Any number you put in front of cyclopentene is actually the same molecule (check it out–draw them!) In a cycloalkene molecule, the double bond is automatically bond #1.

8.101 *There are* cis *and* trans *forms of 2-butene, but not of 1-butene. Why is this?*

In 1-butene, one of the carbons is attached to two identical H atoms–switching them doesn't make any difference to the molecule, so this molecule does not have *cis* and *trans* isomers. But in 2-butene, switching the two groups on one of the double-bonded carbon atoms gives a different molecule, so *cis* and *trans* designations must be made.

1-butene:

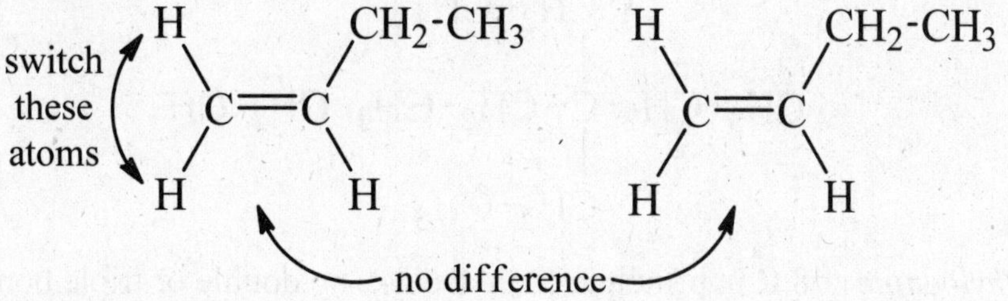

H, H on one C; CH_2-CH_3, H on the other: $H_2C{=}CH{-}CH_2{-}CH_3$ (switch these atoms) → $H_2C{=}CH{-}CH_2{-}CH_3$

no difference

2-butene:

H_3C, H on one C; CH_3, H on the other (switch these groups)

cis: H_3C and CH_3 on the same side   trans: H_3C and CH_3 on opposite sides

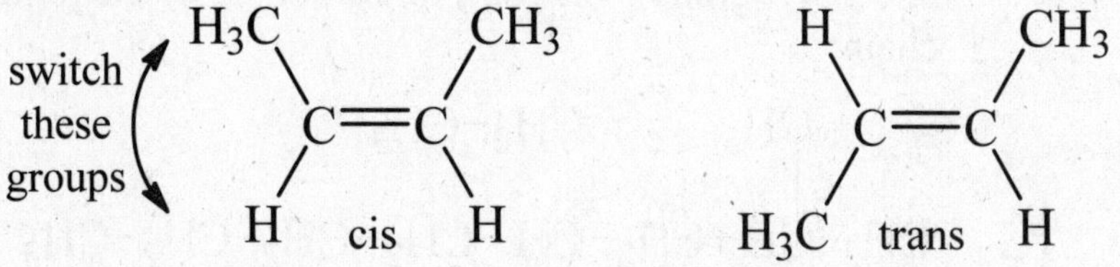

8.103 a) *In the combustion reaction of a hydrocarbon, what additional reactant is necessary?*

Combustion is always, by definition, a reaction with O_2.

b) *What are the products of the combustion reaction?*

In any complete combustion reaction, carbon in the reactants forms carbon dioxide in the products, and hydrogen in the reactants forms water in the products.

Solutions to Summary and Challenge Problems

* indicates more challenging problems.

8.105 *Draw condensed structural formulas for each of the following hydrocarbons. You may draw line structures for any rings.*

For each of these, remember to think of the name as a set of instructions for drawing the molecule. Go to the end of the name first, because that tells you the length of the principle chain; if the name ends in –ene or –yne, add a double or triple bond as appropriate; add the alkyl groups according to their numbered positions; and wait until the last thing to fill in hydrogen atoms, giving each C atom four bonds.

a) pentane 5 C, no double or triple bonds, no branches:

$$CH_3\text{–}CH_2\text{–}CH_2\text{–}CH_2\text{–}CH_3 \text{ or } CH_3(CH_2)_3CH_3$$

b) 2-methylhexane 6 C in principle chain (-hex-), no double or triple bonds (-ane), a $-CH_3$ group (-methyl-) on carbon #2 of the principle chain (2-):

$$\begin{array}{l} \quad\;\; CH_3 \\ \quad\;\; | \\ CH_3\text{–}CH\text{–}CH_2\text{–}CH_2\text{–}CH_2\text{–}CH_3 \end{array}$$

c) 3,3-diethylheptane 7 C in principle chain (-hept-), no double or triple bonds (-ane), two $-CH_2\text{–}CH_3$ groups (-diethyl-), both on carbon #3 of the principle chain (3,3-):

$$\begin{array}{l} \qquad\qquad\quad CH_2\text{–}CH_3 \\ \qquad\qquad\quad | \\ CH_3\text{–}CH_2\text{–}C\text{–}CH_2\text{–}CH_2\text{–}CH_2\text{–}CH_3 \\ \qquad\qquad\quad | \\ \qquad\qquad\quad CH_2\text{–}CH_3 \end{array}$$

d) 4-ethyl-2-methyloctane 8 C in principle chain (-oct-), no double or triple bonds (-ane), one $-CH_2\text{–}CH_3$ group (-ethyl-) on carbon #4, and one $-CH_3$ group (-methyl-) on carbon #2 of the principle chain:

$$\begin{array}{l} \quad\;\; CH_3 \qquad\quad CH_2\text{–}CH_3 \\ \quad\;\; | \qquad\qquad\quad | \\ CH_3\text{–}CH\text{–}CH_2\text{–}CH\text{–}CH_2\text{–}CH_2\text{–}CH_2\text{–}CH_3 \end{array}$$

e) cyclopropane 3 C (-prop-) in a ring (cyclo-), no double or triple bonds (-ane):

f) propylcyclopentane 5 C (-pent-) in a ring (cyclo-), no double or triple bonds (-ane), and a three-carbon alkyl group (propyl-) on the ring (all positions on the ring are equivalent, the propyl group doesn't need a number to tell which carbon it's attached to):

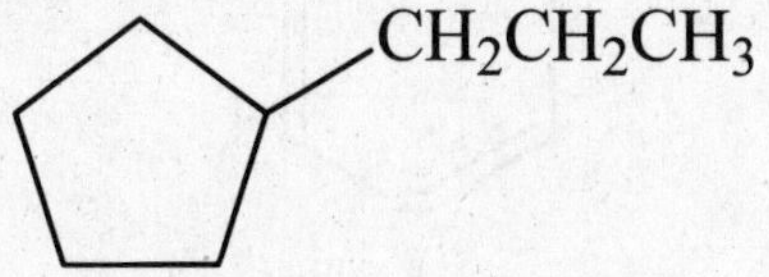

$CH_2CH_2CH_3$

g) 2-heptyne 7 C (-hept-), with a triple bond (-yne) on bond #2 of the principle chain:

$$CH_3{-}C{\equiv}C{-}CH_2{-}CH_2{-}CH_2{-}CH_3$$

h) trans-*2-hexene* 6 C (-hex-), with a trans double bond (trans- -ene) on bond #2 of the principle chain:

H　　　$CH_2{-}CH_2{-}CH_3$

　$C{=}C$

CH_3　　　H

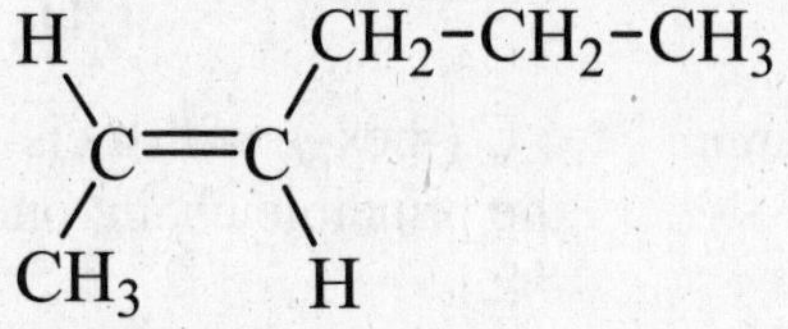

i) ethylene This is the common name for ethene (2 C, double bond): **$CH_2{=}CH_2$**

j) cyclopentene A ring (cyclo-) of 5 C (-pent-) with one double bond (-ene).

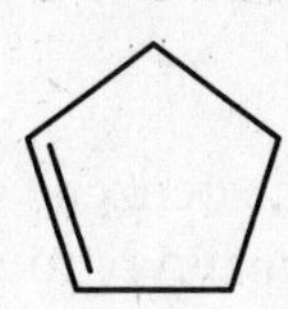

NOTE:
It doesn't matter where you put the double bond, all positions are equivalent:

k) 2-propyl-1-pentene The principle chain has 5 C (-pent-) and a double bond (-ene) on the #1 bond; attached to carbon #2 is a three-carbon alkyl group (-propyl-). Remember that for alkanes and alkynes, the principal chain is not necessarily the longest chain, but **the longest chain that includes the double or triple bond.**

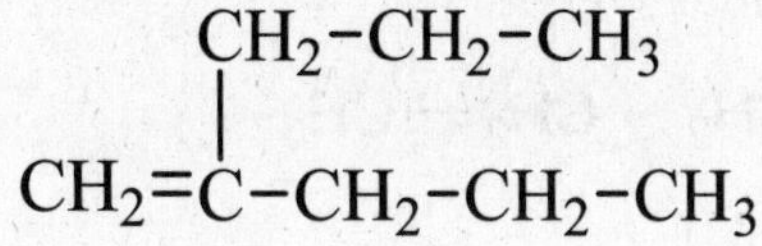

$CH_2{-}CH_2{-}CH_3$

$CH_2{=}C{-}CH_2{-}CH_2{-}CH_3$

l) benzene

This one you have to memorize, because there are no clues in the name; it's a 6-membered ring with alternating double and single bonds all the way around:

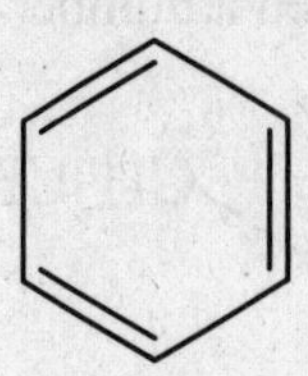

m) 4,4-dimethyl-2-hexyne

6 C (-hex-), with a triple bond (-yne) on bond #2 of the principle chain; two $-CH_3$ groups (-dimethyl-), both on carbon #4 (4,4-):

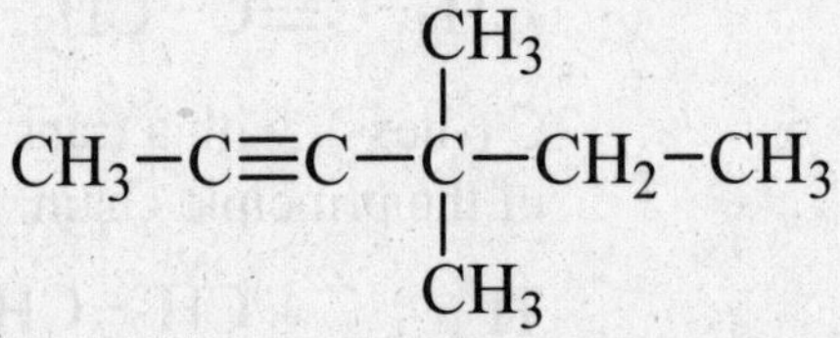

$$
\begin{array}{c}
\qquad\qquad\quad CH_3 \\
\qquad\qquad\quad | \\
CH_3-C\equiv C-C-CH_2-CH_3 \\
\qquad\qquad\quad | \\
\qquad\qquad\quad CH_3
\end{array}
$$

*n) 4-methyl-*cis*-2-hexene*

6 C (-hex-), with a cis- double bond (-ene) on bond #2 of the principle chain; one $-CH_3$ group (-methyl-) on carbon #4:

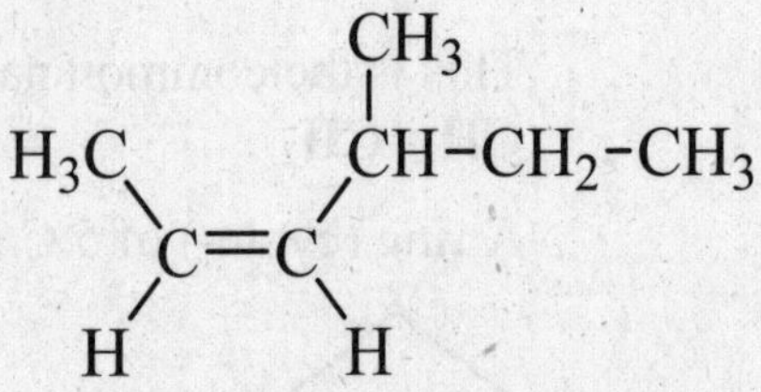

$$
\begin{array}{ccc}
 & & CH_3 \\
 & & | \\
H_3C & & CH-CH_2-CH_3 \\
 & C=C & \\
H & & H
\end{array}
$$

o) propylbenzene

Start with the benzene structure as in *l* above, and add a three-carbon alkyl group (propyl-):

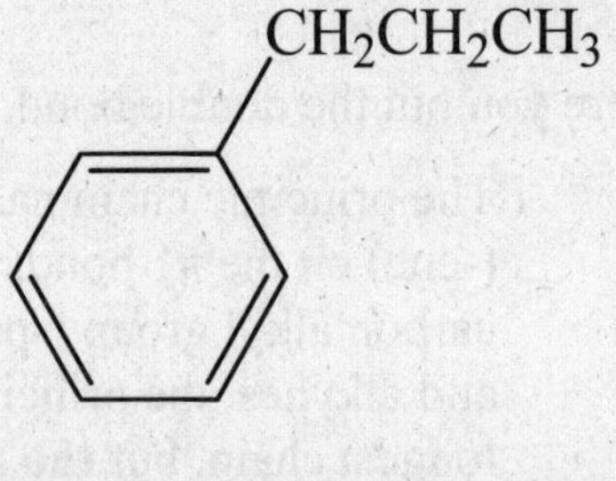

$CH_2CH_2CH_3$

8.107 *Name the following compounds, using the IUPAC rules. Be sure to indicate the* cis *or* trans *isomer where appropriate.*

a) $CH_3-CH_2-CH_2-CH_2-CH_3$

b)
$$
\begin{array}{c}
\qquad\qquad\qquad\qquad CH_2-CH_3 \\
\qquad\qquad\qquad | \\
CH_3-CH_2-CH_2-CH-CH_2-CH_3
\end{array}
$$

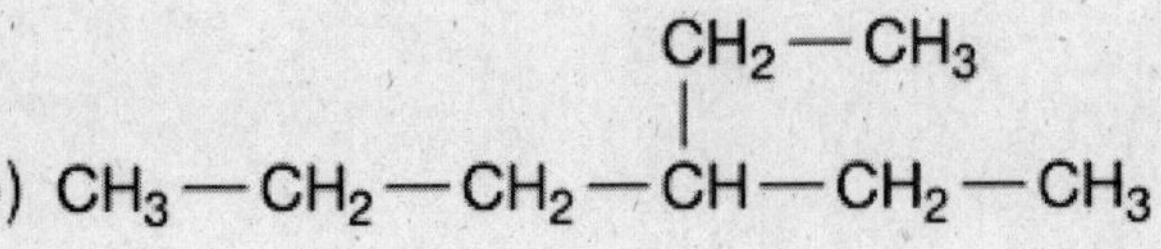

c) 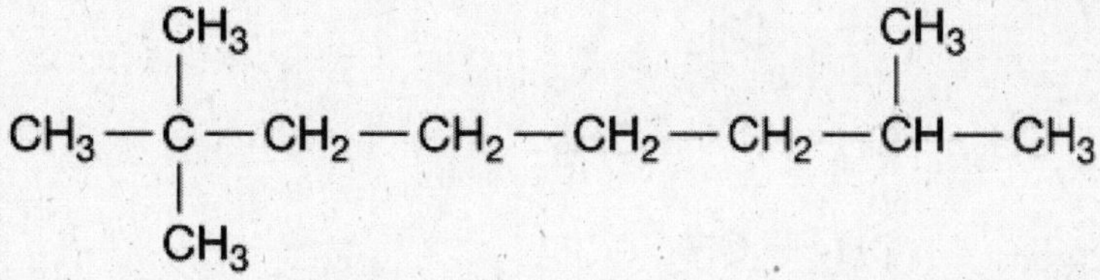

d) 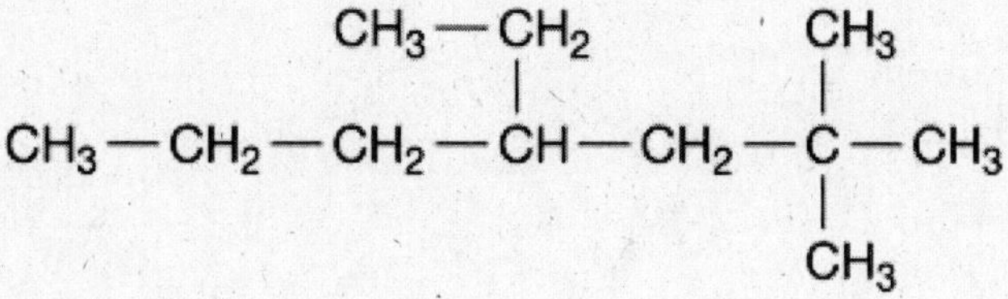

e) 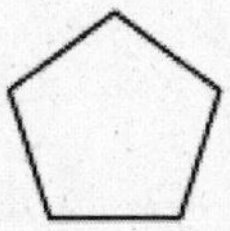

f) 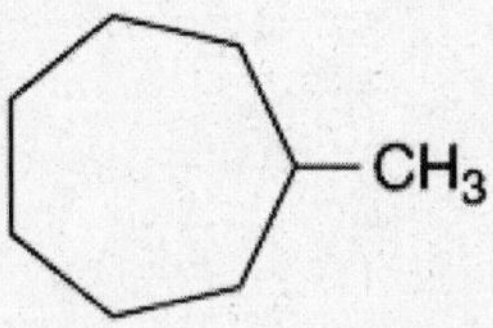

g) $CH_3-CH_2-C\equiv C-CH_3$

h) 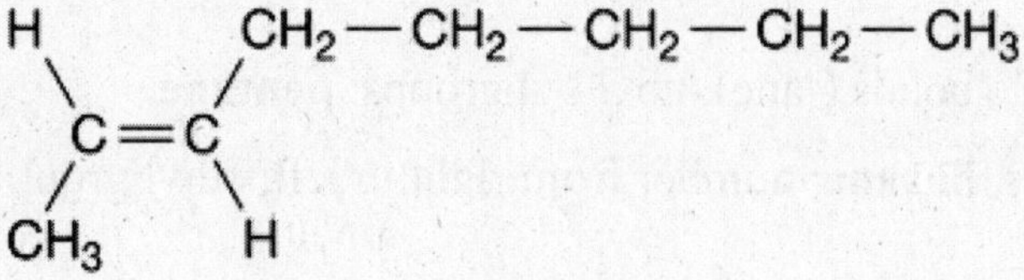

i) 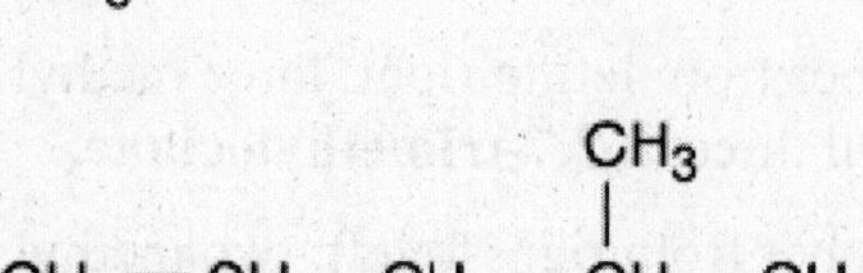

j) 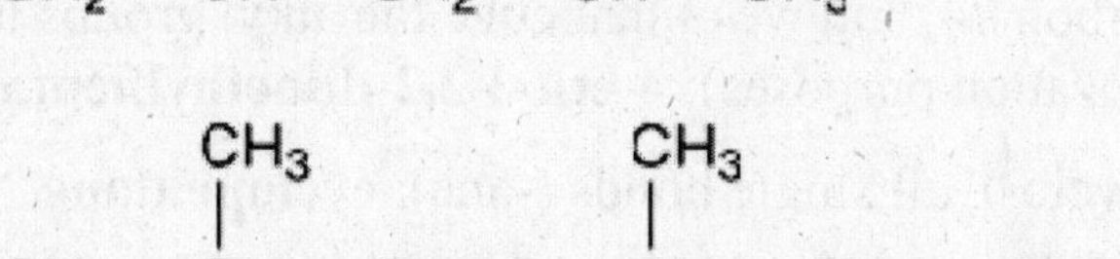

k) 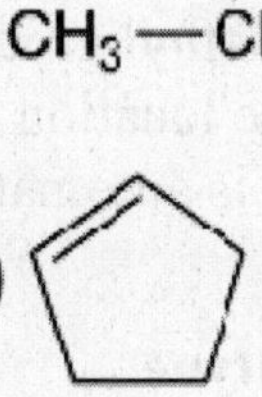

l) 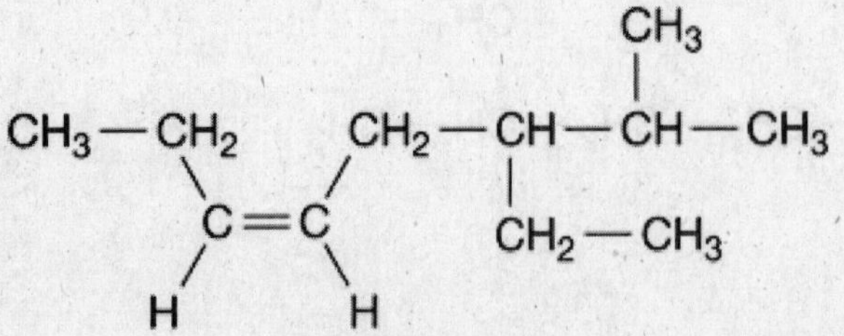 CH_3-CH_2 $\quad$ $CH_2-CH-CH-CH_3$ (with CH_3 on the third carbon and CH_2-CH_3 on the second carbon of that chain)
$C=C$
H $\quad$ H

m) 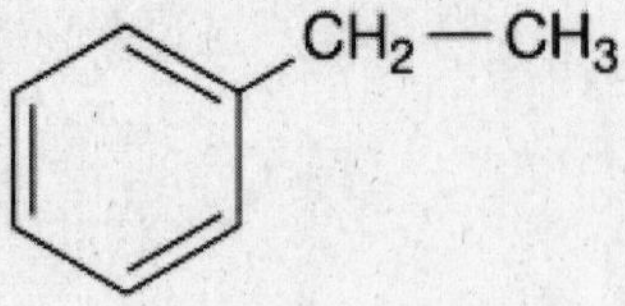 CH_2-CH_3

n)

o)

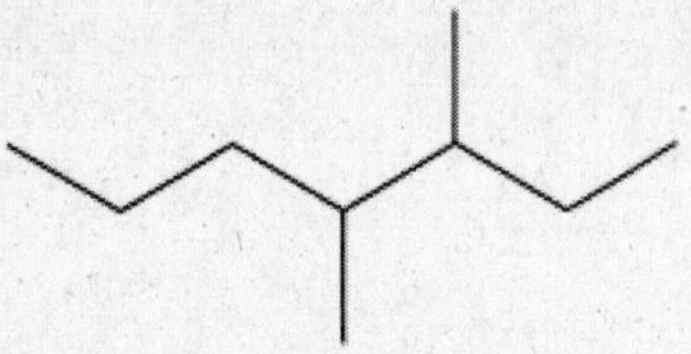

p)

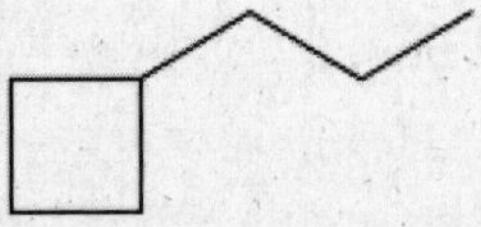

a) 5 carbons (pent-), all single bonds (-ane), no alkyl groups: **pentane.**

b) Longest chain is 6 carbons, **hexane**; number from right to left, ethyl group on carbon #3: **3-ethylhexane.**

c) Longest chain is 8 carbons, octane; number from left to right; three methyl groups, and we have to specify the positions of all three: **2,2,7-trimethyloctane.**

d) Longest chain is 7 carbons, heptane; number from right to left; two methyl groups on carbon #2, ethyl group on carbon #4, and we alphabetize the alkyl groups in the name (ignoring the di- for alphabetization purposes): **4-ethyl-2,2-dimethylheptane.**

e) 5 carbons (pent-) in a ring (cyclo-), all single bonds (-ane): **cyclopentane.**

f) 7 carbons (hept-) in a ring (cyclo-), all single bonds (-ane), one methyl group attached: **methylcycloheptane.** (No number needed to specify the location of the methyl group because the carbon with the methyl group attached is automatically carbon #1.)

g) 5 carbons, number from right to left, second bond is triple: **2-pentyne.**

h) Longest chain is 8 carbons, oct-; number from left to right; second bond is double, 2-octene; and because it makes a difference which direction the alkyl groups and hydrogen atoms are pointing from the double-bonded carbons, we have to specify *cis* or *trans*; in this case, the hydrogen atoms on the double-bond carbons are pointing away from each other across the molecule, so *trans*: ***trans*-2-octene.**

i) Longest chain is 5 carbons, pent-; number left to right; first bond is double, 1-pentene, and the methyl group is on carbon #4: **4-methyl-1-pentene**. (Remember double and triple bonds take precedence over alkyl groups when deciding which direction to number the chain.)

j) 7 carbons, number left to right, third bond is triple, 3-heptyne; methyl groups on carbons #2 and #5: **2,5-dimethyl-3-heptyne**.

k) Ring of 5 carbons, one double bond: **cyclopentene.** (No need to specify the location of the double bond, because in a cycloalkene or cycloalkyne the double or triple bond automatically becomes bond #1.)

l) 8 carbons, number left to right, third bond is double, 3-octene; ethyl group on carbon #6, methyl on carbon #7; and because it makes a difference which direction the alkyl groups and hydrogen atoms are pointing from the double-bonded carbons, we have to specify *cis*: **6-ethyl-7-methyl-*cis*-3-octene**.

NOTE:
Double and triple bonds take precedence over alkyl groups when deciding which direction to number the chain; alkyl groups are alphabetized in the name.

m) Benzene ring with an ethyl group attached: **ethylbenzene**. (No number needed to specify the location of the ethyl group, because the carbon with the ethyl group attached is automatically carbon #1.)

n) 6-carbon alkane: **hexane.**

o) 7-carbon alkane, number from right to left, methyl groups on carbon #3 and #4: **3,4-dimethylheptane.**

p) 4-membered ring, propyl group attached: **propylcyclobutane.** (No number needed to specify the location of the propyl group, because the carbon with the propyl group attached is automatically carbon #1.)

8.109 *What is the molecular formula of the cycloalkane that has the following line structure?*

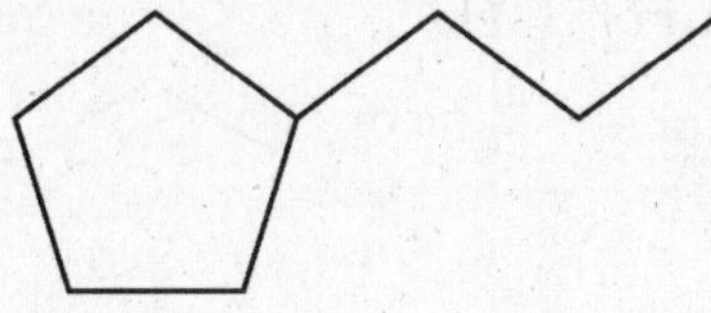

Remember ends and bends are C atoms, and fill in with hydrogen atoms to give each C four bonds: $\mathbf{C_8H_{16}}$

*8.111 *Draw the structure of a compound that fits each of the following descriptions:*

a) a branched alkane that is an isomer of 2-methylhexane.

We need a 7-carbon alkane, and the problem specifies "branched," so we can't use heptane. We could move the methyl group to carbon #3 on hexane (3-methylhexane):

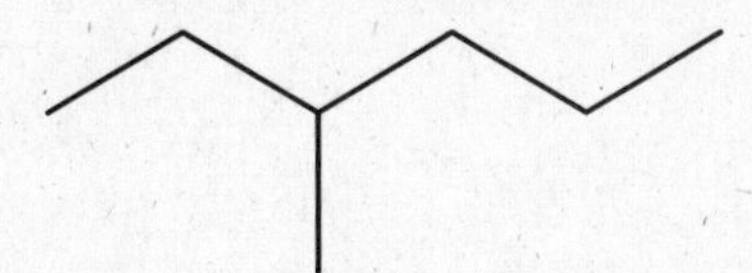

We could also use a shorter chain length and more methyl groups or an ethyl group:

pentane variations:

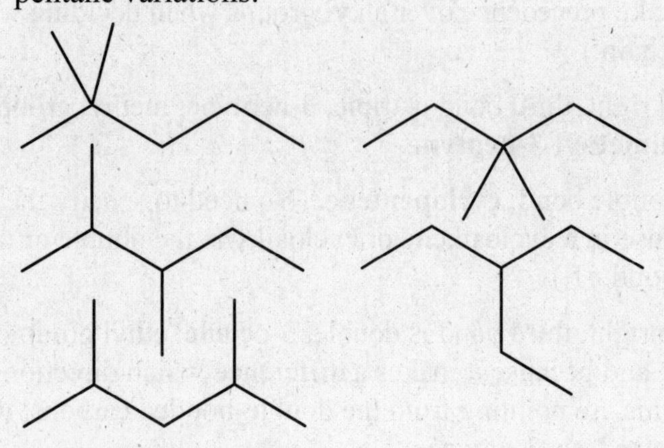

2,2,3-trimethylbutane

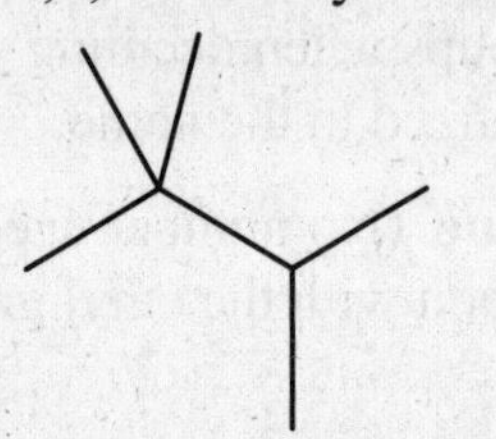

b) a linear alkane that is an isomer of 2-methylhexane.

The only option here, because the problem specifies "linear alkane," is heptane:

c) an alkene that is a stereoisomer of trans*-3-octene.*

Stereoisomers are *cis/trans* isomers, so our only option here is *cis*-3-octene:

$$\begin{array}{ccc} CH_3{-}CH_2 & & CH_2{-}CH_2{-}CH_2{-}CH_3 \\ & C{=}C & \\ H & & H \end{array}$$

or

d) an unbranched alkene that is a constitutional isomer of trans-3-octene.

Constitutional isomers have atoms or double or triple bonds rearranged. Since the problem specified an "unbranched alkene," we have to keep the chain length at 8, so our only option is to move the double bond, to form 1-octene, *cis*- and *trans*-2-octene, and *cis*- and *trans*-4-octene:

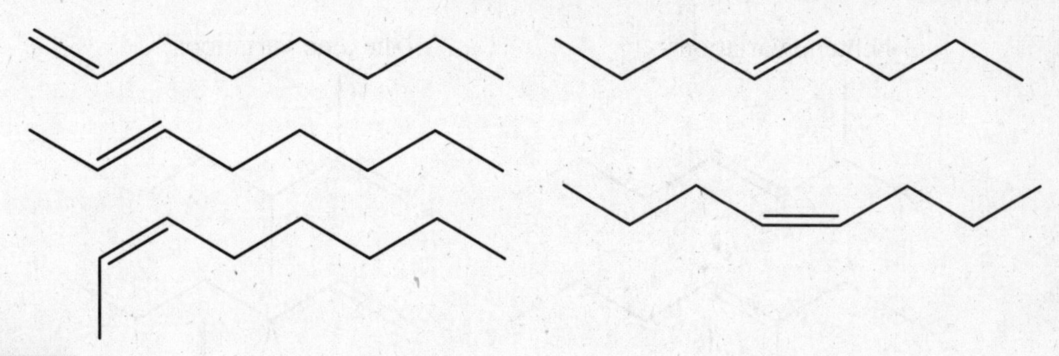

NOTE:
You only have to draw one of these molecules to answer the question.) All other unbranched alkenes are either repeats of these, *trans*-3-octene, or its stereoisomer *cis*-3-octene.

e) *Draw the structure of a branched alkene that is a constitutional isomer of* trans-*3-octene.*

Constitutional isomers can have atoms, double bonds, or triple bonds moved around, so there are many options. This one specified "branched alkene," so we will have to change to a different principal chain length (7 or fewer) and keep a double bond somewhere in the molecule. Here are some options:

1-heptene variations: 2-heptene variations:

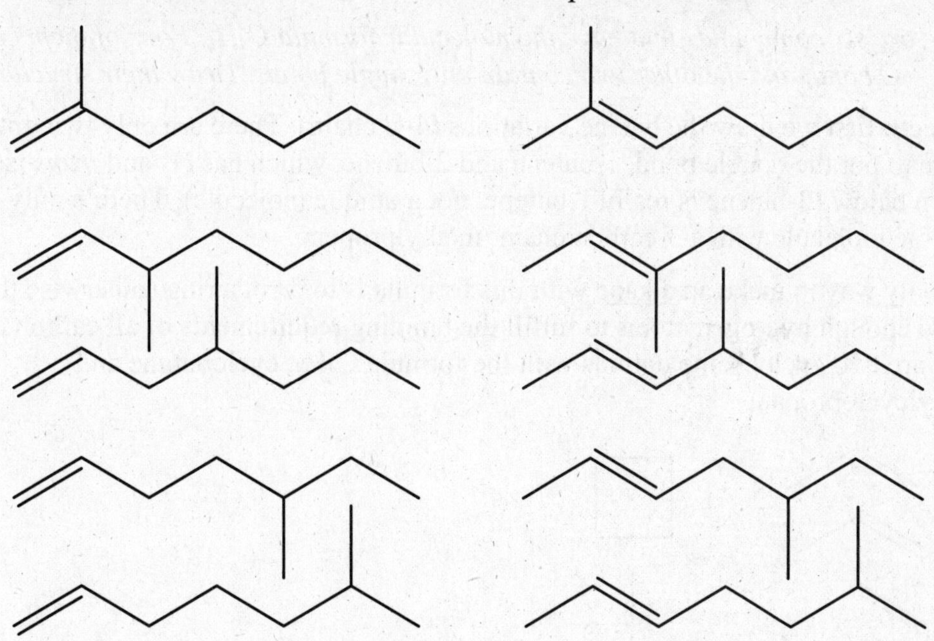

3-heptene variations: 1-hexene variations:

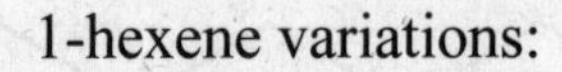

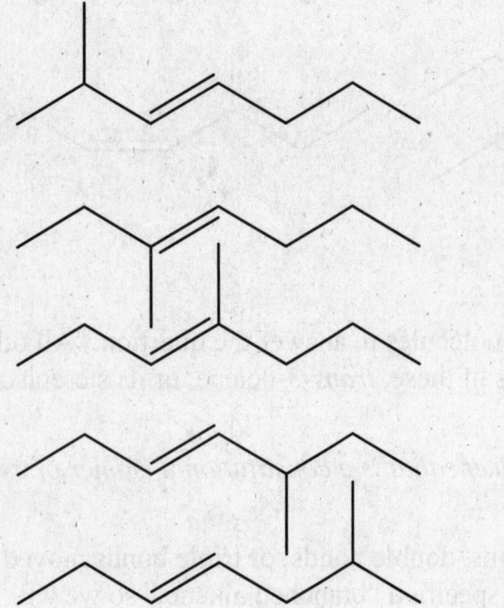

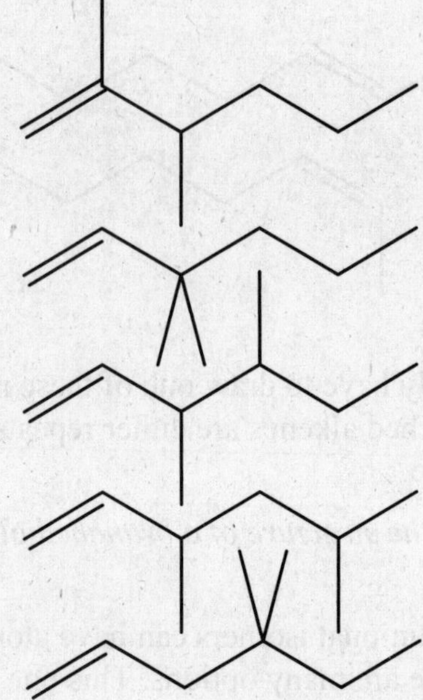

and so on, for a total of more than 60 possible compounds, not including *cis*- and *trans*- isomers (several of the molecules above have *cis/trans* isomers).

*8.113 *There are six compounds that have the molecular formula C_4H_8. Four of them contain one C=C bond, and the other two contain only single bonds. Draw their structures.*

Strategy: first we draw the butene variations (4-C chain). There are only two options for where to put the double bond, 1-butene and 2-butene, which has *cis* and *trans* isomers, shown below (3-butene is really 1-butene, not a unique molecule). There's only one possible molecule with a 3-carbon chain, methylpropene.

The only way to make an <u>alkane</u> with this formula is to form a ring (otherwise there will not be enough hydrogen atoms to fulfill the bonding requirements of all carbon atoms); there are two cycloalkane options with the formula C_4H_8, cyclobutane and methylcyclopropane.

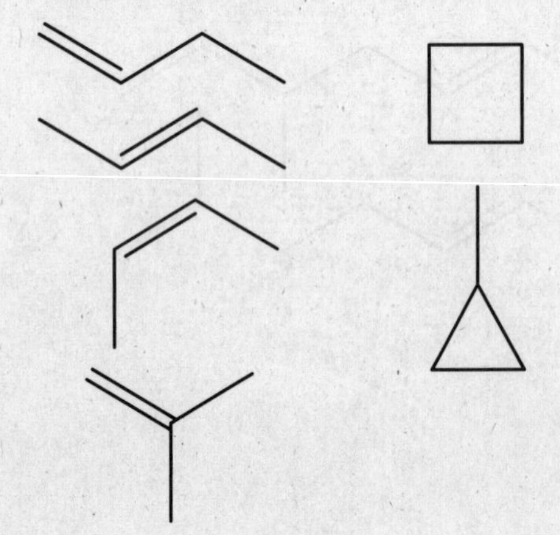

8.115 *Each of the following molecules contains one or more of the hydrocarbon functional groups you have studied in this chapter (alkene, alkyne, and aromatic ring). Circle each hydrocarbon functional group, and tell what type of functional group it is.*

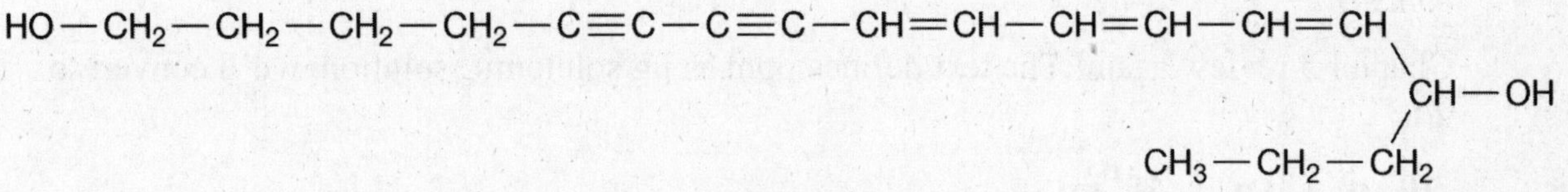

HO—CH$_2$—CH$_2$—CH$_2$—CH$_2$—C≡C—C≡C—CH=CH—CH=CH—CH=CH—CH(OH)—CH$_2$—CH$_2$—CH$_3$

Cicutoxin
(a toxic compound found in water hemlock)

CH$_2$=CH—CH$_2$—C$_6$H$_3$(O—CH$_3$)—O—CH$_2$—C(=O)—N(CH$_2$—CH$_3$)—CH$_2$—CH$_3$

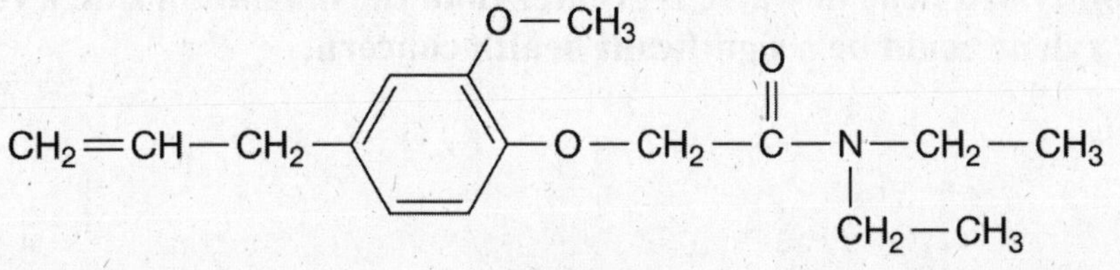

Estil
(an anesthetic)

NOTE:
The problem specifies "*hydrocarbon* functional group," so you are not required to identify the other functional groups in the structures.

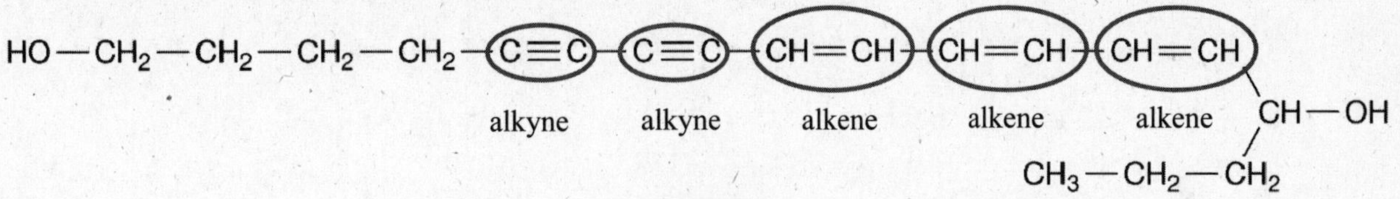

HO—CH$_2$—CH$_2$—CH$_2$—CH$_2$—(C≡C)—(C≡C)—(CH=CH)—(CH=CH)—(CH=CH)—CH(OH)—CH$_2$—CH$_2$—CH$_3$

alkyne, alkyne, alkene, alkene, alkene

Cicutoxin
(a toxic compound found in water hemlock)

(CH$_2$=CH)—CH$_2$—(C$_6$H$_3$)(O—CH$_3$)—O—CH$_2$—C(=O)—N(CH$_2$—CH$_3$)—CH$_2$—CH$_3$

alkene; aromatic

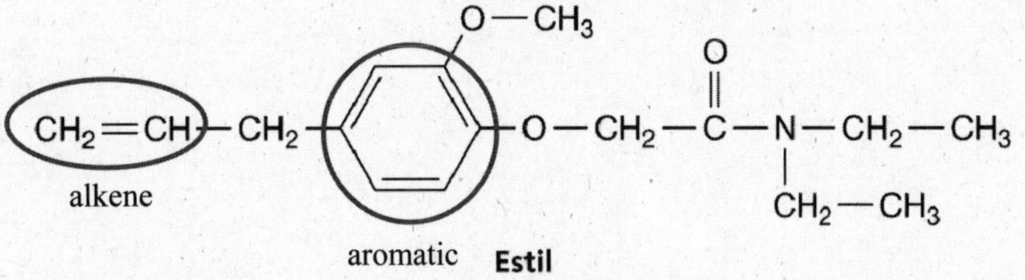

Estil
(an anesthetic)

*8.117 *If you have 2.50 g of pentane, how many moles do you have?*

Chapter 5 review! The conversion between grams and moles requires the molar mass, which requires the molecular formula of the compound–pentane is C_5H_{12}, 72.15 g/mol.

$$\mathbf{2.50\ g \times \frac{1\ mol}{72.15\ g} = 0.0347\ mol}$$

*8.119 *The United States government has set the maximum level of xylene in drinking water at 10 ppm. Is this level greater than or less than the solubility of xylene in water (0.17 g/L)? Based on this comparison, could xylene in drinking water be a significant health concern?*

Chapter 5 review again! The text defines ppm as μg solute/mL solution, we'll convert to g/L:

$$\frac{10\ \mu g}{mL} \times \frac{1\ g}{10^6\ \mu g} \times \frac{10^3\ mL}{1\ L} = 0.01\ g/L$$

The solubility of xylene in water is greater than the maximum safe level in drinking water, so xylene could be a significant health concern.

Chapter 9

Hydration, Dehydration, and Alcohols

Solutions to Section 9.1 Core Problems

9.1 *Draw the structures of the products that are formed when the following alkenes are hydrated. Be sure to tell whether the reaction forms two products or only one.*

a) $CH_2{=}CH{-}CH_2{-}CH_2{-}CH_3$

b) 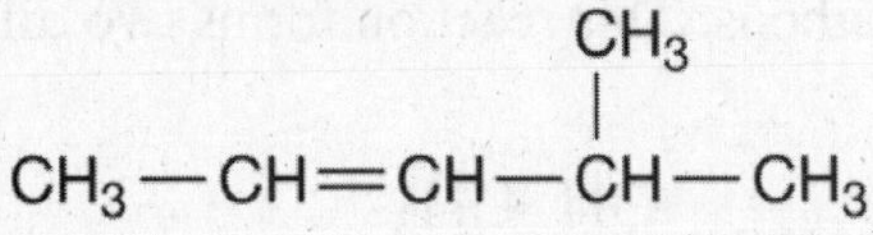 $CH_3{-}CH{=}CH{-}CH(CH_3){-}CH_3$

c) 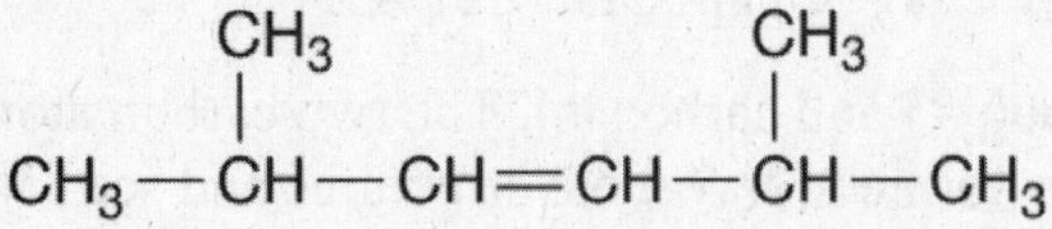 $CH_3{-}CH(CH_3){-}CH{=}CH{-}CH(CH_3){-}CH_3$

d)

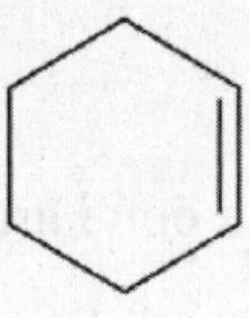

e) 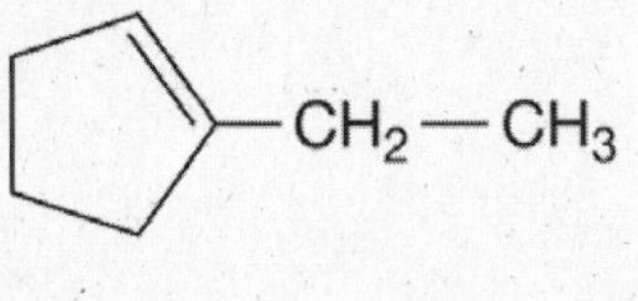 $-CH_2-CH_3$

f) 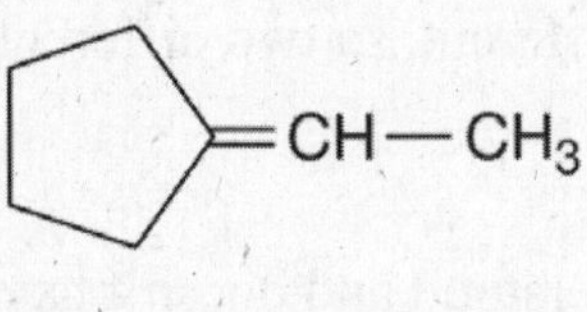 $=CH-CH_3$

g)

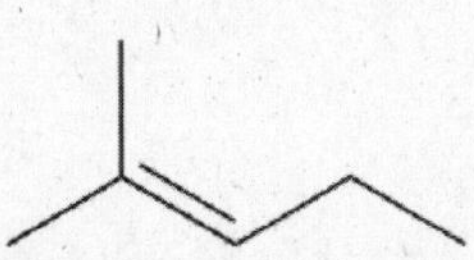

In each of these, we remove the double bond between the two carbon atoms, and put an –H atom on one of the two carbons and an –OH group on the other. Then we draw the structure again, switching the –H and –OH, and see if the second compound is different from the first.

a) The double bond is between the first two carbons, so we can put the –OH group on carbon #1 or carbon #2. This reaction forms two different products:

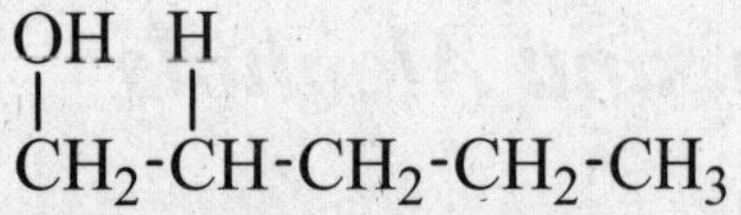

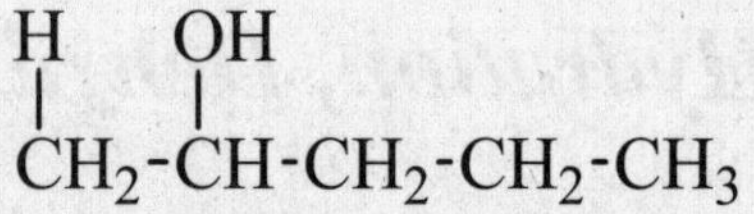

$$\begin{array}{l} \text{OH} \quad \text{H} \\ | \qquad | \\ CH_2\text{-}CH\text{-}CH_2\text{-}CH_2\text{-}CH_3 \end{array} \qquad \begin{array}{l} \text{H} \quad \text{OH} \\ | \qquad | \\ CH_2\text{-}CH\text{-}CH_2\text{-}CH_2\text{-}CH_3 \end{array}$$

We can also clean up these structures a little. Wherever an –H atom has been added to a carbon, we can simplify the structure by combining the new –H with the others on that carbon:

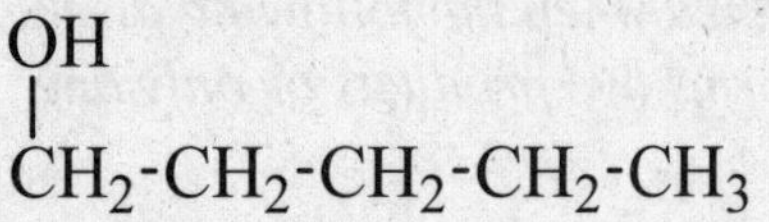

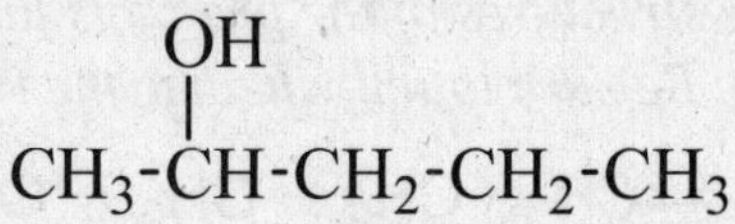

$$\begin{array}{l} \text{OH} \\ | \\ CH_2\text{-}CH_2\text{-}CH_2\text{-}CH_2\text{-}CH_3 \end{array} \qquad \begin{array}{l} \quad\;\; \text{OH} \\ \quad\;\; | \\ CH_3\text{-}CH\text{-}CH_2\text{-}CH_2\text{-}CH_3 \end{array}$$

b) The double bond is between the second and third carbons from the left, so we can put the –OH group on either of those two carbons. This reaction forms two different products:

$$\begin{array}{l} \quad\;\; \text{OH} \qquad\quad\;\; CH_3 \\ \quad\;\; | \qquad\qquad\quad | \\ CH_3\text{–}CH\text{–}CH_2\text{–}CH\text{–}CH_3 \end{array} \qquad \begin{array}{l} \qquad\qquad\;\; \text{OH} \;\; CH_3 \\ \qquad\qquad\;\; | \qquad | \\ CH_3\text{–}CH_2\text{–}CH\text{–}CH\text{–}CH_3 \end{array}$$

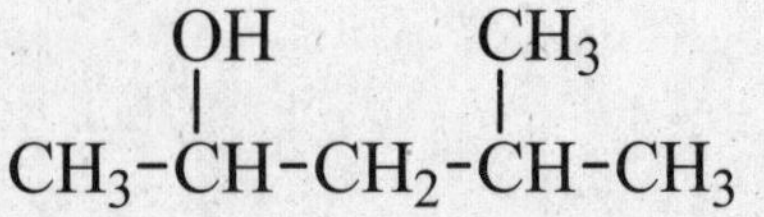

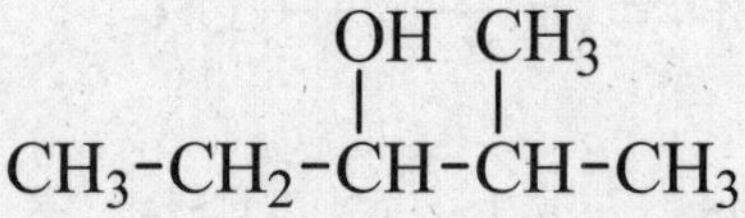

c) The double bond is between carbon #3 and carbon #4. The two carbon atoms are equivalent, so only one product is formed (draw both structures and see for yourself!):

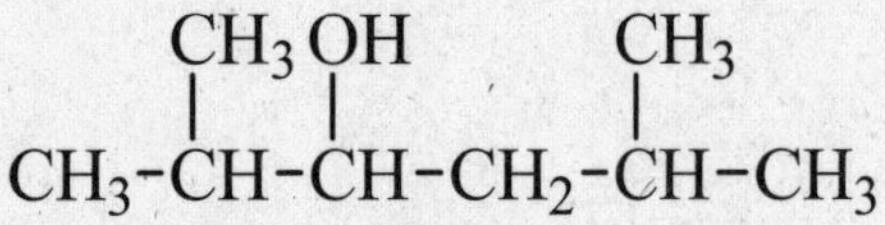

$$\begin{array}{l} \quad\;\; CH_3\,\text{OH} \qquad\; CH_3 \\ \quad\;\; | \qquad | \qquad\qquad | \\ CH_3\text{–}CH\text{–}CH\text{–}CH_2\text{–}CH\text{–}CH_3 \end{array}$$

d) Because this is a cycloalkene, the two carbon atoms in the double bond are equivalent, so only one product is formed (draw and see!):

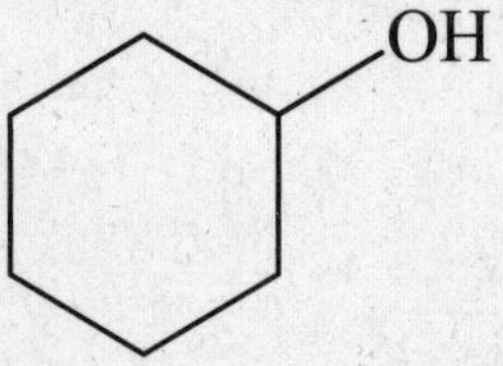

OH

e) The two carbons involved in the double bond are different, so two different products are formed.

NOTE:

In line structures, the H atom that was added is understood and doesn't have to be shown:

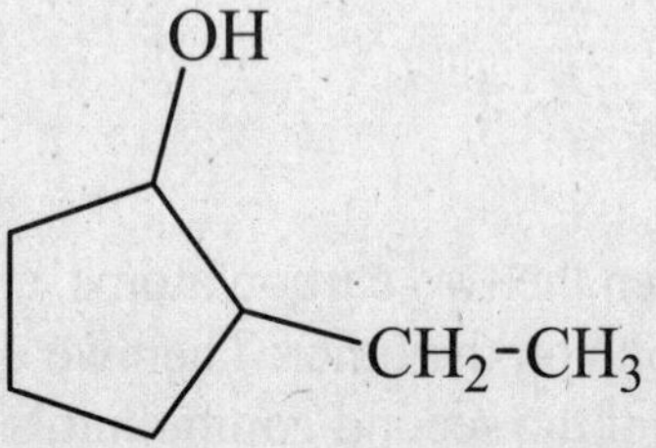

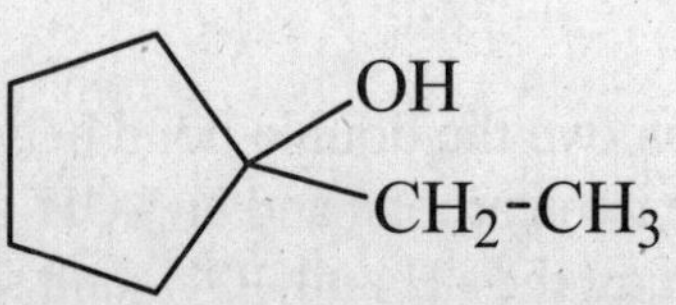

OH

$CH_2\text{–}CH_3$

OH

$CH_2\text{–}CH_3$

f) The two carbons involved in the double bond are different, so two different products are formed.

NOTE:
In the second structure, the H atom that was added is understood and doesn't have to be shown; also note that the first product is the same as one of the products in (e):

g) The two carbons involved in the double bond are different, so two different products are formed:

9.3 *If the following alkene is hydrated in a laboratory, one of the two possible products will be formed in much greater amount than the other, based on Markovnikov's Rule. Draw the structure of this product.*

$$CH_3-\underset{|}{\overset{CH_3}{CH}}-CH{=}CH_2$$

The two structures are:

$$CH_3-\overset{CH_3}{\underset{}{CH}}-CH_2-\overset{OH}{CH_2} \qquad CH_3-\overset{CH_3}{CH}-\overset{OH}{CH}-CH_3$$

According to Markovnikov's Rule, the carbon that has more H to begin with will get the new H, so the major product of this reaction is the **second isomer**:

$$CH_3-\overset{CH_3}{CH}-\overset{OH}{CH}-CH_3$$

9.5 *Which of the following alkenes can form only one product when they react with water? (List all of the correct answers.)*

a) $CH_3-CH{=}CH-CH_2-CH_2CH_3$

b) $CH_3-CH_2-CH{=}CH-CH_2-CH_3$

c) $CH_3-CH_2-CH{=}CH-\overset{CH_3}{CH}-CH_3$

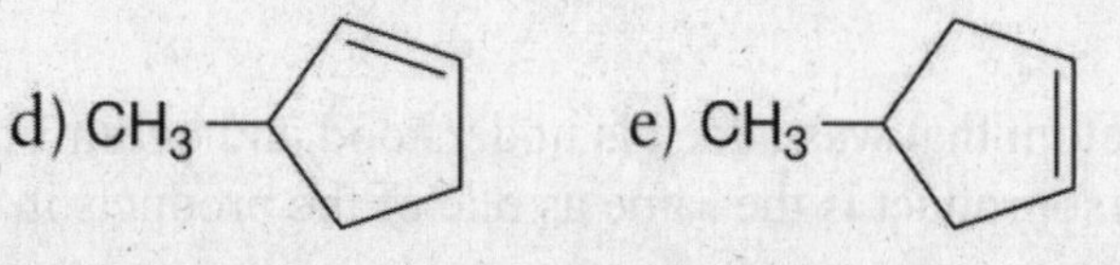

d) CH_3– (cyclopentene)   e) CH_3– (cyclopentene)

What "form only one product" really means is that the two possible places you can put the –OH group turn out to be the same molecule, flipped over or turned around. You can always draw both structures and check, but to answer this question *without* drawing both structures, look at the two carbons of the double bond and decide: are they equivalent, or can you tell them apart? If you flip the molecule so that the positions of the two carbons are reversed, does the molecule look the same? If the carbon atoms are equivalent, then it doesn't matter which one you put the –OH group on, you will get the same product. If the alkene carbon atoms are different, then hydration will make two different products.

a) The two carbon atoms involved in the double bond in this molecule are different. **Hydration will give two different products** (alcohol group on carbon #2 or carbon #3).

b) The two carbon atoms involved in the double bond in this molecule are equivalent. Counting from either direction, the first carbon of the double bond is carbon #3. **Putting the alcohol group on either carbon will give the same product.**

c) The two carbon atoms involved in the double bond in this molecule are different. **Hydration will give two different products** (alcohol group and methyl group on adjacent carbons, or one CH_2 group between).

d) The two carbon atoms involved in the double bond in this molecule are different. **Hydration will give two different products** (the alcohol group could be on carbon #3 or carbon #4).

e) The two carbon atoms involved in the double bond in this molecule are equivalent. **Putting the alcohol group on either carbon will give the same product** (draw them!).

Solutions to Section 9.2 Core Problems

9.7 *What is an enzyme?*

Enzymes are protein molecules that act as catalysts for specific reactions in the body.

9.9 *The following reaction occurs when your body breaks down carbohydrates, fats, and proteins to obtain energy. The reaction is catalyzed by an enzyme. Must this enzyme select one of the possible products in this reaction? If so, what is the other possible product?*

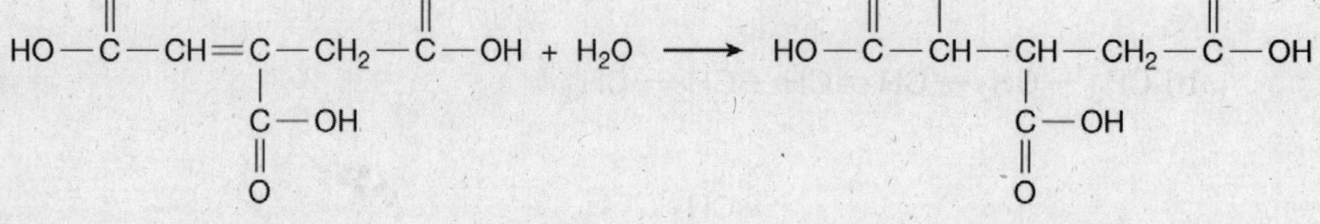

HO–C(=O)–CH=C(–C(=O)–OH)–CH₂–C(=O)–OH + H_2O ⟶ HO–C(=O)–CH(OH)–CH(–C(=O)–OH)–CH₂–C(=O)–OH

Aconitic acid   Isocitric acid

The two carbon atoms in the double bond are different, so the uncatalyzed hydration of the double bond would give two different products. Therefore the enzyme must select the product (isocitric acid). The other possible product would be:

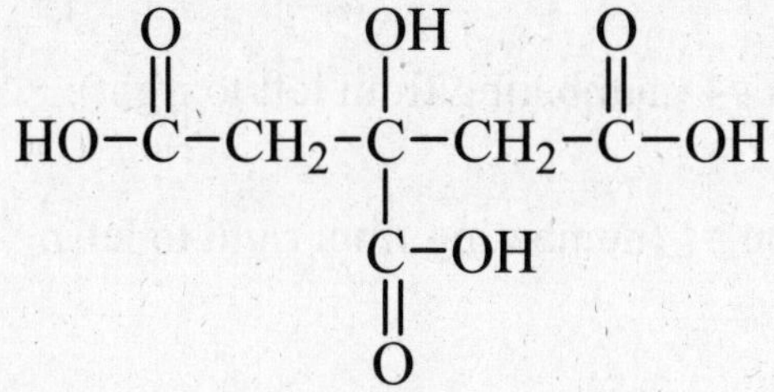

$$HO-\overset{\overset{O}{\|}}{C}-CH_2-\overset{\overset{OH}{|}}{\underset{\underset{\underset{O}{\|}}{C-OH}}{|}}{C}-CH_2-\overset{\overset{O}{\|}}{C}-OH$$

Solutions to Section 9.3 Core Problems

9.11 *Write the IUPAC names for each of the following alcohols:*

a) $CH_3-CH_2-CH_2-CH_2-OH$

b) 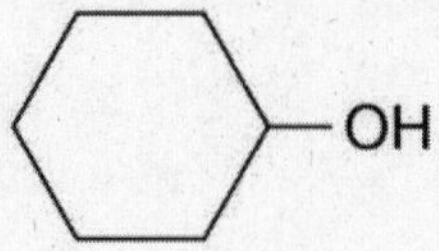 (cyclohexane ring)–OH

c) 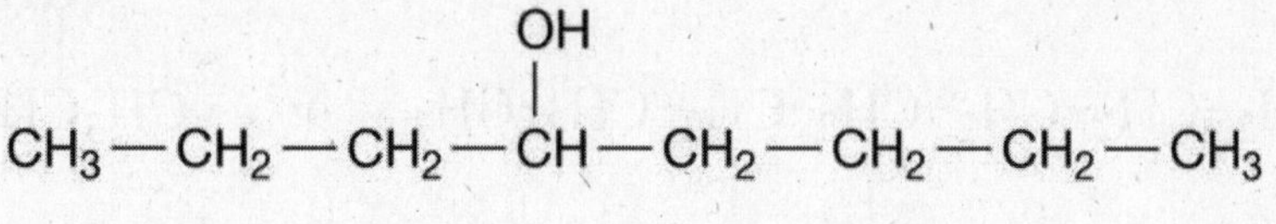 $CH_3-CH_2-CH_2-\overset{\overset{OH}{|}}{CH}-CH_2-CH_2-CH_2-CH_3$

d) 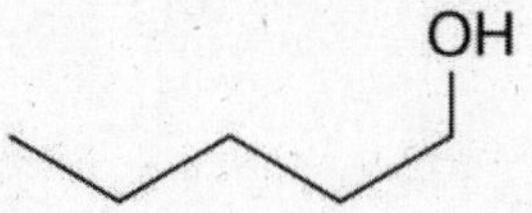 (skeletal structure: five-carbon chain with terminal OH)

a) 4-carbon alkane chain (butan-); number from right to left, alcohol group (-ol) on carbon #1, **1-butanol.**

b) cyclohexane molecule with alcohol group (-ol) attached: **cyclohexanol.** (No position number needed, because the carbon with the alcohol group attached is automatically carbon #1.)

c) octane with alcohol group (-ol) on carbon #4 (numbering from left to right): **4-octanol.**

d) pentane with alcohol group (-ol) on carbon #1 (numbering from right to left): **1-pentanol**

9.13 *Draw the condensed structural formula and the line structure of each of the following alcohols:*

a) 3-pentanol

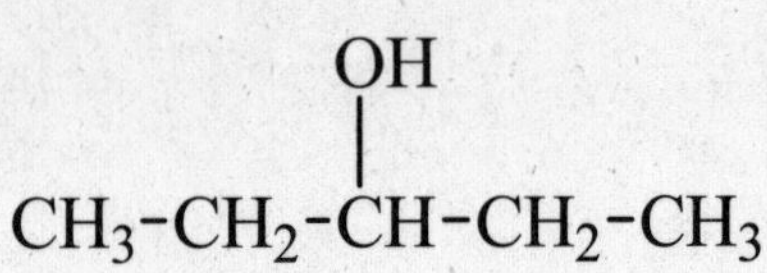

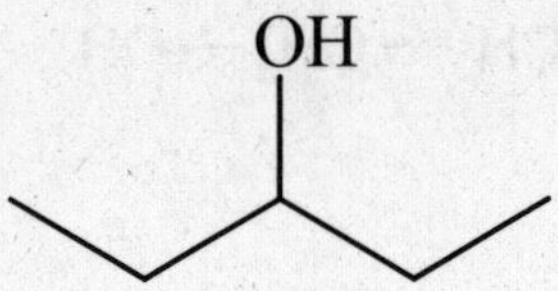

$$CH_3-CH_2-\underset{}{\overset{OH}{\overset{|}{C}}}H-CH_2-CH_3$$

b) 2-nonanol

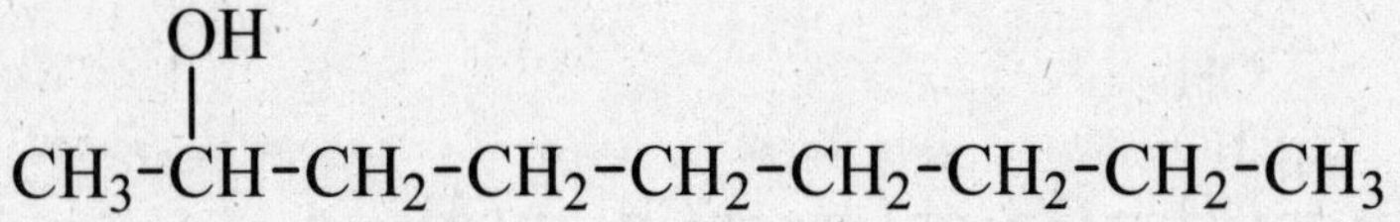

$$CH_3-\overset{OH}{\overset{|}{C}}H-CH_2-CH_2-CH_2-CH_2-CH_2-CH_2-CH_3 \quad \text{or} \quad CH_3\overset{OH}{\overset{|}{C}}H(CH_2)_6CH_3$$

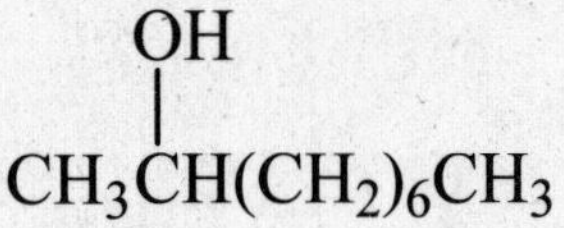

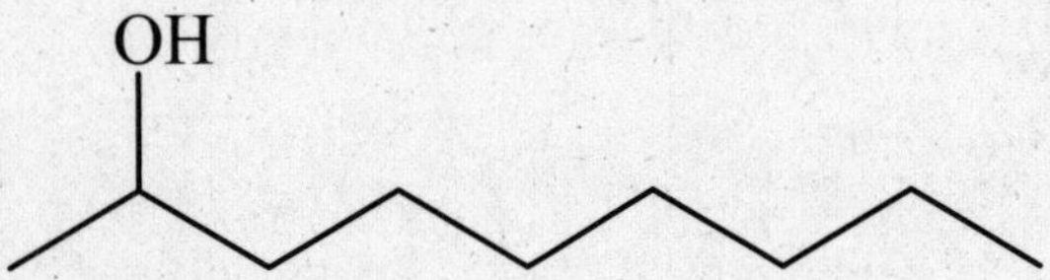

c) cyclobutanol

$$\begin{array}{l} \quad\quad\;\; OH \\ \quad\quad\; / \\ CH_2-CH \\ \;| \quad\quad\; | \\ CH_2-CH_2 \end{array}$$

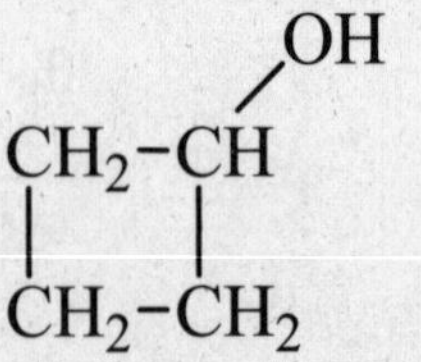

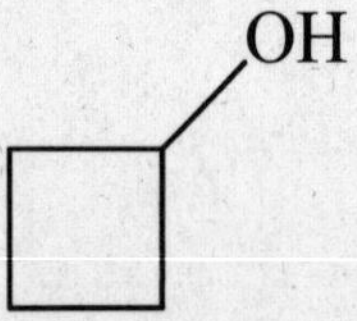

9.15 *Draw the structure of methyl alcohol, and give its IUPAC name.*

The name "methyl alcohol" indicates a methyl group ($-CH_3$) attached to an alcohol group ($-OH$): **CH_3-OH, methanol**

9.17 *Which of the following names are incorrect, according to the IUPAC rules? Explain your answer.*

a) pentanol — **Incorrect.** This name could apply to 3 different molecules (1-pentanol, 2-pentanol, 3-pentanol). A number is needed to specify the position of the alcohol group.

b) 1-pentanol — **Correct.** This name refers to a unique structure, correctly numbered.

c) 2-pentanol — **Correct.** This name refers to a unique structure, correctly numbered.

d) 4-pentanol — **Incorrect.** The molecule should be numbered starting at the end nearest the functional group. The correct name for this molecule is 2-pentanol.

Solutions to Section 9.4 Core Problems

9.19 *Draw a picture that shows the two ways in which a water molecule and a molecule of methanol can form hydrogen bonds with each other.*

For a review of hydrogen bonding, see Section 4.5. Hydrogen bonds are attractions between positively-charged H atoms and lone pairs on O or N atoms.

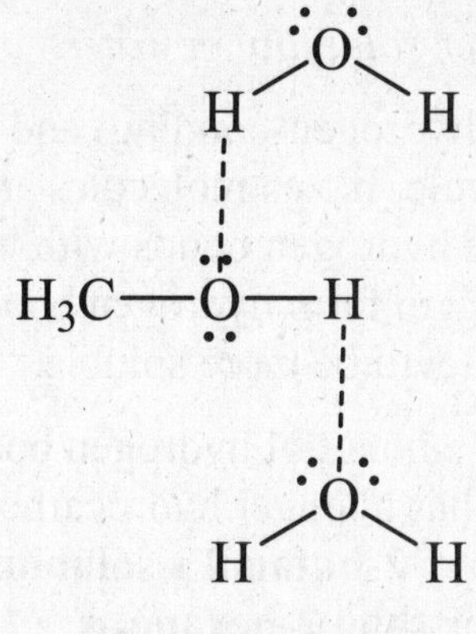

9.21 *From each pair of compounds, select the one with the higher boiling point.*

Boiling point depends on the strength of the attractions between molecules: dispersion forces, which are determined by the number of electrons in the molecules, and hydrogen bonding, which we learned about in Chapter 4. If two molecules are about the same size and therefore have the same dispersion forces, then hydrogen bonding may break the tie; if both molecules have equivalent hydrogen bonding, then look at dispersion forces.

a) 3-pentanol or pentane

3-pentanol is a larger molecule with more electrons and therefore slightly stronger dispersion forces than pentane. It also has hydrogen bonding, which pentane lacks. **3-pentanol should have a significantly higher boiling point than pentane.**

b) 2-butanol or 2-heptanol

Both alcohols have the same ability to form hydrogen bonds between the molecules, but 2-heptanol has stronger dispersion forces because it has more electrons. **2-heptanol should have a significantly higher boiling point than 2-butanol.**

c)

$$CH_3-CH_2-\underset{|}{\overset{OH}{CH}}-CH_2-CH_3$$

or

$$\overset{OH}{\underset{}{CH_2}}-CH_2-\overset{OH}{CH}-CH_2-CH_3$$

The second compound has both stronger dispersion forces (because it has more electrons) and more hydrogen bonding interactions (because it has two –OH groups, in comparison to only one in the first molecule) between its molecules. **The second compound is expected to have a higher boiling point than the first.**

9.23 *From each pair of compounds, select the one with the higher solubility in water.*

Solubility in water depends on the balance of hydrophilic (hydrogen-bonding) and hydrophobic (especially hydrocarbon) regions of the molecule. If two molecules are about the same size, the one with the greater ability to form hydrogen bonds with water will be more soluble; if two molecules have the same ability to form hydrogen bonds with water, the molecule with the smaller hydrophobic area will be more soluble.

a) 2-butanol or 2-hexanol Both alcohols have the same amount of hydrogen bonding (hydrophilic), but 2-hexanol has a longer hydrocarbon chain and is more hydrophobic. **2-butanol's solubility in water is significantly greater than 2-hexanol's.**

b) 2-butanol or hexane Hexane has no ability to hydrogen bond–no hydrophilic areas, the entire molecule is hydrophobic. 2-butanol has a hydrophilic alcohol group, so **2-butanol's solubility in water is much greater than 2-hexanol's.**

c) ethanol or ethyne Ethyne has no ability to hydrogen bond–no hydrophilic areas, the entire molecule is hydrophobic. Ethanol has a hydrophilic alcohol group, so **ethanol's solubility in water is much greater than ethyne's.**

9.25 *Tell whether each of the following solutions is acidic, neutral, or basic.*

a) 0.01 M KOH KOH is an ionic compound and dissociates into K^+ ions and OH^- ions. OH^- ions make a solution **basic**.

b) 0.01 M CH_3OH — CH_3OH is a covalent compound; the alcohol group is neither acidic or basic, so methanol does not dissociate in water. Alcohols are **neutral** in aqueous solution.

9.27 *One of the following compounds does not dissolve in benzene. Which one?*

a) $CH_3–CH_2–CH_2–CH_3$ *b) $CH_3–CH_2–CH_2–OH$* *c) $HO–CH_2–CH_2–OH$*

Benzene is a hydrocarbon with no hydrophilic functional groups. Compounds with a significant amount of hydrophilic nature are likely to be insoluble in benzene, while compounds that are mostly hydrophobic are expected to be highly soluble in benzene. **Of the three, Compound (c) $HO–CH_2–CH_2–OH$, which has two hydrophilic –OH groups, is the most hydrophilic and therefore the least soluble in benzene.**

Solutions to Section 9.5 Core Problems

9.29 *Tell whether each of the following objects is chiral or achiral:*

If an object is chiral, we can tell the difference between the object and its mirror image. If an object fits differently on or in one hand (or foot) than the other, it's probably chiral. If it feels the same in either hand, it's achiral (not chiral).

a) a shoe — A shoe does not fit equally well on either foot–a shoe and its mirror image cannot be superimposed on each other. **A shoe is chiral.**

b) a sock — Most socks fit equally well on either foot–a sock and its mirror image are superimposable. **Socks are not chiral.**

c) a spoon — A spoon and its mirror image are superimposeable. The spoon fits equally well in either hand. **Spoons are not chiral.**

d) a coffee cup — A typical coffee cup and its mirror image are superimposeable. The cup fits equally well in either hand. **Coffee cups are not generally chiral.** (However a particular coffee cup with a curved handle to fit one hand, or a coffee cup with a design on it, might well be distinguishable from its mirror image and therefore **chiral**.)

9.31 *Find and circle each of the chiral carbon atoms in the following molecules:*

a) 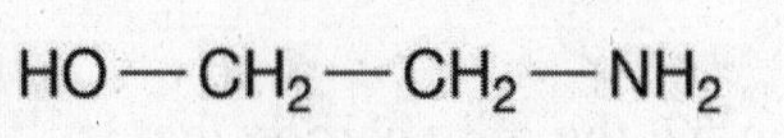 $HO—CH_2—CH_2—NH_2$

Ethanolamine,
a component of
cell membrane lipids

b) 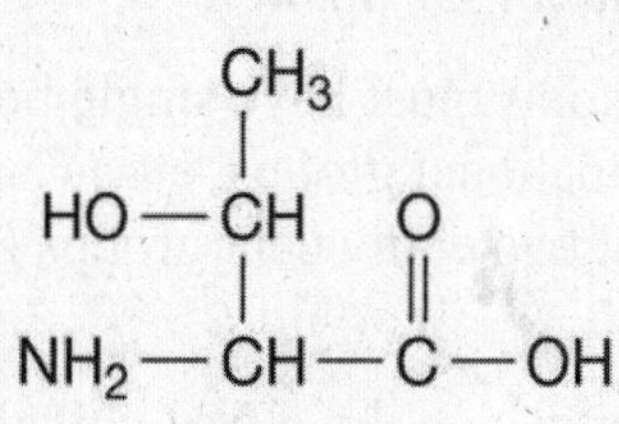 $NH_2—CH(—CH(OH)—CH_3)—C(=O)—OH$

```
          CH3
          |
       HO—CH   O
          |    ||
b) NH2—CH—C—OH
```

Threonine,
an amino acid

c) 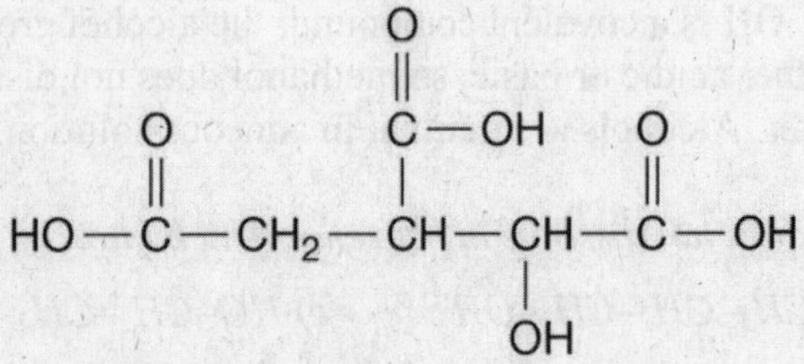 HO—C(=O)—CH_2—CH(—C(=O)—OH)—CH(—OH)—C(=O)—OH

Isocitric acid,
a product of the oxidation of
carbohydrates, fats, and proteins

Chiral carbon atoms must have single bonds to four different groups (hydrogen atoms, alkyl groups, functional groups, etc.) Carbon atoms with double bonds, or those with bonds to at least two equivalent groups (for example, the carbons in CH_2 and CH_3), are not chiral. Ethanolamine has no chiral carbon atoms; each of the two carbon atoms has two equivalent groups (two hydrogen atoms on each carbon). Threonine and isocitric acid each have two chiral carbon atoms (these are circled below)

a) None are chiral

b)

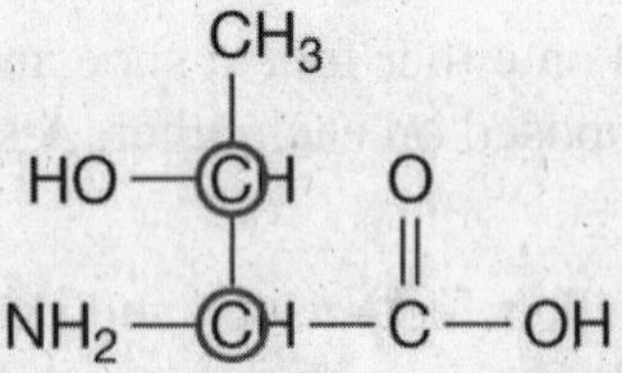

CH_3
HO—(CH)
NH_2—(CH)—C(=O)—OH

c)

HO—C(=O)—CH_2—(CH)(—C(=O)—OH)—(CH)(—OH)—C(=O)—OH

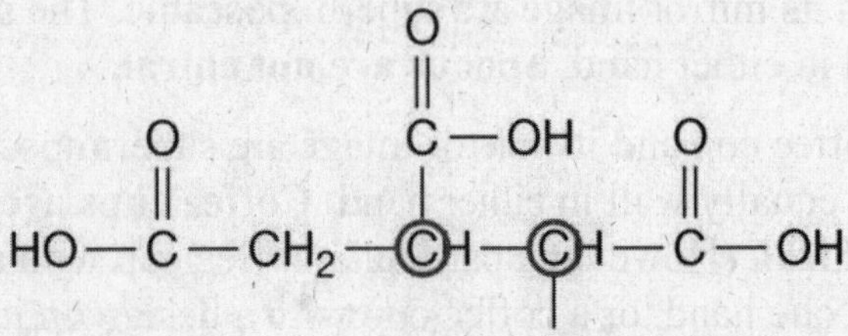

9.33 *One of the following molecules is chiral, but the other two are not. Tell which molecule is chiral, and explain your answer.*

Chiral carbon atoms must have single bonds to four different groups (hydrogen atoms, alkyl groups, functional groups, etc.) Carbon atoms with double bonds, or those with bonds to at least two equivalent groups (for example, the carbons in CH_2 and CH_3), are not chiral.

a) 2-methylheptane — **Not chiral**; Carbon #2, the only carbon that isn't a CH_2 or CH_3, has two identical methyl groups attached:

$$CH_3-CH(CH_3)-CH_2-CH_2-CH_2-CH_2-CH_3$$

b) *3-methylheptane* **Chiral. Carbon #3 (circled) has four different groups attached: one –H, one methyl, one ethyl, and one butyl.**

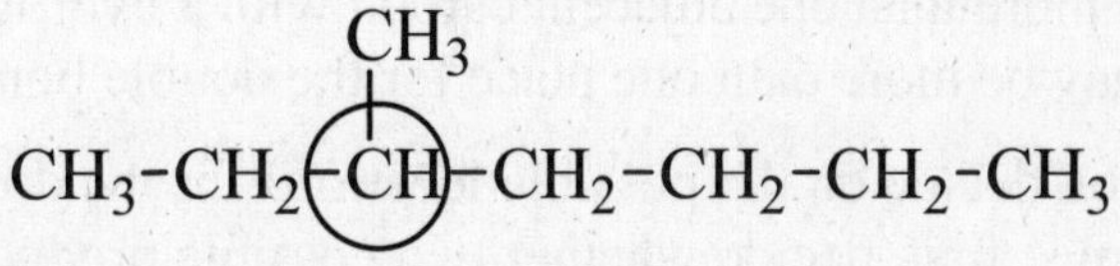

$$CH_3-CH_2-\underset{\text{(circled)}}{\overset{\displaystyle CH_3}{\overset{|}{CH}}}-CH_2-CH_2-CH_2-CH_3$$

c) *4-methylheptane* **Not chiral**; Carbon #4 has two identical propyl groups attached.

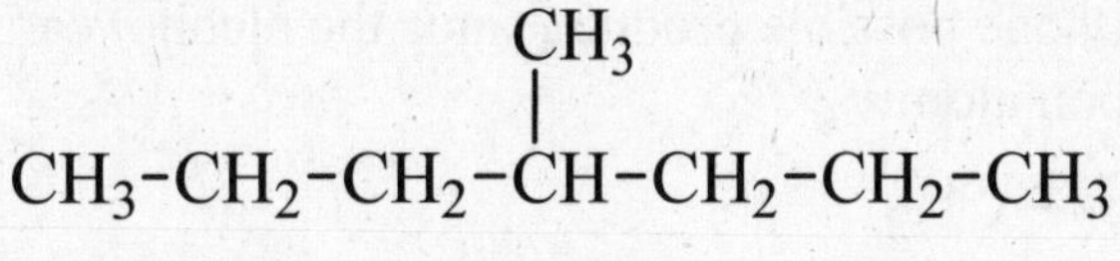

$$CH_3-CH_2-CH_2-\overset{\displaystyle CH_3}{\overset{|}{CH}}-CH_2-CH_2-CH_3$$

9.35 *Why is it important that enzymes are chiral molecules?*

Carbohydrates and proteins are chiral, so to make, use, or otherwise do chemistry on these substances, chiral enzymes are required; enzymes must have structures that match up with the molecules whose reactions they catalyze.

Solutions to Section 9.6 Core Problems

9.37 *Draw the structures of the products that are formed when the following alcohols are dehydrated. Be sure to draw all of the possible products for each reaction.*

a) CH_3-CH_2-OH

b) 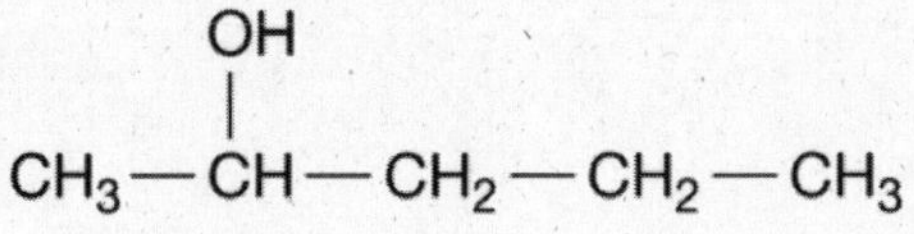 $CH_3-\overset{\displaystyle OH}{\overset{|}{CH}}-CH_2-CH_2-CH_3$

c) 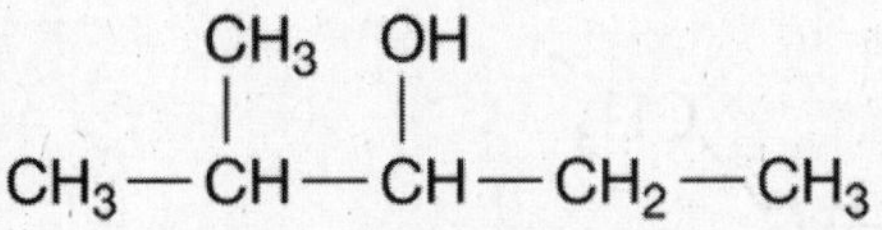 $CH_3-\overset{\displaystyle CH_3}{\overset{|}{CH}}-\overset{\displaystyle OH}{\overset{|}{CH}}-CH_2-CH_3$

d) 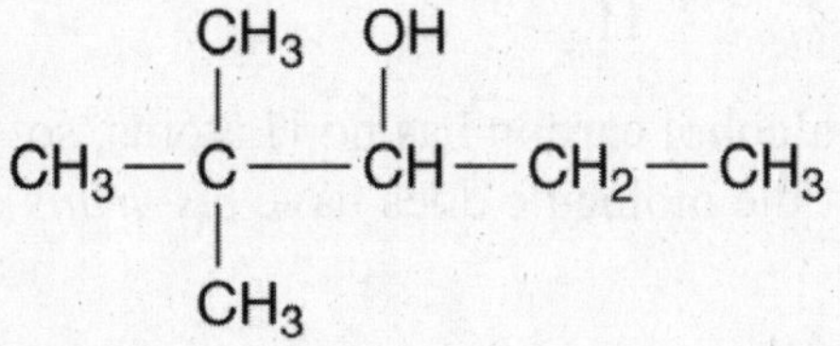 $CH_3-\underset{\underset{\displaystyle CH_3}{|}}{\overset{\displaystyle CH_3}{\overset{|}{C}}}-\overset{\displaystyle OH}{\overset{|}{CH}}-CH_2-CH_3$

e) 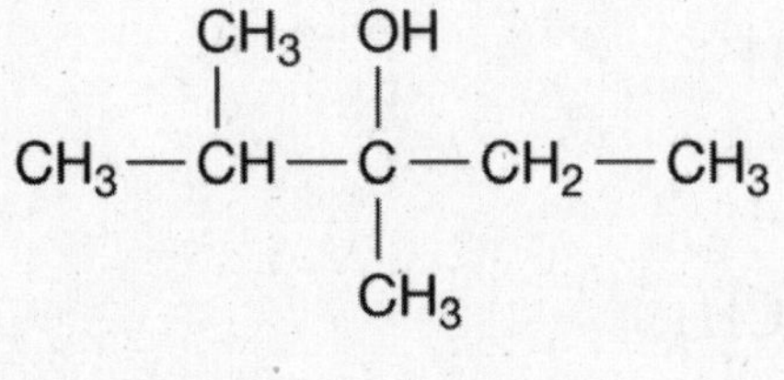 $CH_3-\overset{\displaystyle CH_3}{\overset{|}{CH}}-\underset{\underset{\displaystyle CH_3}{|}}{\overset{\displaystyle OH}{\overset{|}{C}}}-CH_2-CH_3$

f) 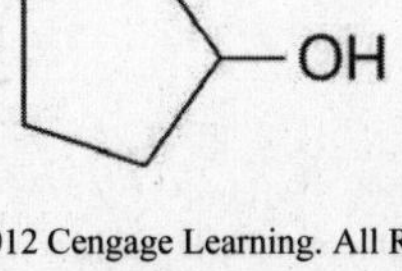 cyclopentane ring –OH

Dehydration removes an alcohol group (–OH) from one carbon and a hydrogen atom (–H) from an adjacent carbon, forming a double bond between the two carbon atoms. If there is more than one adjacent carbon with a hydrogen atom that can be removed, then there may be more than one place for the double bond to go in the molecule.

Once you decide on the possible locations for the double bond, you still have to check on two things: first, decide whether two possible products actually turn out to be the same molecule; and second, since we're dealing with alkenes, look to see if *cis–trans* isomers are possible.

a) Only one possible product, since the alcohol carbon is only connected to one other carbon atom:

$CH_2{=}CH_2$

b) Two possible locations for the double bond, one of which has *cis–trans* isomers:

$CH_2{=}CH{-}CH_2{-}CH_2{-}CH_3$

H, H / C=C / CH_3, $CH_2{-}CH_3$ CH_3, H / C=C / H, $CH_2{-}CH_3$

c) Two possible locations for the double bond, one of which gives a molecule with *cis–trans* isomers:

$CH_3{-}C(CH_3){=}CH{-}CH_2{-}CH_3$

$CH_3{-}CH(CH_3)$, H / C=C / H, CH_3 $CH_3{-}CH(CH_3)$, CH_3 / C=C / H, H

d) One of the two carbon atoms connected to the alcohol carbon has no H atoms, so there is only one place for the double bond, but the molecule does have *cis–trans* isomers:

$CH_3{-}C(CH_3)(CH_3)$, CH_3 / C=C / H, H $CH_3{-}C(CH_3)(CH_3)$, H / C=C / H, CH_3

e) There are two possible locations for the double bond formed in the dehydration reaction:

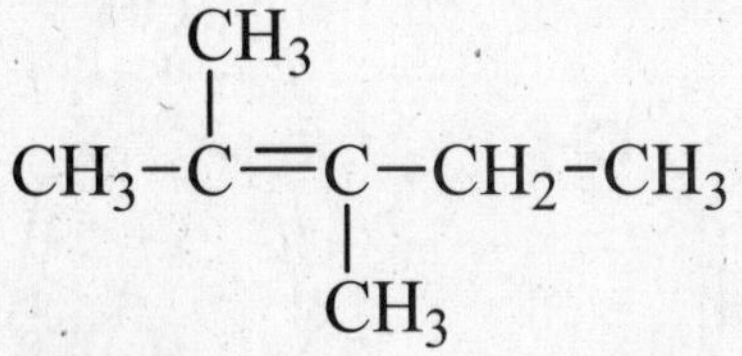

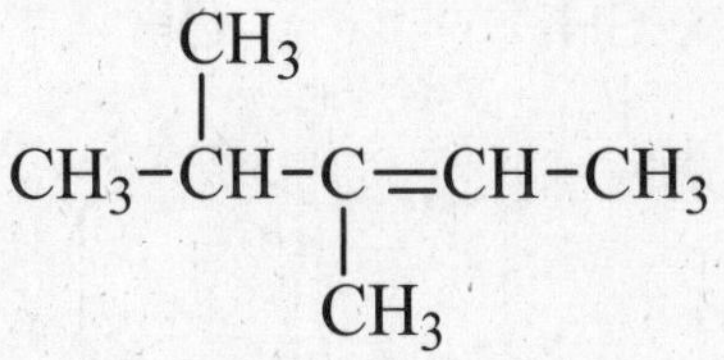

$$CH_3-\underset{\displaystyle CH_3}{\overset{\displaystyle CH_3}{C}}{=}\!\!\!\!\!\!\quad C-CH_2-CH_3 \qquad\qquad CH_3-\overset{\displaystyle CH_3}{CH}-\underset{\displaystyle CH_3}{C}=CH-CH_3$$

NOTE:
The second molecule actually has *cis–trans* isomers, but it's beyond the scope of this text to determine which isomer is which.)

f) While there are two directions the double bond can go from the alcohol carbon, they both actually give the same molecule, so there's only one possible product:

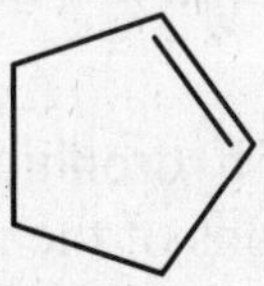

9.39 *The following alcohol cannot be dehydrated. Explain.*

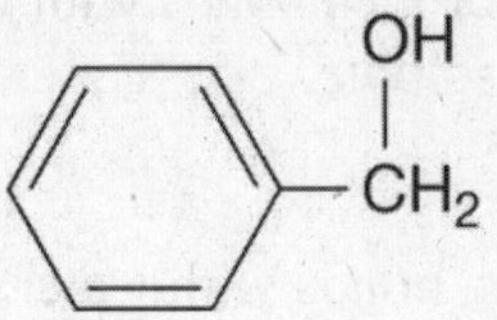

OH
CH2

In a dehydration reaction, an alcohol group from one carbon and a hydrogen atom from an adjacent carbon are removed, forming a double bond between the two carbon atoms. In this molecule, the carbon atom next to the alcohol carbon has no hydrogen atoms, so dehydration cannot occur. Or, to look at it another way, the product of the dehydration would have a double bond between the alcohol carbon and the next carbon, in the ring; this would give the carbon in the ring 5 bonds, so it can't happen.

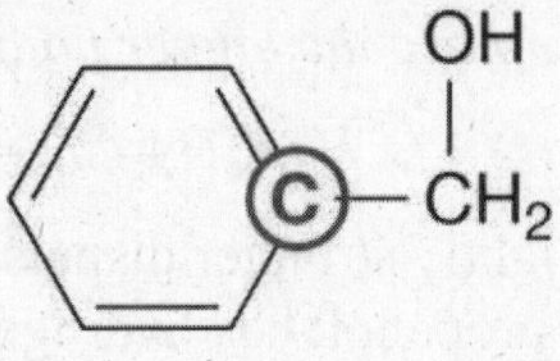

OH
C
CH2

Solutions to Section 9.7 Core Problems

9.41 *Classify each of the following molecules as an alcohol, a phenol, or a thiol:*

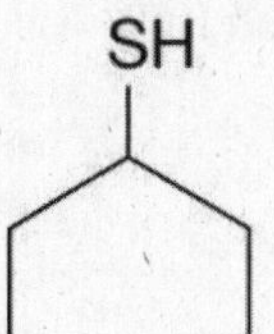

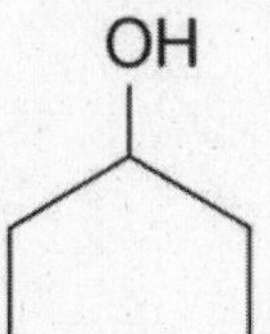

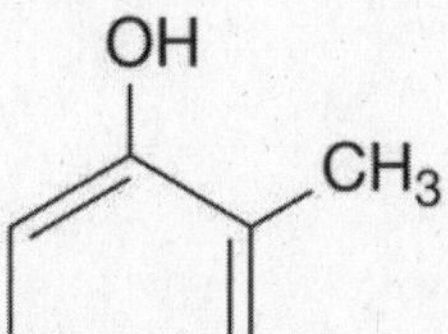

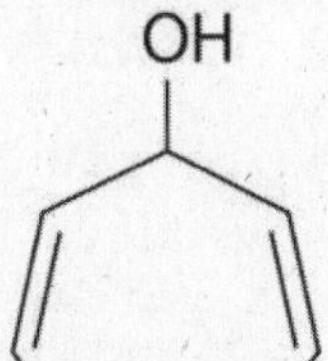

SH OH OH CH3 OH

An alcohol has an –OH group on an alkane, alkene or alkyne carbon; a phenol has an –OH group on an aromatic ring carbon; and a thiol has an –SH group.

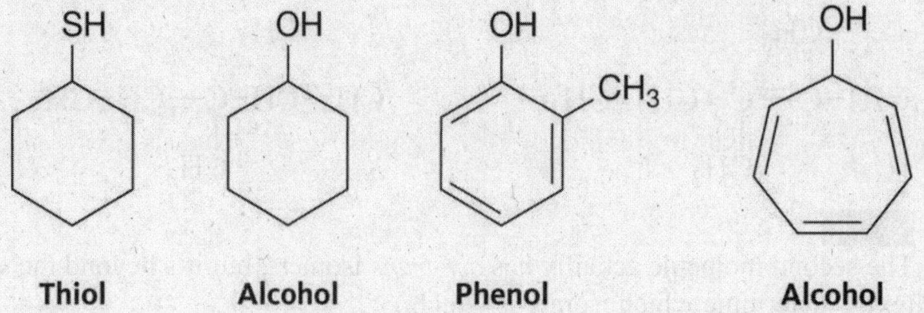

SH | OH | OH CH_3 | OH

Thiol | Alcohol (ring is not aromatic) | Phenol (ring is aromatic) | Alcohol (ring is not aromatic)

9.43 *From each of the following pairs of molecules, select the compound that has the higher solubility in water.*

Solubility of organic compounds in water depends on the balance of hydrophilic (hydrogen-bonding) and hydrophobic (especially hydrocarbon) regions of the molecule. If two molecules are about the same size, the one with the greater ability to form hydrogen bonds with water is more hydrophilic and will be more soluble; if two molecules have the same ability to form hydrogen bonds with water, the molecule with the smaller hydrophobic area will be more soluble.

a) *phenol or benzene*

Benzene has no ability to form hydrogen bonds with water; phenol, with its –OH group, can. **Phenol is much more soluble in water than benzene is.**

b) $CH_3–CH_2–CH_2–SH$ *or* $CH_3–CH_2–CH_2–OH$

The thiol group in the first molecule, propanethiol, is hydrophobic and it cannot participate in hydrogen bonding with water. The alcohol group in the second molecule, propanol, is hydrophilic. **Propanol is much more soluble in water than propanethiol is.**

9.45 *Which of the following compounds should have the higher boiling point?*

$CH_3–CH_2–CH_2–SH$ *or* $CH_3–CH_2–CH_2–OH$

$CH_3–CH_2–CH_2–SH$, propanethiol, has slightly stronger dispersion forces, but no hydrogen bonding. $CH_3–CH_2–CH_2–OH$, propanol, has hydrogen bonding attractions between its molecules, in addition to dispersion forces. **The total attractions between molecules are greater in propanol, and it has a higher boiling point than propanethiol.**

Solutions to Concept Questions

* indicates more challenging problems.

9.47 a) *When water reacts with an alkene, what type of compound is formed?*

The reaction of water with an alkene is a hydration reaction, which forms an **alcohol**.

b) *When an alcohol is dehydrated, what type of organic compound is formed?*

The dehydration of an alcohol removes an –OH group and an –H from adjacent carbon atoms, forming a double bond between the two carbon atoms: an **alkene**.

9.49 *Most reactions in our bodies require an enzyme, even if they have only one possible product. Why is this?*

Most of the reactions that happen in our bodies would be so slow, without a catalyst, that they wouldn't be fast enough to do us any good; to happen at an appreciable rate they require enzyme catalysis.

9.51 *Both of the following compounds are polar, but only one of them dissolves well in water. Which one is the water-soluble compound? Explain your reasoning.*

$F-CH_2-CH_2-F$ $HO-CH_2-CH_2-OH$

Ethylene difluoride   **Ethylene glycol**

The major attraction between molecules in water is hydrogen bonding. Other molecules that can participate in hydrogen bonding are often soluble in water, while molecules that cannot participate in hydrogen bonding (and therefore interrupt, rather than add to, the hydrogen-bonding network in liquid water) are usually insoluble.

Ethylene glycol contains –OH groups that can form hydrogen bonds with water. Ethylene difluoride cannot participate in hydrogen bonding (F can only participate in hydrogen bonding in the specific compound HF; fluorine in organic compounds does not form hydrogen bonds.) Therefore **ethylene glycol is soluble in water, while ethylene difluoride is not.**

9.53 *Why is ethanol more soluble in water than ethane is?*

The major attraction between molecules in water is hydrogen bonding. Other molecules that can participate in hydrogen bonding are often soluble in water, while molecules that cannot participate in hydrogen bonding (and therefore interrupt, rather than add to, the hydrogen-bonding network in liquid water) are usually insoluble. **Ethane, a hydrocarbon, has no ability to form hydrogen bonds with water, while ethanol has a hydrophilic –OH group and can participate in hydrogen bonding with water.**

9.55 *How can you identify a chiral carbon atom in a molecule?*

A chiral carbon atom must have four single bonds to four different groups (hydrogen atoms, alkyl groups, or functional groups.) If the carbon atom has one or more multiple bonds, or if any two groups attached to it are the same, then the carbon atom is not chiral.

9.57 *What is the maximum number of alkenes that could be formed by dehydrating an alcohol? Explain your answer.*

In the dehydration of an alcohol, an –OH group from one carbon and an –H from an adjacent carbon are removed, forming a double bond between the two carbon atoms. In principle, there could be as many as three different possible products, assuming that the –OH carbon had three carbon atoms attached to it, each with at least one hydrogen atom. For example:

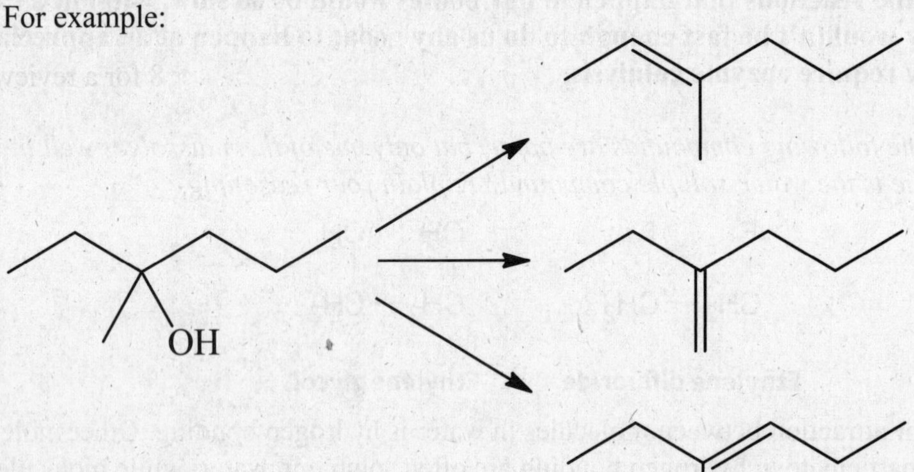

OH

No more than three options are possible (not counting *cis/trans* isomers), because carbon can only make 4 bonds—there can't be more than three different carbons attached to the alcohol carbon.

9.59 *What is the difference between an alcohol and a phenol?*

An alcohol has an –OH group attached to an alkene, alkane or alkyne carbon. A phenol has an –OH group attached directly to one of the carbon atoms in an aromatic ring.

Solutions to Summary and Challenge Problems

* indicates more challenging problems.

9.61 *Draw structures of molecules that fit each of the following descriptions:*

a) *a cyclic alcohol*

Choose any cyclic alkane and add an –OH group. Two options (of many possible structures):

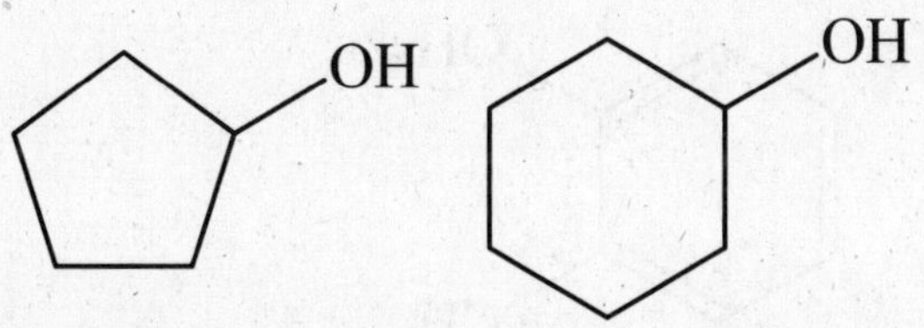

OH OH

b) *a thiol that contains three carbon atoms*

Make a propane chain and add an –SH group. This time there are only two possible answers:

$$CH_3\text{-}CH_2\text{-}CH_2\text{-}SH \qquad \begin{array}{c} SH \\ | \\ CH_3\text{-}CH\text{-}CH_3 \end{array}$$

c) *an alcohol that contains a branched carbon chain*

For this one there are an infinite number of answers. See Chapter 8 for a review of branched carbon chains if you need to. Here are two:

$$\begin{array}{l} \quad CH_3 \\ \quad\ | \\ CH_3\text{-}CH\text{-}CH_2\text{-}OH \end{array} \qquad \begin{array}{c} OH \\ | \\ CH_3\text{-}C\text{-}CH_3 \\ | \\ CH_3 \end{array}$$

d) *a constitutional isomer of 1-pentanol*

The molecular formula of 1-pentanol is $C_5H_{12}O$. We can move the alcohol group around:

$$\begin{array}{l} \qquad OH \\ \qquad\ | \\ CH_3\text{-}CH\text{-}CH_2\,CH_2\text{-}CH_3 \end{array} \qquad \begin{array}{l} \qquad\qquad\quad OH \\ \qquad\qquad\quad\ | \\ CH_3\text{-}CH_2\text{-}CH\text{-}CH_2\text{-}CH_3 \end{array}$$

or branch the chain, changing the carbon connectivity:

$$\begin{array}{l} OH \\ | \\ CH_2\text{-}CH\text{-}CH_2\,CH_3 \\ \qquad | \\ \quad\ CH_3 \end{array} \qquad \begin{array}{l} \qquad\ OH \\ \qquad\ \ | \\ CH_3\text{-}C\text{—}CH_2\,CH_3 \\ \qquad\ \ | \\ \qquad CH_3 \end{array}$$

(+ many other options)

We can also give the O atom two bonds to carbon instead of only one (it's no longer an alcohol then):

$CH_3\text{-}CH_2\text{-}O\text{-}CH_2\text{-}CH_2\text{-}CH_3$ (+ several other options)

e) *a phenol that contains more than six carbon atoms*

Since we have to start with an aromatic ring with an –OH group attached, we're limited to adding carbon atoms elsewhere on the ring somehow. Here are two options of many:

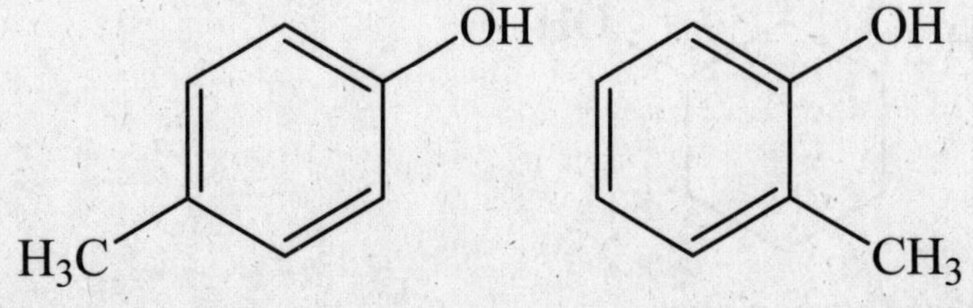

$$\text{4-methylphenol (OH, } H_3C\text{ para)} \qquad \text{2-methylphenol (OH, } CH_3\text{ ortho)}$$

*9.63 *a) Two possible alcohols can be dehydrated to form the alkene shown here. Draw the structures of these two alcohols.*

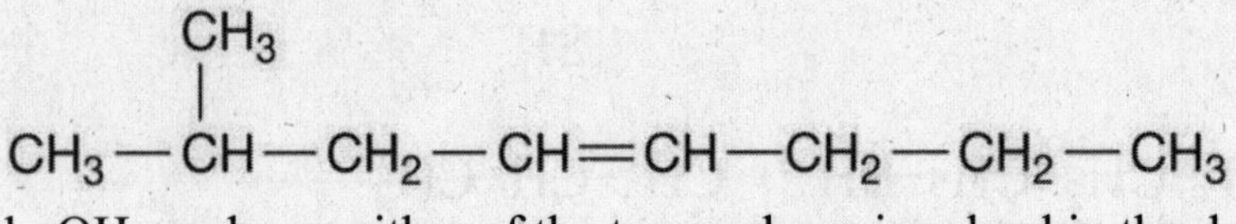

$$CH_3-\underset{}{\overset{CH_3}{\overset{|}{C}}}H-CH_2-CH{=}CH-CH_2-CH_2-CH_3$$

The alcohol –OH can be on either of the two carbons involved in the double bond:

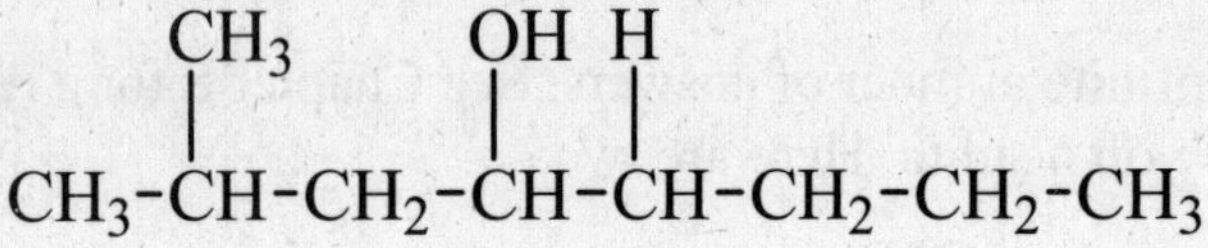

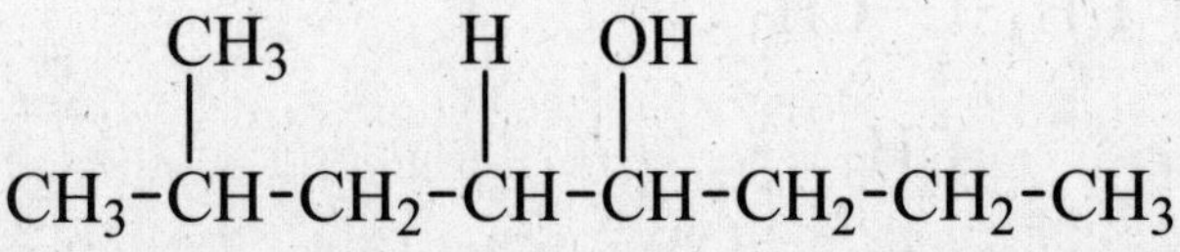

$$CH_3\text{-}\overset{CH_3}{\overset{|}{C}}H\text{-}CH_2\text{-}\overset{OH}{\overset{|}{C}}H\text{-}\overset{H}{\overset{|}{C}}H\text{-}CH_2\text{-}CH_2\text{-}CH_3$$

$$CH_3\text{-}\overset{CH_3}{\overset{|}{C}}H\text{-}CH_2\text{-}\overset{H}{\overset{|}{C}}H\text{-}\overset{OH}{\overset{|}{C}}H\text{-}CH_2\text{-}CH_2\text{-}CH_3$$

b) Each of the alcohols you drew in part a can make a second alkene. Draw the structures of the other alkenes that you could make by dehydrating these alcohols.

In each case, remove the –OH and an –H from the other adjacent carbon atom:

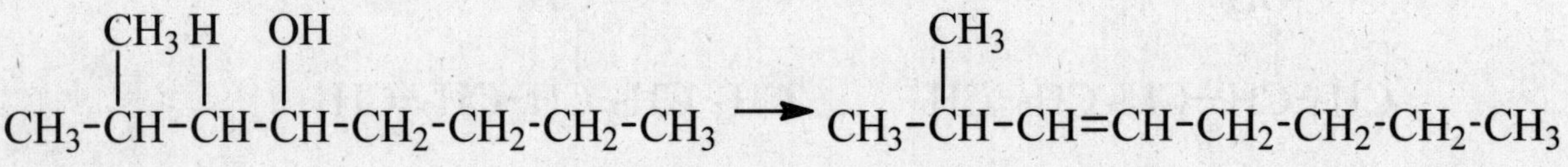

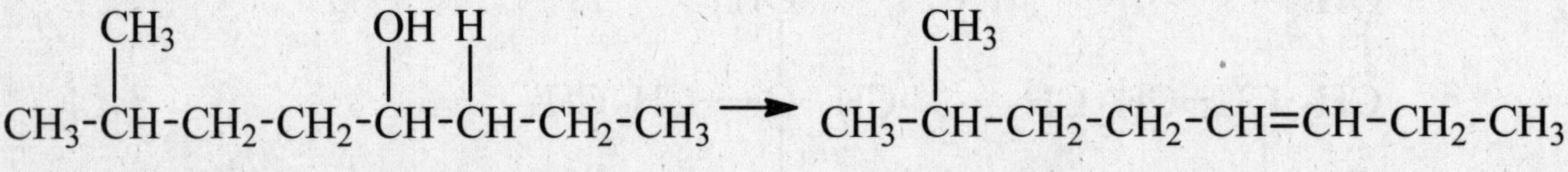

$$CH_3\text{-}\overset{CH_3}{\overset{|}{C}}H\text{-}\overset{H}{\overset{|}{C}}H\text{-}\overset{OH}{\overset{|}{C}}H\text{-}CH_2\text{-}CH_2\text{-}CH_2\text{-}CH_3 \longrightarrow CH_3\text{-}\overset{CH_3}{\overset{|}{C}}H\text{-}CH{=}CH\text{-}CH_2\text{-}CH_2\text{-}CH_2\text{-}CH_3$$

$$CH_3\text{-}\overset{CH_3}{\overset{|}{C}}H\text{-}CH_2\text{-}CH_2\text{-}\overset{OH}{\overset{|}{C}}H\text{-}\overset{H}{\overset{|}{C}}H\text{-}CH_2\text{-}CH_3 \longrightarrow CH_3\text{-}\overset{CH_3}{\overset{|}{C}}H\text{-}CH_2\text{-}CH_2\text{-}CH{=}CH\text{-}CH_2\text{-}CH_3$$

9.65 *Name the following compounds:*

a) CH_3-OH

b) 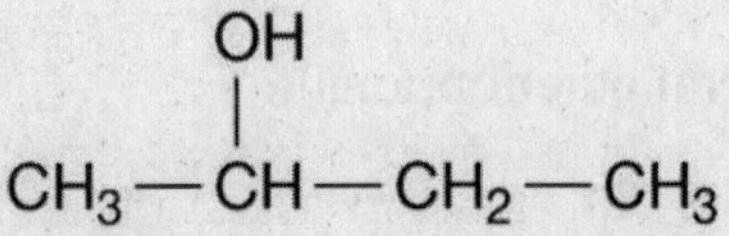 $CH_3-\overset{OH}{\overset{|}{C}}H-CH_2-CH_3$

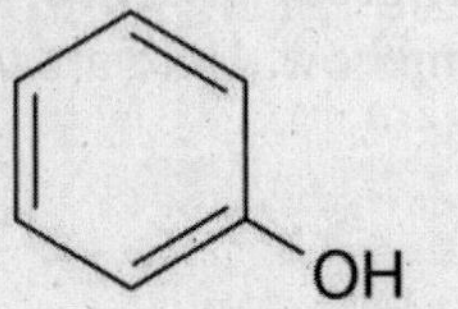

c) (benzene ring with OH)

d) $CH_3—CH_2—CH_2—CH_2—CH_2—CH(OH)—CH_2—CH_3$

e) (cyclohexane ring with OH)

a) one-carbon alcohol: **methanol**

b) four-carbon alcohol, –OH group on second carbon: **2-butanol**

c) alcohol group on a benzene ring: **phenol**

d) eight-carbon alcohol, –OH group on third carbon (number this molecule from right to sleft): **3-octanol**

e) alcohol group on six-carbon ring: **cyclohexanol**

9.67 *Write the molecular formula of each of the following molecules:*

If you need to draw the structure of the molecule first and count up C and H atoms, do so! But there's a shortcut here: any time you draw the structure of an alcohol, you remove an –H atom and add an –OH. (Think about this and make sure you see why.) So in the molecular formula, you would only have to add an O. So methane CH_4 becomes methanol, CH_3OH or CH_4O; ethane C_2H_6 becomes ethanol, C_2H_5OH or C_2H_6O; etc.

a) 2-propanol — Propane is C_3H_8, so propanol is **C_3H_8O**.

b) cyclopropanol — Cyclopropane is C_3H_6, so propanol is **C_3H_6O**.

9.69 *Answer the following questions by selecting compounds from the following list:*

Compound A (phenol, –OH) Compound B (cyclohexanol, –OH) Compound C (cyclohexanethiol, –SH)

a) Which compound has the lowest solubility in water?

Solubility in water depends on the balance of hydrophilic (hydrogen-bonding) and hydrophobic (especially hydrocarbon) regions of the molecule. If two molecules are about the same size, the one with the greater ability to form hydrogen bonds with water will be more soluble. Since thiol groups cannot participate in hydrogen bonding, **compound C is the least soluble**.

b) Which compounds can be dehydrated?

In a dehydration reaction, an alcohol group from one carbon and a hydrogen atom from an adjacent carbon are removed, forming a double bond between the two carbon atoms. In Compound A, dehydration would disrupt the aromatic ring (the alternating pattern of double and single bonds), so dehydration cannot occur (phenols do not undergo dehydration). Compound C doesn't have an alcohol group, so dehydration cannot occur. **Compound B can be dehydrated**: the carbon atoms adjacent to the alcohol carbon have hydrogen atoms that can be removed.

c) *Which compounds can form hydrogen bonds?*

See Chapter 4 if you need a review on hydrogen bonding. Hydrogen bonds are attractions between H atoms covalently bonded to N or O atoms, and N or O atoms with lone pairs and δ- charges. **Compounds A and B can form hydrogen bonds,** both as pure substances and with water in aqueous solution. Compound C does not contain N or O atoms and cannot form hydrogen bonds.

9.71 *Examine the following pairs of structures. In each case, are the two molecules constitutional isomers of one another? If not, explain why not.*

a) $CH_2(OH)-CH_2-CH_2-CH_3$ and $CH_3-CH(OH)-CH_2-CH_3$

b) $CH_3-CH(OH)-CH_2-CH_3$ and $CH_3-CH_2-CH(OH)-CH_3$

c) $CH_3-CH(OH)-CH_2-CH_3$ and $CH_3-CH(OH)-CH(CH_3)-CH_3$

d) $CH_3-CH(OH)-CH_2-CH_3$ and $CH_3-C(OH)(CH_3)-CH_3$

e) $CH_3-CH(OH)-CH_2-CH_2-CH_3$ and $CH_3-O-C(CH_3)_2-CH_3$

f) $CH_3-CH_2-CH_2-CH(OH)-CH_3$ and cyclopentanol (cyclopentane ring with —OH)

g) 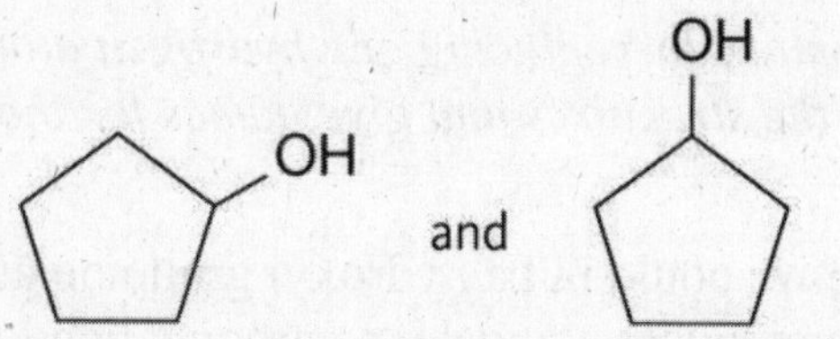 OH and OH

h) 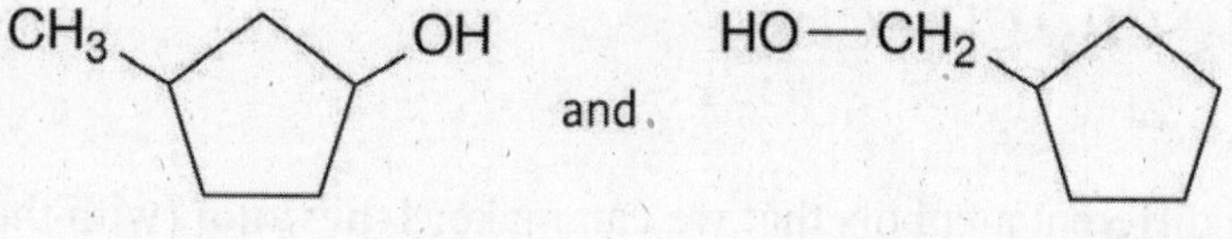 CH_3 OH and HO—CH_2

i) 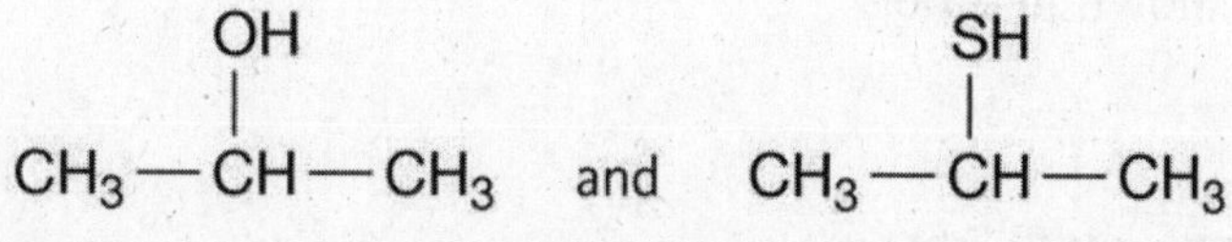 $CH_3—CH(OH)—CH_3$ and $CH_3—CH(SH)—CH_3$

j) 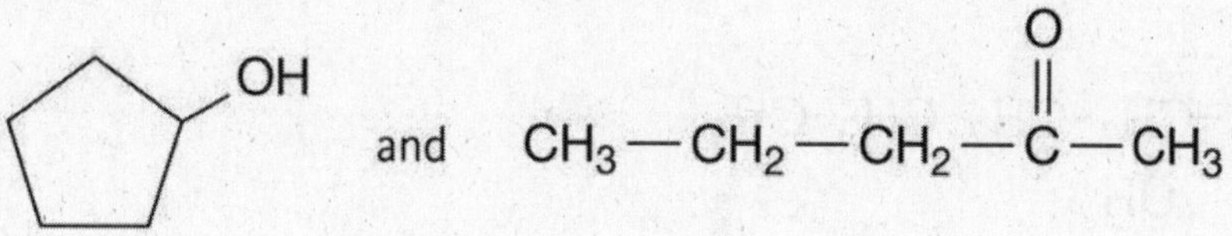 OH and $CH_3—CH_2—CH_2—C(=O)—CH_3$

k) 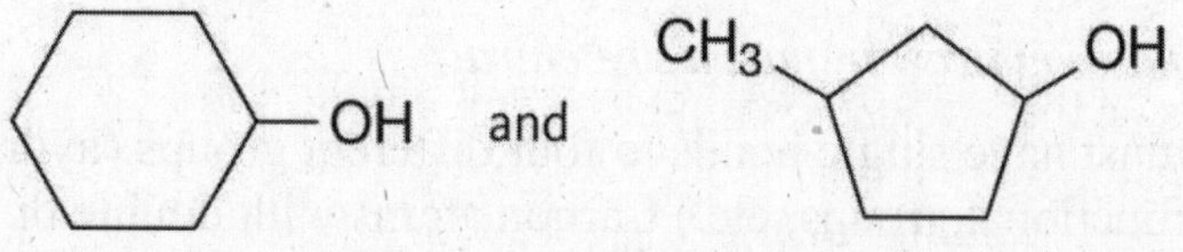 —OH and CH_3 OH

Constitutional isomers have the same molecular formula, but the atoms are connected in a different way. If the molecular formulas of two substances are different, they are not isomers; if the only difference between two molecules is that one is *cis* and one is *trans*, then they are stereoisomers, not constitutional isomers; and if the two structures represent different ways of drawing the same molecule, then they aren't isomers.

a) Two molecules with the same molecular formula ($C_4H_{10}O$) and different connectivity: **constitutional isomers.**

b) **Same molecule, not isomers**: remember that the principal chain is numbered starting from the end closer to the functional group; both structures represent 2-butanol.

c) **Not isomers**. These molecules have different formulas ($C_4H_{10}O$ and $C_5H_{12}O$).

d) Same molecular formula ($C_4H_{10}O$) and different connectivity: **constitutional isomers.**

e) Same molecular formula ($C_5H_{12}O$) and different connectivity: **constitutional isomers.**

f) **Not isomers**. These molecules have different formulas ($C_5H_{12}O$ and $C_5H_{10}O$).

*9.73 a) *How many different alcohols can you make by replacing one hydrogen atom in hexane with a hydroxyl group? Draw the structures and give names for these compounds.*

First, let's look at hexane. In principle we could put a hydroxyl group on any of the six carbons, but in reality, carbon #1 and #6 are equivalent, carbon #2 and #5 are equivalent, and carbon #3 and #4 are equivalent:

$$CH_3-CH_2-CH_2-CH_2-CH_2-CH_3$$
1 2 3 4 5 6

So there are three different alcohols that we can make: **1-hexanol** (with the –OH group on carbon #1), **2-hexanol** (–OH on carbon #2) and **3-hexanol** (–OH on carbon #3). The structures are shown in order:

$$\overset{OH}{\overset{|}{CH_2}}-CH_2-CH_2-CH_2-CH_2-CH_3$$

$$CH_3-\overset{OH}{\overset{|}{CH}}-CH_2-CH_2-CH_2-CH_3$$

$$CH_3-CH_2-\overset{OH}{\overset{|}{CH}}-CH_2-CH_2-CH_3$$

b) *Which of the compounds you drew in part a are chiral?*

Chiral carbon atoms must have single bonds to four different groups (hydrogen atoms, alkyl groups, functional groups, etc.) Carbon atoms with double or triple bonds, or those with bonds to at least two equivalent groups (for example, the carbons in CH_2 and CH_3 groups), are not chiral.

Only 2-hexanol and 3-hexanol are chiral (the chiral carbon atoms are circled in the structures below); in 1-hexanol, every carbon atom has at least two hydrogen atoms attached to it.

$$CH_3-\overset{OH}{\overset{|}{\textcircled{CH}}}-CH_2-CH_2-CH_2-CH_3 \qquad CH_3-CH_2-\overset{OH}{\overset{|}{\textcircled{CH}}}-CH_2-CH_2-CH_3$$

9.75 *In one of the reactions in Problem 9.74, the enzyme selected one of the two possible products. Draw the structure of the product that is* <u>not</u> *formed in this reaction.*

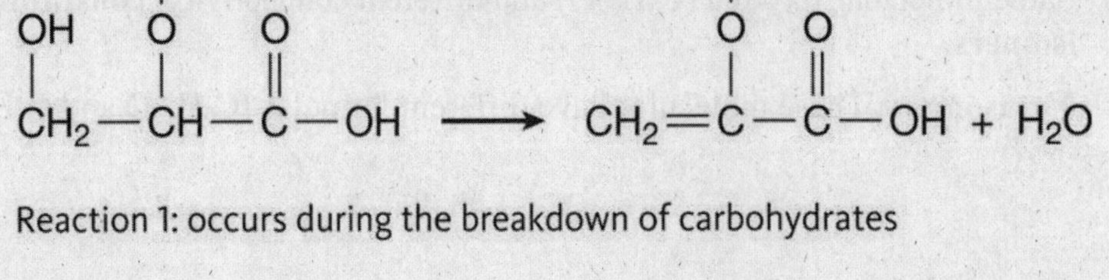

$$\overset{OH}{\overset{|}{CH_2}}-\overset{\overset{PO_3^{2-}}{|}}{\overset{O}{\overset{|}{CH}}}-\overset{O}{\overset{\|}{C}}-OH \longrightarrow CH_2=\overset{\overset{PO_3^{2-}}{|}}{\overset{O}{\overset{|}{C}}}-\overset{O}{\overset{\|}{C}}-OH + H_2O$$

Reaction 1: occurs during the breakdown of carbohydrates

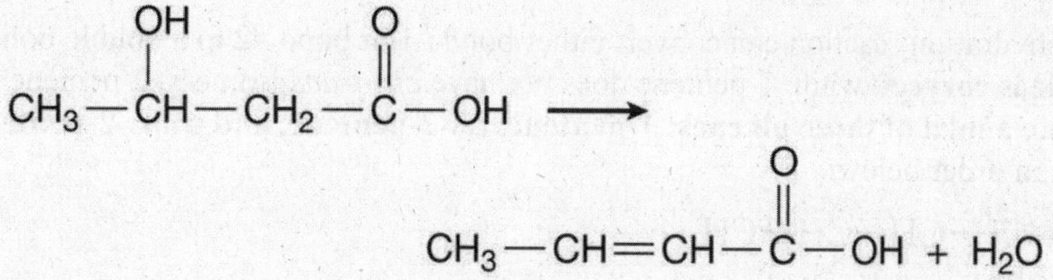

$$CH_3-\underset{OH}{CH}-CH_2-\overset{O}{\overset{\|}{C}}-OH \longrightarrow CH_3-CH=CH-\overset{O}{\overset{\|}{C}}-OH + H_2O$$

Reaction 2: occurs during the formation of fatty acids

The alcohol in Reaction 1 can only give one dehydration product, because the alcohol carbon is only connected to one other carbon atom. The alcohol in Reaction 2 can give two possible dehydration products, and the enzyme only gives the product with the double bond as bond #2. In the other possible product, bond #1 is the double bond:

$$CH_2=CH-CH_2-\overset{O}{\overset{\|}{C}}-OH$$

NOTE:
The carboxylic acid functional group is not involved in the reaction.

*9.77 *The following molecule can lose two molecules of water. Draw the structures of all of the possible products of this double dehydration. (You may ignore cis–trans isomerism.)*

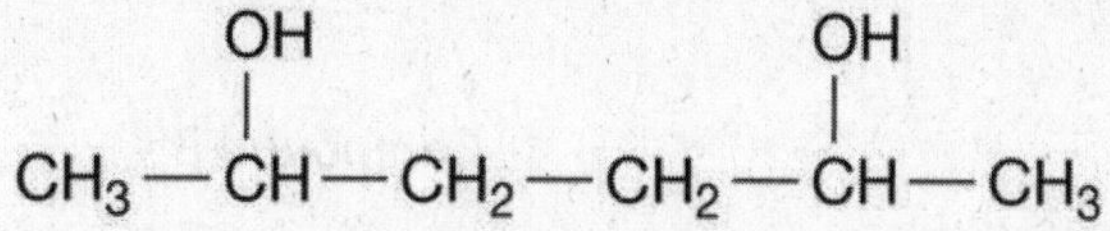

$$CH_3-\underset{OH}{CH}-CH_2-CH_2-\underset{OH}{CH}-CH_3$$

For each dehydration, the double bond formed can go either to the left or to the right of the alcohol carbon, giving double bonds on bond 1 or 2 from carbon #2, and bond 4 or 5 from carbon #5. We draw all possible combinations and check for repeats:

$CH_2=CH-CH_2-CH=CH-CH_3$ — Bonds 1&4 (or 2&5)

$CH_2=CH-CH_2-CH_2-CH=CH_2$ — Bonds 1&5

$CH_3-CH=CH-CH=CH-CH_3$ — Bonds 2&4

*9.79 *Joshua says, "When you dehydrate 2-pentanol, you make two alkenes, 1-pentene and 2-pentene." Rayelle answers, "No, there are more than two, because you've forgotten about cis–trans isomers." Is Rayelle correct? If so, how many alkenes can you make, and what are their names and structures?*

Start by looking at 2-pentanol:

$$CH_3-\underset{OH}{CH}-CH_2-CH_2-CH_2-CH_3$$

The dehydration reaction can convert either bond #1 or bond #2 to a double bond. **Rayelle is correct**: while 1-pentene does not have *cis–trans* isomers, 2-pentene does, so there are a total of **three alkenes: 1-pentene, *cis*-2-pentene, and *trans*-2-pentene**, shown in order below:

$CH_2{=}CH{-}CH_2{-}CH_2{-}CH_3$

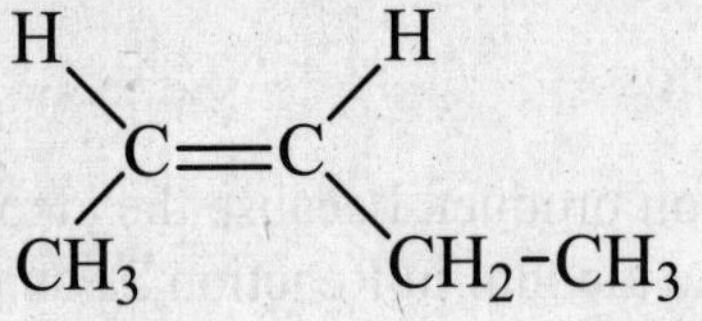

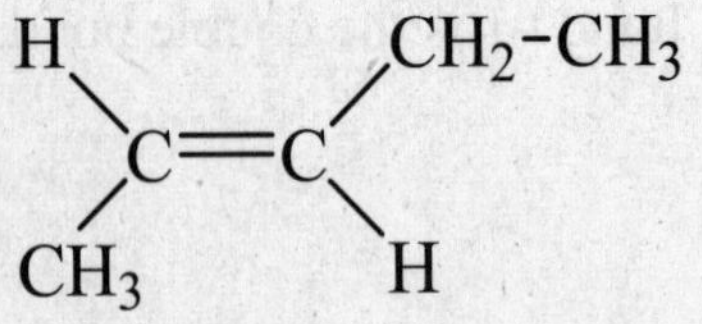

H H
C=C
CH_3 $CH_2{-}CH_3$

H $CH_2{-}CH_3$
C=C
CH_3 H

9.81 *Answer the following questions by selecting compounds from the following list:*

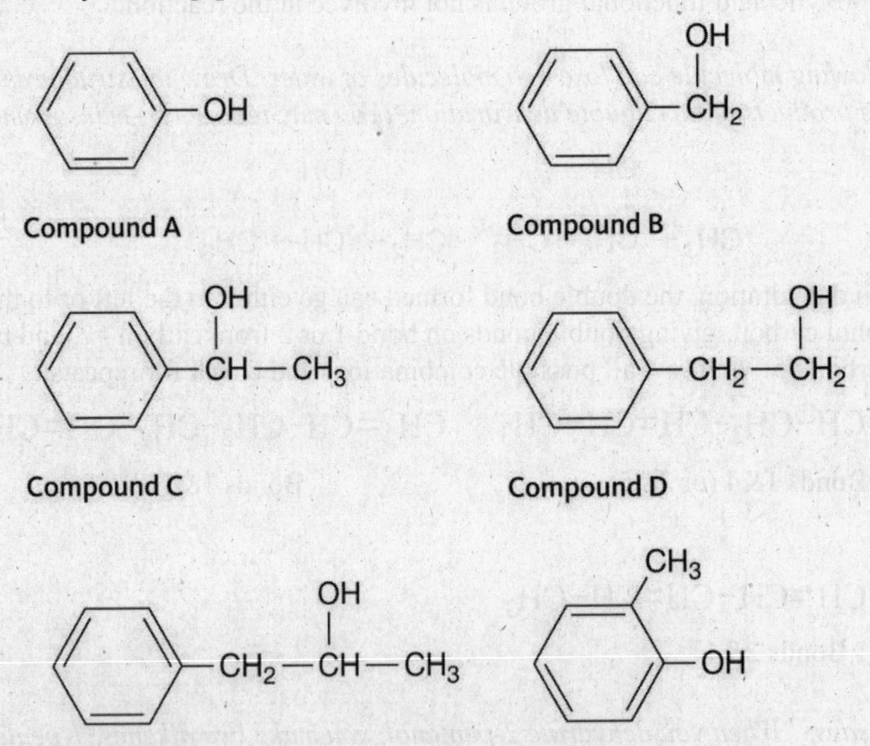

a) *Which of these compounds are phenols?*

A phenol has an –OH group attached directly to one of the carbon atoms in an aromatic ring. **Compounds A and F are phenols.**

b) *Which of these compounds cannot be dehydrated?*

In a dehydration reaction, an alcohol group from one carbon and a hydrogen atom from an adjacent carbon are removed, forming a double bond between the two carbon atoms. In Compounds A and F, dehydration would disrupt the aromatic ring (the alternating pattern of double and single bonds), so dehydration cannot occur (phenols do not undergo dehydration). In Compound B, there is no hydrogen on the carbon atom adjacent to the alcohol carbon, so dehydration cannot occur (phenols do not undergo dehydration). **Compounds C, D and E can be dehydrated**: the carbon atoms adjacent to the alcohol carbon have hydrogen atoms that can be removed.

c) *Which of these compounds form only one product when dehydrated?*

In Compounds C and D, there is only one possible way to take an –H from a carbon adjacent to the alcohol carbon, and only one product will be formed in the dehydration. Interestingly, the dehydration product is the same for both C and D:

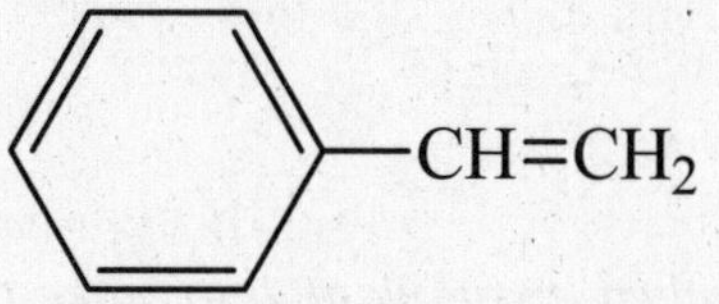

$-CH{=}CH_2$

d) *Which of these compounds form two different products when dehydrated?*

Only Compound E. The products are:

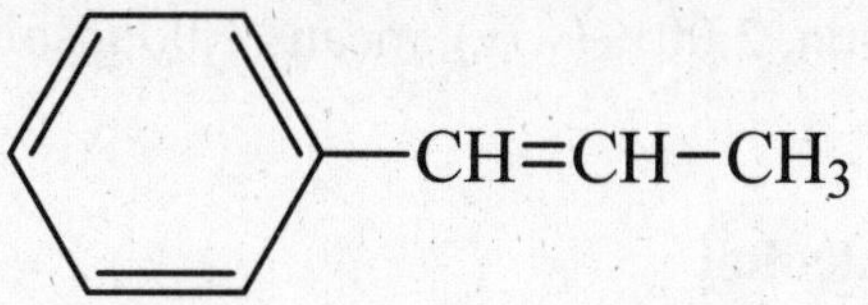

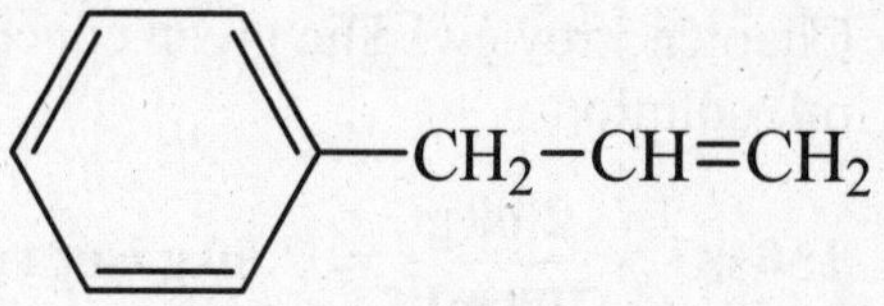

$-CH{=}CH{-}CH_3$ $\qquad$ $-CH_2{-}CH{=}CH_2$

9.83 *Would you expect ethynylestradiol (see Problem 9.82) to be soluble in water? Explain your answer.*

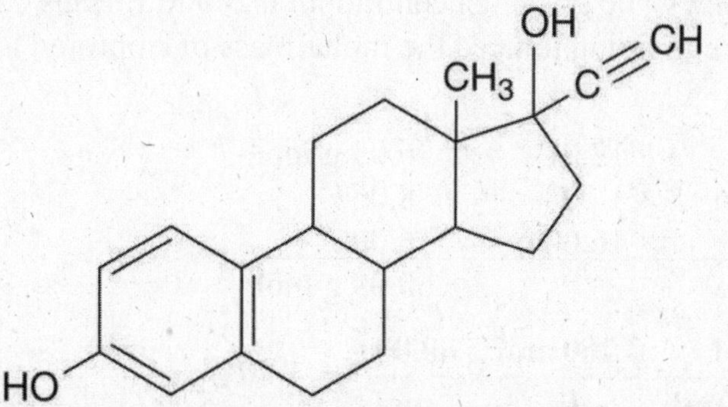

OH
CH
CH_3
C
HO

The molecule has only a modest ability to form hydrogen bonds (with one alcohol group and one phenol group), and has large hydrophobic hydrocarbon regions; it is probably not very soluble in water. (We've seen that it takes about one –OH group per 3-4 carbons in smaller alcohols to give appreciable solubility in water; this molecule falls far short of that ratio.)

*9.85 a) *How much would 2.50 moles of 2-butanol weigh?*

Chapter 5 review! Add up the molar mass of 2-butanol, $C_4H_{10}O$ (draw the molecule to determine its molecular formula, if you need to):

4 carbons:	4 × 12.01	=	48.04 g/mol
10 hydrogens:	10 × 1.008	=	10.08
+ 1 oxygen:	1 × 16.00	=	16.00
			74.12 g/mol

$$\mathbf{2.50\ moles \times \frac{74.12\ g}{1\ mol} = 185\ g}$$

b) *2-butanol is a liquid with a density of 0.802 g/mL. What would be the volume of 2.00 moles of 2-butanol?*

You can review density, and its use as a conversion factor, in Chapter 1.

$$\mathbf{2.00\ moles \times \frac{74.12\ g}{mol} \times \frac{1\ mL}{0.802\ g} = 185\ mL}$$

*9.87 *How many grams of isopropyl alcohol would you need if you wanted to make the following?*

a) *150 mL of a 2.00% (w/v) solution of isopropyl alcohol in water*

Chapter 5 review! The given concentration, 2.00% (w/v), means 2.00 g solute in 100 mL solution:

$$\mathbf{150\ mL \times \frac{2.00\ g}{100\ mL} = 3.00\ g\ isopropyl\ alcohol}$$

b) *150 mL of a 0.200 M solution of isopropyl alcohol in water*

Chapter 5 review! The given concentration, 0.200 M, means 0.200 mol solute per L of solution. We're going to need the molar mass of isopropyl alcohol (C_3H_8O) for this one:

3 carbons:	3 × 12.01	=	36.03 g/mol
8 hydrogens:	8 × 1.008	=	8.064
+ 1 oxygen:	1 × 16.00	=	16.00
			60.09 g/mol

$$\mathbf{150\ mL \times \frac{1\ L}{1000\ mL} \frac{0.200\ mol}{L} \times \frac{60.09\ g}{mol} = 1.80\ g}$$

*9.89 *If you dissolve 4.0 grams of glycerol in enough water to make 100 mL of solution, will this solution be isotonic, hypertonic, or hypotonic? Show your reasoning.*

Chapter 5 review! Tonicity is based on the molarity (mol solute/L solution) of the solution, so we need to calculate the molar mass of glycerol:

3 carbons:	3 × 12.01	=	36.03 g/mol
8 hydrogens:	8 × 1.008	=	8.064
+ 3 oxygen:	3 ×48.00	=	48.00
			92.09 g/mol

Then we can calculate the molarity of the solution:

$$\frac{\mathbf{4.0\ g}}{\mathbf{100\ mL}} \times \frac{\mathbf{1\ mol}}{\mathbf{92.09\ g}} \times \frac{\mathbf{1000\ mL}}{\mathbf{L}} = \mathbf{0.43\ M}$$

This molarity is greater than the isotonic concentration (0.28 M), so this solution is **hypertonic**.

*9.91 *Alcohols can burn, just as hydrocarbons can. The combustion of alcohols follows the same general scheme as that of hydrocarbons:*

$$alcohol + O_2 \rightarrow CO_2 + H_2O$$

Using this information, write the balanced equation for the combustion of ethanol.

See Chapters 6 and 8 for a review of combustion. Ethanol is C_2H_6O (we don't need a structural formula to write a balanced equation, just a molecular formula.) Balance the equation as in Chapter 6:

$$\mathbf{C_2H_6O + 3\ O_2 \rightarrow 2\ CO_2 + 3\ H_2O}$$

*9.93 *Ethanol is used in some rubbing alcohol formulas, because it evaporates readily and cools the skin. The specific heat of ethanol is 0.58 cal/g·°C and the heat of vaporization is 201 cal/g.*

Chapter 4 review!

a) *How much heat is needed to raise the temperature of 25.0 g of ethanol from 20°C to 37°C (body temperature)?*

The temperature change is 17°C.

Heat = mass × specific heat × ΔT
= 25.0 g × 0.58 cal/g·°C × 17°C
= **250 cal** (rounded to 2 significant figures)

b) *How much heat is needed to evaporate 25.0 g of ethanol?*

Heat required for evaporation = mass × heat of vaporization
= 25.0 g × 201 cal/g
= **5030 cal** (rounded to 3 significant figures)
= **5.03 kcal**

Carbonyl Compounds and Redox Reactions

Solutions to Section 10.1 Core Problems

10.1 *The following compound is dehydrogenated during the breakdown of lysine (one of the components of proteins). The portion of the molecule that is dehydrogenated is circled. Draw the structure of the product of this reaction.*

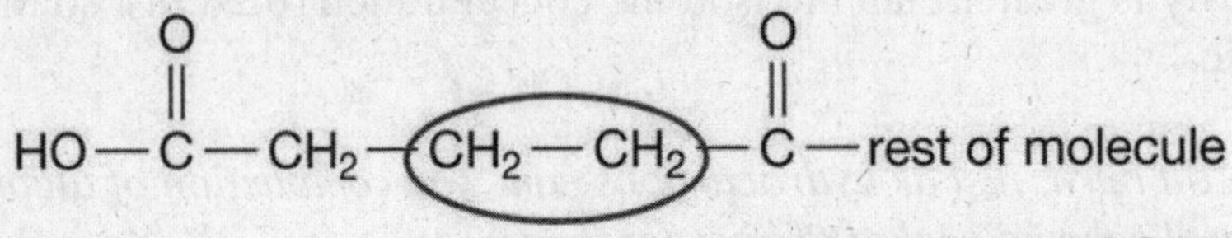

$$HO-\overset{\overset{O}{\|}}{C}-CH_2-\boxed{CH_2-CH_2}-\overset{\overset{O}{\|}}{C}-\text{rest of molecule}$$

In a dehydrogenation reaction, **two hydrogen atoms are removed from adjacent single-bonded carbon atoms, forming a carbon-carbon double bond and a molecule of H_2.**

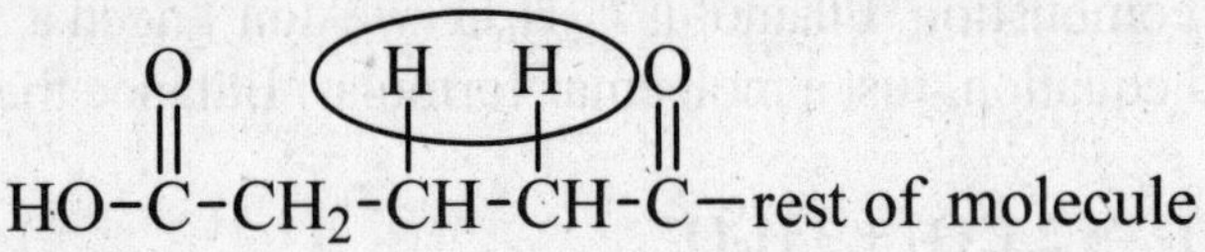

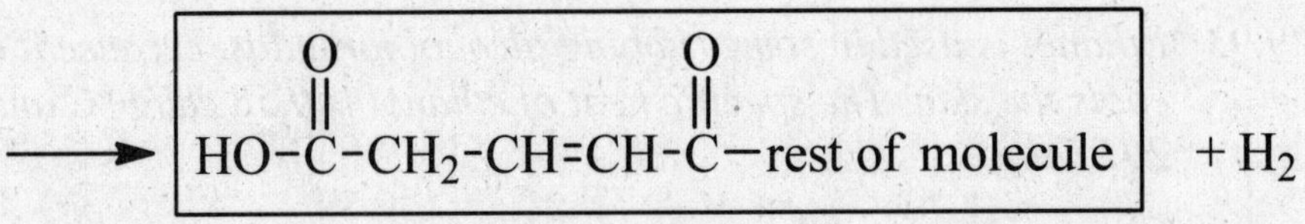

$$HO-\overset{\overset{O}{\|}}{C}-CH_2-\overset{\overset{H}{|}}{C}H-\overset{\overset{H}{|}}{C}H-\overset{\overset{O}{\|}}{C}-\text{rest of molecule}$$

$$\longrightarrow \boxed{HO-\overset{\overset{O}{\|}}{C}-CH_2-CH=CH-\overset{\overset{O}{\|}}{C}-\text{rest of molecule}} + H_2$$

10.3 *Draw the structures of the products that are formed when each of the following alkenes is hydrogenated:*

In a hydrogenation reaction, a double bond is converted to a single bond, and a hydrogen atom is added to each carbon of the double bond involved.

a) *1-pentene* Hydrogenation will form pentane:

$CH_3-CH_2-CH_2-CH_2-CH_3$

b)

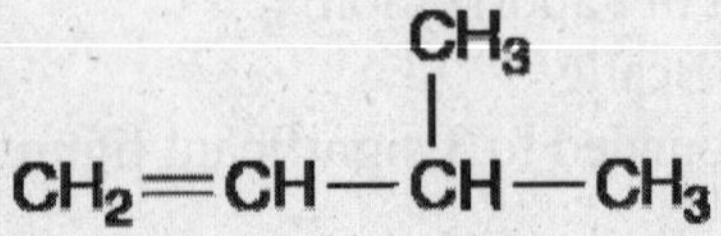

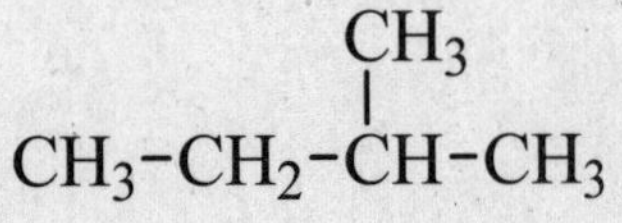

$$CH_2=CH-\underset{}{\overset{\overset{CH_3}{|}}{C}H}-CH_3$$

$$CH_3-CH_2-\overset{\overset{CH_3}{|}}{C}H-CH_3$$

10.5 *Is the following reaction an oxidation, or is it a reduction? Explain your answer.*

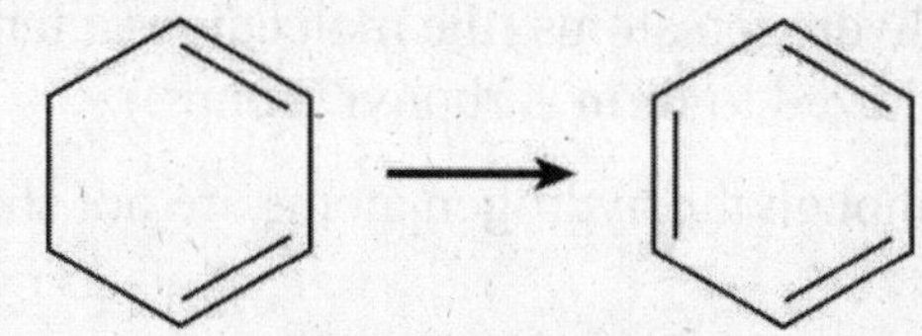

In an oxidation, oxygen atoms are added or hydrogen atoms are removed. In a reduction, oxygen atoms are removed or hydrogen atoms are added. **In the reaction shown, hydrogen atoms are removed, so it is an oxidation reaction.** In line structures, the hydrogen atoms are not shown but are understood to be present; if you have trouble seeing what's going on in the reaction, try drawing out a full or condensed structure (Chapter 8).

Solutions to Section 10.2 Core Problems

10.7 *Draw the structures of the carbonyl compounds that are formed when the following alcohols are oxidized. If the alcohol cannot be oxidized, write "no reaction."*

a) 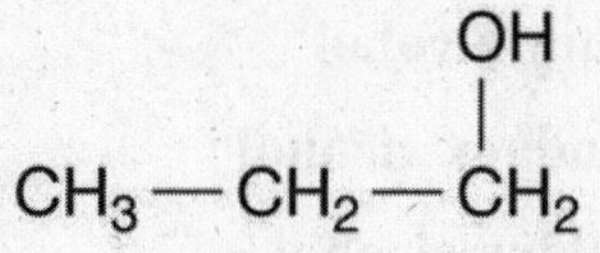 $CH_3—CH_2—CH_2$ (OH on the terminal CH_2)

b) 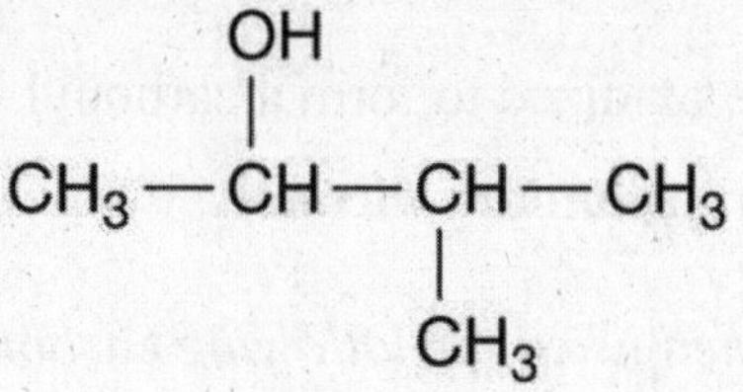 $CH_3—CH—CH—CH_3$ (OH on the first CH; CH_3 on the second CH)

c) $CH_3—C—CH_2—CH_2—CH_3$ (OH and CH_3 on C)

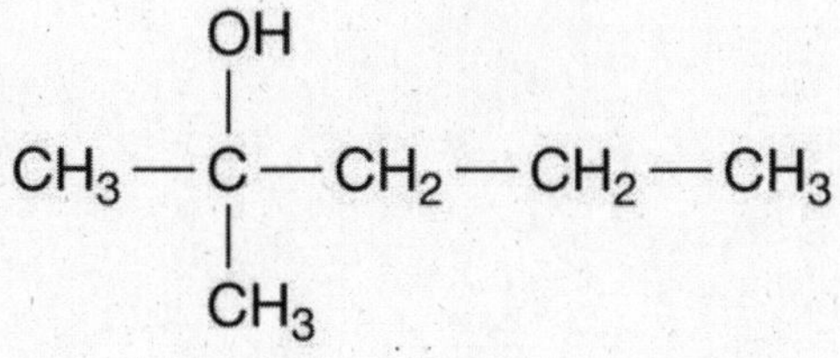

d) 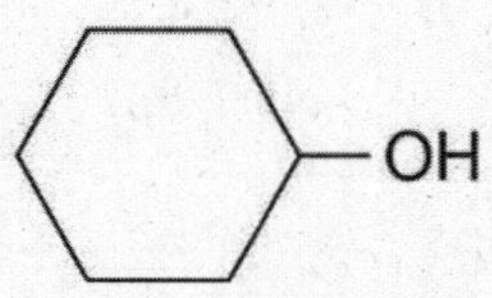 cyclohexane ring—OH

Oxidation of an alcohol removes two H atoms, one from the alcohol O and one from the carbon attached to the alcohol O, forming a C=O double bond.

a) 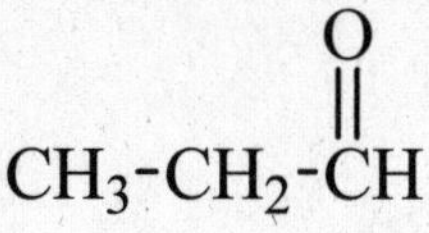 $CH_3\text{-}CH_2\text{-}\overset{\overset{\large O}{||}}{C}H$ b) 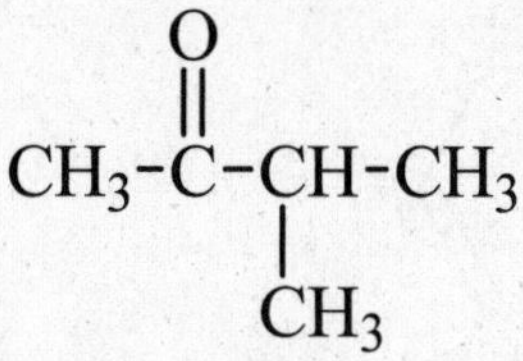 $CH_3\text{-}\overset{\overset{\large O}{||}}{C}\text{-}\underset{\underset{\large CH_3}{|}}{CH}\text{-}CH_3$

c) **No** reaction. This molecule cannot be oxidized; the carbon atom attached to the alcohol group has no hydrogen atoms (the molecule is a tertiary alcohol, and tertiary alcohols cannot be oxidized to form carbonyl groups.)

d) Remember that even though the hydrogen atoms are not shown in line structures, they're still there:

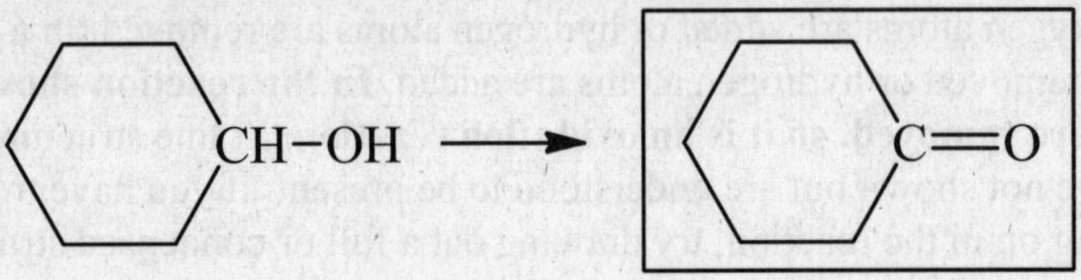

(cyclohexyl)CH−OH ⟶ (cyclohexyl)C=O

10.9 *Classify each of the alcohols in Problem 10.7 as a primary, secondary, or tertiary alcohol.*

Alcohols are classified according to the number of carbon atoms attached to the alcohol carbon.

a) one carbon atom attached to alcohol carbon: **primary alcohol**

b) two carbon atoms attached to alcohol carbon: **secondary alcohol**

c) three carbon atoms attached to alcohol carbon: **tertiary alcohol**

NOTE:
As shown in 10.7, tertiary alcohols cannot be oxidized to form a carbonyl.

d) two carbon atoms attached to alcohol carbon: **secondary alcohol**

10.11 *Draw the structures of the alcohols that are formed when the following carbonyl compounds are reduced:*

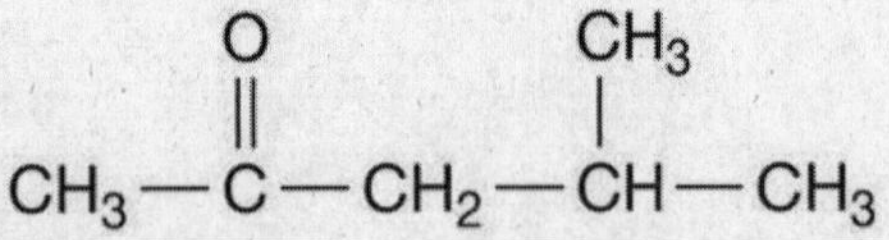

a) 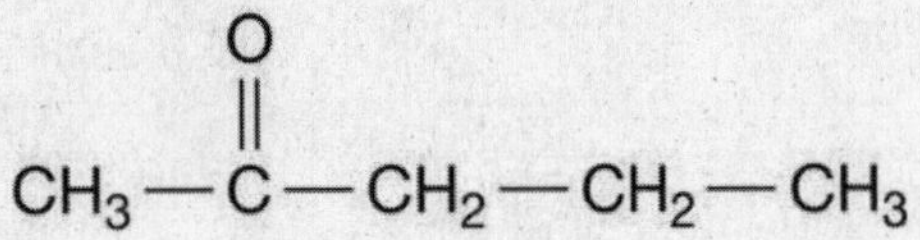 $CH_3-C(=O)-CH_2-CH_2-CH_3$

b) $CH_3-C(=O)-CH_2-CH(CH_3)-CH_3$

c) (cyclopentyl)=O

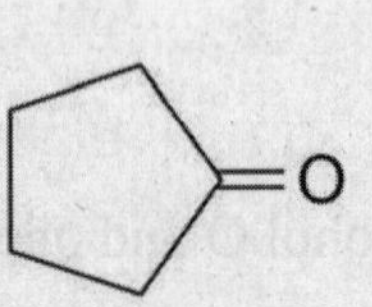

d) 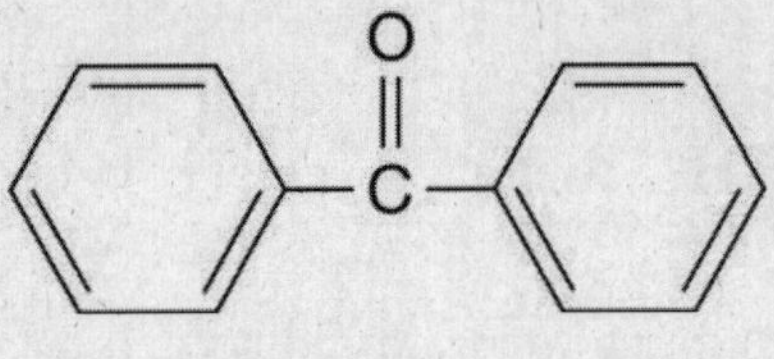 (phenyl)−C(=O)−(phenyl)

Reduction of a carbonyl (C=O) adds two hydrogen atoms, one to the O and one to the C of the carbonyl, to give a C–O single bond in the alcohol produced.

a) $CH_3\text{-}CH(OH)\text{-}CH_2\text{-}CH_2\text{-}CH_3$

b) $CH_3\text{-}CH(OH)\text{-}CH_2\text{-}CH(CH_3)\text{-}CH_3$

c) (cyclopentyl)CH–OH or (cyclopentyl)–OH

d) (phenyl)–CH(OH)–(phenyl)

NOTE:
In c) you can choose to show the alcohol carbon and its hydrogen atom, or not.

10.13 *Both of the following compounds can be reduced, but one produces a chiral alcohol while the other does not. Which is which? Explain your answer.*

$CH_3-CH_2-C(=O)-CH_2-CH_3$ **3-Pentanone**

$CH_3-C(=O)-CH_2-CH_2-CH_3$ **2-Pentanone**

Review the concept of chirality in Chapter 9. **Because 3-butanone is symmetric, after reduction the alcohol carbon will have two identical ethyl groups attached. A chiral carbon atom must have four *different* groups attached, so the reduction product of 3-butanone is not chiral. After reduction, the alcohol carbon in 2-butanone will have 4 different groups attached: an alcohol, a hydrogen, a methyl group and a propyl group, so the reduction product of 2-butanone is chiral.**

$$CH_3\text{-}CH_2\text{-}C(=O)\text{-}CH_2\text{-}CH_3 \longrightarrow CH_3\text{-}CH_2\text{-}CH(OH)\text{-}CH_2\text{-}CH_3$$

$$CH_3\text{-}C(=O)\text{-}CH_2\text{-}CH_2\text{-}CH_3 \longrightarrow CH_3\text{-}CH(OH)\text{-}CH_2\text{-}CH_2\text{-}CH_3$$

Solutions to Section 10.3 Core Problems

10.15 *Name the compounds below, using the IUPAC rules:*

a) 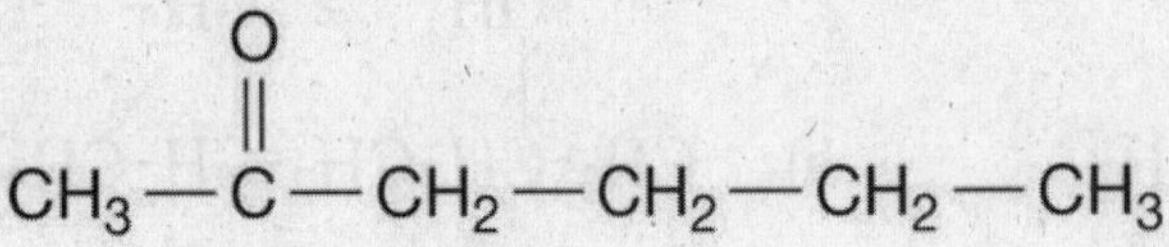 $CH_3-C(=O)-CH_2-CH_2-CH_2-CH_3$

b) 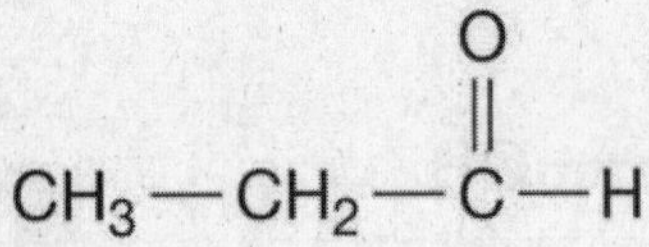 $CH_3-CH_2-C(=O)-H$

c) $CH_3-CH_2-CH_2-CH_2-CHO$

d) 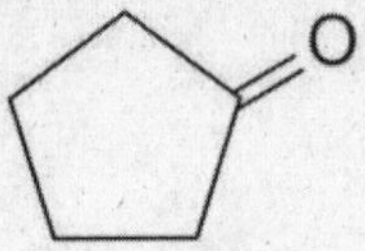 (cyclopentane ring with =O)

e) 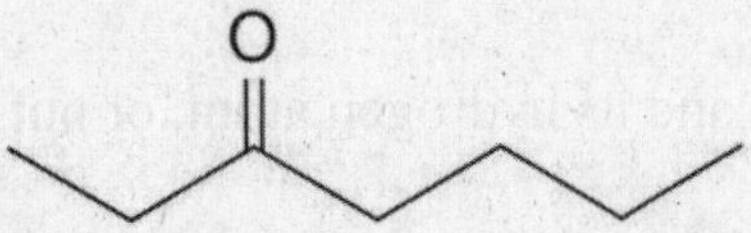 (skeletal structure with =O)

In each case, identify the principle carbon chain, identify the functional group, and assemble the name using the suffix -one for ketones and -al for aldehydes. Because aldehydes must always be at the end of a carbon chain, no number is required to specify the position of an aldehyde group, but with ketones it's important to check and see whether an address is needed (by approximately the same method used in Chapter 8 to determine whether a number was needed to specify the position of an alkyl group in a branched alkane or cyclic alkane).

a) 6-carbon chain with a ketone group on the #2 carbon: **2-hexanone.** (To see whether the number for the position of a ketone group is actually needed, just try other numbers and see if they give different compounds. 1-hexanone can't exist, because a ketone group must always be in the interior of the molecule, but 3-hexanone does exist and is a different molecule from 2-hexanone. Therefore the 2- is required.)

b) 3-carbon chain with aldehyde group: **propanal.**

c) 5-carbon chain with aldehyde group: **pentanal.**

NOTE:
Aldehyde groups are often shown in condensed form as –CHO.)

d) 5-carbon ring with ketone group: **cyclopentanone**. Just as with alkyl groups on cyclic compounds in Chapter 8, no number is needed because all positions on the ring are equivalent.

e) 7-carbon chain with ketone group on #3 carbon: **3-heptanone.**

10.17 *Give the trivial name of the following compound:*

$$CH_3-\overset{\overset{\large O}{\|}}{C}-CH_3$$

No way around this one but to memorize it: **acetone.**

10.19 *Draw the structures of each of the following compounds:*
a) 3-pentanone *b) octanal* *c) formaldehyde* *d) cyclobutanone*

a) 5-carbon chain with ketone group on carbon #3:

$$CH_3\text{-}CH_2\text{-}\overset{\overset{O}{\|}}{C}\text{-}CH_2\text{-}CH_3$$ or

b) 8-carbon chain with aldehyde group:

$$CH_3\text{-}CH_2\text{-}CH_2\text{-}CH_2\text{-}CH_2\text{-}CH_2\text{-}CH_2\text{-}\overset{\overset{O}{\|}}{C}H$$ or $\overset{\overset{O}{\|}}{C}H$

c) As with other common or trivial names, this one has to be memorized, but it's helpful to know that the "form-" root in trivial names is often equivalent to "meth-" in systematic names, so this is the one-carbon aldehyde:

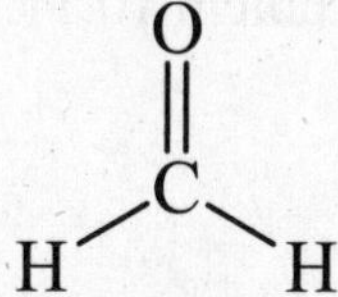

d) 4-carbon ring with ketone group:

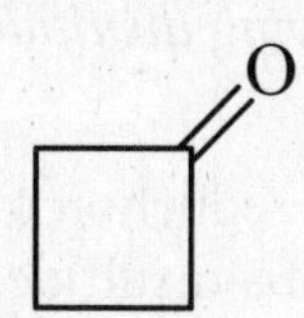

10.21 *One of the compounds below boils at 36°C, one boils at 75°C, and one boils at 99°C. Match each compound with its boiling point.*

$CH_3{-}CH_2{-}CH_2{-}CH{=}O$	$CH_3{-}CH_2{-}CH_2{-}CH_2{-}OH$	$CH_3{-}CH_2{-}CH_2{-}CH_2{-}CH_3$
butanal	*1-butanol*	*pentane*

Chapter 4 review! For this we look at the attractions between molecules in each compound. All four molecules are about the same size and have about the same total number of electrons, so the dispersion forces in each compound are about the same. **Pentane, with only dispersion forces, will have the lowest boiling point, 36°C.** In addition to the dispersion forces, 1-butanol has hydrogen bonding between its molecules, while butanal has dipole-dipole forces. Hydrogen bonding is generally a stronger attraction than dipole-dipole forces, so **1-butanol will have the highest boiling point of the three, 99°C, and butanal will be in between at 75°C.**

10.23 *The solubilities of the following three compounds are 12 g/L, 1.2 g/L, and 0.1 g/L. Match each compound with its solubility.*

$CH_3CH_2CH_2CH_2CH_3$	$CH_3CH_2CH_2CH_2COH$	$CH_3CH_2CH_2CH_2CH_2CH_2CHO$
pentane	*pentanal*	*heptanal*

Chapter 5 review! For this we compare the hydrophobic and hydrophilic areas of the molecules. **Pentane, with no ability to form hydrogen bonds with water, will have the lowest solubility, 0.1 g/L.** Pentanal and heptanal have equivalent ability to accept hydrogen bonds from water, and pentanal has a smaller hydrophobic area, so **pentanal will have the greatest solubility, 12 g/L, and heptanal will be in between at 1.2 g/L.**

Solutions to Section 10.4 Core Problems

10.25 *Draw the structure of the compound that is formed when the following thiol is oxidized. (Hint: You need two molecules of the thiol.)*

$$CH_3-CH_2-CH_2-\underset{\displaystyle CH_3}{\underset{|}{CH}}-SH$$

While oxidation of an alcohol forms a product with a C=O double bond, oxidation of a thiol couples two molecules together with an S–S single bond. Draw two molecules of the thiol, and turn one around, lining the two molecules up so that the S atoms point toward each other. Then, in the oxidation reaction, the hydrogen atoms are removed and the two sulfur atoms are linked with a single bond. See Sample Problem 10.11.

$$CH_3\text{-}CH_2\text{-}CH_2\text{-}\underset{\displaystyle CH_3}{\underset{|}{CH}}-S-S-\underset{\displaystyle CH_3}{\underset{|}{CH}}\text{-}CH_2\text{-}CH_2\text{-}CH_3$$

10.27 *Draw the structures of the products that are formed when the following disulfide is reduced.*

As reduction of a carbonyl is the reverse of oxidation of an alcohol, reduction of a disulfide is the reverse of the oxidation of a thiol. In the reduction of a disulfide, the S–S bond is broken and hydrogen atoms are added to the sulfur atoms, forming two thiols

$$CH_3\text{-}CH_2\text{-}S-S\text{-}CH_2\text{-}CH_3 \ + \ 2\,[H] \longrightarrow \boxed{CH_3\text{-}CH_2\text{-}SH \ + \ HS\text{-}CH_2\text{-}CH_3}$$

10.29 *Ascorbic acid (vitamin C) reacts with many substances. In these reactions, the ascorbic acid is converted into dehydroascorbic acid, as shown here. Is this reaction an oxidation, or is it a reduction? Explain your answer.*

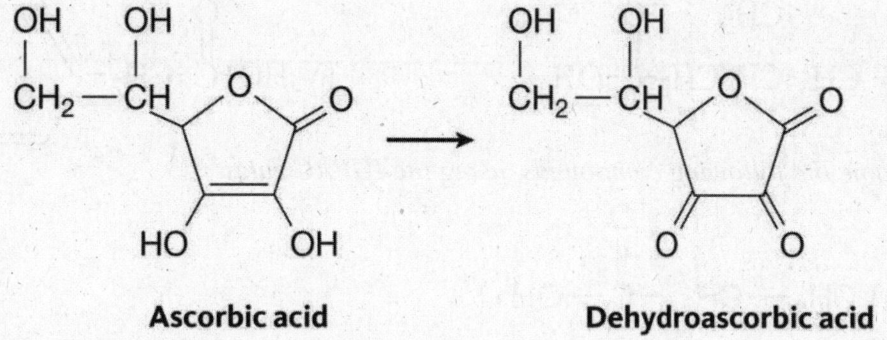

OH OH | | CH_2—CH ... O ... O ... HO OH → OH OH | | CH_2—CH ... O ... O ... O O

Ascorbic acid → Dehydroascorbic acid

This reaction reflects the removal of two H atoms from the ascorbic acid molecule, so it is an oxidation. To make this easier to see, let's look at the structure of just the part of the molecule that's involved in the reaction:

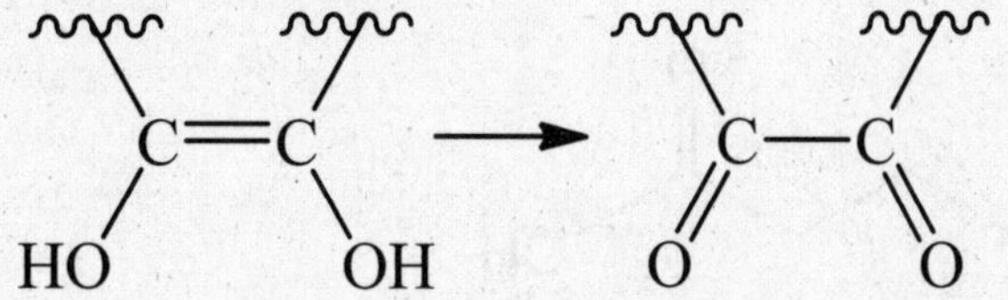

HO—C=C—OH → O=C—C=O

In reduction reactions, the number of C–H bonds *increases*, or the number of C–O bonds *decreases* (or both). Oxidation reactions do the opposite, *increasing* the number of C–O bonds or *decreasing* the number of C–H bonds. In this reaction, the affected carbon atoms end up with more bonds to oxygen, so the reaction is an oxidation.

NOTE:
Even though the alkene group in ascorbic acid is converted to a C—C single bond in dehydroascorbic acid, this is accomplished by forming more bonds to O atoms, not by adding H atoms.

Solutions to Section 10.5 Core Problems

10.31 *Draw the structures of the products that are formed when the following aldehydes are oxidized:*

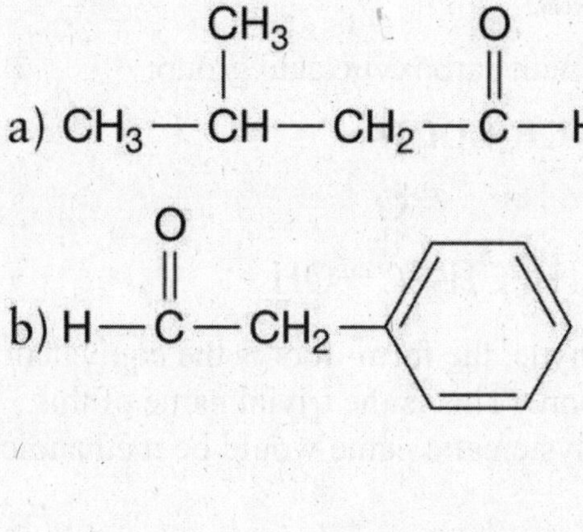

a) CH_3—CH(CH_3)—CH_2—C(=O)—H

b) H—C(=O)—CH_2—C_6H_5

Oxidation of an aldehyde forms a carboxylic acid. To draw the structure, we simply insert an O between the carbonyl and its hydrogen (although the complete chemical pathway is a lot more complex than this!):

a) $CH_3-CH(CH_3)-CH_2-C(=O)-OH$ b) 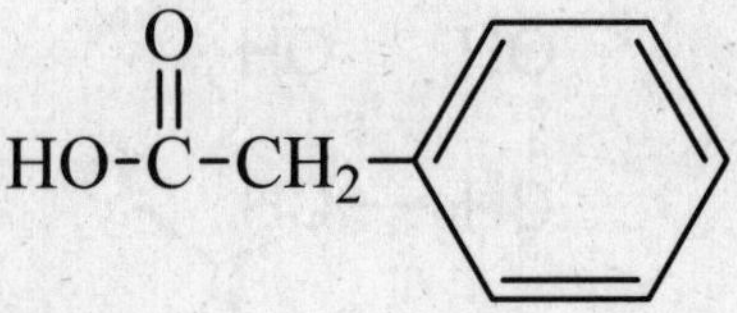 $HO-C(=O)-CH_2-C_6H_5$

10.33 *Name the following compounds, using the IUPAC rules:*

a) $CH_3-CH_2-C(=O)-OH$

b) $CH_3-CH_2-CH_2-CH_2-CH_2-CH_2-COOH$

c) 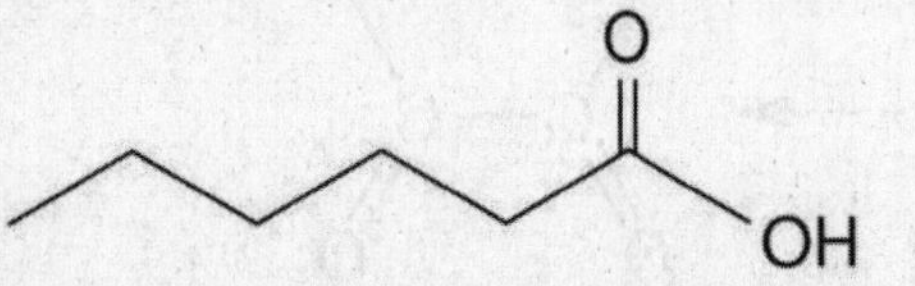 (line structure: six-carbon chain ending in C(=O)OH)

Since carboxylic acid groups, like aldehyde groups, are always at the end of the chain, we don't have to use a number to specify the position of the group in the chain. Determine the principal chain as in Chapter 8, then add the -oic acid suffix for the carboxylic acid group.

a) 3-carbon carboxylic acid: **propanoic acid**

b) Don't be fooled by the condensed representation of the carboxylic acid group: $-COOH$ and $-CO_2H$ are common ways to write this group in condensed structures. 7-carbon carboxylic acid: **heptanoic acid.**

c) Remember that in line structures, ends and bends are carbon atoms. 6-carbon carboxylic acid: **hexanoic acid.**

10.35 *Draw the structures of the following compounds:*

a) pentanoic acid 5-carbon chain with carboxylic acid group:

$CH_3-CH_2-CH_2-CH_2COOH$ or

$CH_3-CH_2-CH_2-CH_2-C(=O)-OH$

b) formic acid As in formaldehyde, the form- root is the equivalent of meth-, one carbon. (This is the trivial name of this compound; its systematic name would be methanoic acid.)

$$\mathbf{HCOOH} \text{ or } \quad H-\overset{\overset{\displaystyle O}{\|}}{C}-OH$$

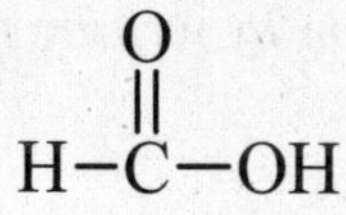

10.37 *One of the following compounds is much more soluble in water than the other one. Which compound has the higher solubility?*

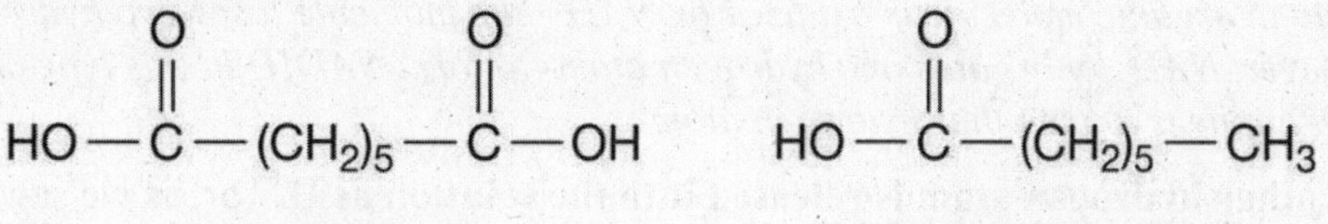

$$HO-\overset{\overset{\displaystyle O}{\|}}{C}-(CH_2)_5-\overset{\overset{\displaystyle O}{\|}}{C}-OH \qquad HO-\overset{\overset{\displaystyle O}{\|}}{C}-(CH_2)_5-CH_3$$

Pimelic acid **Heptanoic acid**

To compare water solubilities of substances, we compare their hydrophilic and hydrophobic nature. **Pimelic acid has two hydrogen bonding functional groups, and is therefore much more soluble in water than heptanoic acid, with its one hydrogen bonding group and long hydrophobic chain.**

10.39 *One of the compounds in Problem 10.37 is a solid at room temperature, and one is a liquid. Which is which? Explain your answer.*

To compare melting points of substances, we compare the strengths of the attractions between their molecules. Both molecules have similar dispersion forces, but **pimelic acid has two hydrogen bonding functional groups, stronger total attractions between its molecules than heptanoic acid, and therefore is likely to have a higher melting point. Pimelic acid is a solid at room temperature, while heptanoic acid (with weaker attractions between its molecules) is a liquid.**

Solutions to Section 10.6 Core Problems

10.41 *Which of the redox coenzymes is normally used to supply the hydrogen atoms for biological reduction reactions?*

Nicotinamide adenine dinucleotide phosphate, or NADP$^+$, supplies the hydrogen atoms in biological reduction reactions.

10.43 *The following reaction occurs in many kinds of plants:*

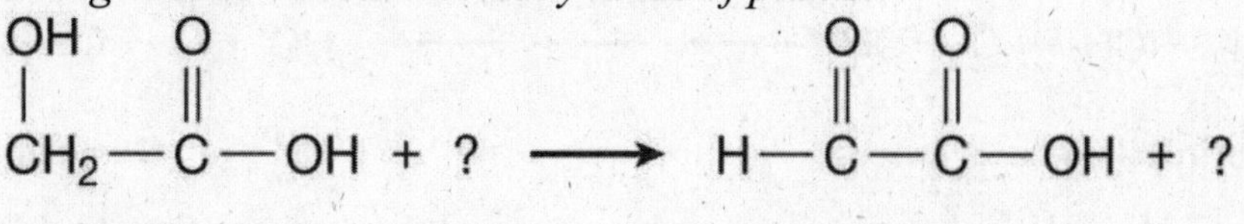

$$\overset{\overset{\displaystyle OH}{|}}{CH_2}-\overset{\overset{\displaystyle O}{\|}}{C}-OH + ? \longrightarrow H-\overset{\overset{\displaystyle O}{\|}}{C}-\overset{\overset{\displaystyle O}{\|}}{C}-OH + ?$$

Glycolic acid **Glyoxylic acid**

a) *Which of the redox coenzymes is most likely to be involved in this reaction?*

The reaction converts a primary alcohol to an aldehyde: an oxidation reaction. **The active redox coenzyme in alcohol oxidations is nicotinamide adenine dinucleotide, or NAD$^+$.** It removes two H atoms; one adds to the NAD$^+$ as H$^-$, forming NADH, and the other is released into the solution as H$^+$ (or more precisely, as we saw in Chapter 7, attaches to a water molecule to form H_3O^+).

b) *Complete the reaction by writing the correct form of the redox coenzyme on each side of the equation.*

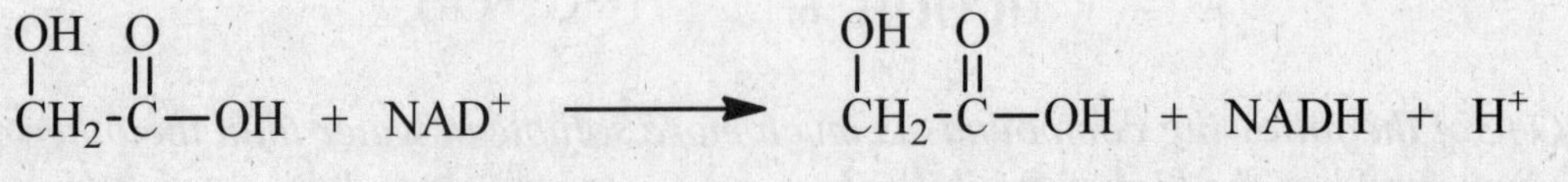

$$CH_2(OH)\text{-}C(=O)\text{—}OH + NAD^+ \longrightarrow CH_2(OH)\text{-}C(=O)\text{—}OH + NADH + H^+$$

10.45 *When an organic molecule is oxidized by NAD^+, the molecule loses two hydrogen atoms. However, NAD^+ only gains one hydrogen atom (to form NADH) in this type of reaction. What happens to the other hydrogen atom?*

The other hydrogen atom is released into the solution as H^+ (or, as we saw in Chapter 7, adds to water to form H_3O^+).

Solutions to Section 10.7 Core Problems

10.47 *The following reactions occur in plants. (These are not balanced equations.)*

Reaction #1: glucose → glucose-6-phosphate
Reaction #2: glucose-6-phosphate → glucose-1-phosphate
Reaction #3: glucose-1-phosphate → ADP-glucose
Reaction #4: ADP-glucose → starch

Do these reactions make up a metabolic pathway? If so, what are the starting material and the final product of this pathway?

Yes, these reactions do make up a metabolic pathway: a metabolic pathway is a sequence of reactions that changes one important biological molecule into another. Each step in the reaction uses the product of the previous step as its staring material. **The starting material in this pathway is glucose; the final product is starch.**

10.49 *One of the most important metabolic pathways in animals and plants is the citric acid cycle (or Krebs cycle), a series of eight reactions that is involved in the breakdown of all types of nutrients to produce energy. Part of this pathway involves the conversion of succinic acid into oxaloacetic acid, using the sequence of three reactions described in this section. Draw the structures of each of the compounds that are formed during this sequence of reactions.*

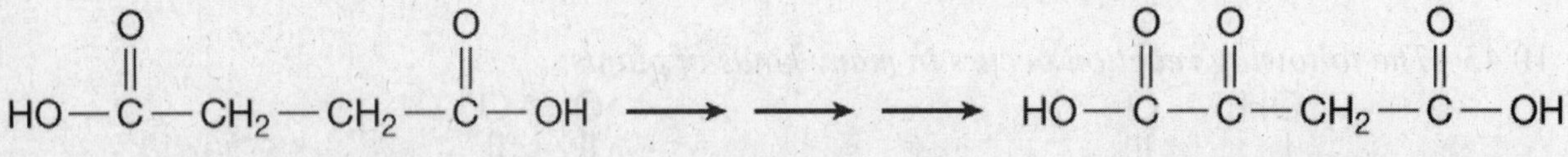

$$HO\text{—}C(=O)\text{—}CH_2\text{—}CH_2\text{—}C(=O)\text{—}OH \longrightarrow \longrightarrow \longrightarrow HO\text{—}C(=O)\text{—}C(=O)\text{—}CH_2\text{—}C(=O)\text{—}OH$$

Succinic acid **Oxaloacetic acid**

This reaction involves the common sequence of **a dehydrogenation (oxidation) to form an alkene, hydration of the alkene to form an alcohol, and oxidation of the alcohol to form a carbonyl.** (This is the three-reaction sequence described in this section.)

$$\mathrm{HO{-}\overset{O}{\overset{\|}{C}}{-}\overset{H}{\overset{|}{C}}H{-}\overset{H}{\overset{|}{C}}H{-}\overset{O}{\overset{\|}{C}}{-}OH \xrightarrow[\text{(oxidation)}]{\text{dehydrogenation}} \boxed{HO{-}\overset{O}{\overset{\|}{C}}{-}CH{=}CH{-}\overset{O}{\overset{\|}{C}}{-}OH}}$$

$$\mathrm{\xrightarrow{\text{hydration}} \boxed{HO{-}\overset{O}{\overset{\|}{C}}{-}\overset{OH}{\overset{|}{C}}H{-}\overset{H}{\overset{|}{C}}H{-}\overset{O}{\overset{\|}{C}}{-}OH}}$$

$$\mathrm{\xrightarrow{\text{oxidation}} \boxed{HO{-}\overset{O}{\overset{\|}{C}}{-}\overset{O}{\overset{\|}{C}}{-}\overset{H}{\overset{|}{C}}H{-}\overset{O}{\overset{\|}{C}}{-}OH}}$$

Solutions to Concept Questions

* indicates more challenging problems.

10.51 *What do most biological oxidation reactions have in common with one another?*

Most biological oxidation reactions use the same redox coenzyme, nicotinamide adenine dinucleotide or NAD^+.

10.53 *In a carbonyl group, both the carbon atom and the oxygen atom are electrically charged.*

a) Which atom is positively charged, and which is negatively charged?

b) Using electronegativities, explain your answer to part a.

Chapter 3 review! **Oxygen is more electronegative than carbon and attracts the electrons in the bond, so it has a δ– charge, and C has a δ+ charge.**

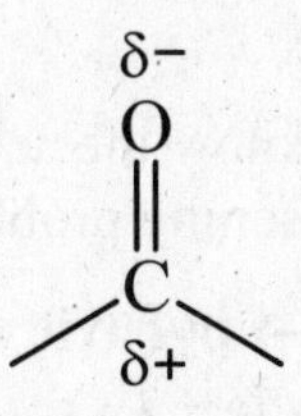

10.55 *Each of the following names violates the IUPAC rules, but each does so for a different reason. Explain why each name is incorrect.*

a) 1-pentanal — **Because the aldehyde group must always be at the end of a molecule, aldehydes do not require a number to specify the location of the aldehyde group. The aldehyde carbon is automatically carbon #1. This compound should be named "pentanal."**

b) 2-pentanal — **An aldehyde group can never be on an interior carbon atom; aldehydes can only be at the end of a chain. 2-pentanal cannot exist. If there's a carbonyl on carbon #2, the compound is 2-pentanone, not 2-pentanal.**

10.57 *Explain why the boiling points of aldehydes and ketones are substantially lower than the boiling points of similar-sized alcohols.*

Aldehyde and ketone compounds cannot form hydrogen bonds between molecules (as pure substances), so the attractions between their molecules are due only to dispersion and dipole-dipole forces. Similar-sized alcohols would have equivalent dispersion forces, but alcohols are also capable of forming hydrogen bond attractions between their molecules, giving alcohols generally higher melting and boiling points than aldehydes and ketones.

10.59 *What are the three common redox coenzymes, and what do our bodies use each of them for?*

NAD^+ (nicotinamide adenine dinucleotide) is the hydrogen acceptor in most oxidations. FAD (flavin adenine dinucleotide) accepts hydrogen atoms in dehydrogenation reactions that produce alkene functional groups. $NADP^+$ (nicotinamide adenine dinucleotide phosphate) supplies the hydrogen atoms in reduction reactions.

10.61 *The following reactions occur in some types of yeast. Put the reactions in order so that they form a metabolic pathway.*

pyruvic acid → acetaldehyde + CO_2
phosphoenolpyruvic acid → pyruvic acid
acetaldehyde → ethanol

In a metabolic pathway, the product of each step becomes the reactant in the following step. We look at the reactions for any species that appears as both a reactant and a product, and put the step in which it's a product before the step in which it's a reactant. The compounds that appear twice in this set of reactions are pyruvic acid and acetaldehyde.

NOTE:
In this problem, you never need to know the actual structures of the compounds involved to answer the question; this is a reasoning problem.)

phosphoenolpyruvic acid → pyruvic acid
pyruvic acid → acetaldehyde + CO_2
acetaldehyde → ethanol

Solutions to Summary and Challenge Problems

* indicates more challenging problems.

10.63 *Name the following compounds, using the IUPAC rules.*

a) 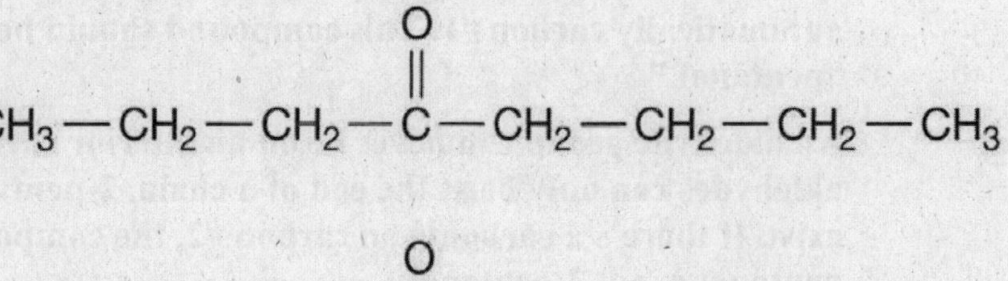 $CH_3-CH_2-CH_2-\overset{\overset{\displaystyle O}{\|}}{C}-CH_2-CH_2-CH_2-CH_3$

b) 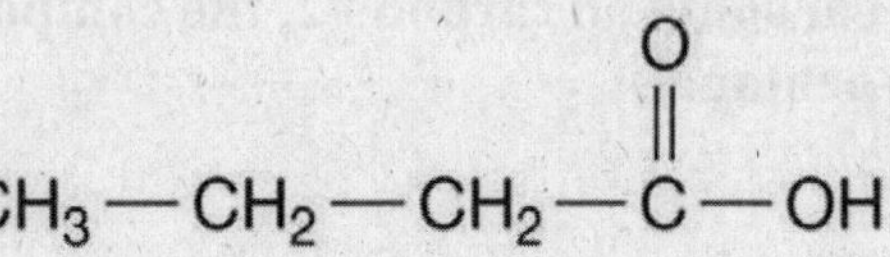 $CH_3-CH_2-CH_2-\overset{\overset{\displaystyle O}{\|}}{C}-OH$

c) 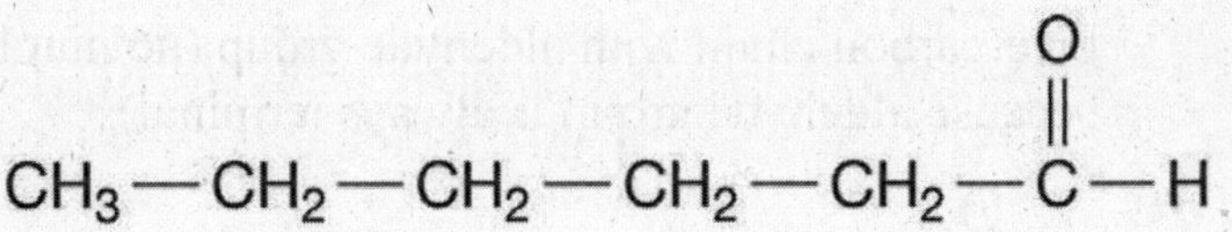 $CH_3-CH_2-CH_2-CH_2-CH_2-\overset{\large O}{\overset{\|}{C}}-H$

d) 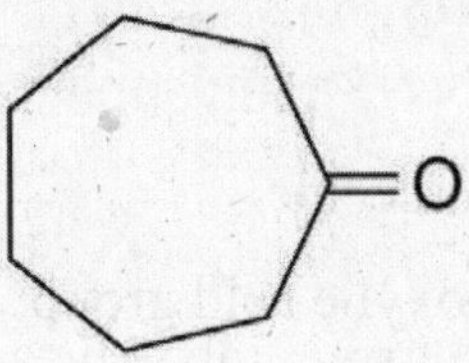 (cycloheptanone ring structure, =O)

e) 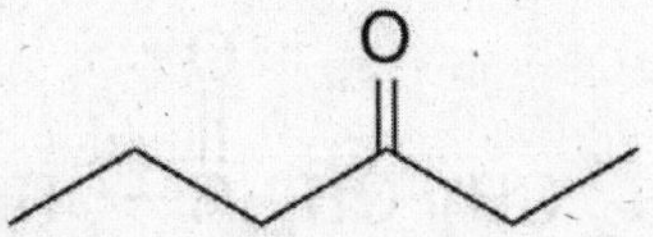 (skeletal structure with O)

f)  COOH

g)  CHO

a) 8-carbon chain, ketone on carbon #4: **4-octanone**

b) 4-carbon carboxylic acid: **butanoic acid**

c) 6-carbon aldehyde: **hexanal**

d) 7-carbon ring with ketone group: **cycloheptanone**

e) 6-carbon chain, ketone on carbon #3 (numbering starts at the end closer to the functional group, in this case right to left): **3-hexanone**

f) 3-carbon carboxylic acid: **propanoic acid** (don't be fooled by the condensed representation of the carboxylic acid group)

g) 6-carbon aldehyde: **hexanal**

NOTE:
This is another representation of the same compound in 10.63c.

10.65 *Draw condensed structural formulas for each of the following compounds:*

a) 2-hexanone Six-carbon chain, ketone group on carbon #2:

$$CH_3-\overset{\large O}{\overset{\|}{C}}-CH_2-CH_2-CH_2-CH_3$$

b) pentanal Five-carbon chain with aldehyde group (no number needed, because aldehyde group is always terminal).

$$\overset{\displaystyle O}{\overset{\|}{H-C}}-CH_2-CH_2-CH_2-CH_3$$

NOTE:
The aldehyde group can be shown on either end of the molecule.

c) heptanoic acid Seven-carbon chain with carboxylic acid group. Again, no number needed, and the carboxylic acid group can be shown on either end:

$$CH_3-CH_2-CH_2-CH_2-CH_2-CH_2-\overset{\displaystyle O}{\overset{\|}{C}}-OH$$

d) cyclobutanone Four-carbon ring with ketone group:

$$\begin{array}{ccc} CH_2 & - & C{=}O \\ | & & | \\ CH_2 & - & CH_2 \end{array}$$

e) acetaldehyde This is the trivial name of ethanal, the two-carbon aldehyde:

$$\overset{\displaystyle O}{\overset{\|}{H-C}}-CH_3$$

f) formic acid This is the trivial name of methanoic acid.

$$\overset{\displaystyle O}{\overset{\|}{H-C}}-OH$$

NOTE:
The "form" root is often equivalent to "meth," one carbon, and the "-ic acid" ending is a clue to the carboxylic acid group.

*10.67 *Draw the structure of a molecule that fits each of the following descriptions:*

a) An isomer of 1-butanol

We can move the alcohol group, change to a branched chain, or both. (We could also move the O to between C atoms rather than keeping the alcohol group; this would create a functional group called an "ether," which will be introduced in a later chapter.) The three alcohol compounds are shown:

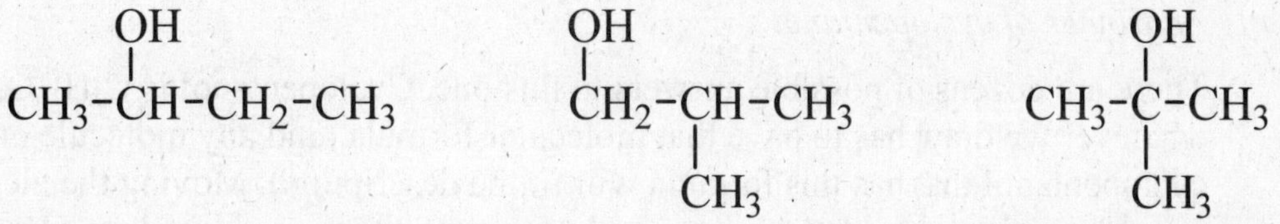

$$\begin{array}{c} OH \\ | \\ CH_3-CH-CH_2-CH_3 \end{array} \qquad \begin{array}{l} OH \\ | \\ CH_2-CH-CH_3 \\ \quad\;\; | \\ \quad\;\; CH_3 \end{array} \qquad \begin{array}{c} OH \\ | \\ CH_3-C-CH_3 \\ | \\ CH_3 \end{array}$$

b) *An isomer of butanal*

Again, we can move the carbonyl to a different carbon, change to a branched chain, or both. We could also change from having a C=O double bond to having a C=C double bond. There are many possible compounds; here are three:

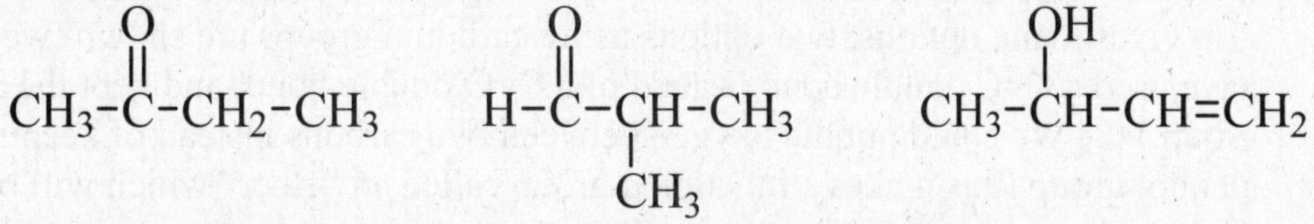

$$\begin{array}{c} O \\ \| \\ CH_3-C-CH_2-CH_3 \end{array} \qquad \begin{array}{c} O \\ \| \\ H-C-CH-CH_3 \\ | \\ CH_3 \end{array} \qquad \begin{array}{c} OH \\ | \\ CH_3-CH-CH=CH_2 \end{array}$$

c) *A ketone that has the same carbon skeleton as 2-methylpentane*

Start by drawing the 2-methylpentane skeleton (with only carbon atoms, no hydrogen or other atoms yet), and assess where a ketone group can be added.

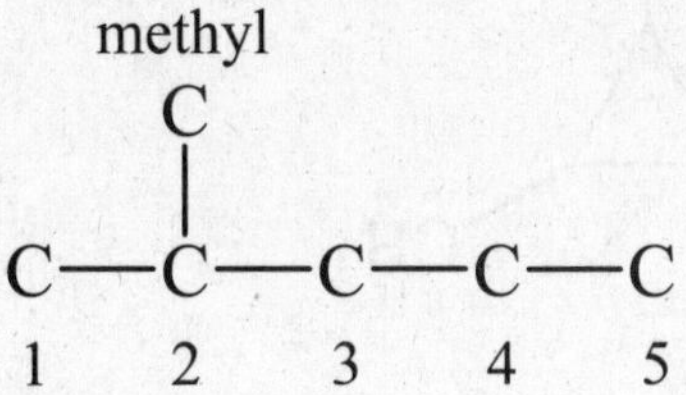

$$\begin{array}{ccccc} & \text{methyl} & & & \\ & C & & & \\ & | & & & \\ C- & C- & C- & C- & C \\ 1 & 2 & 3 & 4 & 5 \end{array}$$

The ketone group **cannot** be placed on carbon #2, because that carbon already has three of its four bonds taken up. A carbonyl on carbon #1 or #5, or on the methyl group, would form an aldehyde, not a ketone. The ketone group can therefore only be placed on carbon #3 or #4:

$$\begin{array}{c} CH_3 \qquad\quad O \\ | \qquad\qquad\; \| \\ CH_3-CH-CH_2-C-CH_3 \end{array} \qquad \begin{array}{c} CH_3\;\; O \\ | \quad\;\; \| \\ CH_3-CH-C-CH_2-CH_3 \end{array}$$

d) *A carboxylic acid that contains a cyclohexane ring*

Since the carboxylic acid group always has to be at the end of a chain, the group cannot be on one of the carbon atoms in the ring. The carboxylic acid group has to be outside the ring, but with that one limit, you could draw *any* compound that contained both a carboxylic acid group and a cyclohexane ring. The smallest possible molecule that fits this description is shown:

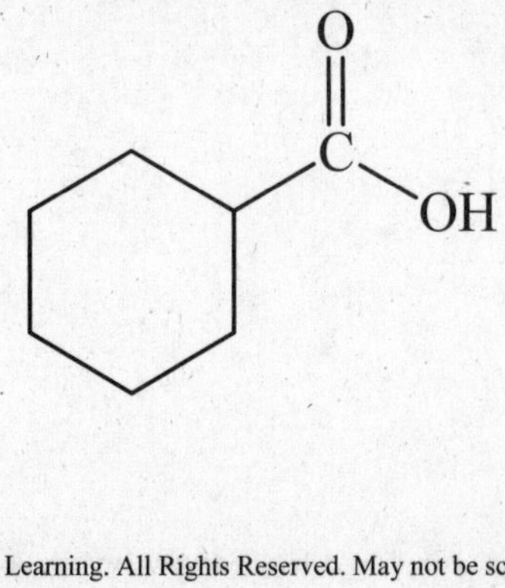

O

C

OH

e) An isomer of cyclopentanol

There are dozens of possible answers to this one. Cyclopentanol is $C_5H_{10}O$, so whatever we draw has to have that molecular formula (and any molecule other than cyclopentanol that has this formula will fit the description). Moving the alcohol group around the ring won't help, because all of the positions are equivalent. Here are three possible approaches: (1) If we want to keep a cyclic alcohol, we'll have to make the ring smaller (this gives many options; three are shown.) (2) We could also notice that closing a ring, like forming a double bond, requires a molecular formula with two fewer H atoms than a linear or branched alkane. So we can draw a linear or branched molecule with a double bond (alkene or carbonyl), rather than a cyclic alkane. This also gives many options; two options using carbonyl groups are shown (we could also have used a C=C double bond instead of a C=O double bond, and kept the alcohol group.) (3) We could put the oxygen between two carbons instead of keeping the alcohol group (this makes a functional group called an "ether" which will be discussed in a later chapter.) Again, there are many possible structures, and this option can also be combined with either of the others.

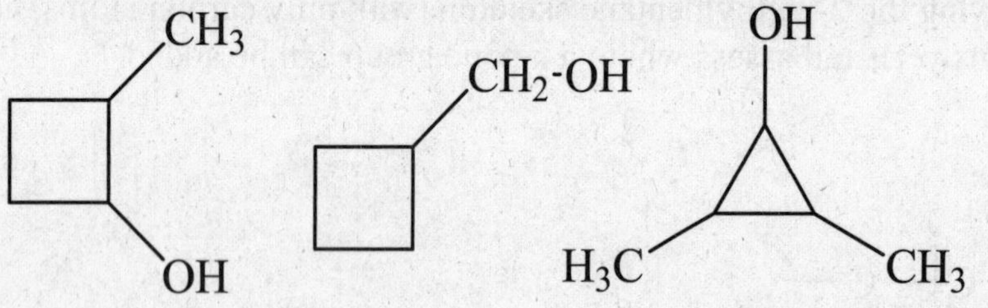

CH_3 / OH (cyclobutane ring) CH_2-OH (cyclobutane ring) OH / H_3C / CH_3 (cyclopropane ring)

$$H-\overset{\overset{\large O}{\|}}{C}-CH_2-CH_2-CH_2-CH_3 \qquad CH_3-\overset{\overset{\large O}{\|}}{C}-CH_2-CH_2-CH_3$$

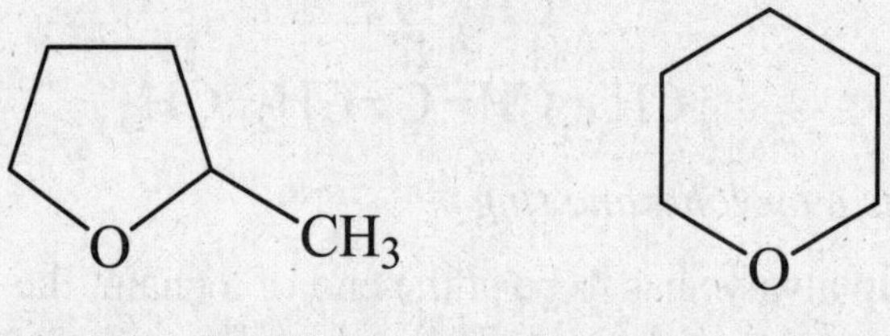

O / CH_3 (five-membered ring) O (six-membered ring)

10.69 *Draw the structure of the product that is formed when each of the following molecules is hydrogenated:*

a) $CH_3-CH=CH-CH_2-CH_3$

b) 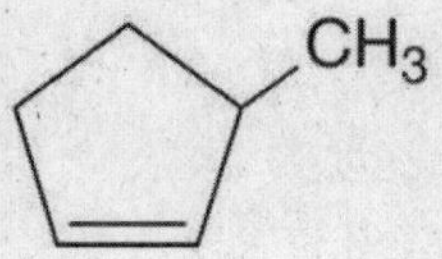 CH_3 (cyclopentene ring)

In each case, we identify the alkene present and turn it into a carbon-carbon single bond by adding hydrogen atoms:

a) $CH_3\text{-}CH_2\text{-}CH_2\text{-}CH_2\text{-}CH_3$

b) 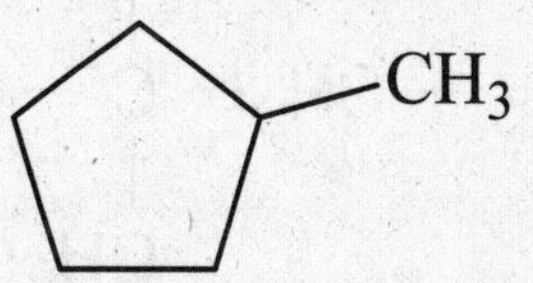 (cyclopentane ring)–CH_3

10.71 *The burning of fats to produce energy involves many reactions, one of which is the dehydrogenation of the following compound. The portion of the molecule that reacts is circled. Draw the product of this reaction.*

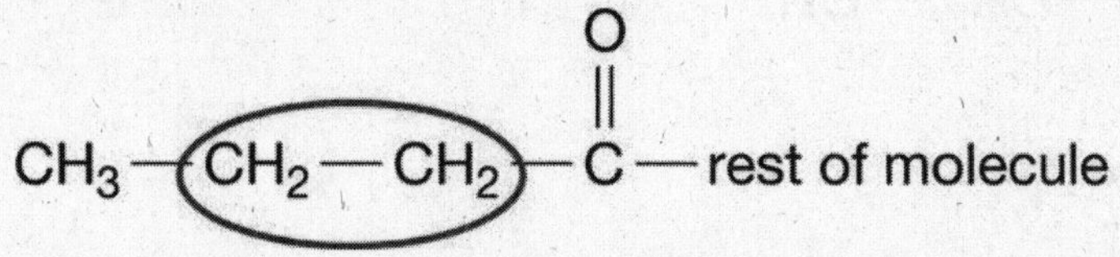

$$CH_3\text{—}(CH_2\text{—}CH_2)\text{—}\overset{\overset{\displaystyle O}{\|}}{C}\text{—rest of molecule}$$

Dehydrogenation removes two H atoms from adjacent carbons, forming a C=C double bond. (The complete reaction is shown below, with the hydrogen atoms that are removed circled.

NOTE:
You would only need to draw the product molecule.)

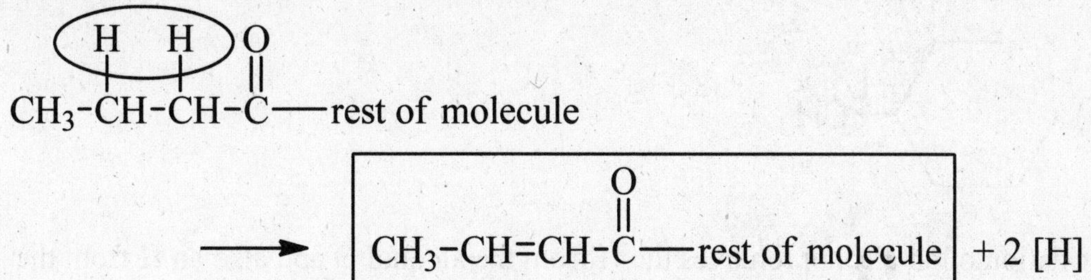

$$CH_3\text{-}\underset{}{\overset{H}{CH}}\text{-}\overset{H}{CH}\text{-}\overset{\overset{\displaystyle O}{\|}}{C}\text{—rest of molecule}$$

$$\longrightarrow \boxed{CH_3\text{-}CH\text{=}CH\text{-}\overset{\overset{\displaystyle O}{\|}}{C}\text{—rest of molecule}} + 2\,[H]$$

10.73 *Which of the alcohols in Problem 10.72 can be oxidized?*

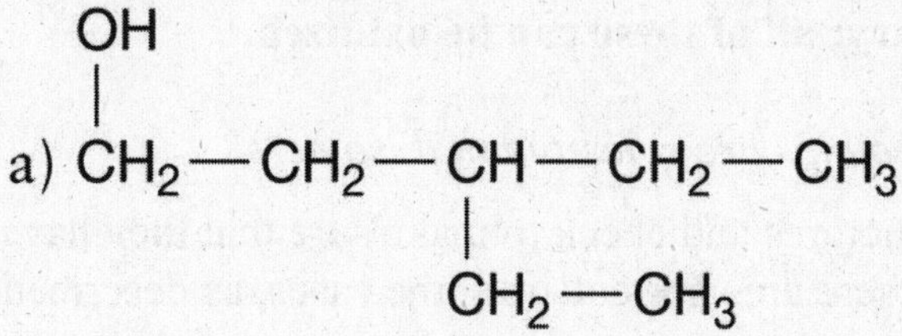

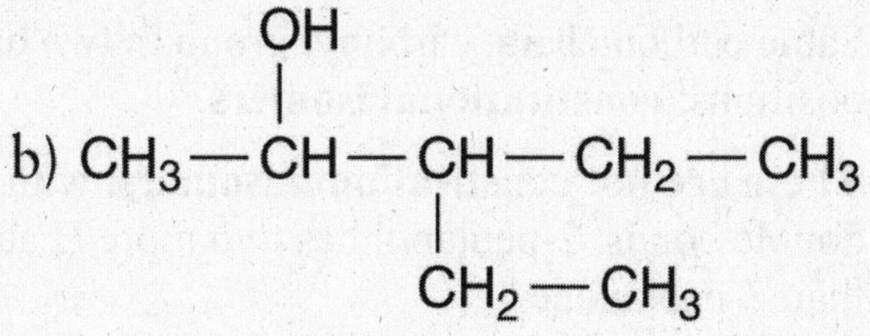

a)
$$\overset{\overset{\displaystyle OH}{|}}{CH_2}\text{—}CH_2\text{—}\underset{\underset{\displaystyle CH_2\text{—}CH_3}{|}}{CH}\text{—}CH_2\text{—}CH_3$$

b)
$$CH_3\text{—}\overset{\overset{\displaystyle OH}{|}}{CH}\text{—}\underset{\underset{\displaystyle CH_2\text{—}CH_3}{|}}{CH}\text{—}CH_2\text{—}CH_3$$

c) $CH_3-CH_2-C(OH)(CH_2-CH_3)-CH_2-CH_3$

d) (cyclohexane ring)$-CH_2-OH$

e) (cyclohexane ring with OH and CH_3 on the same carbon)

f) (cyclohexane ring with OH and CH_3 on adjacent carbons)

Oxidation of an alcohol removes the H from the alcohol O and also an H from the alcohol C. If the carbon atom attached to the alcohol group does not have any H atoms, then the alcohol cannot be oxidized (that is, tertiary alcohols cannot be oxidized.)

Options c) and e) are tertiary alcohols and cannot be oxidized. **Options a) and d) are primary, and b) and f) are secondary; all of these can be oxidized.**

10.75 *Which of the following pairs of compounds are constitutional isomers?*

The safest thing to do is draw the structures and check to make sure that they have the same molecular formula. However, there are shortcuts in some cases, as described below.

a) 2-pentanone and 3-pentanone — Same carbon chain, same functional group in different locations: **constitutional isomers.**

b) 2-pentanone and pentanal — Same carbon chain, carbonyl group in two different positions: **constitutional isomers.**

c) 2-pentanone and 2-pentanol — **These are not constitutional isomers**; with no double bonds, 2-pentanol has two more H atoms than 2-pentanone.

d) 2-pentanone and cyclopentanol — Same molecular formula, different structure. This is the hardest one to shortcut, and the safest thing is to draw the structures and count up the molecular formulas ($C_5H_{10}O$): **Constitutional isomers.**

10.77 *From each of the following pairs of compounds, select the compound that should have the higher boiling point:*

Chapter 4 review! For this we look at the attractions between molecules in each compound: dispersion forces, dipole-dipole forces, and hydrogen bonding.

a) *propane or propanal*

These molecules have similar dispersion forces; propanal adds the effect of dipole-dipole forces because of the polar carbonyl group, so **propanal will have a higher boiling point.**

b) *2-butanol or 2-butanone*

The two molecules have similar dispersion forces and dipole-dipole forces; 2-butanol adds hydrogen bonding, while 2-butanone cannot form hydrogen bonds between its molecules, so **2-butanol will have a higher boiling point.**

c) *acetic acid or pentanoic acid*

These molecules have the same functional group and therefore have approximately the same dipole-dipole forces and hydrogen bonding attractions; therefore the compound with stronger dispersion forces will have the higher boiling point. **Pentanoic acid, a larger molecule with more electrons, will have stronger dispersion forces and a higher boiling point.**

*10.79 *Phenylalanine is a component of proteins. It is made by bacteria, using a sequence of reactions that includes the reduction of the following compound. Only the ketone group of this molecule is reduced. Draw the structure of the product of this reduction.*

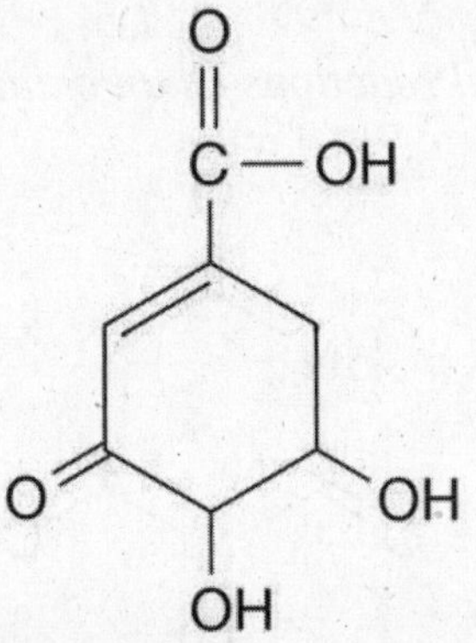

O
C—OH
O
OH
OH

3-Dehydroshikimic acid

Reduction of a carbonyl adds hydrogen atoms to both the O and C of the carbonyl, converting the carbonyl to an alcohol group. We identify the ketone group (lower left) and leave the rest of the molecule alone as we draw the reduction product:

$$HOOC\text{-cyclohexene ring with HO, OH, OH substituents}$$

10.81 *When 1-butanol is oxidized, it is converted into compound A. Compound A can also be oxidized, and the product of this second oxidation is compound B. Draw the structures of compounds A and B.*

$$CH_3-CH_2-CH_2-CH_2(OH) \longrightarrow \text{Compound A} \longrightarrow \text{Compound B}$$

Oxidation of an alcohol removes two H atoms, one from the alcohol O and one from the alcohol H, forming a C=O double bond. In this case, oxidation of the primary alcohol converts it to an aldehyde. In the second step, oxidation of the aldehyde group adds an O between the aldehyde C and H, converting it to a carboxylic acid.

$$CH_3\text{-}CH_2\text{-}CH_2\text{-}\overset{O}{\overset{\|}{C}}\text{-}H \longrightarrow CH_3\text{-}CH_2\text{-}CH_2\text{-}\overset{O}{\overset{\|}{C}}\text{-}OH$$

compound A compound B

10.83 *Identify each of the following biochemical reactions as an oxidation or a reduction:*

a) [dihydroorotic acid structure: HN, O, N-H, COOH] ⟶ [orotic acid structure: HN, O, N-H, COOH]

b) [cyclic imine ring with N and COOH] ⟶ [piperidine ring with N and COOH]

c) 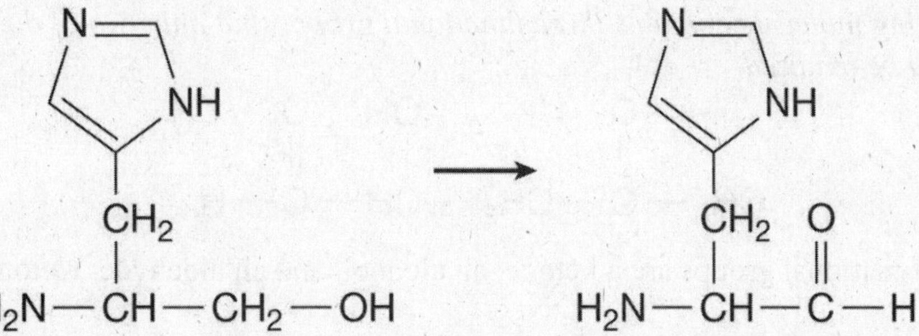 (imidazol-4-yl)–CH_2–CH(NH_2)–CH_2–OH $\longrightarrow$ (imidazol-4-yl)–CH_2–CH(NH_2)–C(=O)–H

$$H_2N-CH(CH_2\text{-imidazolyl})-CH_2-OH \longrightarrow H_2N-CH(CH_2\text{-imidazolyl})-\overset{O}{\overset{\|}{C}}-H$$

d)

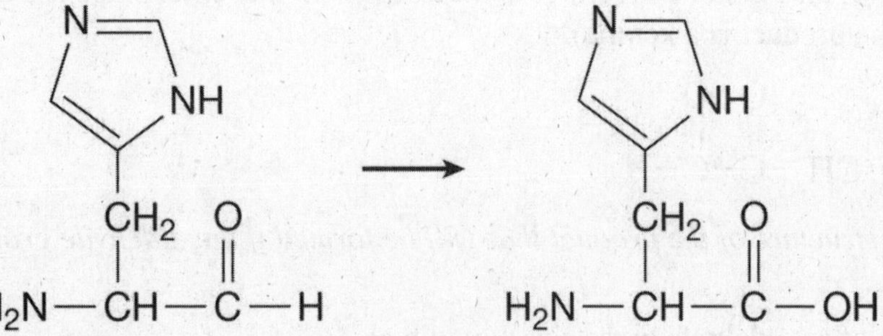

$$H_2N-CH(CH_2\text{-imidazolyl})-\overset{O}{\overset{\|}{C}}-H \longrightarrow H_2N-CH(CH_2\text{-imidazolyl})-\overset{O}{\overset{\|}{C}}-OH$$

e)

$$HO-\overset{O}{\overset{\|}{C}}-CH_2-\overset{O}{\overset{\|}{C}}-\overset{O}{\overset{\|}{C}}-OH \longrightarrow HO-\overset{O}{\overset{\|}{C}}-CH_2-\overset{OH}{\overset{|}{C}H}-\overset{O}{\overset{\|}{C}}-OH$$

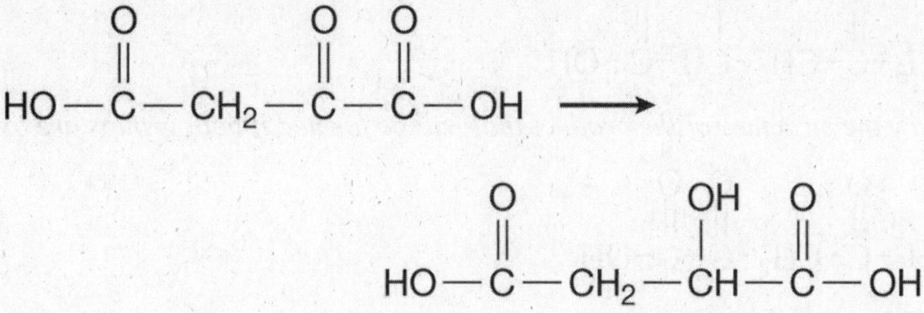

Oxidation removes H or adds O (or increases the number of bonds to O). Reduction adds H or decreases the number of bonds to O. In each reaction, try to identify what's really changing between the reactant and product molecules, and ignore everything else—anything that isn't changing in the reaction is just a distraction.

a) In this reaction, a C–C single bond is converted into a C=C double bond; this requires removal of two H atoms from the carbon atoms (dehydrogenation), an **oxidation** reaction.

b) In this reaction, a C=N double bond is converted to a C–N single bond; this requires the addition of two H atoms and is a **reduction** reaction.

c) In this reaction, an alcohol group is converted to an aldehyde (carbonyl) group. Two hydrogen atoms are removed, the number of C–O bonds is increased, and the reaction is an **oxidation**.

d) This reaction converts an aldehyde to a carboxylic acid. An oxygen is added to the molecule and the number of C–O bonds is increased: **oxidation**.

e) In this reaction, the middle carbonyl group, the ketone, is converted to an alcohol with the addition of two H atoms: **reduction**.

*10.85 *The following molecule contains three functional groups, but only two of the functional groups can be oxidized:*

$$CH_3-\overset{\overset{\displaystyle O}{\|}}{C}-CH_2-\overset{\overset{\displaystyle OH}{|}}{CH}-\overset{\overset{\displaystyle O}{\|}}{C}-H$$

The three functional groups are a ketone, an alcohol, and an aldehyde. Ketones cannot be oxidized.

a) *Draw the structure of the product that will be formed if the alcohol group is oxidized.*

Oxidation of an alcohol converts it to a carbonyl (in this case, since it's a secondary alcohol, the product is a ketone).

$$CH_3-\overset{\overset{\displaystyle O}{\|}}{C}-CH_2-\overset{\overset{\displaystyle O}{\|}}{C}-\overset{\overset{\displaystyle O}{\|}}{C}-H$$

b) *Draw the structure of the product that will be formed if the aldehyde group is oxidized.*

Oxidation of an aldehyde gives a carboxylic acid:

$$CH_3-\overset{\overset{\displaystyle O}{\|}}{C}-CH_2-\overset{\overset{\displaystyle OH}{|}}{CH}-\overset{\overset{\displaystyle O}{\|}}{C}-OH$$

c) *Draw the structure of the product that will be formed if both groups are oxidized.*

$$CH_3-\overset{\overset{\displaystyle O}{\|}}{C}-CH_2-\overset{\overset{\displaystyle O}{\|}}{C}-\overset{\overset{\displaystyle O}{\|}}{C}-OH$$

10.87 *Which of the redox coenzymes is most likely to be involved in each of the following types of reactions?*

NAD^+ (nicotinamide adenine dinucleotide) is the hydrogen acceptor in most oxidations. FAD (flavin adenine dinucleotide) accepts hydrogen atoms in dehydrogenation reactions that produce alkene functional groups. $NADP^+$ (nicotinamide adenine dinucleotide phosphate) supplies the hydrogen atoms in reduction reactions. We identify each reaction as an oxidation (which will involve NAD^+, or in the particular case of dehydrogenation to form an alkene, FAD) or a reduction (which will involve $NADP^+$).

a) *The conversion of an alcohol into a ketone.*

This is an oxidation reaction, in which two hydrogen atoms are removed; the redox coenzyme is likely to be **NAD^+**.

b) *The conversion of an alkene into an alkane.*

This is a hydrogenation reaction, a type of reduction. **$NADP^+$** supplies the hydrogen atoms in reduction reactions.

c) *The conversion of a thiol into a disulfide.*

This is an oxidation reaction; even if you haven't memorized that the 2 thiol → disulfide reaction is an oxidation, you can identify it because the reaction requires removing two H atoms: **NAD^+**.

d) *The conversion of an aldehyde into a carboxylic acid.*

This is an oxidation reaction: **NAD^+**.

*10.89 *If you have 13.5 g of cyclopentanol, how many moles do you have?*

Chapter 5 review! First, figure out the molecular formula of cyclopentanol ($C_5H_{10}O$), so that you can calculate its molar mass:

5 C	=	5 × 12.01	=	60.05 g/mol
10 H	=	10 × 1.008	=	10.08
+ 1 O	=	1 × 16.00	=	16.00
			=	86.13 g/mol

Then, use the molar mass to convert from grams to moles:

$$\mathbf{13.5\ g \times \frac{1\ mol}{86.13\ g} = 0.157\ mol}$$

10.91 *A student is asked to draw the structure of the product that is formed when ethanol is oxidized. The student writes the reaction shown below. Explain why this answer is not reasonable, and draw the actual product of this oxidation.*

$$\underset{\text{ethanol}}{CH_3\text{–}CH_2\text{–}OH} \rightarrow \underset{\text{oxidation product?}}{CH_2\text{=}CH\text{–}OH}$$

Oxidation of alcohols is easier than dehydrogenation—in fact, dehydrogenation is the most difficult of the oxidations we've looked at. So the dehydrogenation reaction written by the student is not going to happen if there is an alcohol present that can be oxidized. In fact, the oxidation product of the alcohol will be a carbonyl, in this case giving **ethanal** (trivial name, **acetaldehyde**):

$$CH_3\text{–}\overset{\overset{\displaystyle O}{||}}{C}\text{–}H$$

NOTE:
The aldehyde group could also be oxidized further, to produce a carboxylic acid group; the compound formed in this oxidation would be acetic acid:

$$CH_3\text{–}\overset{\overset{\displaystyle O}{||}}{C}\text{–}OH$$

*10.93 *Isoleucine is one of the components of proteins. It can be burned to obtain energy by a metabolic pathway that involves a sequence of nine reactions. Part of this pathway involves the conversion shown here, using the sequence of three reactions described in Section 10.7. Draw the structures of each of the compounds that are formed during this sequence of reactions.*

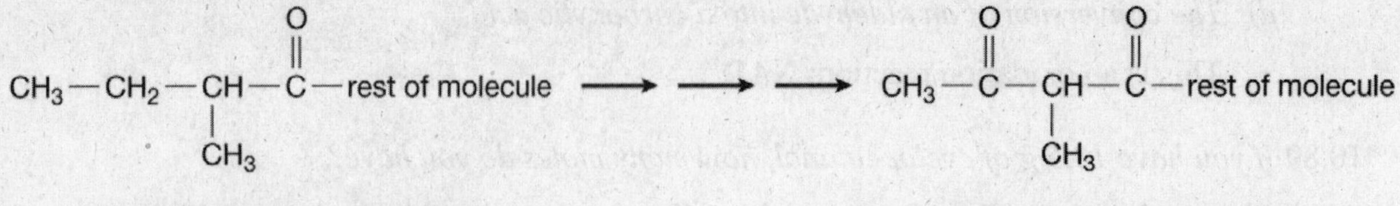

$$CH_3-CH_2-\underset{\displaystyle CH_3}{\underset{|}{CH}}-\overset{\displaystyle O}{\overset{\|}{C}}-\text{rest of molecule} \longrightarrow \longrightarrow \longrightarrow CH_3-\overset{\displaystyle O}{\overset{\|}{C}}-\underset{\displaystyle CH_3}{\underset{|}{CH}}-\overset{\displaystyle O}{\overset{\|}{C}}-\text{rest of molecule}$$

α-Methylbutyryl-coenzyme A **α-Methylacetoacetyl-coenzyme A**

The sequence mentioned in Section 10.7 is a three-step oxidation/hydration/oxidation pattern. In this case, the **first (oxidation) step converts the alkane to an alkene, the second (hydration) step converts the alkene to an alcohol, and the third step (the second oxidation) converts the alcohol to a carbonyl** (in this case a ketone.)

NOTE:

In the hydration step, you have to make a choice of where to put the alcohol group; only by putting the alcohol on the second carbon from the left can you get to the product shown. (This doesn't follow Markovnikov's rule; many enzyme-controlled reactions do not. Since we have the structure of the final product, we take the pathway that gets us there.)

Though the problem only asks for the two intermediate products, the complete reaction sequence is shown on the following page:s

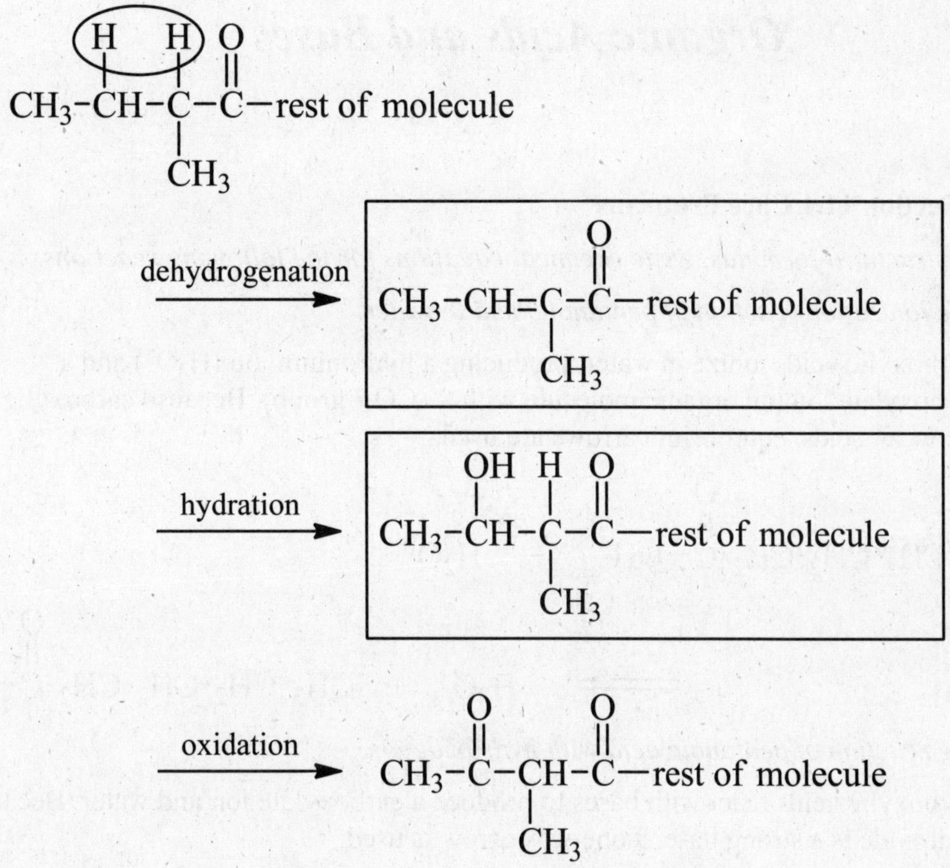

$$CH_3-CH-\underset{\underset{CH_3}{|}}{\overset{H}{C}}-\overset{O}{\overset{\|}{C}}-\text{rest of molecule}$$

(the H on the second carbon and the H on the third carbon are circled)

$$\xrightarrow{\text{dehydrogenation}} CH_3-CH=\underset{\underset{CH_3}{|}}{C}-\overset{O}{\overset{\|}{C}}-\text{rest of molecule}$$

$$\xrightarrow{\text{hydration}} CH_3-\overset{OH}{\overset{|}{CH}}-\underset{\underset{CH_3}{|}}{\overset{H}{\overset{|}{C}}}-\overset{O}{\overset{\|}{C}}-\text{rest of molecule}$$

$$\xrightarrow{\text{oxidation}} CH_3-\overset{O}{\overset{\|}{C}}-\underset{\underset{CH_3}{|}}{CH}-\overset{O}{\overset{\|}{C}}-\text{rest of molecule}$$

If you have trouble seeing this when working the problem out for yourself, you can always just work backward from the products: the reverse of an oxidation/hydration/oxidation sequence would be reduction/dehydration/reduction (stop a moment here and make sure you see why). Then working *backward from the product* (and noticing that only the left ketone is affected—the ketone on the right is unchanged in the reaction), reduction of the left ketone gives an alcohol; dehydration of the alcohol forms an alkene; and reduction (hydrogenation) of the alkene forms an alkane.

11

Chapter 11

Organic Acids and Bases

Solutions to Section 11.1 Core Problems

11.1 *Using structural formulas, write chemical equations for the following reactions:*

a) The ionization reaction of pentanoic acid in water.

Carboxylic acids ionize in water, producing a hydronium ion (H_3O^+) and a carboxylate ion (an organic molecule with a $-CO_2^-$ group). Because carboxylic acids are weak acids, equilibrium arrows are used:

$$CH_3\text{-}CH_2\text{-}CH_2\text{-}CH_2\text{-}\overset{\overset{\large O}{\|}}{C}-OH \quad + \quad H_2O$$

$$\rightleftharpoons \quad H_3O^+ \quad + \quad CH_3\text{-}CH_2\text{-}CH_2\text{-}CH_2\text{-}\overset{\overset{\large O}{\|}}{C}-O^-$$

b) The reaction of pentanoic acid with hydroxide ion.

Carboxylic acids react with bases to produce a carboxylate ion and water. Because hydroxide is a strong base, a one-way arrow is used:

$$CH_3\text{-}CH_2\text{-}CH_2\text{-}CH_2\text{-}\overset{\overset{\large O}{\|}}{C}-OH \ + \ OH^-$$

$$\longrightarrow \quad CH_3\text{-}CH_2\text{-}CH_2\text{-}CH_2\text{-}\overset{\overset{\large O}{\|}}{C}-O^- \quad + \quad H_2O$$

11.3 *Draw structural formulas for the following:*

a) butanoate ion Four-carbon chain with a carboxylate ion group (a $-CO_2^-$ group):

$$CH_3\text{-}CH_2\text{-}CH_2\text{-}\overset{\overset{\large O}{\|}}{C}-O^-$$

b) potassium butanoate The butanoate ion (four-carbon chain, carboxylate ion group) combines in a 1:1 ratio with potassium ion to form an ionic compound:

$$CH_3\text{-}CH_2\text{-}CH_2\text{-}\overset{\overset{\large O}{\|}}{C}-O^- \ K^+$$

11.5 *Write the chemical formula of calcium butanoate.*

The butanoate ion has a four-carbon chain with a carboxylate ion group (see Problem 11.3). Since the butanoate ion has a −1 charge, it combines in a 2:1 ratio with calcium ion:

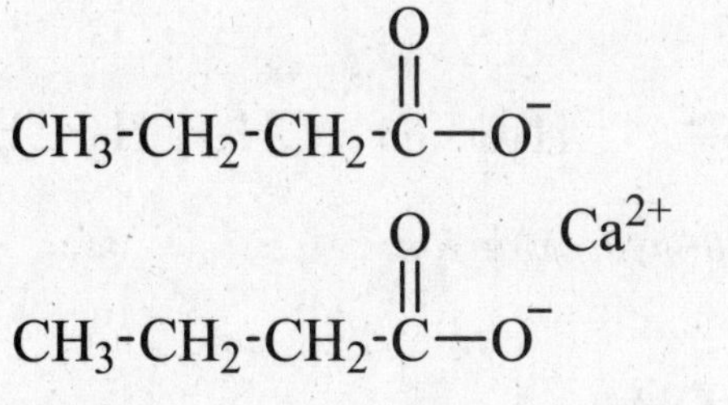

$$CH_3\text{-}CH_2\text{-}CH_2\text{-}\overset{\displaystyle O}{\overset{\|}{C}}\text{—}O^-$$

$$Ca^{2+}$$

$$CH_3\text{-}CH_2\text{-}CH_2\text{-}\overset{\displaystyle O}{\overset{\|}{C}}\text{—}O^-$$

The chemical formula of calcium butanoate is $Ca(C_4H_7O_2)_2$.

11.7 *Name the following substances:*

a) 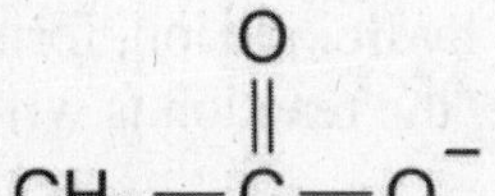 $CH_3\text{—}\overset{\displaystyle O}{\overset{\|}{C}}\text{—}O^-$

b) 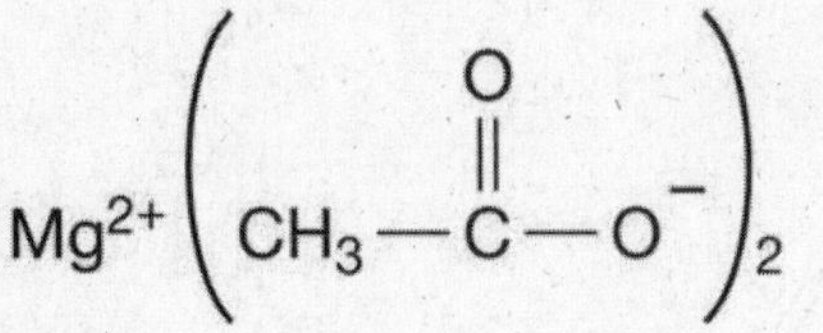 $Mg^{2+}\left(CH_3\text{—}\overset{\displaystyle O}{\overset{\|}{C}}\text{—}O^-\right)_2$

a) This ion is the conjugate base of acetic acid

NOTE:
Acetic acid is almost never called by its systematic name, ethanoic acid; this structure and name must be memorized). To name a carboxylate ion we remove the "–ic acid" ending from the name of the acid and add the suffix "-ate": **acetate ion**.

b) As in Chapter 3, ionic compounds are named by giving the name of the cation, then the anion. This compound of magnesium ion and acetate ion is named **magnesium acetate.**

11.9 *Which of the following compounds should have the higher solubility in water? Explain your reasoning.*

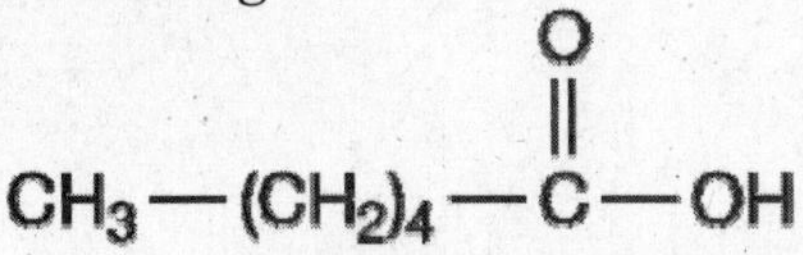

$$CH_3\text{—}(CH_2)_4\text{—}\overset{\displaystyle O}{\overset{\|}{C}}\text{—}OH \qquad CH_3\text{—}(CH_2)_4\text{—}\overset{\displaystyle O}{\overset{\|}{C}}\text{—}O^-\ Na^+$$

Hexanoic acid 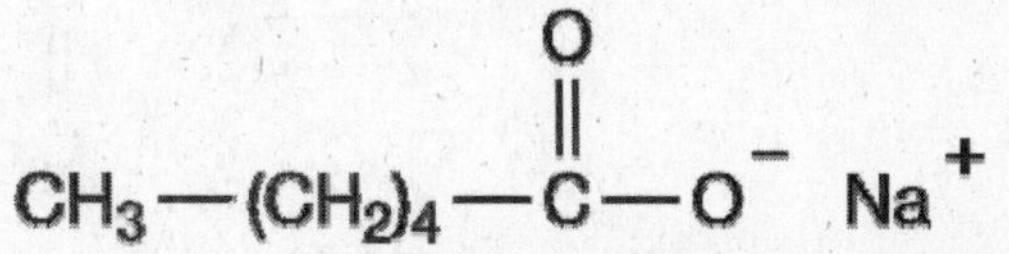 **Sodium hexanoate**

Most sodium and potassium salts of carboxylate ions are more soluble than the corresponding carboxylic acids, so sodium hexanoate should be more soluble than hexanoic acid.

11.11 *Using structural formulas, write chemical equations for the following reactions:*

a) The ionization reaction of CH_3–CH_2–CH_2–SH in water:

Thiols are weak acids, so in the ionization reaction of this thiol, the acidic hydrogen is transferred from the thiol S to a water molecule. Because thiols are weak acids, the reaction is written with equilibrium arrows:

$$CH_3\text{-}CH_2\text{-}CH_2\text{-}SH + H_2O \rightleftharpoons H_3O^+ + CH_3\text{-}CH_2\text{-}CH_2\text{-}S^-$$

b) The reaction of the following compound with hydroxide ion:

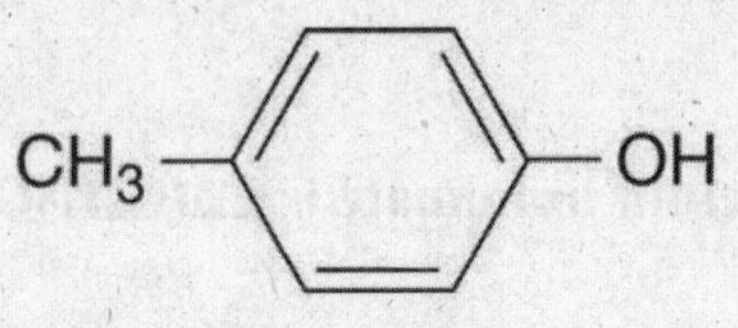

CH_3–C_6H_4–OH

Phenols are weak acids, so when this molecule reacts with OH^-, a strong base, the acidic hydrogen is transferred from the phenol O to the hydroxide ion, forming a water molecule. Because hydroxide ion is a strong base, the reaction is written with a one-way arrow:

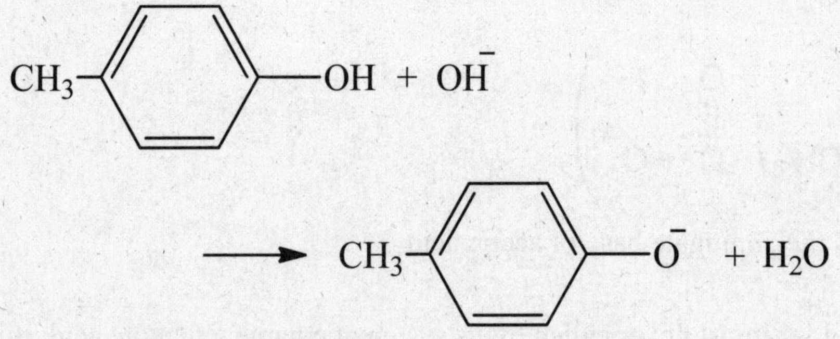

$$CH_3\text{–}C_6H_4\text{–}OH + OH^- \longrightarrow CH_3\text{–}C_6H_4\text{–}O^- + H_2O$$

Solutions to Section 11.2 Core Problems

11.13 *What small molecule is formed in all decarboxylation reactions?*

In a decarboxylation reaction, a carboxylate group leaves the organic molecule as a molecule of **carbon dioxide, CO_2.**

11.15 *a) Identify the α- and β- carbon atoms in the following carboxylic acid:*

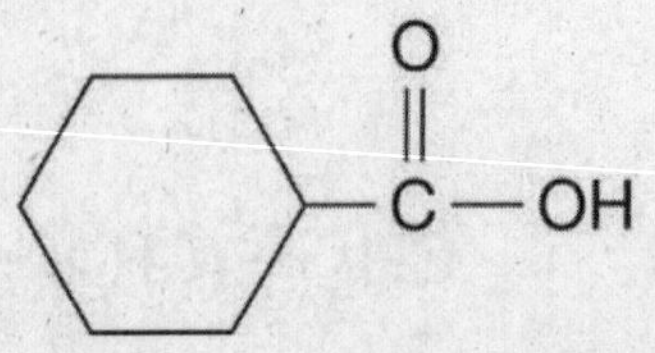

C_6H_{11}–C(=O)–OH

The alpha (α) carbon is the carbon next to the carboxylic acid, and the beta (β) carbon is next to the alpha carbon:

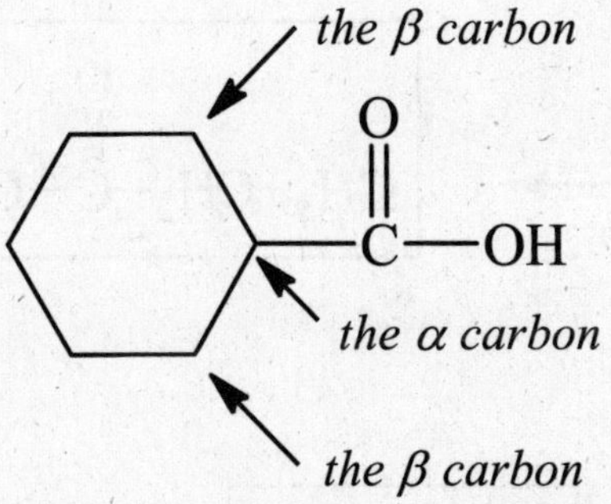

the β carbon

$$\text{C}_6\text{H}_{11}-\overset{\displaystyle O}{\overset{\|}{C}}-OH$$

the α carbon

the β carbon

b) *Can this compound be decarboxylated? Explain how you can tell.*

In practice, a decarboxylation reaction will only occur if there is another functional group (usually a carbonyl group, C=O) on the α- or β-carbon atoms. This molecule has no functional groups on the α- or β-carbons, so it cannot be decarboxylated.

11.17 *Draw the structures of the organic compounds that will be formed when each of the following molecules is decarboxylated:*

a) 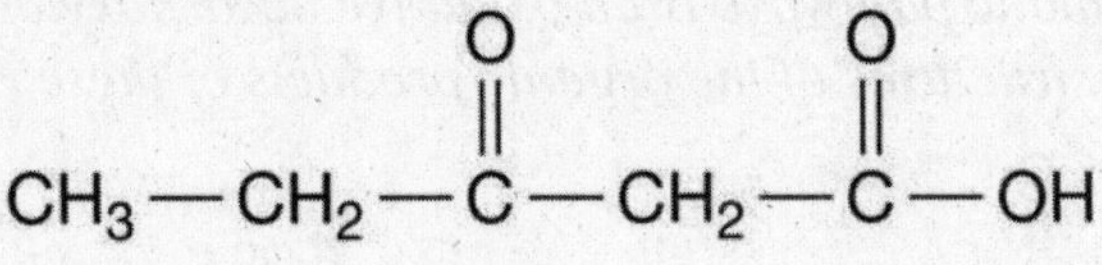 $CH_3-CH_2-\overset{\displaystyle O}{\overset{\|}{C}}-CH_2-\overset{\displaystyle O}{\overset{\|}{C}}-OH$

b) 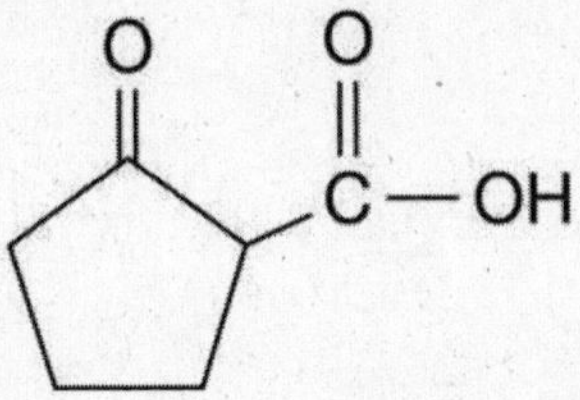 2-oxocyclopentane ring with $-\overset{\displaystyle O}{\overset{\|}{C}}-OH$

In a normal decarboxylation of a carboxylic acid with a β-ketone group, we remove the carboxylic acid group as CO_2 and replace the H on the α-carbon. Though the question only asks for the products (in boxes below), the complete reactions are shown below, with the carboxylic acid groups circled:

a) $CH_3-CH_2-\overset{O}{\overset{\|}{C}}-CH_2-\overset{O}{\overset{\|}{C}}-OH \longrightarrow CH_3-CH_2-\overset{O}{\overset{\|}{C}}-CH_3 + O=C=O$

b) (cyclopentanone with $-\overset{O}{\overset{\|}{C}}-OH$ on the CH) $\longrightarrow$ (cyclopentanone, CH_2) + $O=C=O$

11.19 *Each of the following organic acids is formed during the breakdown of an amino acid. The next step in each of these metabolic pathways is an oxidative decarboxylation, using coenzyme A as the thiol. Draw the structures of the organic products of these reactions.*

a) $CH_3-CH_2-\underset{}{\overset{CH_3}{\overset{|}{CH}}}-\overset{O}{\overset{\|}{C}}-\overset{O}{\overset{\|}{C}}-OH$

Formed when isoleucine is broken down

b) $HO-\overset{O}{\overset{\|}{C}}-CH_2-CH_2-CH_2-\overset{O}{\overset{\|}{C}}-\overset{O}{\overset{\|}{C}}-OH$

Formed when lysine is broken down

In all oxidative decarboxylations, we remove the carboxylic acid group and replace it with the –S–CoA group to make a thioester. The rest of the molecule remains unchanged.

a) The organic product of the decarboxylation is:

$CH_3-CH_2-\overset{CH_3}{\overset{|}{CH}}-\overset{O}{\overset{\|}{C}}-S-CoA$

b) There are two carboxylic acid groups in this molecule. Oxidative decarboxylation removes a carboxylic acid with an α-ketone, so only the left-hand carboxylic acid is affected:

$$HO-\overset{\overset{\large O}{\|}}{C}-CH_2-CH_2-CH_2-\overset{\overset{\large O}{\|}}{C}-S-CoA$$

Solutions to Section 11.3 Core Problems

11.21 *Classify each of the following compounds as a primary, secondary, or tertiary amine:*

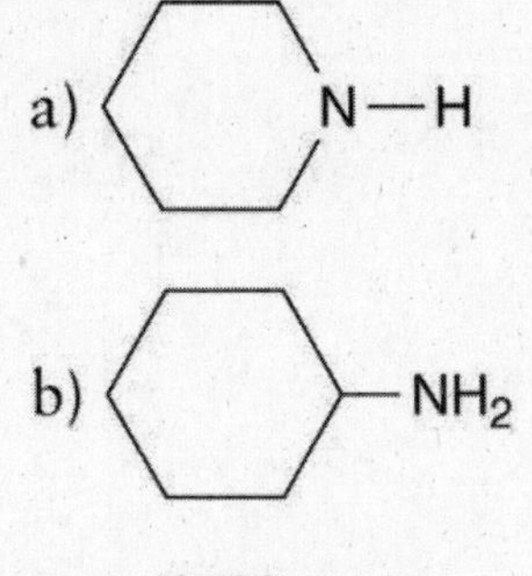

c) 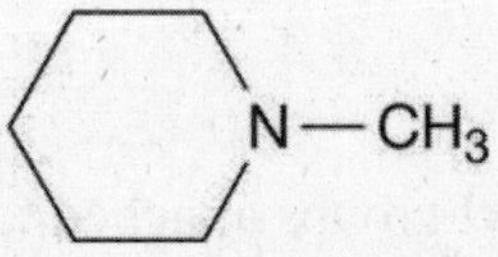 N—CH_3

Amines are classified as primary, secondary or tertiary based on the number of carbon atoms bonded to the amine nitrogen atom.

a) Two carbon atoms attached to amine nitrogen: **secondary amine**

b) One carbon atom attached to amine nitrogen: **primary amine**

c) Three carbon atoms attached to amine nitrogen: **tertiary amine**

11.23 *One of the compounds in Problem 11.21 contains an amino group. Which compound is it?*

An "amino group" is the $-NH_2$ part of a primary amine. Only **b)** has this group.

11.25 *Draw the structures of the following amines:*

Amines are named by listing, in alphabetical order, the alkyl groups attached to the amine nitrogen, and adding the word "amine."

a) pentylamine 5-carbon chain attached to amine group:

CH_3-CH_2-CH_2-CH_2-CH_2-NH_2 or NH_2

b) dipropylamine Two 3-carbon chains attached to amine group:

CH_3-CH_2-CH_2-NH-CH_2-CH_2-CH_3 or  NH

c) *cyclohexylmethylamine* 6-carbon ring and methyl group attached to amine group:

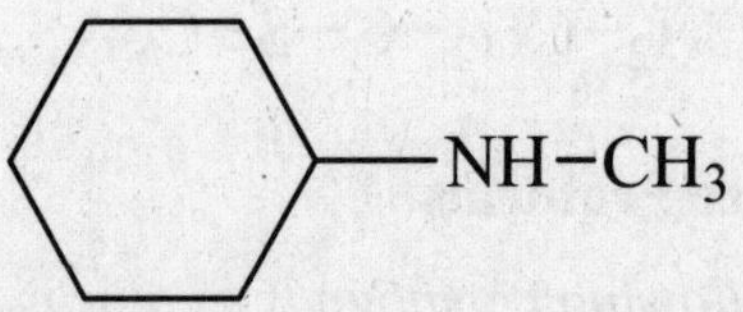

$$\text{C}_6\text{H}_{11}-\text{NH}-\text{CH}_3$$

11.27 *Name the following amines:*

a) 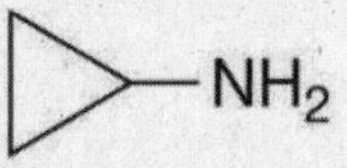 (cyclopropyl)$-NH_2$

b) $CH_3-CH_2-NH-CH_2-CH_2-CH_2-CH_3$

c) 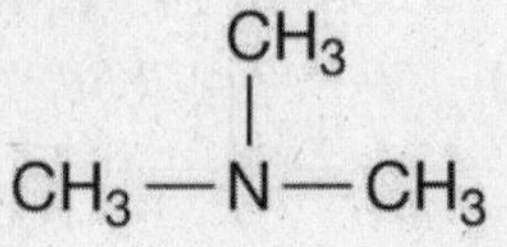 $CH_3-N(CH_3)-CH_3$

Amines are named by listing, in alphabetical order, the alkyl groups attached to the amine nitrogen, and adding the word "amine."

a) The 3-membered ring is a cyclopropyl group: **cyclopropylamine.**

b) This molecule contains an ethyl group (2-carbon chain) and a butyl group (4-carbon chain), listed alphabetically: **butylethylamine.**

c) Three methyl groups attached to amine: **trimethylamine.**

11.29 *Using structural formulas, show how hydrogen bonding can occur between the following:*

a) *two molecules of ethylamine*

The hydrogen atom in a primary or secondary amine is a hydrogen bond donor; the nitrogen atom in an amine is a hydrogen bond acceptor. If you need a review of hydrogen bonding, see Chapter 4 (hydrogen bonding will continue to be an important concept throughout the remaining chapters, so make sure to know it!).

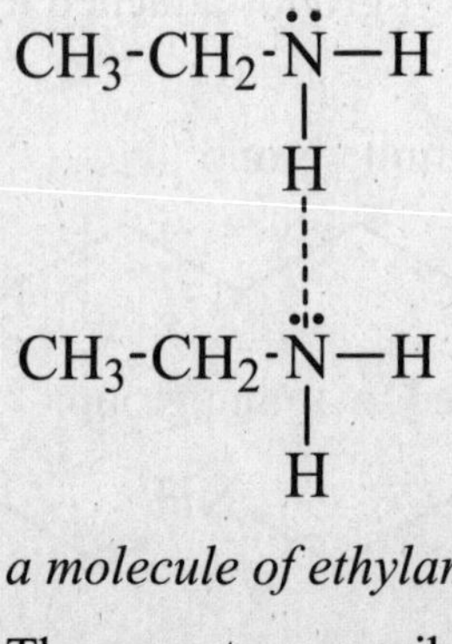

$$CH_3\text{-}CH_2\text{-}\ddot{N}(-H)-H \cdots \ddot{N}(-H)(-H)\text{-}CH_2\text{-}CH_3$$

b) *a molecule of ethylamine and a molecule of water*

There are two possible modes of hydrogen bonding: one in which the hydrogen atom in water is a hydrogen bond donor and the amine nitrogen atom is the acceptor (top water molecule, possible for all amines), and one in which the amine is the donor and water is the acceptor (lower water molecule, only possible for primary and secondary amines).

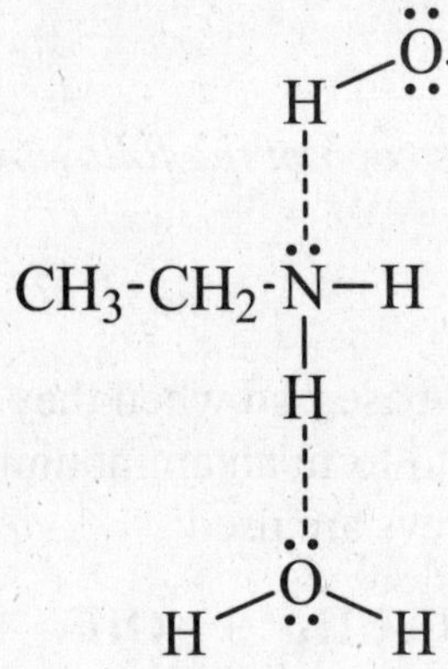

$$\begin{array}{c} H-\ddot{O}-H \\ \vdots \\ CH_3\text{-}CH_2\text{-}\ddot{N}-H \\ | \\ H \\ \vdots \\ H-\ddot{O}-H \end{array}$$

11.31 *The following two compounds are constitutional isomers. One boils at 145°C, while the other boils at 185°C. Match each structure with its boiling point, and explain your answer.*

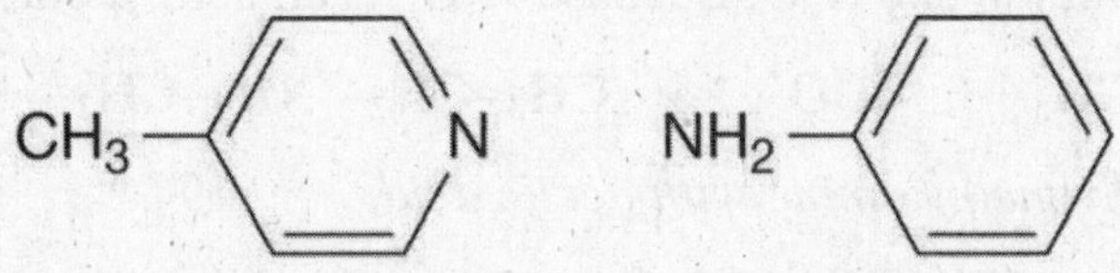

4-Methylpyridine **Aniline**

Since both molecules have the same formula and the same number of electrons, the dispersion forces between the molecules in each compound are about equal. The difference in their boiling points must come from hydrogen bonding interactions. Tertiary amines do not have a hydrogen atom bonded to the nitrogen, so they cannot be hydrogen bond donors. As a result, the attraction between molecules of a tertiary amine is weaker than the attraction between molecules of other amines. **4-methylpyridine cannot form hydrogen bonds between its molecules as a pure substance, so it has the weaker attractions and the lower boiling point, 145°C. Aniline molecules can form hydrogen bonds to each other, so aniline has the stronger attractions and the higher boiling point, 185°C.**

11.33 *Which of the following compounds should have the higher solubility in water? Explain your answer.*

$$CH_3-CH_2-NH_2$$

Ethylamine

$$CH_3-CH_2-CH_2-CH_2-CH_2-CH_2-NH_2$$

Hexylamine

As always, in comparing water solubilities of organic compounds, we look at the relative sizes of the hydrophobic and hydrophilic areas of the molecules. **These two primary amines have the same ability to form hydrogen bonds with water (the same**

hydrophilic area), but hexylamine has a much larger hydrophobic area and will be less soluble in water than ethylamine.

Solutions to Section 11.4 Core Problems

11.35 *Using condensed structures, write a chemical equation for each of the following reactions:*

a) *the ionization of ethylmethylamine in water.*

Ethylmethylamine is CH_3–CH_2–NH–CH_3. Amines are bases, so when they ionize in water, the amine N accepts H^+ from a water molecule to form an ammonium ion and OH^-. Because amines are weak bases, equilibrium arrows are used:

$$\mathbf{CH_3–CH_2–NH–CH_3 + H_2O \rightleftarrows CH_3–CH_2–{}^{+}NH_2–CH_3 + OH^-}$$

b) *the reaction of ethylmethylamine with H_3O^+.*

Ethylmethylamine is CH_3–CH_2–NH–CH_3. The amine N accepts H^+ from H_3O^+ to form an ammonium ion and H_2O. Because H_3O^+ is an acid, a one-way arrow is used:

$$\mathbf{CH_3–CH_2–NH–CH_3 + H_3O^+ \rightarrow CH_3–CH_2–{}^{+}NH_2–CH_3 + H_2O}$$

c) *the reaction of ethylmethylamine with acetic acid.*

Ethylmethylamine is CH_3–CH_2–NH–CH_3. The amine N accepts H^+ from CH_3COOH to form an ammonium ion and CH_3COO^-. Because this is an acid-base reaction, a one-way arrow is used:

$$\mathbf{CH_3–CH_2–NH–CH_3 + CH_3COOH \rightarrow}$$
$$\mathbf{CH_3–CH_2–{}^{+}NH_2–CH_3 + CH_3COO^-}$$

11.37 *The following amino acid forms a zwitterion. Draw the structure of the zwitterion.*

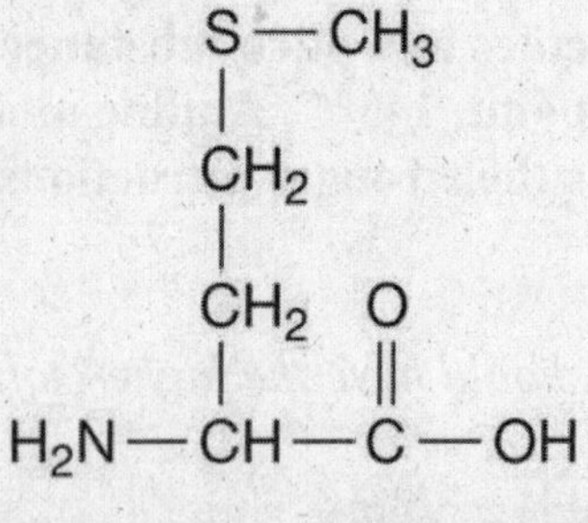

```
S—CH3
|
CH2
|
CH2  O
|    ||
H2N—CH—C—OH
```

Methionine
An amino acid

Because amino acids contain both acid (H^+ donor, the carboxylic acid group) and base (H^+ acceptor, the amine group) groups, they self-ionize: the H^+ leaves the carboxylic acid group and attaches to the amine group, leaving a carboxylate ion group and an ammonium ion group. It's like any other reaction between an amine and a carboxylic acid, except that the two groups are attached to the same molecule:

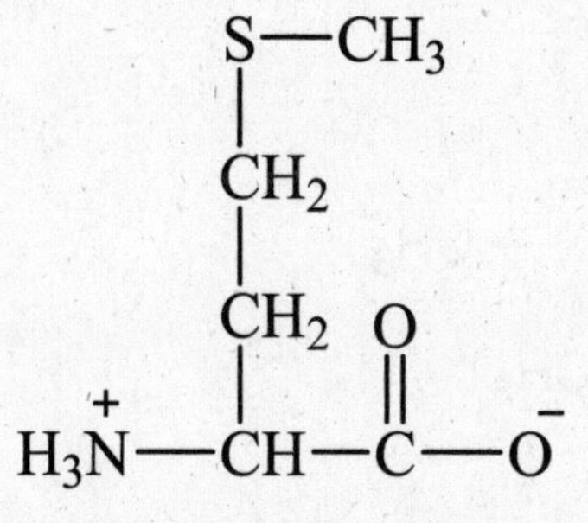

$$\begin{array}{l} S{-}CH_3 \\ | \\ CH_2 \\ | \\ CH_2 \quad O \\ | \qquad \| \\ H_3\overset{+}{N}{-}CH{-}C{-}O^- \end{array}$$

11.39 *Draw the structures of the following:*

a) *the conjugate acid of propylamine*

Chapter 7 review! The conjugate acid of any species is what you get when you add an H^+ to it. Propylamine is $CH_3{-}CH_2{-}CH_2{-}NH_2$, and its conjugate acid is

$CH_3{-}CH_2{-}CH_2{-}NH_3^+$

b) *butylammonium bromide*

Butylamine is $CH_3{-}CH_2{-}CH_2{-}CH_2{-}NH_2$, its conjugate acid is $CH_3{-}CH_2{-}CH_2{-}CH_2{-}NH_3^+$, and the compound formed by butylammonium ion with bromide ion is

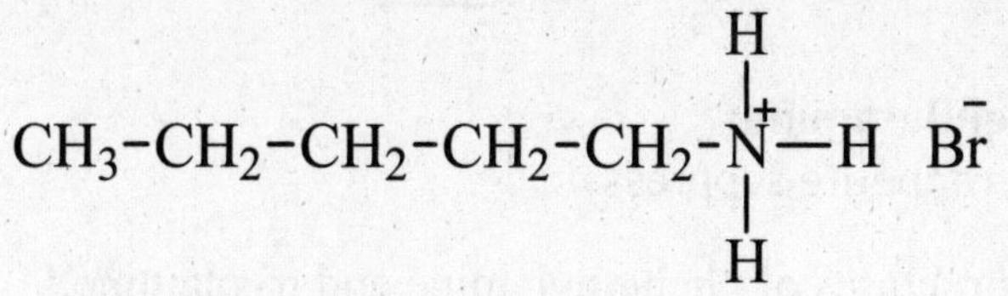

$$CH_3{-}CH_2{-}CH_2{-}CH_2{-}CH_2{-}\overset{+}{N}H_2{-}H \quad Br^-$$

c) *hexylmethylammonium formate*

Hexylmethylammonium ion is the conjugate acid of hexylmethylamine,

$CH_3{-}CH_2{-}CH_2{-}CH_2{-}CH_2{-}CH_2{-}NH{-}CH_3$. Formate ion is the conjugate base of formic acid (the trivial name for methanoic acid), HCOOH (review carboxylate ions, Section 11.1). The two ions combine in a 1:1 ratio:

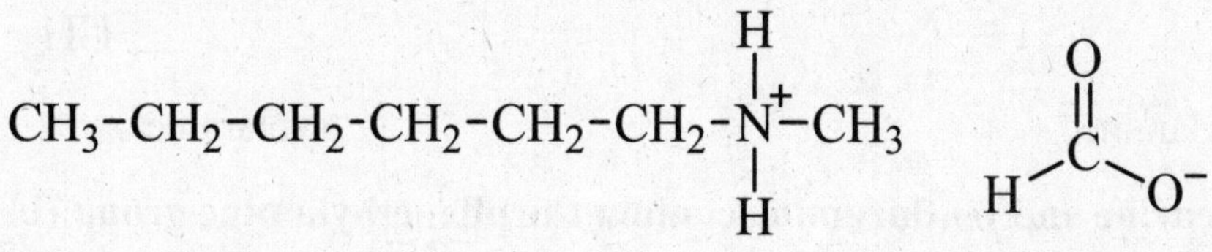

$$CH_3{-}CH_2{-}CH_2{-}CH_2{-}CH_2{-}CH_2{-}\overset{+}{N}H_2{-}CH_3 \qquad H{-}\overset{O}{\overset{\|}{C}}{-}O^-$$

Solutions to Section 11.5 Core Problems

11.41 *Which of the following substances are structurally related to phenethylamine, and which are structurally related to tryptamine?*

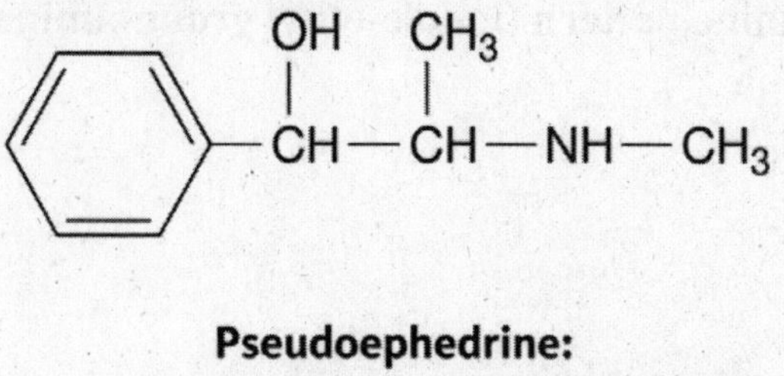

$$C_6H_5{-}\underset{}{\overset{OH}{\overset{|}{C}H}}{-}\overset{CH_3}{\overset{|}{C}H}{-}NH{-}CH_3$$

Pseudoephedrine:
a nasal decongestant found
in many cold remedies

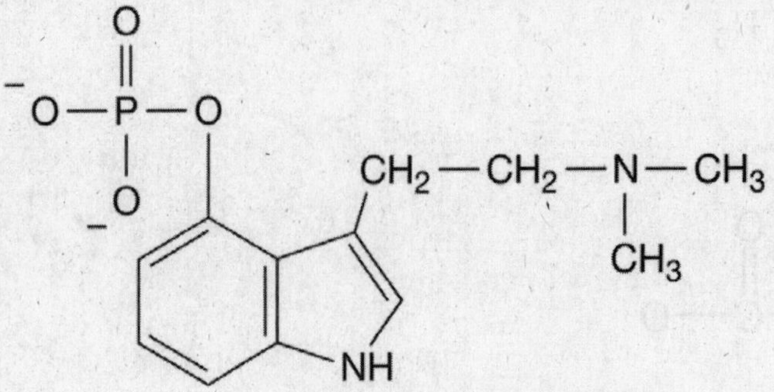

$$^{-}O-\overset{\overset{O}{\|}}{\underset{\underset{O^{-}}{|}}{P}}-O$$

$CH_2-CH_2-N(CH_3)-CH_3$

NH

Psilocybin:
a hallucinogen found in
certain species of mushrooms

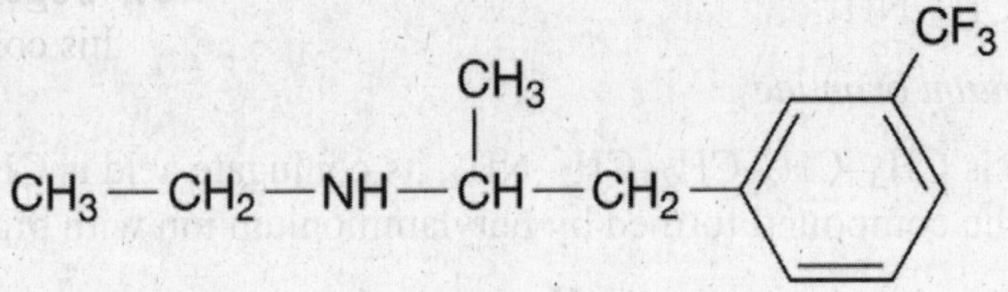

CF_3

$CH_3-CH_2-NH-CH(CH_3)-CH_2-$

Fenfluramine:
an appetite suppressant

Compare each structure to the structures of phenethylamine and tryptamine.

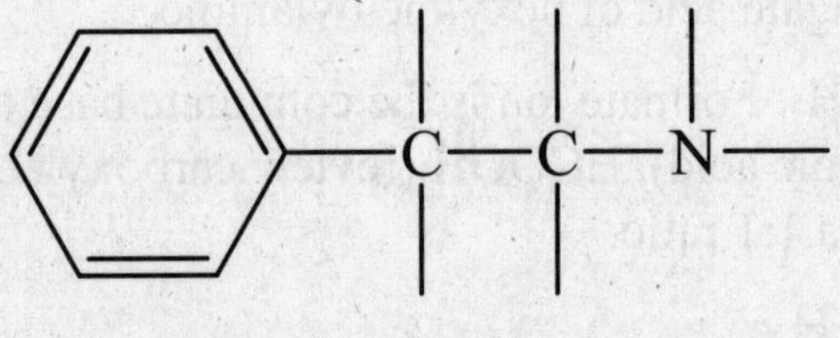

H
N

C—C—N

$CH_2\text{-}CH_2\text{-}NH_2$

phenethylamine 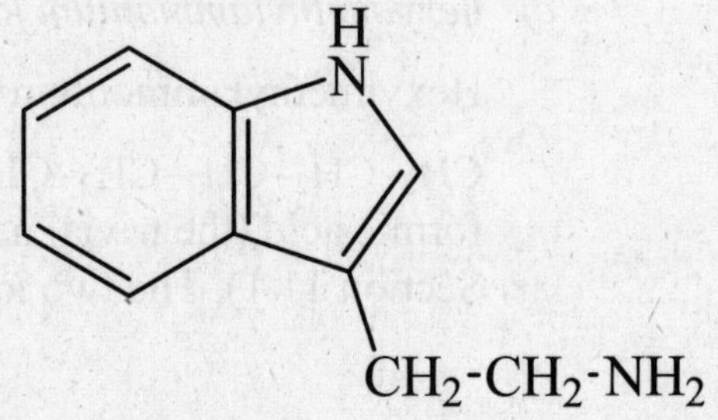   tryptamine

Pseudoephedrine and fenfluramine contain the phenethylamine group (benzene ring-ethyl group-amine).

NOTE:
The fenfluramine structure shown is flipped over from the phenethylamine structure, but the pattern is still there.

Psilocybin has the tryptamine pattern (indole-ethyl group-amine).

11.43 *The active ingredient in the antidepressant medication Zoloft is sertraline hydrochloride, which is formed when sertraline reacts with HCl. The structure of sertraline is shown below.*

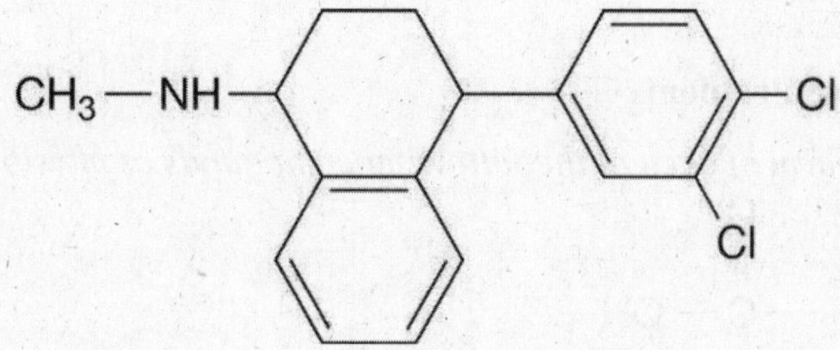

CH_3—NH— ... —Cl

Cl

Sertraline

a) *Draw the structure of sertraline hydrochloride.*

Adding "hydrochloride" to the name of an amine tells us that the actual structure is a salt (ionic compound) of the ammonium ion (the conjugate acid of the amine) with chloride ion. Attaching the hydrogen ion from HCl to the nitrogen atom in sertraline gives us the conjugate acid of sertraline. The salt contains this conjugate acid and a chloride ion, as shown below.

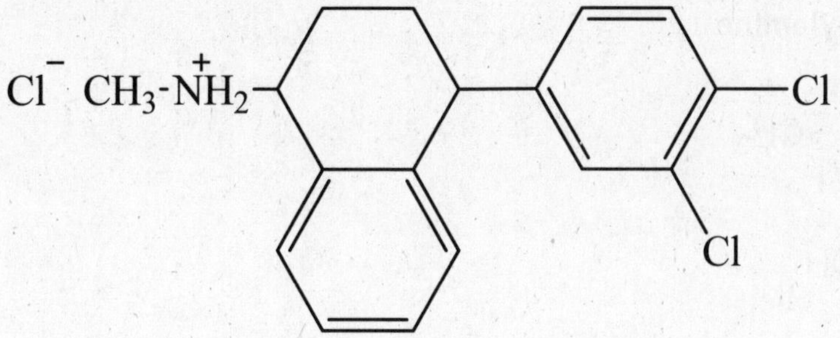

Cl^- CH_3-$\overset{+}{N}H_2$— ... —Cl

Cl

b) *Why do you think that the manufacturers of Zoloft use this salt, rather than sertraline itself?*

Organic amines are usually insoluble in water and may be liquids or oily, low-melting solids at room temperature. The salts of amine drugs are usually much more soluble in water and have higher melting points than the original amines, making them solid for storing and dispensing at room temperature. Therefore many amine drugs are actually sold as their hydrochloride (or other) salts.

11.45 *Tetrahydrocannabinol (THC) is the physiologically active compound in marijuana, which is made from the dried seed pods of the hemp plant (Cannabis sativa). The structure of THC is shown here. Is THC an alkaloid? Why or why not?*

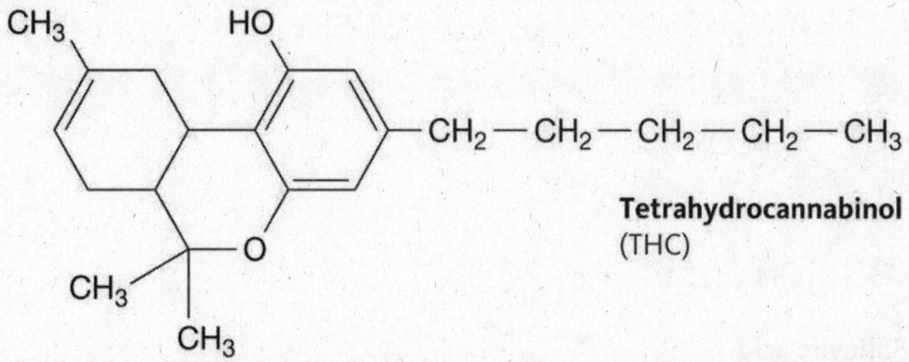

CH_3 HO

—CH_2—CH_2—CH_2—CH_2—CH_3

O

CH_3

CH_3

Tetrahydrocannabinol
(THC)

Alkaloids are organic amines that are produced by plants or fungi and that have some sort of physiological activity. **THC does not contain an amine group, so it is not an alkaloid.**

Solutions to Section 11.6 Core Problems

11.47 *What is the dominant form of each of the following compounds at physiological pH?*

a) 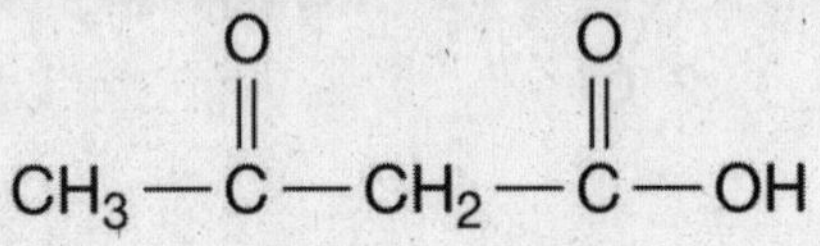 $CH_3-C(=O)-CH_2-C(=O)-OH$

Acetoacetic acid

b) 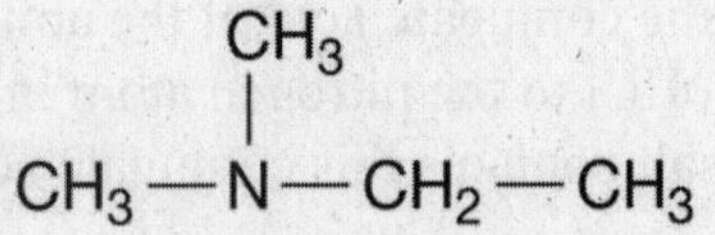 $CH_3-N(CH_3)-CH_2-CH_3$

Ethyldimethylamine

c) 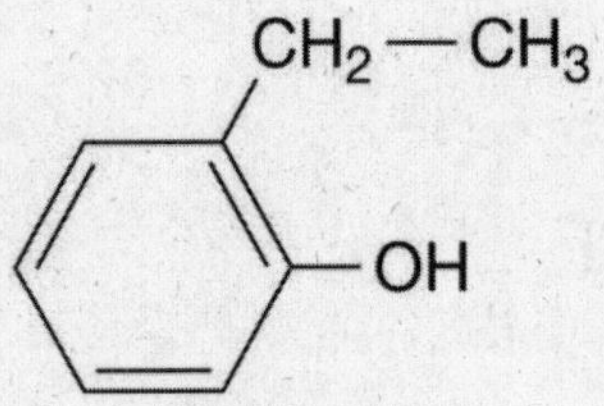 (benzene ring with CH_2-CH_3 and adjacent OH)

Phlorol

d) 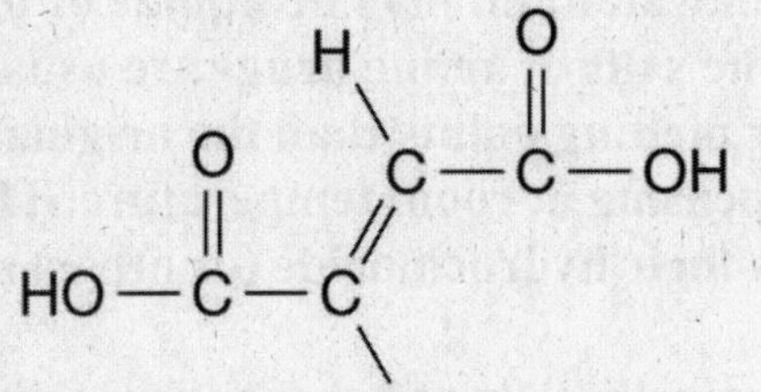 $HO-C(=O)-C(H)=C(H)-C(=O)-OH$

Fumaric acid

e) 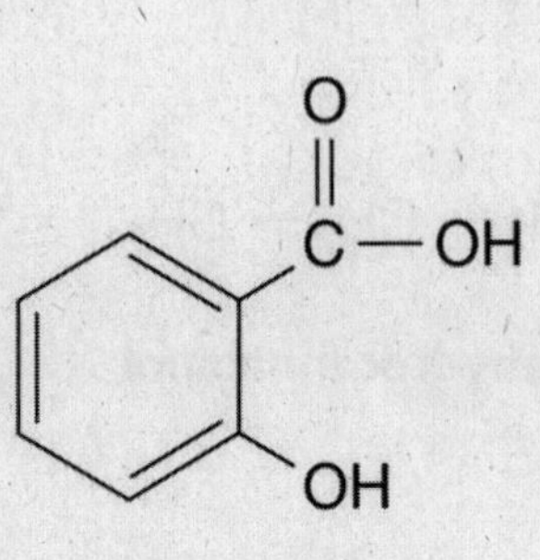 (benzene ring with $C(=O)-OH$ and adjacent OH)

Salicylic acid

f) $NH_2-CH_2-CH_2-CH_2-CH_2-CH_2-NH_2$

Cadaverine

In solutions near pH 7, carboxylic acids are ionized to the carboxylate ions, and most amines are ionized to their ammonium ions. All other functional groups (including the acidic phenol and thiol groups, because they are too weak to be ionized at physiological pH) are shown in their normal form.

a) This molecule contains a carboxylic acid group, which will be ionized to a **carboxylate ion** at physiological pH, and a ketone group, which is neither acidic nor basic and appears normally:

$$CH_3-\overset{\overset{O}{\|}}{C}-CH_2-\overset{\overset{O}{\|}}{C}-O^-$$

b) This molecule contains an amine group, which will be ionized to an **ammonium ion** at physiological pH:

$$CH_3-\overset{\overset{CH_3}{|}}{\underset{\underset{H}{|}}{N^+}}-CH_2-CH_3$$

c) This molecule contains a phenol group, so **no change**: phenols do not ionize at physiological pH.

d) This molecule contains two carboxylic acid groups, both of which will be ionized to form **carboxylate ions** at physiological pH

$$^-O-\overset{\overset{O}{\|}}{C}-\underset{\underset{H}{\diagdown}}{C}=\overset{\overset{H}{\diagdown}}{C}-\overset{\overset{O}{\|}}{C}-O^-$$

e) This molecule contains a carboxylic acid group, which will be ionized to a **carboxylate ion** at physiological pH, and a phenol group, which is such a weak acid that it is not ionized at physiological pH:

$$C_6H_4(OH)-\overset{\overset{O}{\|}}{C}-O^-$$

f) This molecule contains two amine groups, which will be ionized to form **ammonium ions** at physiological pH:

$\overset{+}{N}H_3\text{-}CH_2\text{-}CH_2\text{-}CH_2\text{-}CH_2\text{-}CH_2\text{-}\overset{+}{N}H_3$

11.49 *Glyceraldehyde-3-phosphate is formed in the metabolic pathway that breaks down sugars to produce energy. How will this molecule actually appear at physiological pH?*

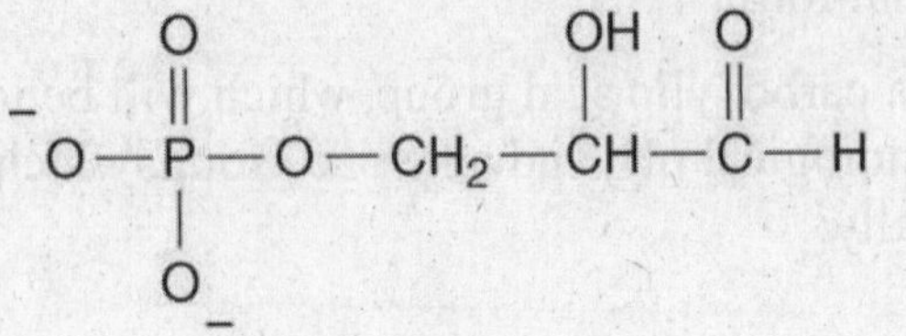

$$^{-}O-\overset{\overset{O}{\|}}{\underset{\underset{O^{-}}{|}}{P}}-O-CH_2-\overset{\overset{OH}{|}}{CH}-\overset{\overset{O}{\|}}{C}-H$$

Glyceraldehyde-3-phosphate

Alcohols and aldehydes appear in their normal (neutral) forms at physiological pH, so we don't have to show any changes to those groups. Organic phosphates appear in a mixture of two forms at pH 7. One form has no hydrogen atoms bonded to the oxygen atoms of the phosphate group, as shown above, so the given structure is actually an acceptable answer to the question. In the other form, one of the single-bonded phosphate oxygen atoms has a hydrogen atom attached. We can draw the hydrogen atom attached to either of the two singly-bonded oxygen atoms in the phosphate group.
The two forms of this ion, at equilibrium at physiological pH, are:

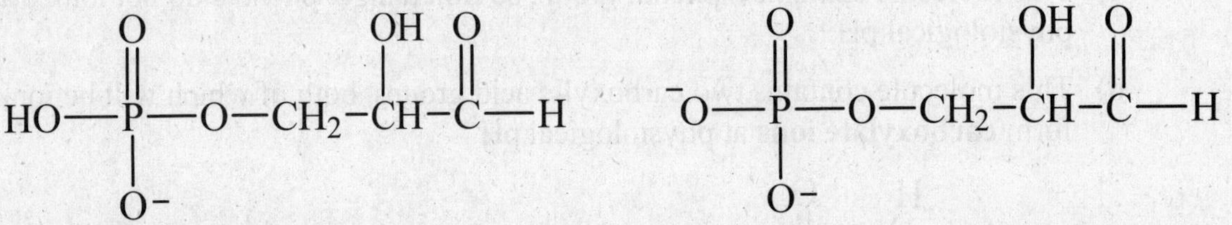

$$HO-\overset{\overset{O}{\|}}{\underset{\underset{O^{-}}{|}}{P}}-O-CH_2-\overset{\overset{OH}{|}}{CH}-\overset{\overset{O}{\|}}{C}-H \qquad ^{-}O-\overset{\overset{O}{\|}}{\underset{\underset{O^{-}}{|}}{P}}-O-CH_2-\overset{\overset{OH}{|}}{CH}-\overset{\overset{O}{\|}}{C}-H$$

Solutions to Concept Questions

* indicates more challenging problems.

11.51 *What three types of organic compounds produce an acidic solution when they dissolve in water?*

The acidic functional groups are carboxylic acid, phenol, and thiol, so any organic compound that contains one of these functional groups will produce an acidic solution.

11.53 *When we write the equation for the reaction of an organic acid with hydroxide ion, why do we use a single arrow?*

Hydroxide ion is a strong base. Reactions that involve a strong acid or strong base go entirely to products and are not equilibrium processes.

11.55 *All oxidative decarboxylation reactions require NAD^+. What is the function of the NAD^+?*

Chapter 10 review:
NAD^+ accepts the hydrogen atoms that are removed in most types of oxidation reactions.

11.57 *How many bonds does a nitrogen atom usually form in an organic compound? Explain your answer.*

Nitrogen atoms have five valence electrons, and need three to complete their octet, so the typical bonding pattern for a nitrogen atom is 3 bonds and 1 lone pair. However, nitrogen atoms can also form four bonds in ammonium ions, where they will have a +1 charge. In an ammonium ion, the two electrons in the lone pair of the amine are converted into a bonding pair.

11.59 *Why do tertiary amines have lower boiling points than primary and secondary amines that contain the same numbers of atoms?*

"The same number of atoms" is a hint to remind us that the dispersion forces in these compounds are about the same. When dispersion forces are equivalent in two substances, we look to hydrogen bonding to decide which of two or more compounds will have the strongest total attractions between molecules, and therefore the highest boiling point. Primary and secondary amines have, by definition, at least one H atom directly bonded to the amine N, and therefore can form hydrogen bonds between the molecules of the pure substance. Tertiary amines, by definition, have no H atoms bonded to N, and therefore cannot form hydrogen bonds between the molecules of the pure substance. **Hydrogen bonding interactions between the molecules of primary and secondary amines give them stronger total attractions between molecules, and therefore higher boiling points, than tertiary amines that contain the same numbers of atoms.**

11.61 *When an amine dissolves in water, the resulting solution has a pH above 7. Why is this?*

Like ammonia, amines are able to accept an H^+ ion from water, forming an ammonium ion and a hydroxide ion. Substances that accept H^+ ions are (by definition) bases, and the formation of OH^- in aqueous solution causes the solution to have a pH above 7.

11.63 *Many metabolic pathways produce carboxylic acids. Under normal physiological conditions, these acids immediately lose H^+. Where do these hydrogen ions go? Give a specific example of a substance that is present in body fluids and can accept a hydrogen ion.*

Chapter 7 review: see Section 7.8 for a review of the buffers in human physiology. We aren't ready yet to show the specific substances involved in the protein buffer system, but we can easily answer the question in terms of the $H_2PO_4^-$/ HPO_4^{2-} buffer system in intracellular fluid and the H_2CO_3/ HCO_3^- buffer in the bloodstream. **In all buffer systems, any added H^+ ions react with the conjugate base in the buffer. Within cells, H^+ ions react with HPO_4^{2-}; in the bloodstream, H^+ ions react with HCO_3^-.**

11.65 *What are the two main forms of phosphate ion at physiological pH? Draw their structures and write their chemical formulas.*

As we saw in Chapter 7, at pH 7 phosphate forms a buffer that contains a mixture of $H_2PO_4^-$ and HPO_4^{2-} ions. H_3PO_4 is a sufficiently strong acid that the phosphate ions do not bond to three hydrogen ions at physiological pH.

NOTE:
Organic phosphates behave very similarly; the only difference is that one hydrogen in each form of the phosphate ion would be replaced with an organic group.

HO—P(=O)(—O⁻)—O⁻   HO—P(=O)(—O⁻)—OH

Solutions to Summary and Challenge Problems

* indicates more challenging problems.

11.67 *Complete the following chemical equations. Use condensed structures or line structures for organic compounds.*

a) $CH_3-CH(CH_3)-CH_2-C(=O)-OH + H_2O \xrightleftharpoons{\text{Ionization}}$

b) $CH_3-C_6H_4-OH + OH^- \longrightarrow$

c) $CH_3-CH_2-CH(SH)-CH_3 + OH^- \longrightarrow$

d) $C_5H_9-C(=O)-OH + CH_3-NH-CH_3 \longrightarrow$ (cyclopentyl)

e) piperidine (ring NH) $+ H_2O \xrightleftharpoons{\text{Ionization}}$

f) 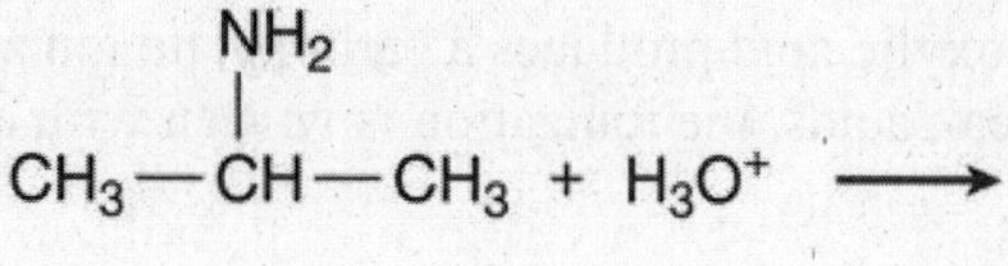 $CH_3-CH(NH_2)-CH_3 + H_3O^+ \longrightarrow$

g) 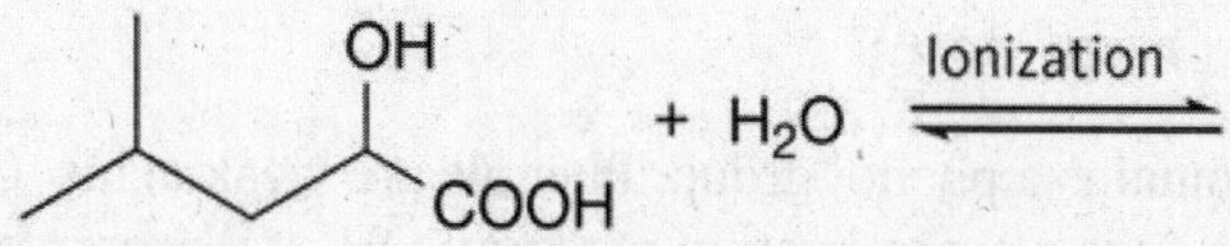 (3-methyl-2-hydroxybutanoic acid, OH, COOH) $+ H_2O \xrightleftharpoons{\text{Ionization}}$

h) $C_6H_5-CH_2-\overset{O}{\overset{\|}{C}}-CH_2-\overset{O}{\overset{\|}{C}}-OH \xrightarrow{\text{Decarboxylation}}$

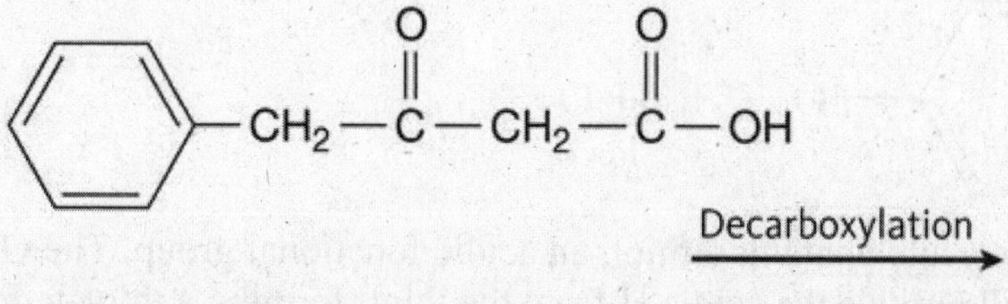

i) 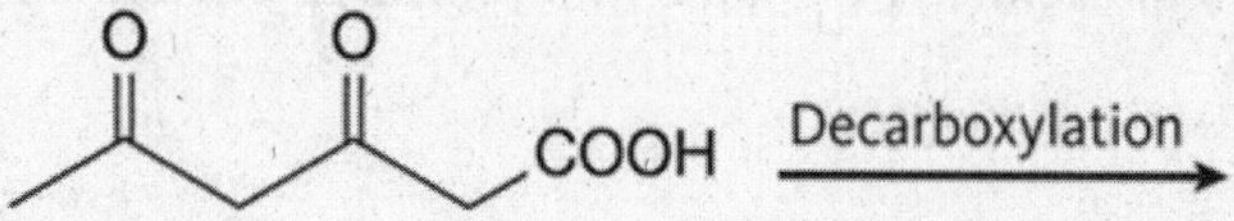 (O, O, COOH) $\xrightarrow{\text{Decarboxylation}}$

j) 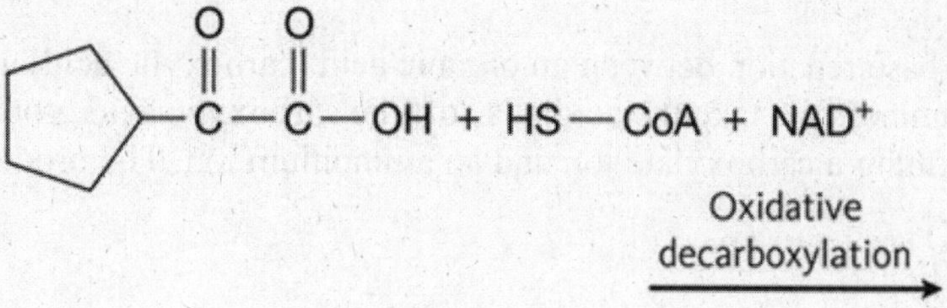 $C_5H_9-\overset{O}{\overset{\|}{C}}-\overset{O}{\overset{\|}{C}}-OH + HS-CoA + NAD^+ \xrightarrow{\text{Oxidative decarboxylation}}$

k) 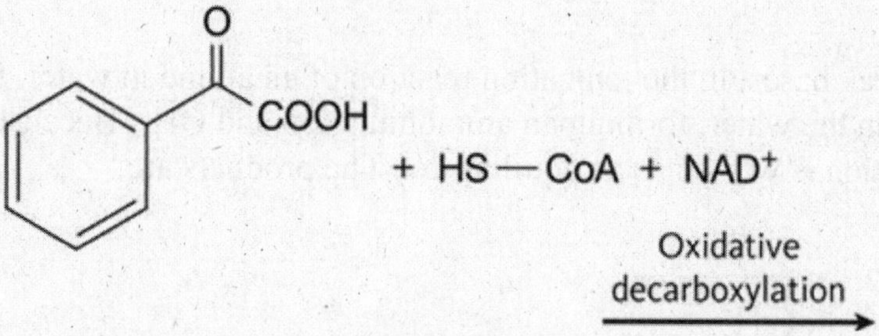 (phenyl, O, COOH) $+ HS-CoA + NAD^+ \xrightarrow{\text{Oxidative decarboxylation}}$

In each case, we need to identify the type of reaction that is occurring, identify the functional groups that will be involved, and draw the product molecule(s) or ion(s). While only the products of each reaction are shown here.

NOTE:

The problem asks for the complete equations, so be prepared to write out both the reactants (as given in the problem) and products.

a) The ionization reaction of a carboxylic acid produces a carboxylate ion and H_3O^+. (Because carboxylic acids are weak acids, the ionization is written with equilibrium arrows.) The products are:

$$CH_3-\underset{|}{\overset{CH_3}{}}CH-CH_2-\overset{O}{\overset{\|}{C}}-O^- \quad + \quad H_3O^+$$

b) The organic molecule contains a phenol group. Phenols are weak acids. OH^- is a strong base, so this reaction is a neutralization reaction: OH^- will remove the acidic H from the phenol, forming a phenolate ion and water. The products are:

$$CH_3-C_6H_4-O^- \quad + \quad H_2O$$

c) The organic molecule contains a thiol, an acidic functional group. The OH^- ion, a strong base, will remove the acidic H from the thiol, forming a thiolate ion and water. The products are:

$$CH_3-CH_2-\overset{S^-}{\overset{|}{C}}H-CH_3 \quad + \quad H_2O$$

d) This is an acid-base reaction between an organic acid (carboxylic acid) and an organic base (amine). We take the acidic H^+ off the carboxylic acid group and put in on the base, forming a carboxylate ion and an ammonium ion. The products are:

$$C_5H_9-\overset{O}{\overset{\|}{C}}-O^- \quad + \quad CH_3-\overset{+}{N}H_2-CH_3$$

e) Amines are weak bases. In the ionization reaction of an amine in water, the amine accepts H^+ from the water, forming an ammonium ion and OH^-. Because the base is weak, the reaction is written as an equilibrium. The products are:

$$C_5H_{10}\overset{+}{N}H_2 \quad + \quad OH^-$$

f) This is a reaction between an amine (a weak base) and H_3O^+ (a strong acid). The amine accepts H^+ from the acid, forming an ammonium ion and H_2O. The products are:

$$CH_3-\overset{\overset{+}{N}H_3}{\overset{|}{C}}H-CH_3 \quad + \quad H_2O$$

g) In the ionization reaction of a carboxylic acid in water, the acid donates H^+ to the water, forming a carboxylate ion and H_3O^+. Because carboxylic acids are weak acids, the reaction is shown as an equilibrium. The alcohol group is not involved (alcohols do not participate in normal acid-base reactions). The products are:

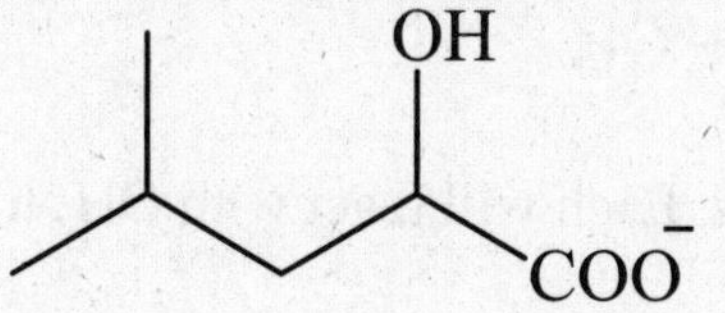

OH

COO^- + H_3O^+

h) This carboxylic acid has a ketone group in the β (beta) position, so it can undergo a normal decarboxylation reaction. We remove the carboxylic acid group, rewrite it as CO_2, and put the hydrogen back on the α (alpha) carbon. The products are:

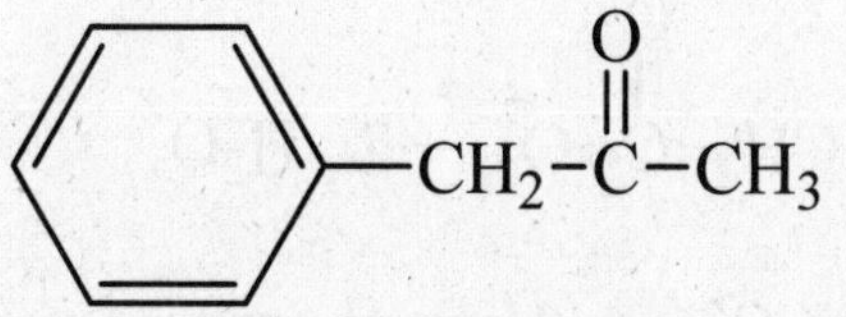

$C_6H_5-CH_2-C(=O)-CH_3$ + CO_2

i) It's a little harder to see the carboxylic acid here, as it's written in the condensed –COOH form. This carboxylic acid has a ketone group in the β (beta) position, so it can undergo a normal decarboxylation reaction. We remove the carboxylic acid group, rewrite it as CO_2, and put the hydrogen back on the α (alpha) carbon. The products are:

O O

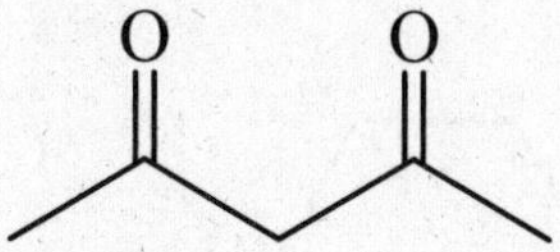

+ CO_2

j) In an oxidative decarboxylation, we remove the carboxylic acid group, replace it with the S—CoA group to make a thioester, and distribute the two hydrogen atoms produced as NADH and H^+. The products are:

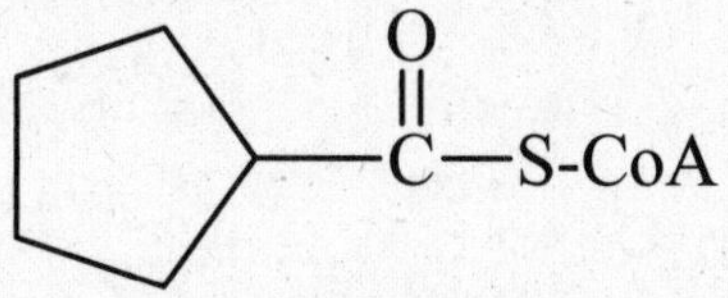

O

C—S-CoA + CO_2 + NADH + H^+

k) Notice that the carboxylic acid is written in the condensed –COOH form. In an oxidative decarboxylation, we remove the carboxylic acid group, replace it with the S—CoA group to make a thioester, and distribute the two hydrogen atoms produced as NADH and H^+. The products are:

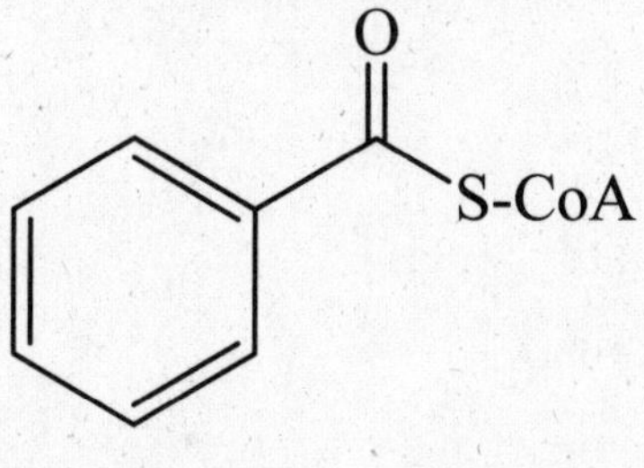

O

S-CoA

+ CO_2 + NADH + H^+

11.69 *Succinic acid can react with two hydroxide ions. Using structural formulas, write a chemical equation for this reaction.*

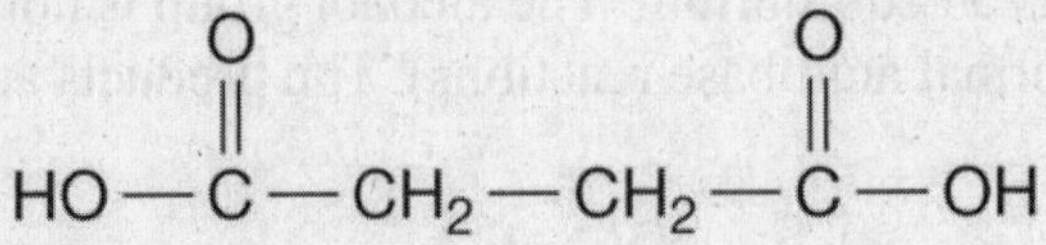

$$HO-\overset{\overset{O}{\|}}{C}-CH_2-CH_2-\overset{\overset{O}{\|}}{C}-OH \quad \textbf{Succinic acid}$$

Succinic acid contains two carboxylic acid groups. Each will react with OH^- to form carboxylate ions and water:

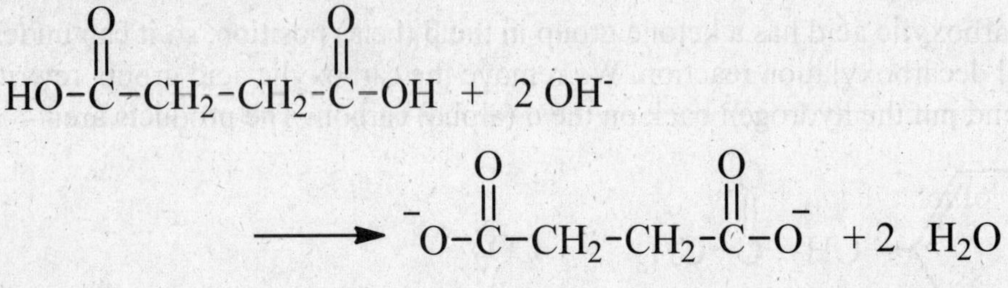

$$HO-\overset{\overset{O}{\|}}{C}-CH_2-CH_2-\overset{\overset{O}{\|}}{C}-OH + 2\ OH^-$$

$$\longrightarrow {}^-O-\overset{\overset{O}{\|}}{C}-CH_2-CH_2-\overset{\overset{O}{\|}}{C}-O^- + 2\ H_2O$$

11.71 *The following compound can react with two H_3O^+ ions. Using structures for the organic molecules, write a balanced equation for this reaction.*

HN NH (piperazine ring)

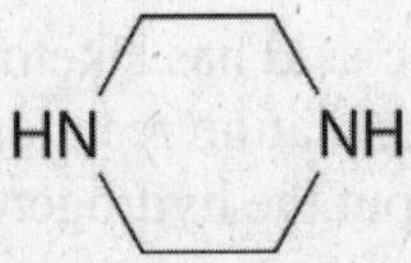

The compound shown contains two amine groups. Each will react with H_3O^+ to form ammonium ions and water:

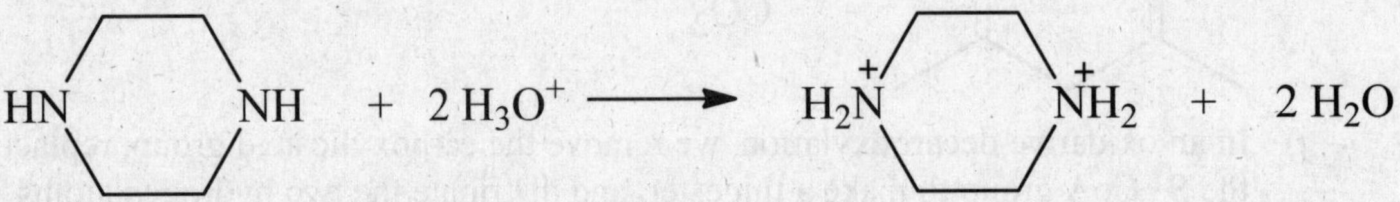

$$HN\langle\rangle NH + 2\ H_3O^+ \longrightarrow H_2\overset{+}{N}\langle\rangle\overset{+}{N}H_2 + 2\ H_2O$$

11.73 *Name the following compounds and ions:*

a) 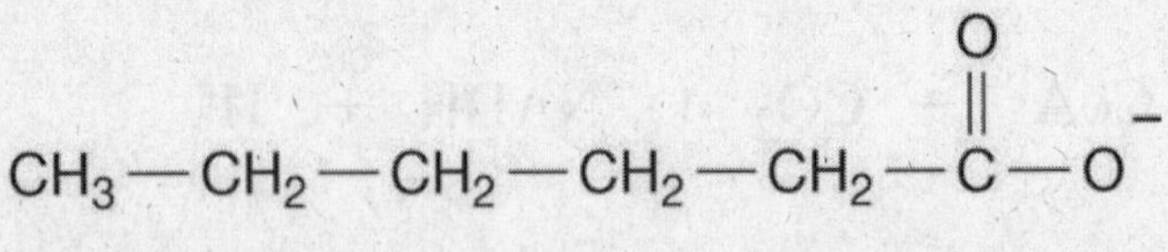 $CH_3-CH_2-CH_2-CH_2-CH_2-\overset{\overset{O}{\|}}{C}-O^-$

b)  (skeletal structure) $\overset{\overset{O}{\|}}{C}-O^-\ Na^+$

c) $CH_3-CH_2-CH_2-CH_2-NH_2$

d) 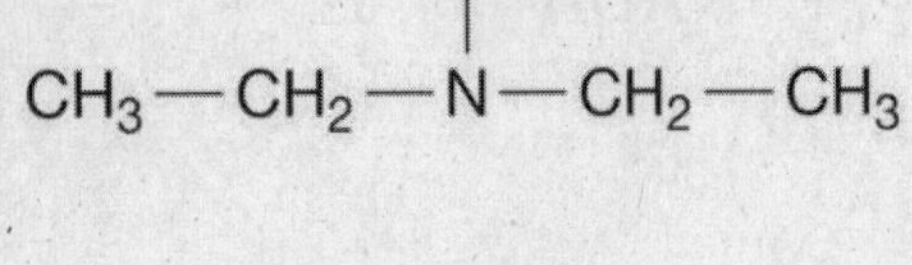 $CH_3-CH_2-\underset{}{\overset{\overset{CH_2-CH_3}{|}}{N}}-CH_2-CH_3$

$$\text{e)}\ CH_3-\overset{\overset{\large CH_3}{|}}{\underset{\underset{\large H}{|}}{N^+}}-CH_2-CH_2-CH_3$$

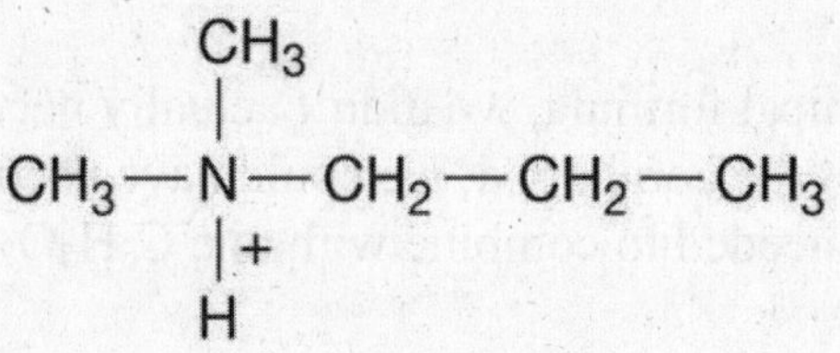

$$\text{f)}\ C_6H_{11}-\overset{+}{N}H_3\ \ Cl^-$$

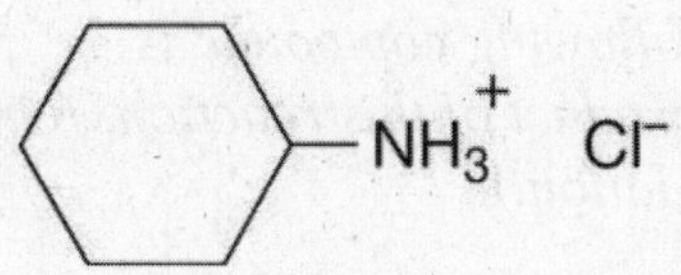

a) Six-carbon chain with carboxylate ion group: **hexanoate ion.**

b) Seven-carbon chain with carboxylate ion group, combined with sodium ion to form an ionic compound (salt): **sodium hexanoate** (remember that as in Chapter 3, the names of ionic compounds always begin with the cation).

c) An amine with a four-carbon chain: **butylamine.**

d) An amine with three chains of two carbons each: **triethylamine.**

e) An ammonium ion with two methyl groups (1 carbon each) and a propyl group (3-carbon chain): **dimethylpropylammonium ion.**

f) An ammonium ion with a six-carbon ring (cyclohexyl group) attached, combined with chloride ion to form an ionic compound (salt): **cyclohexylammonium chloride** (again, the cation is named first in ionic compounds).

*11.75 *The citrate ion has the chemical formula $C_6H_5O_7^{3-}$ and the structure shown here. Write the chemical formula of potassium citrate.*

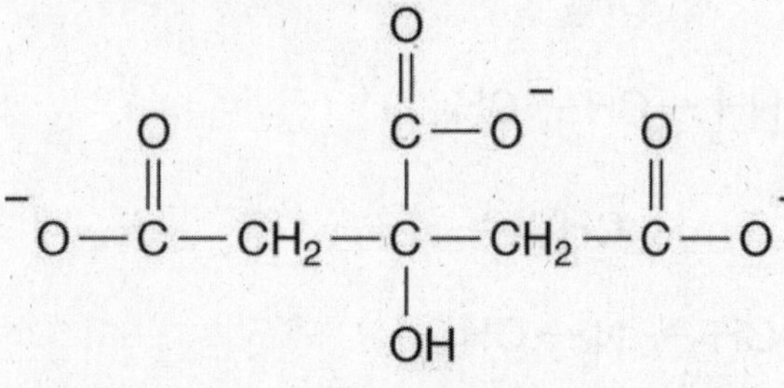

$$^-O-\overset{\overset{\large O}{||}}{C}-CH_2-\overset{\overset{\large O=C-O^-}{|}}{\underset{\underset{\large OH}{|}}{C}}-CH_2-\overset{\overset{\large O}{||}}{C}-O^-$$

To form the ionic compound, we match each negative charge on the citrate ion with a positive potassium ion, so that the compound will have a zero charge overall:

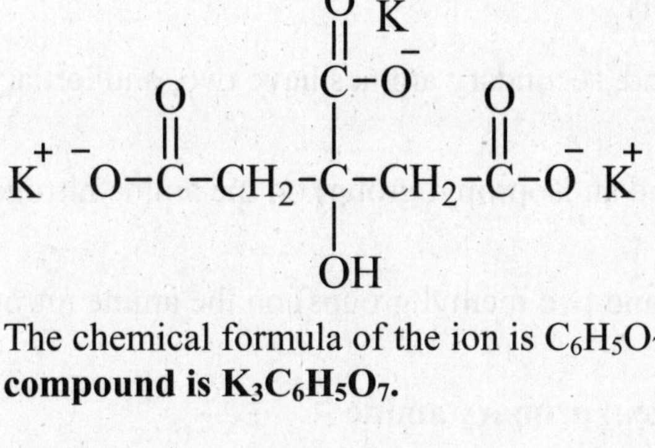

$$K^+\ ^-O-\overset{\overset{\large O}{||}}{C}-CH_2-\overset{\overset{\large O=C-O^-\ K^+}{|}}{\underset{\underset{\large OH}{|}}{C}}-CH_2-\overset{\overset{\large O}{||}}{C}-O^-\ K^+$$

The chemical formula of the ion is $C_6H_5O_7^{3-}$, so **the chemical formula of the compound is $K_3C_6H_5O_7$.**

NOTE:
Since the problem only asked for the chemical formula, we didn't actually need to see the structure of the ion or draw the structure of the compound; we could have used Chapter 3 principles to realize that three K^+ ions are needed to combine with one $C_6H_5O_7^{3-}$.)

11.77 *The amino acid lysine is a component of all proteins. Our bodies can burn lysine to obtain energy. During this metabolic pathway, the following compound is decarboxylated. Draw the structure of the organic product of this reaction. (Only the circled carboxyl group is involved in this decarboxylation.)*

$$\mathrm{HO{-}\overset{\overset{\displaystyle O}{\|}}{C}{-}CH_2{-}CH_2{-}\underset{}{\overset{\overset{\displaystyle HO-\overset{\overset{\displaystyle O}{\|}}{C}}{|}}{CH}}{-}\overset{\overset{\displaystyle O}{\|}}{C}{-}\overset{\overset{\displaystyle O}{\|}}{C}{-}OH}$$

This carboxylic acid group has a ketone group on the β-carbon, so it is a normal decarboxylation, not an oxidative decarboxylation; NAD^+ and CoA are not required. To write the product of a decarboxylation reaction, remove the entire carboxylic acid group and replace it with a hydrogen atom:

$$\mathrm{HO{-}\overset{\overset{\displaystyle O}{\|}}{C}{-}CH_2{-}CH_2{-}\overset{\overset{\displaystyle H}{|}}{CH}{-}\overset{\overset{\displaystyle O}{\|}}{C}{-}\overset{\overset{\displaystyle O}{\|}}{C}{-}OH}$$

11.79 *Classify each of the following compounds as a primary, secondary, or tertiary amine:*

a) $\mathrm{CH_3{-}CH_2{-}NH{-}\overset{\overset{\displaystyle CH_3}{|}}{CH}{-}CH_3}$

b) $\mathrm{CH_3{-}CH_2{-}CH_2{-}\overset{\overset{\displaystyle CH_3}{|}}{N}{-}CH_3}$

c) $\mathrm{CH_3{-}CH_2{-}CH_2{-}\overset{\overset{\displaystyle CH_3}{|}}{CH}{-}NH_2}$

Primary amines have only one N—C bond, secondary amines have two, and tertiary amines have three.

a) Two alkyl groups (an ethyl group and an isopropyl group) on the amine nitrogen: **secondary amine**

b) Three alkyl groups (a propyl group and two methyl groups) on the amine nitrogen: **tertiary amine**

c) One alkyl group on the amine nitrogen: **primary amine**

11.81 *The following two compounds are constitutional isomers. One of them boils at 36°C while the other boils at 69°C. Match each compound with its boiling point, and explain your reasoning.*

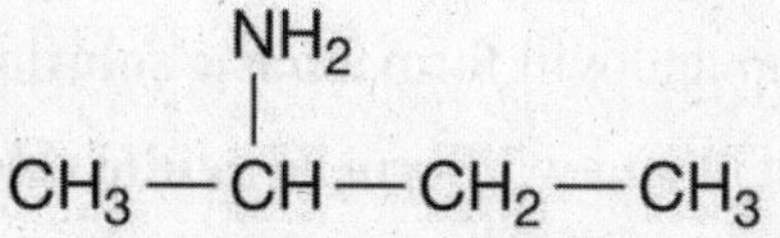

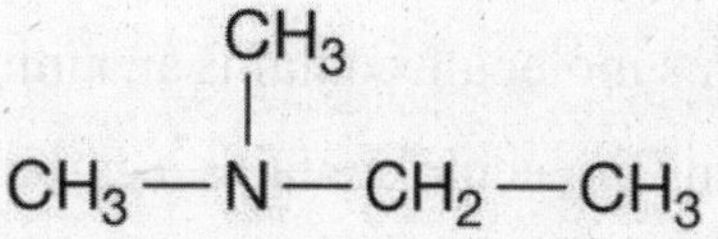

$$CH_3-CH(NH_2)-CH_2-CH_3 \qquad CH_3-N(CH_3)-CH_2-CH_3$$

The compound with the higher boiling point will always be the one with the stronger total attractions between molecules. The two molecules shown have the same chemical formula, the same number of electrons, and therefore equivalent dispersion forces. When the dispersion forces in two compounds are about the same, we look to hydrogen bonding interactions: **the left-hand molecule, a primary amine, has two hydrogen atoms on the amine, so the compound has hydrogen bonding interactions between its molecules. The right-hand compound, a tertiary amine, cannot form hydrogen bonds between its molecules; it will therefore have the lower boiling point, 36°C, while the primary amine will have the higher boiling point, 69°C.**

11.83 *Rank the following compounds in order of solubility in water. Start with the most soluble compound.*

$CH_3–(CH_2)_{11}–NH_2$ $\qquad$ $CH_3–(CH_2)_7–NH_2$ $\qquad$ $CH_3–(CH_2)_3–NH_2$

To decide on the water solubility of organic compounds, we compare the hydrophobic and hydrophilic areas of the molecule. These three molecules all have the same hydrophilic group, a primary amine, but the length of the hydrophobic hydrocarbon chain is different in all of them. The most soluble compound will be the one with the shortest hydrocarbon chain, and the least soluble will be the one with the longest chain:

$CH_3–(CH_2)_3–NH_2$ $\qquad$ **$CH_3–(CH_2)_7–NH_2$** $\qquad$ **$CH_3–(CH_2)_{11}–NH_2$**

least hydrophobic ⟵⟶ **most hydrophobic**
most soluble ⟵⟶ **least soluble**

11.85 *Each of the following compounds is soluble in water. If they are dissolved in water, which of the resulting solutions will be acidic, which will be basic, and which will be neutral?*

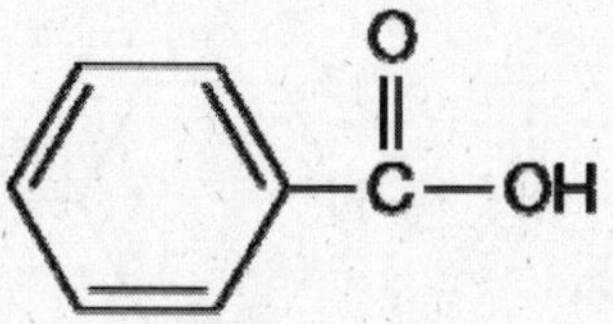

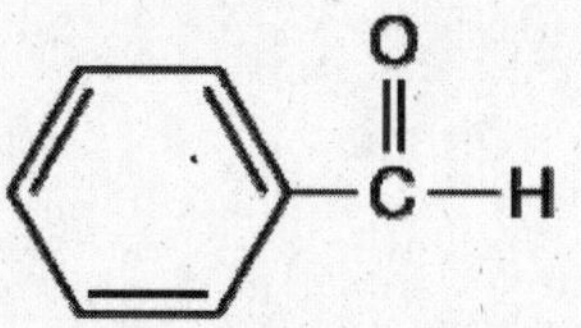

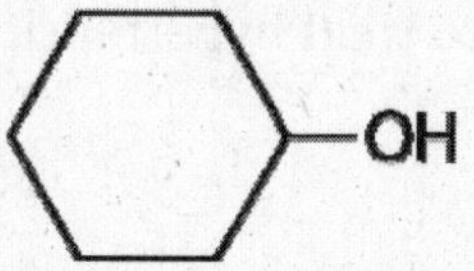

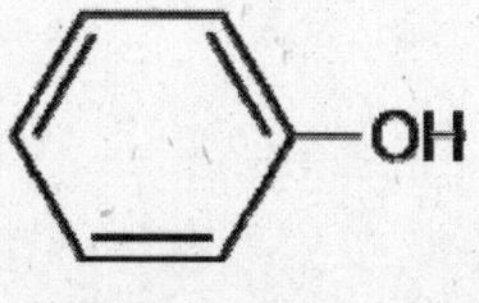

a) 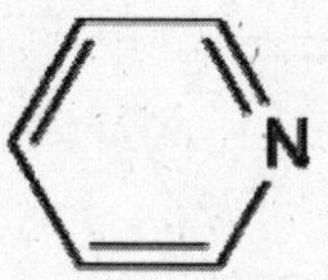 pyridine (N in ring)

b) benzene ring–C(=O)–OH

c) benzene ring–C(=O)–H

d) cyclohexane ring–OH

e) benzene ring–OH

The acidic functional groups are carboxylic acids, phenols, thiols, and ammonium ions; the basic functional groups are the conjugate bases of these—carboxylate ions, phenolate ions, thiolate ions, and amines. All other functional groups are neutral.

a) This molecule contains an amine group and will form a **basic solution**.

b) This molecule contains a carboxylic acid and will form an **acidic solution**.

c) This molecule contains an aldehyde group, which neither donates nor accepts a proton. It will form a **neutral solution**.

d) This molecule contains an alcohol group.

NOTE:
This is **not** phenol—the ring is a cyclohexane ring, not an aromatic ring.) Alcohols form **neutral solutions** in water.

e) This molecule is phenol (an alcohol group on an aromatic ring). Phenol and phenol compounds form **acidic solutions** in water.

11.87 *Is each of the following amines related to phenethylamine, to tryptamine, or to neither?*

HO—CH₂ / OH / CH₃
HO—(ring)—CH—CH₂—NH—CH—CH₃

Albuterol
(dilates the respiratory passages; also used to prevent pre-term labor)

H
N
CH₂—CH₂—N—(ring)—NH—C(=O)—(ring)

Indoramin
(used to treat hypertension)

Compare each structure to the structures of phenethylamine and tryptamine:

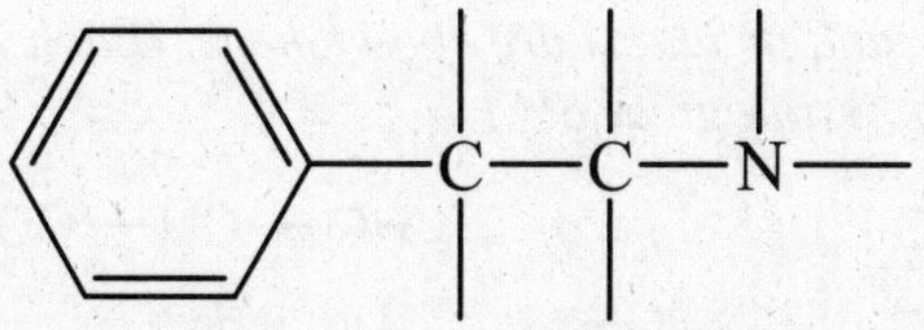

C—C—N—   H N CH_2-CH_2-NH_2

phenethylamine 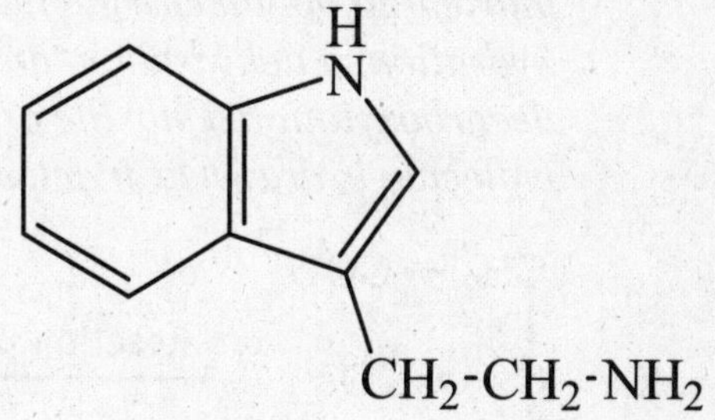   tryptamine

Albuterol is built on the phenethylamine structure, with the aromatic ring-carbon-carbon-amine pattern; **indoramin is based on the tryptamine structure**, with the indole-carbon-carbon-amine pattern.

*11.89 *When your body breaks down carbohydrates to obtain energy, one of the compounds that it forms during the metabolic pathway is phosphoenolpyruvic acid. However, at physiological pH this compound actually exists as a mixture of two ions. One ion has a –2 charge and the other has a –3 charge. Draw the structures of these two ions.*

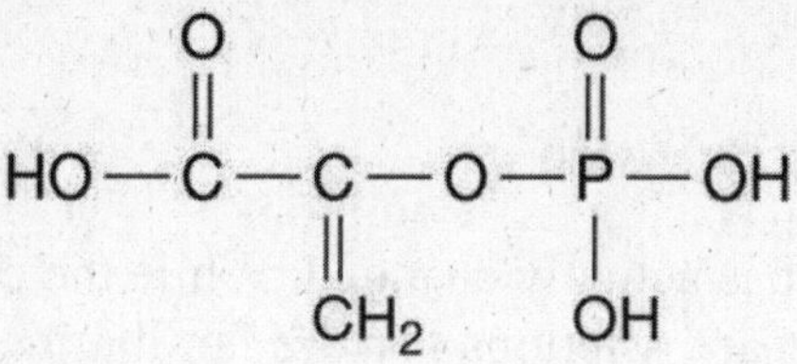

HO—C(=O)—C(=CH_2)—O—P(=O)(OH)—OH **Phosphoenolpyruvic acid**

Around pH 7,

- carboxylic acids are converted to carboxylate ions;
- amines are converted to ammonium ions;
- organic phosphates form buffers that contain a mixture of two ions: one with a single ionizable H, and the other with no ionizable H.

The molecule shown contains a carboxylic acid, which will be converted to the carboxylic acid, and a phosphate group, which will have either of two possible forms at physiological pH:

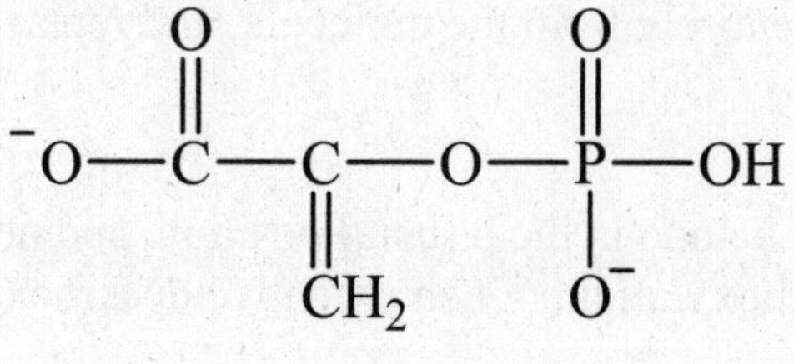

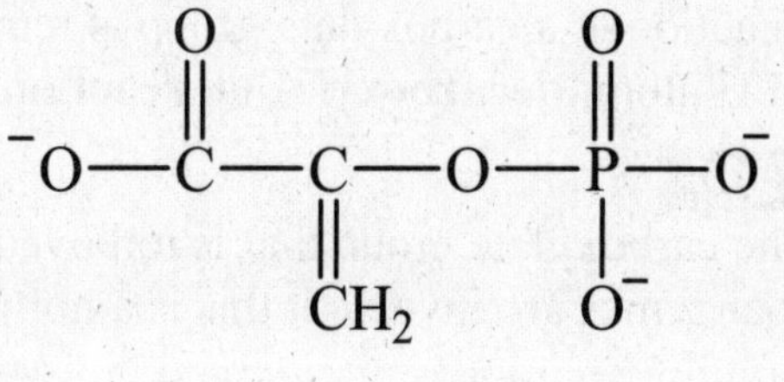

$^{-}$O—C(=O)—C(=CH_2)—O—P(=O)(O^{-})—OH $^{-}$O—C(=O)—C(=CH_2)—O—P(=O)(O^{-})—O^{-}

*11.91 *The following sequence of reactions occurs whenever your body breaks down any kind of nutrient to obtain energy. Identify each of these reactions as an oxidation, a reduction, a hydration, a dehydration, an acid–base reaction, a decarboxylation, or an oxidative decarboxylation. Only the organic reactants and products are shown here. (Each molecule is drawn as it actually appears at physiological pH.)*

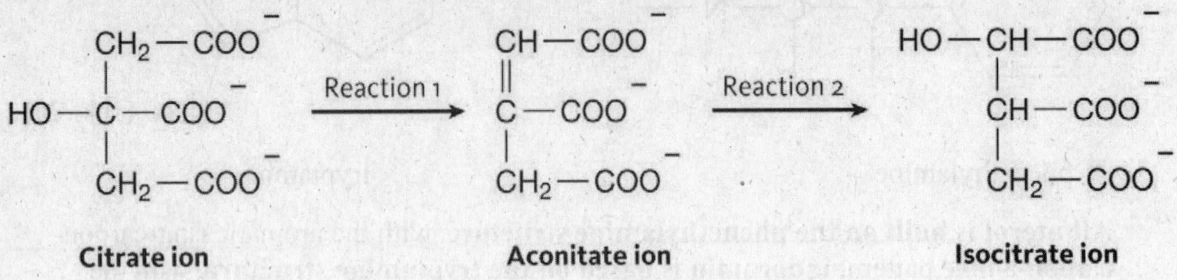

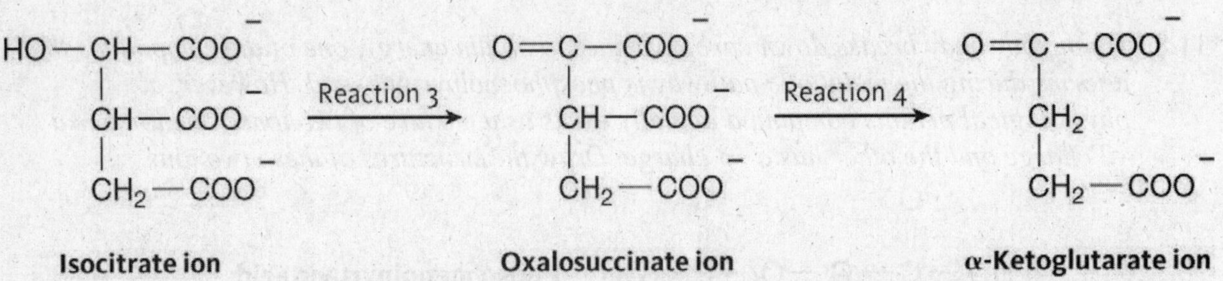

$$\underset{\textbf{Citrate ion}}{\begin{matrix} CH_2-COO^- \\ | \\ HO-C-COO^- \\ | \\ CH_2-COO^- \end{matrix}} \xrightarrow{\text{Reaction 1}} \underset{\textbf{Aconitate ion}}{\begin{matrix} CH-COO^- \\ \| \\ C-COO^- \\ | \\ CH_2-COO^- \end{matrix}} \xrightarrow{\text{Reaction 2}} \underset{\textbf{Isocitrate ion}}{\begin{matrix} HO-CH-COO^- \\ | \\ CH-COO^- \\ | \\ CH_2-COO^- \end{matrix}}$$

$$\underset{\textbf{Isocitrate ion}}{\begin{matrix} HO-CH-COO^- \\ | \\ CH-COO^- \\ | \\ CH_2-COO^- \end{matrix}} \xrightarrow{\text{Reaction 3}} \underset{\textbf{Oxalosuccinate ion}}{\begin{matrix} O=C-COO^- \\ | \\ CH-COO^- \\ | \\ CH_2-COO^- \end{matrix}} \xrightarrow{\text{Reaction 4}} \underset{\boldsymbol{\alpha}\textbf{-Ketoglutarate ion}}{\begin{matrix} O=C-COO^- \\ | \\ CH_2 \\ | \\ CH_2-COO^- \end{matrix}}$$

In each step, look carefully to see what has actually changed. While the molecules may be large, the changes are small, and they are reactions that are familiar to us.

Reaction 1: an OH group and an H atom are removed from adjacent carbon atoms, forming a C=C double bond: **dehydration reaction**.

Reaction 2: an OH group and an H atom are added to adjacent, double-bonded carbon atoms, forming a C–C single bond: **hydration reaction**.

NOTE:
The net effect of the first two reactions is to switch the positions of the H and OH.

Reaction 3: two H atoms are removed from the O and C of an alcohol, forming a C=O double bond: **oxidation reaction**.

Reaction 4: a carboxylate group is removed entirely from the molecule, and replaced with an H atom: **decarboxylation reaction**.

NOTE:
The carboxylate group that is removed has a ketone in the β [beta] position, and no redox coenzymes are involved; this is a normal carboxylation, not an oxidative decarboxylation.

*11.93 *Draw the structures of compounds A through E in the sequence of reactions shown here.*

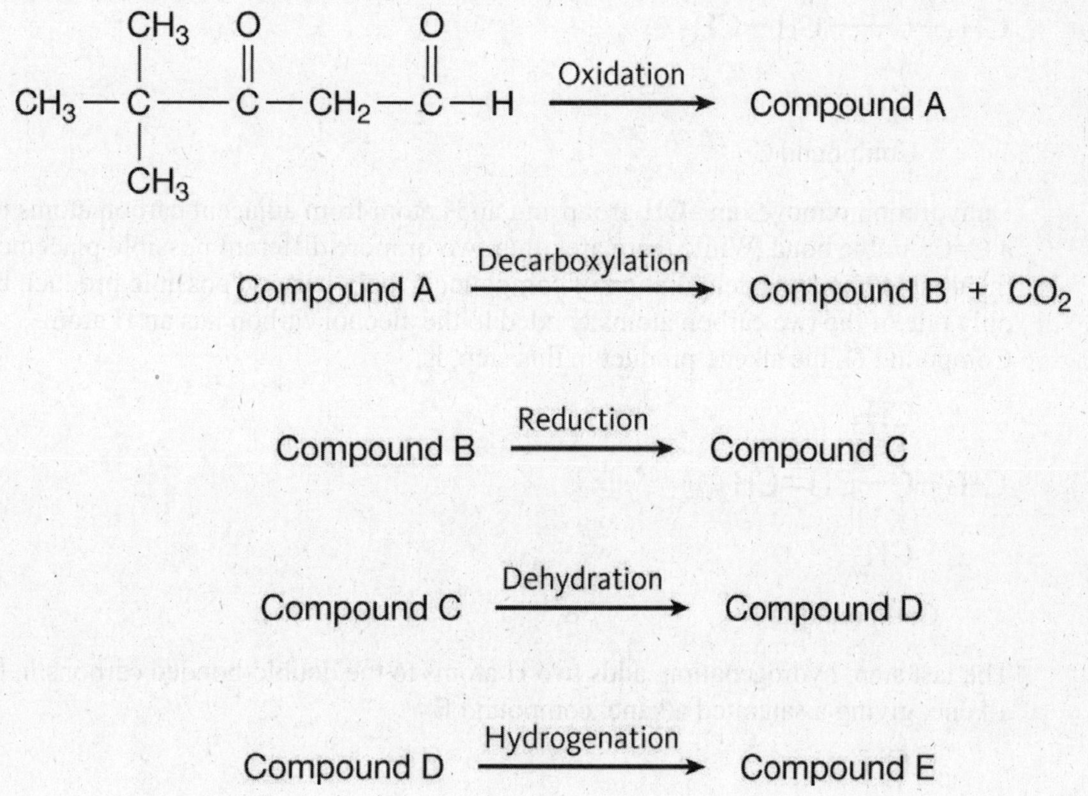

$$CH_3-\overset{\overset{\displaystyle CH_3}{|}}{\underset{\underset{\displaystyle CH_3}{|}}{C}}-\overset{\overset{\displaystyle O}{\|}}{C}-CH_2-\overset{\overset{\displaystyle O}{\|}}{C}-H \xrightarrow{\text{Oxidation}} \text{Compound A}$$

$$\text{Compound A} \xrightarrow{\text{Decarboxylation}} \text{Compound B} + CO_2$$

$$\text{Compound B} \xrightarrow{\text{Reduction}} \text{Compound C}$$

$$\text{Compound C} \xrightarrow{\text{Dehydration}} \text{Compound D}$$

$$\text{Compound D} \xrightarrow{\text{Hydrogenation}} \text{Compound E}$$

Step 1: the only oxidizable group in the original molecule is the aldehyde group (review oxidation reactions in Chapter 10); oxidation of an aldehyde produces a carboxylic acid (Section 10.5), so compound A is:

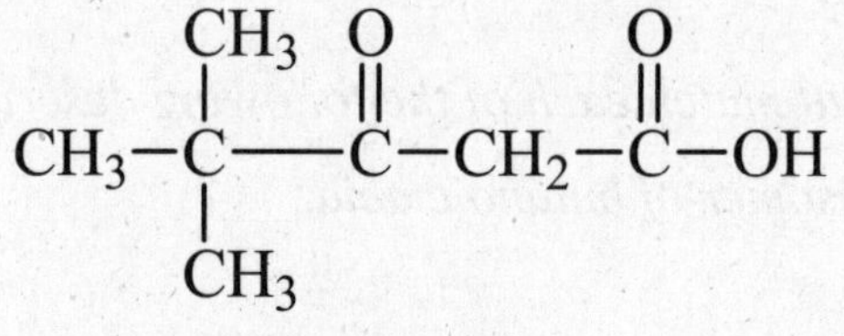

$$CH_3-\overset{\overset{\displaystyle CH_3}{|}}{\underset{\underset{\displaystyle CH_3}{|}}{C}}-\overset{\overset{\displaystyle O}{\|}}{C}-CH_2-\overset{\overset{\displaystyle O}{\|}}{C}-OH$$

Compound A

In the decarboxylation of compound A, the carboxylic acid group is removed (as CO_2) and replaced with an H atom to give compound B:

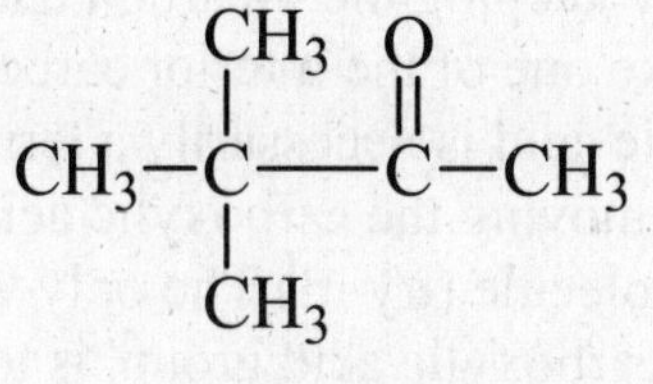

$$CH_3-\overset{\overset{\displaystyle CH_3}{|}}{\underset{\underset{\displaystyle CH_3}{|}}{C}}-\overset{\overset{\displaystyle O}{\|}}{C}-CH_3$$

Compound B

Compound B contains a carbonyl (ketone) group. Reduction of a carbonyl adds two H atoms, one to the carbonyl O and one to the carbonyl C, forming an alcohol, compound C:

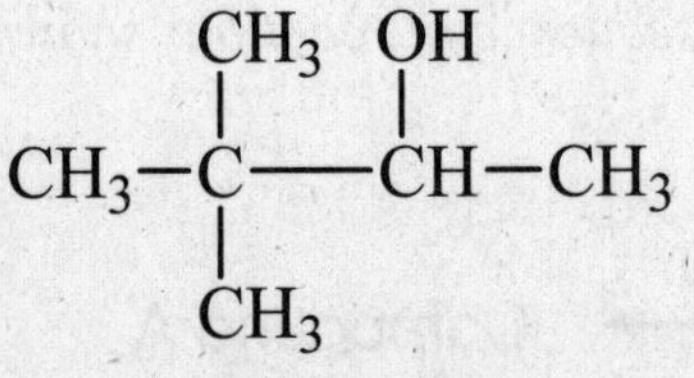

$$\begin{array}{l} \quad\;\; CH_3 \;\; OH \\ \quad\;\;\; | \qquad\;\; | \\ CH_3-C-\!\!-CH-CH_3 \\ \quad\;\;\; | \\ \quad\;\; CH_3 \end{array}$$

Compound C

Dehydration removes an –OH group and an H atom from adjacent carbon atoms to form a C=C double bond. While there are often two or more different possible placements of the double bond, the dehydration of compound C only has one possible product, because only one of the two carbon atoms bonded to the alcohol carbon has an H atom. Compound D, the alkene product in this step, is:

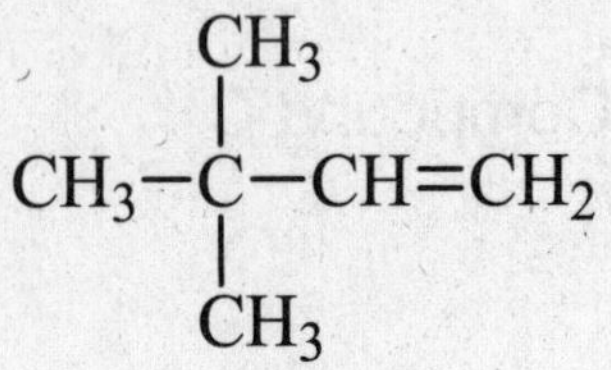

$$\begin{array}{l} \quad\;\; CH_3 \\ \quad\;\;\; | \\ CH_3-C-CH=CH_2 \\ \quad\;\;\; | \\ \quad\;\; CH_3 \end{array}$$

Compound D

The last step, hydrogenation, adds two H atoms to the double-bonded carbons in the alkene, giving a saturated alkane, compound E:

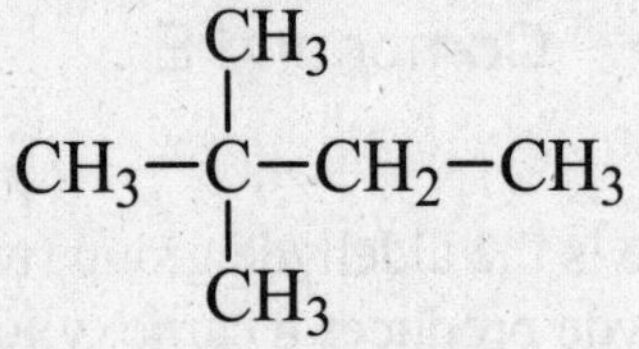

$$\begin{array}{l} \quad\;\; CH_3 \\ \quad\;\;\; | \\ CH_3-C-CH_2-CH_3 \\ \quad\;\;\; | \\ \quad\;\; CH_3 \end{array}$$

Compound E

*11.95 *Draw the structures of organic compounds that match each of the following descriptions:*

a) *A carboxylic acid that is a constitutional isomer of butanoic acid.*

Butanoic acid has the following structure:

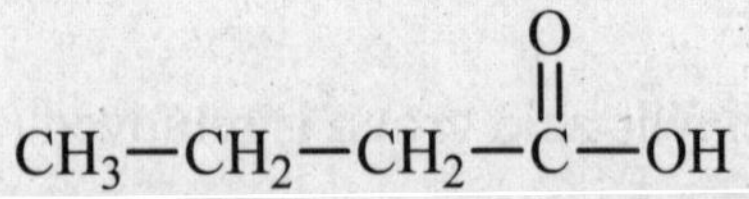

$$CH_3-CH_2-CH_2-\overset{\overset{\displaystyle O}{\|}}{C}-OH$$

In this case, we can't make a constitutional isomer by keeping the 4-carbon chain and just moving the carboxylic acid group. We can't make one of the interior carbon atoms a carboxylic acid group, because the carboxylic acid is necessarily a terminal group (has to be at the end of the carbon chain), and moving the carboxylic acid to the opposite end of the 4-carbon chain gives the same molecule (try it!) The only way we can draw a constitutional isomer, while keeping the carboxylic acid group, is to rearrange the carbons to form a branched structure:

$$CH_3-\underset{\displaystyle CH_3}{\underset{|}{CH}}-\overset{\displaystyle O}{\overset{||}{C}}-OH$$

b) *A primary amine that is a constitutional isomer of methylpropylamine.*

First let's take a look at the structure of methylpropylamine:

$$CH_3-CH_2-CH_2-NH-CH_3$$

Methylpropylamine is a secondary amine (it has two alkyl groups attached to the amine N atom). If we're going to make a primary amine (one with only one alkyl group attached to an $-NH_2$ group), we have to put all 4 carbons into one alkyl group. That gives us four options: two with a four-carbon chain (with the $-NH_2$ group either on carbon #1 or carbon #2), and two with a branched structure for the carbon chain:

$$CH_3-CH_2-CH_2-CH_2-NH_2 \qquad CH_3-CH_2-\overset{\displaystyle NH_2}{\overset{|}{CH}}-CH_3$$

$$CH_3-\underset{\displaystyle CH_3}{\underset{|}{CH}}-CH_2-NH_2 \qquad CH_3-\underset{\displaystyle CH_3}{\underset{|}{\overset{\displaystyle CH_3}{\overset{|}{C}}}}-NH_2$$

c) *A compound that is a constitutional isomer of butanoic acid and that does not contain a carboxylic acid group.*

Butanoic acid is already shown in part a) above, but here we have many more options, because we are not restricted to keeping the carboxylic acid group intact. One option is to put the –COO– group in the middle of the molecule (this gives a functional group called an "ester" that will be introduced in Chapter 12):

$$CH_3-CH_2-\overset{\displaystyle O}{\overset{||}{C}}-O-CH_3 \qquad CH_3-\overset{\displaystyle O}{\overset{||}{C}}-O-CH_2-CH_3$$

We could also split the O atoms onto different carbon atoms on the molecule (two of several possible options are shown):

$$CH_3-\overset{\displaystyle OH}{\overset{|}{CH}}-\overset{\displaystyle O}{\overset{||}{C}}-CH_3 \qquad CH_3-\overset{\displaystyle OH}{\overset{|}{CH}}-CH_2-\overset{\displaystyle O}{\overset{||}{C}}-H$$

Or instead of having a C=O double bond, we could have a C=C double bond, or make a cyclo compound (three of several possible options are shown):

OH
|
$CH_3-C-CH=CH_2$
|
OH

OH
OH

O
OH

*11.97 *You need to make 10.0 mL of a 0.50 M solution of triethylamine. How many grams of triethylamine must you use?*

Chapter 5 review! The setup for the problem is:

10.0 mL solution = ? g triethylamine

0.50 M is the same as 0.50 mol/L, a compound unit we will use as a conversion factor in our problem. Our strategy: convert mL to L, use the concentration to convert L to moles, and use the molar mass of triethylamine to convert moles to grams.

The molecular formula of triethylamine is $C_6H_{15}N$ (draw the structure and add it up.) To add up the molar mass:

6 C	=	6 × 12.01	=	72.06 g/mol
15 H	=	15 × 1.008	=	15.12
+ 1 N	=	1 × 14.01	=	14.01
total				101.19 g/mole

Plugging it all in and rounding to two sig figs:

$$\mathbf{10.0\ mL\ solution \times \frac{1\ L}{1000\ mL\ soln} \times \frac{0.50\ mol}{1\ L\ soln} \times \frac{101.19\ g\ triethylamine}{mol} = 0.51\ g\ triethylamine}$$

*11.99 a) *If you add 30.0 mL of water to 15.0 mL of 1.50% (w/v) sodium acetate, what will be the concentration of sodium acetate in the resulting solution?*

Chapter 5 review!

NOTE:

Here that, as in many dilution problems, we don't need to figure out the formula or the molar mass of the solute. The formula for dilution calculations is: $C_1 \times V_1 = C_2 \times V_2$

In this problem, $C_1 = 1.50\%$ (w/v), $V_1 = 15.0$ mL, and C_2 is unknown. The final volume V_2 is 30.0 + 15.0 mL = 45.0 mL (assuming volumes are additive, which is usually true for dilute solutions).

$$(1.50\%) \times (15.0\ \text{mL}) = C_2 \times (45.0\ \text{mL})$$

$$\mathbf{C_2 = 0.50\%\ (w/v)}$$

b) *How much water must you add to 15.0 mL of 1.50% (w/v) sodium acetate in order to dilute the solution to 0.30% (w/v)?*

$$C_1 \times V_1 = C_2 \times V_2$$

In this problem, $C_1 = 1.50\%$ (w/v), $V_1 = 15.0$ mL, $C_2 = 0.30\%$ (w/v), and V_2 is unknown:

$$(1.50\%) \times (15.0\ \text{mL}) = (0.30\%) \times V_2$$

$$V_2 = 75 \text{ mL}$$

V_2 is the *total* volume needed, 75 mL, so the amount of water that must be added (assuming volumes are additive, which is usually true for dilute solutions) is

75 – 15.0 = 60 mL water

*11.101 *How many grams of NaOH can react with 6.22 g of succinic acid? The structure of succinic acid is shown below.*

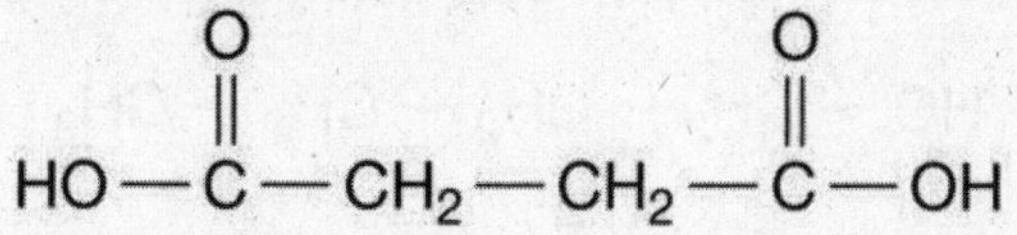

$$HO-\overset{\overset{\displaystyle O}{\|}}{C}-CH_2-CH_2-\overset{\overset{\displaystyle O}{\|}}{C}-OH$$

Succinic acid

Chapter 6 review! To determine mass relationships in reactions, we need a balanced reaction equation. NaOH is a strong base, so this is an acid/base neutralization reaction. Succinic acid has two carboxylic acid groups and therefore two ionizable hydrogen atoms, so it takes 2 moles of NaOH to react with 1 mole of succinic acid. We write the molecular formula of succinic acid with two initial H to reflect the fact that it is diprotic:

$$H_2C_4H_4O_4 + 2\ NaOH \rightarrow C_4H_4O_4^{2-} + 2\ H_2O + 2\ Na^+$$

We add up the molar mass of succinic acid:

4 C	=	4 × 12.01	=	48.04 g/mol
6 H	=	6 × 1.008	=	6.048
+ 4 O	=	4 × 16.00	=	64.00
total				118.09 g/mole

And the molar mass of sodium hydroxide, NaOH:

Na	=	22.99 g/mol
O	=	16.00
H	=	1.008
total		40.00 g/mole

Since 2 moles of NaOH react with 1 mole of succinic acid, the mass relationship is 2(40.00 g) = 80.00 g NaOH for every 118.09 g succinic acid.

$$\mathbf{6.22\ g\ succinic\ acid \times \frac{80.00\ g\ NaOH}{118.09\ g\ succinic\ acid} = 4.21\ g\ NaOH}$$

12

Chapter 12

Condensation and Hydrolysis Reactions

Solutions to Section 12.1 Core Problems

12.1 *Draw the structures of the products of the following condensation reactions:*

a) $CH_3-CH_2-OH + HO-CH_2-CH_2-CH_2-CH_3 \longrightarrow$

b) $CH_3-CH_2-CH_2-OH + CH_3-CH_2-CH_2-OH \longrightarrow$

c) $CH_3-CH_2-OH + CH_3-CH_2-CH_2-CH(OH)-CH_3 \longrightarrow$

In condensation reactions, we first write the molecules with the condensation functional groups (in these problems, alcohols) pointing toward each other. Then we remove an H from one (remembering that this H must be bonded to O or N) and an OH from the other to form a molecule of H_2O, and attach the remaining parts to each other to form the product (in this case, an ether–a functional group in which two alkyl groups are attached to an oxygen atom).

NOTE:
We get the same product, regardless of which reactant supplies the hydroxyl group (in fact, in the actual reaction, it is possible to determine which molecule supplies the –H and which supplies the –OH, but it's well beyond the scope of this text; for purposes of drawing the products it doesn't matter).

a) The two functional groups are already shown pointing toward each other, so drawing the structure of the product is easy:

$CH_3-CH_2-O-CH_2-CH_2-CH_2-CH_3$

b) First, turn the second molecule around so the alcohol groups point toward each other. The atoms that are removed to form an H_2O molecule are circled:

$CH_3-CH_2-CH_2-OH \quad HO-CH_2-CH_2-CH_3$

Then take H from one, OH from the other, and attach the two remaining parts together:

$CH_3-CH_2-CH_2-O-CH_2-CH_2-CH_3$

c) Show the molecules with their alcohol groups in range of each other. The atoms that are removed to form an H_2O molecule are circled:

$$\begin{array}{l} CH_3\text{-}CH_2\text{-}OH \quad H\text{-}O \\ \qquad\qquad\qquad\quad\;\; | \\ \quad CH_3\text{-}CH_2\text{-}CH_2\text{-}CH-CH_3 \end{array}$$

Then draw the condensation product:

$$\begin{array}{l} \quad CH_3\text{-}CH_2\text{-}O \\ \qquad\qquad\qquad\quad | \\ CH_3\text{-}CH_2\text{-}CH_2\text{-}CH-CH_3 \end{array}$$

We could also draw the product more conventionally as:

$$\begin{array}{l} \qquad\qquad\qquad\quad CH_3 \\ \qquad\qquad\qquad\quad\;\; | \\ CH_3-CH_2-O-CH-CH_2\text{-}CH_2\text{-}CH_3 \end{array}$$

12.3 *Name the following ethers:*

a) $CH_3\text{–}CH_2\text{–}CH_2\text{–}CH_2\text{–}CH_2\text{–}O\text{–}CH_2\text{–}CH_3$ *b)* $CH_3\text{–}O\text{–}CH_3$

Ethers are named by listing the two alkyl groups bonded to the oxygen as separate words in alphabetical order, and adding the word "ether."

a) This ether has a pentyl group (5 carbons) and an ethyl group (2 carbons) attached to the ether O: **ethyl pentyl ether.**

b) This ether has two methyl groups (1 carbon each) attached to the ether O: **dimethyl ether**.

Solutions to Section 12.2 Core Problems

12.5 *Draw the structures of the products of the following esterification reactions:*

a) $CH_3-\overset{\overset{\displaystyle O}{\|}}{C}-OH \; + \; HO-C_6H_5 \longrightarrow$

b) $CH_3-\overset{\overset{\displaystyle OH}{|}}{CH}-CH_3 \; + \; CH_3-CH_2-\overset{\overset{\displaystyle O}{\|}}{C}-OH \longrightarrow$

Esterification reactions form esters. In esterification reactions, as in other condensation reactions, we first write the molecules with the condensation functional groups (in these problems, carboxylic acids and alcohols) pointing toward each other. Then we remove an H from one hydroxyl group and an OH from the other molecule, and attach the remaining fragments to each other to form an ester functional group.

NOTE:
We get the same product, regardless of which reactant supplies the hydroxyl group (again, sorting out which molecule actually supplies the hydroxyl group is beyond the scope of this text; for purposes of drawing the product it doesn't matter).

a) The two functional groups are already shown the pointing toward each other, so drawing the structure of the product is easy:

$$CH_3\text{-}\overset{\overset{\large O}{\|}}{C}\text{—O—}C_6H_5$$

b) To make this easier to see, let's rearrange the molecules to show their –OH groups pointing toward each other. The atoms that are removed to form an H_2O molecule are circled:

$$CH_3\text{-}CH_2\text{-}\overset{\overset{\large O}{\|}}{C}\text{—}\boxed{OH\quad H}O\text{—}\overset{\overset{\large CH_3}{|}}{CH}\text{—}CH_3$$

Then draw the condensation product:

$$CH_3\text{-}CH_2\text{-}\overset{\overset{\large O}{\|}}{C}\text{—O—}\overset{\overset{\large CH_3}{|}}{CH}\text{—}CH_3$$

12.7 *Draw the structures of the products of the following amidation reactions:*

a) $CH_3\text{—}CH_2\text{—}\overset{\overset{\large CH_3}{|}}{NH} + HO\text{—}\overset{\overset{\large O}{\|}}{C}\text{—}H \longrightarrow$

b) $CH_3\text{—}CH_2\text{—}NH\text{—}CH_2\text{—}CH_3 \quad +$

$$C_6H_5\text{—}\overset{\overset{\large O}{\|}}{C}\text{—}OH \longrightarrow$$

In amidations, as in other condensation reactions, we first write the molecules with the condensation functional groups (in these problems, carboxylic acids and amines) pointing toward each other. Then we remove an H from one (remembering that this H must be bonded to O or N) and an OH from the other, and attach the remaining parts to each other to form an amide functional group.

NOTE:
This time only one of the reactants has a hydroxyl group, so the H atom must come from the amine N.

a) The two functional groups are already shown pointing toward each other, so drawing the structure of the product is easy:

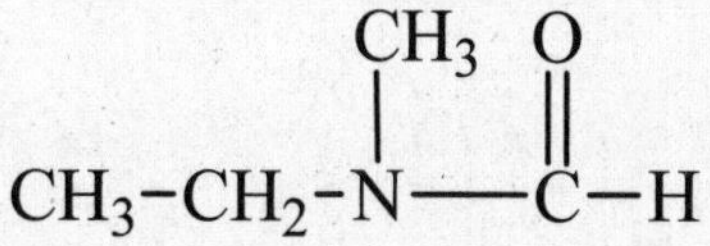

$$\begin{array}{c} \quad\quad\quad\;\; CH_3\;\; O \\ \quad\quad\quad\;\; |\quad\;\; \| \\ CH_3{-}CH_2{-}N{-}C{-}H \end{array}$$

b) Let's first redraw the molecules to show the amine and the carboxylic acid pointing toward each other. The atoms that are removed to form an H_2O molecule are circled:

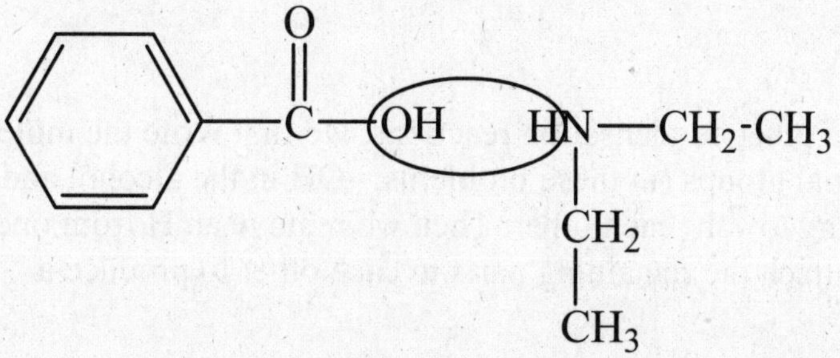

$$C_6H_5{-}\overset{\displaystyle O}{\overset{\|}{C}}{-}\boxed{OH\quad H}N(CH_2{-}CH_3){-}CH_2{-}CH_3$$

Then we can draw the condensation product:

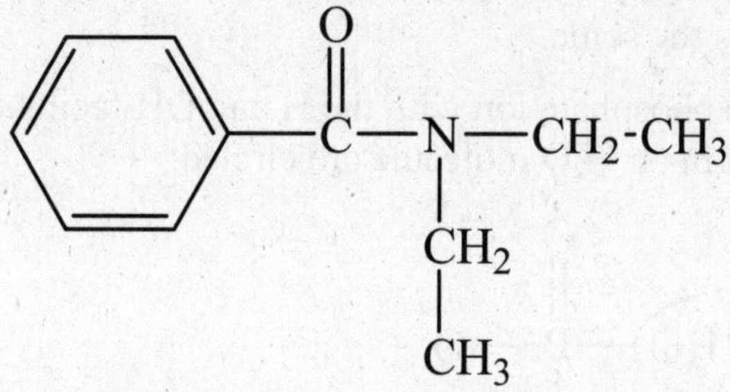

$$C_6H_5{-}\overset{\displaystyle O}{\overset{\|}{C}}{-}N(CH_2{-}CH_3){-}CH_2{-}CH_3$$

12.9 *Which of the following molecules will give a pH higher than 7 when it dissolves in water?*

a) $CH_3{-}CH_2{-}NH{-}CH_2{-}CH_3$

b) $CH_3{-}CH_2{-}NH{-}\overset{\displaystyle O}{\overset{\|}{C}}{-}CH_3$

Chapter 7 & 11 review: pH higher than 7 is a basic solution, which means we are looking for a basic functional group. Amines and anions are basic. Other functional groups are acidic or neutral.

a) **This is an amine and will form a basic solution (pH higher than 7) when it dissolves in water.**

b) This molecule is an amide. Amides are neutral, so **this compound will form a neutral (pH=7) solution when it dissolves in water.**

12.11 *Draw the structures of the phosphoesters that are formed when each of the following compounds reacts with a phosphate ion:*

a) $CH_3—CH_2—CH_2—CH_2—OH$

b) 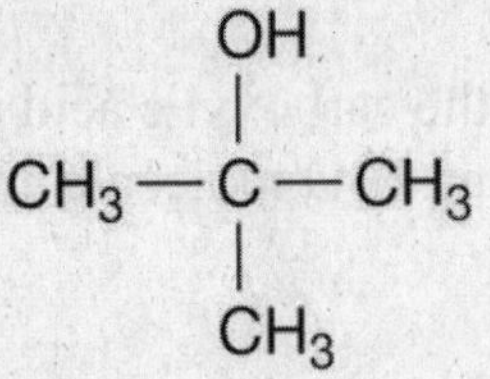 $CH_3—C(OH)(CH_3)—CH_3$

In phosphorylations, as in other condensation reactions, we first write the molecules with the condensation functional groups (in these problems, –OH in the alcohol and –OH in the phosphate ion) pointing toward each other. Then we remove an H from one and an OH from the other, and attach the remaining parts to each other to produce a phosphoester.

NOTE:

For purposes of drawing the product, it doesn't matter which molecule supplies the H and which supplies the OH; the product is the same.

a) First we draw the alcohol and the phosphate ion with the H and OH facing each other. The atoms that are removed to form an H_2O molecule are circled:

$$CH_3\text{-}CH_2\text{-}CH_2\text{-}CH_2\text{-}OH \quad HO—P(=O)(O^-)—O^-$$

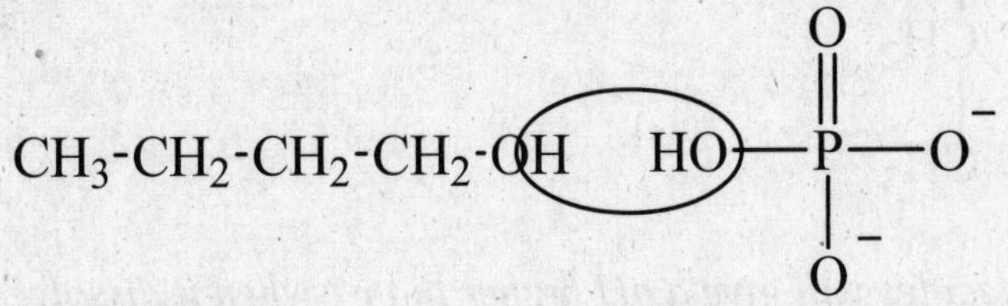

Then we remove OH from the phosphate ion and H from the alcohol, and combine the remaining fragments into a single molecule:

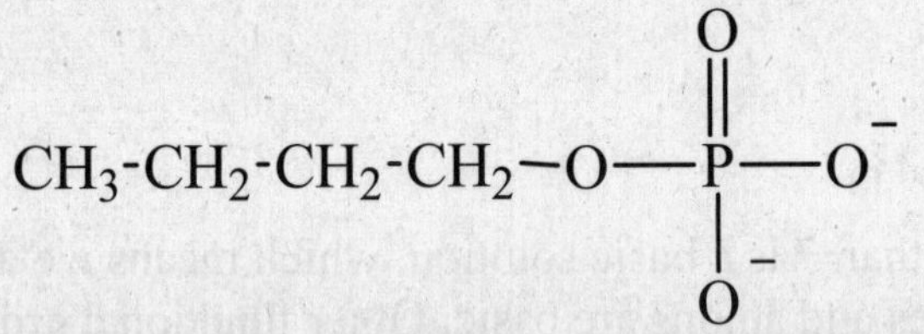

$$CH_3\text{-}CH_2\text{-}CH_2\text{-}CH_2—O—P(=O)(O^-)—O^-$$

b) First we draw the alcohol and the phosphate ion with the H and OH facing each other (for a more convenient arrangement, we rotate the alcohol 90°). The atoms that are removed to form an H_2O molecule are circled:

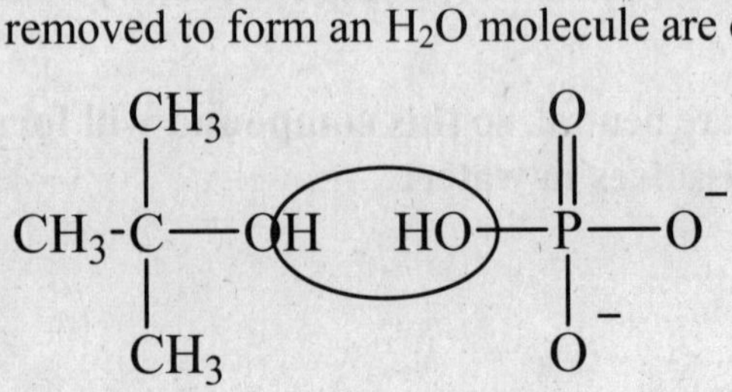

$$CH_3\text{-}C(CH_3)_2—OH \quad HO—P(=O)(O^-)—O^-$$

Then we remove OH from the phosphate ion and H from the alcohol, and combine the remaining fragments into a single molecule:

$$CH_3\text{-}C(CH_3)_2\text{—O—}P(=O)(O^-)\text{—}O^-$$

Solutions to Section 12.3 Core Problems

12.13 *Draw the structure of the organic product that is formed when three molecules of the following compound condense to form a single molecule:*

HO—C_6H_4—OH

Since this molecule has two alcohol groups that can participate in condensations to form ether groups, we can make long strings by repeated condensation reactions. First, line up three molecules:

HO—C_6H_4—OH HO—C_6H_4—OH HO—C_6H_4—OH

Then perform **two** condensation reactions to connect the three molecules. The participating –OH and –H atoms are circled:

HO—C_6H_4—(OH H)O—C_6H_4—(OH H)O—C_6H_4—OH

The final product is:

HO—C_6H_4—O—C_6H_4—O—C_6H_4—OH

12.15 *Lactic acid is responsible for the unpleasant taste and aroma of spoiled milk. Draw the structure of the product that is formed when three molecules of lactic acid condense to form a single molecule.*

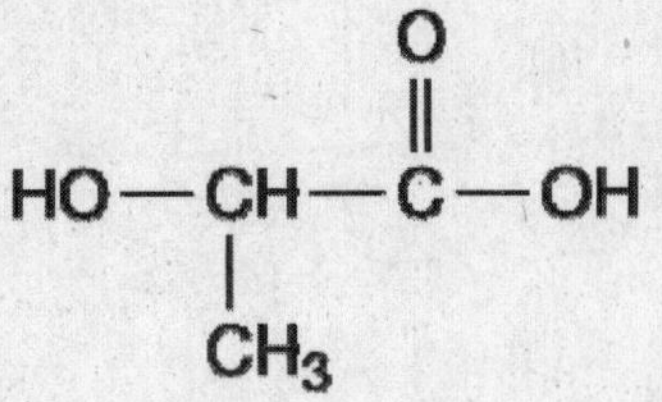

```
              O
              ||
HO—CH—C—OH    Lactic acid
    |
    CH3
```

Since this molecule has two functional groups (an alcohol and a carboxylic acid) that can participate in condensations, we can make long strings by repeated condensation reactions. First, line up three molecules:

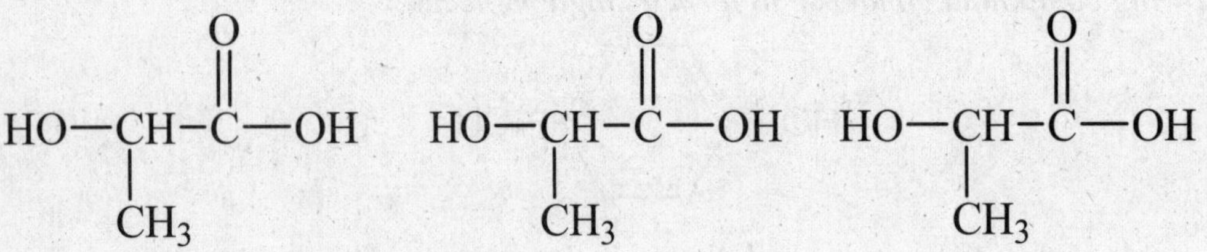

```
         O                  O                  O
         ||                 ||                 ||
HO—CH—C—OH      HO—CH—C—OH      HO—CH—C—OH
    |                  |                  |
    CH3                CH3                CH3
```

Then perform **two** condensation reactions to connect the three molecules. The participating –OH and –H atoms are circled:

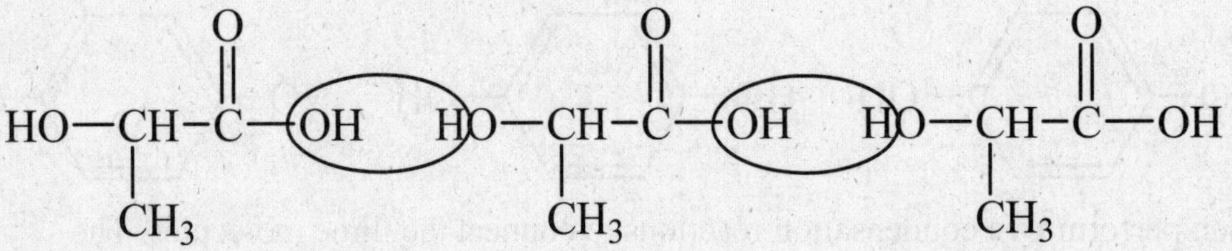

```
         O                  O                  O
         ||                 ||                 ||
HO—CH—C—(OH      H)O—CH—C—(OH      H)O—CH—C—OH
    |                  |                  |
    CH3                CH3                CH3
```

The final product is:

```
         O              O              O
         ||             ||             ||
HO—CH—C—O—CH—C—O—CH—C—OH
    |          |          |
    CH3        CH3        CH3
```

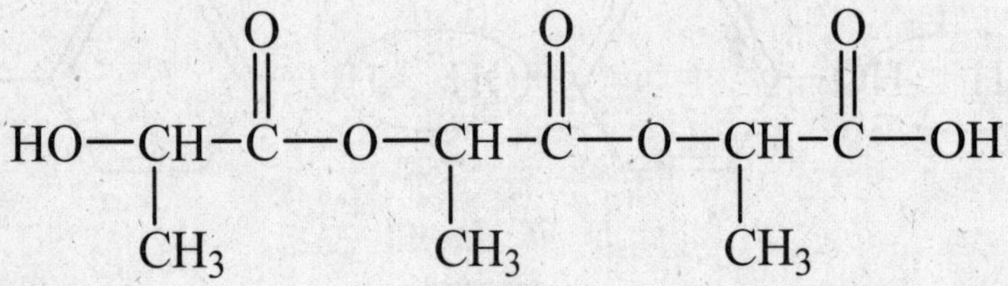

12.17 *Isoleucine is one of the naturally occurring amino acids, and it is a required nutrient in our diet. Draw the structure of the organic product that is formed when three molecules of isoleucine condense.*

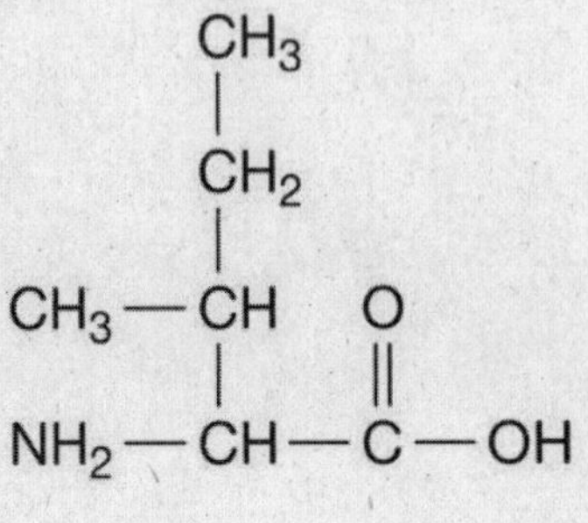

```
          CH3
          |
          CH2
          |                 Isoleucine
    CH3—CH    O
          |    ||
   NH2—CH—C—OH
```

This molecule has two functional groups (an amine and a carboxylic acid) that can participate in condensations, so we can make long strings by repeated condensation reactions. First, line up three molecules (since the amine H atoms will be involved in the condensation, they're shown in full structure form). The participating H and OH atoms are circled:

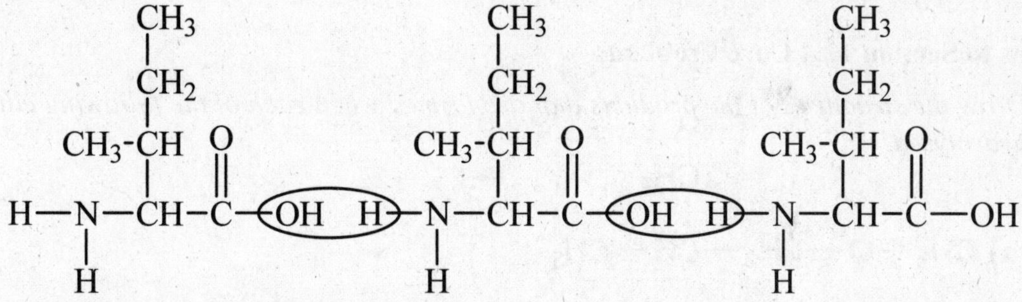

```
          CH3                          CH3                          CH3
          |                            |                            |
          CH2                          CH2                          CH2
          |                            |                            |
     CH3-CH  O                    CH3-CH  O                    CH3-CH  O
          |  ||                        |  ||                        |  ||
H—N—CH-C—(OH   H)—N—CH-C—(OH   H)—N—CH-C—OH
  |                  |                  |
  H                  H                  H
```

Then perform **two** condensation reactions to connect the three molecules, forming amide functional groups. The final product is:

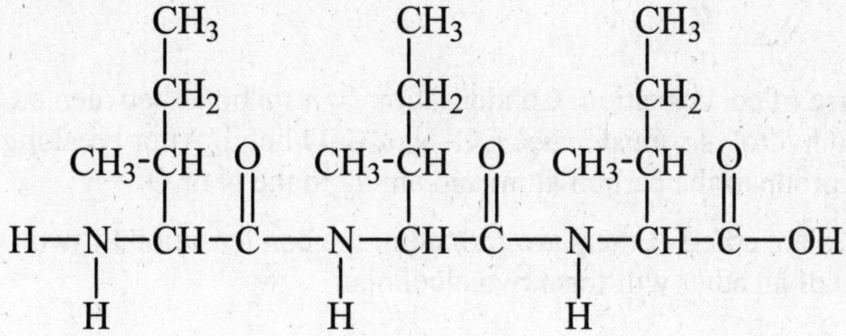

```
          CH3          CH3          CH3
          |            |            |
          CH2          CH2          CH2
          |            |            |
     CH3-CH  O    CH3-CH  O    CH3-CH  O
          |  ||        |  ||        |  ||
H—N—CH-C——N—CH-C——N—CH-C—OH
  |          |          |
  H          H          H
```

12.19 *Ethylenediamine and oxalic acid can form a copolymer. Show how two molecules of each compound can condense to form the beginning of a copolymer.*

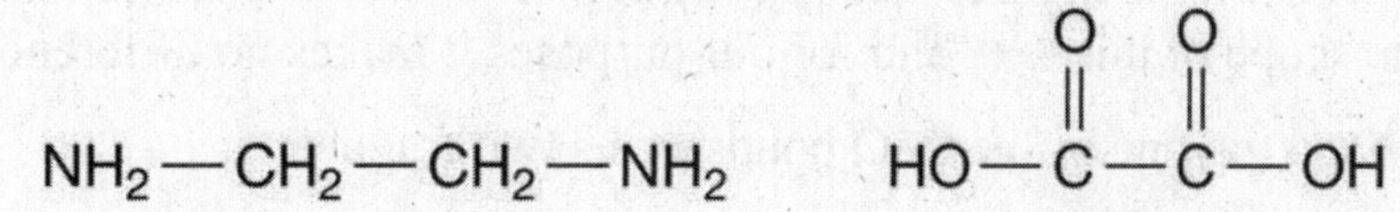

```
                                      O   O
                                      ||  ||
NH2—CH2—CH2—NH2               HO—C—C—OH
```

Ethylenediamine **Oxalic acid**

See Figure 12.7 for help getting started on this one. In a *copolymer*, molecules of two different starting compounds are condensed together to form long strings; the two compounds alternate with each other. In this combination, we'll condense the amine groups on one reactant with the carboxylic acid groups on the other, forming amide functional groups.

First, we draw two molecules of each compound, alternating them and identifying the H atoms and OH groups involved in the condensation (since the amine H atoms will be involved in the condensation, they're shown in full structure form):

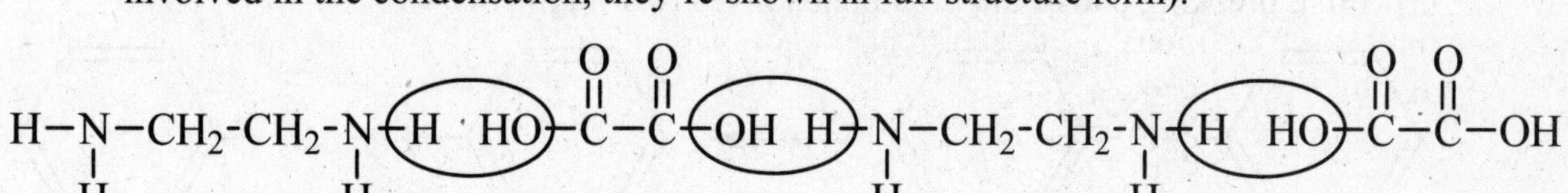

```
                            O  O                                         O  O
                            || ||                                        || ||
H-N-CH2-CH2-N(H  HO)-C-C(OH  H)-N-CH2-CH2-N(H  HO)-C-C-OH
  |          |                     |          |
  H          H                     H          H
```


Then we remove the H and OH, and attach the remaining fragments together:

$$\mathrm{H{-}\underset{\displaystyle H}{\underset{|}{N}}{-}CH_2{-}CH_2{-}\underset{\displaystyle H}{\underset{|}{N}}{-}\overset{\displaystyle O}{\overset{\|}{C}}{-}\overset{\displaystyle O}{\overset{\|}{C}}{-}\underset{\displaystyle H}{\underset{|}{N}}{-}CH_2{-}CH_2{-}\underset{\displaystyle H}{\underset{|}{N}}{-}\overset{\displaystyle O}{\overset{\|}{C}}{-}\overset{\displaystyle O}{\overset{\|}{C}}{-}OH}$$

Solutions to Section 12.4 Core Problems

12.21 *Draw the structures of the products that are formed when each of the following ethers is hydrolyzed:*

a) $\mathrm{CH_3{-}O{-}CH_2{-}\overset{\displaystyle CH_3}{\overset{|}{CH}}{-}CH_3}$

b) $C_6H_5{-}O{-}C_6H_5$ (diphenyl ether)

Hydrolysis is the reverse of condensation. Condensation *forms* a bond between a C atom and an N or O atom, so hydrolysis must *break* a C–N or C–O bond. After breaking the bond, we add an –OH group to the carbon atom and an –H to the N or O.

Both of these molecules are ethers. Ethers are formed in the condensation of two alcohols, so hydrolysis of an ether will form two alcohols.

NOTE:
For ethers, it doesn't matter which C–O bond is broken in the hydrolysis (for the same reason that, in the condensation of two alcohols, it doesn't matter which molecule loses the H and which loses the OH; again, the question of which bond actually breaks is beyond the scope of this text, and for our purposes it makes no difference).

a) First, breaking one of the C–O bonds gives two fragments:

$$\mathrm{CH_3{-}O{\wr}CH_2{-}\overset{\displaystyle CH_3}{\overset{|}{CH}}{-}CH_3 \longrightarrow CH_3{-}O{-} \quad + \quad {-}CH_2{-}\overset{\displaystyle CH_3}{\overset{|}{CH}}{-}CH_3}$$

Then we add an H to the O and an OH to the C, giving two alcohols:

$$\mathrm{CH_3{-}OH \qquad HO{-}CH_2{-}\overset{\displaystyle CH_3}{\overset{|}{CH}}{-}CH_3}$$

b) First, breaking one of the C–O bonds gives two fragments:

$$C_6H_5{\wr}O{-}C_6H_5 \longrightarrow C_6H_5{-}O{-} \quad + \quad {-}C_6H_5$$

Then we add an H to the O and an OH to the C, giving two alcohols:

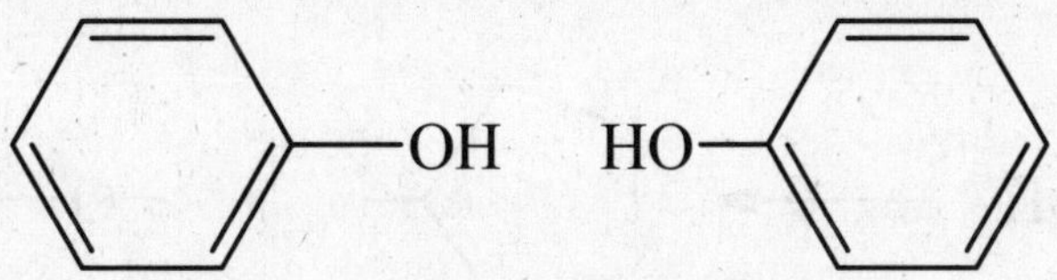

(benzene ring)–OH   HO–(benzene ring)

12.23 *Draw the structures of the products that are formed when each of the following esters is hydrolyzed:*

a) $CH_3-CH_2-CH_2-C(=O)-O-CH_3$

b) 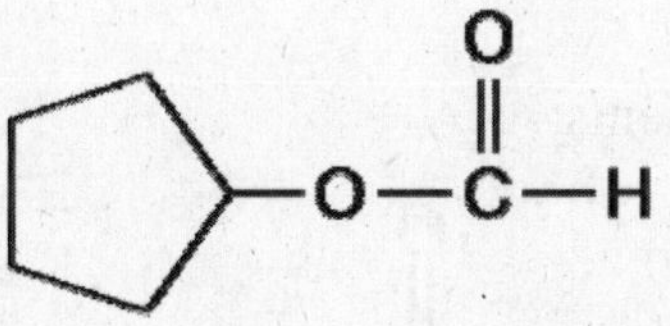 (cyclopentyl)$-O-C(=O)-H$

c) (skeletal structure: propanoate ester with O-isopropyl group)

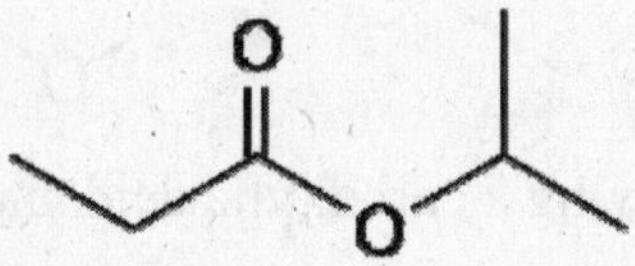

Hydrolysis is the reverse of condensation. Condensation *forms* a bond between a C atom and an N or O atom, so hydrolysis must *break* a C–N or C–O bond. After breaking the bond, we add an –OH group to the carbon atom and an –H to the N or O.

Both of these molecules are esters. Esters are formed in the condensation of a carboxylic acid and an alcohol, so hydrolysis of an ester will form a carboxylic acid and an alcohol.

NOTE:
For esters, as for ethers, it doesn't matter which C–O bond is broken in the hydrolysis (for the same reason that, in the condensation of a carboxylic acid and an alcohol, it doesn't matter which molecule loses the H and which loses the OH).

a) First, breaking one of the C–O bonds gives two fragments:

$$CH_3\text{-}CH_2\text{-}CH_2\text{-}C(=O)\text{-}\wr\text{-}O-CH_3 \longrightarrow CH_3\text{-}CH_2\text{-}CH_2\text{-}C(=O)- \quad + \quad -O-CH_3$$

Then we add an H to the O and an OH to the C, giving a carboxylic acid and an alcohol:

$$CH_3\text{-}CH_2\text{-}CH_2\text{-}C(=O)\text{-}OH \qquad HO-CH_3$$

b) First, breaking one of the C–O bonds gives two fragments:

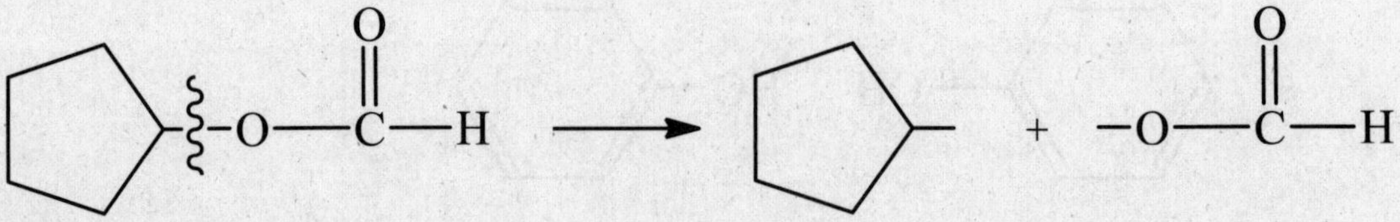

O
‖
–O—C—H ⟶ + –O—C—H

Then we add an H to the O and an OH to the C, giving a carboxylic acid and an alcohol:

–OH   HO—C—H

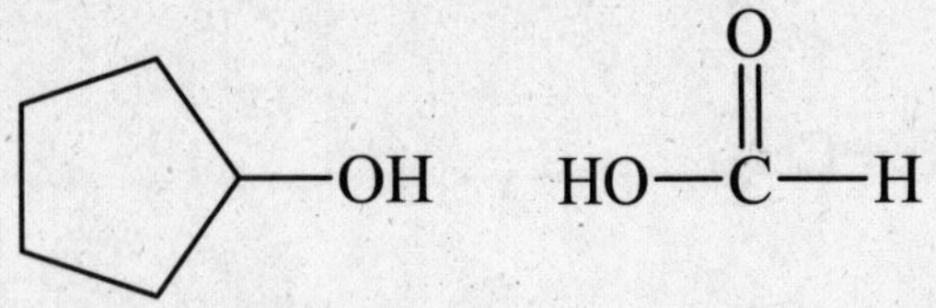

c) Breaking one of the C–O bonds gives two fragments:

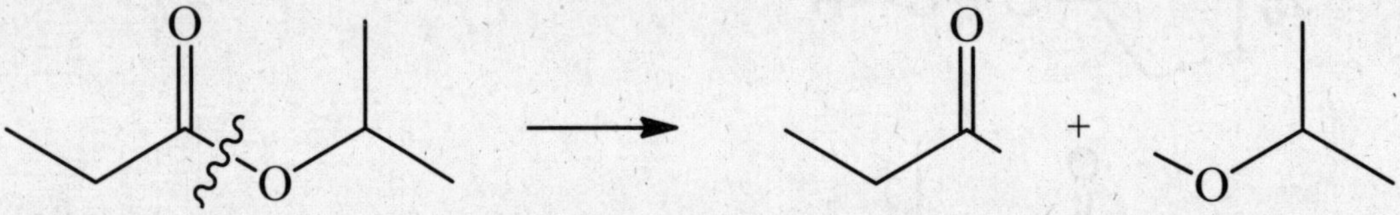

O ⟶ O + O

Then we add an H to the O and an OH to the C, giving a carboxylic acid and an alcohol:

O
OH   HO

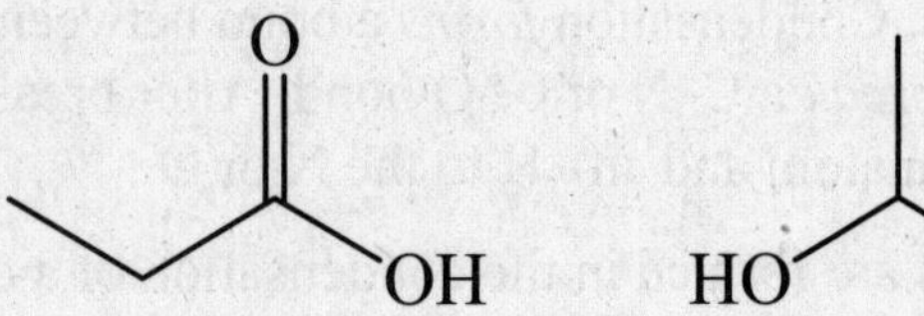

12.25 *Draw the structures of the products that are formed when each of the following amides is hydrolyzed:*

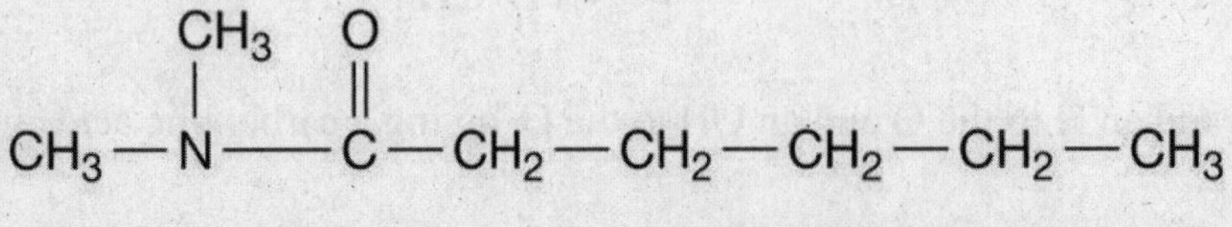

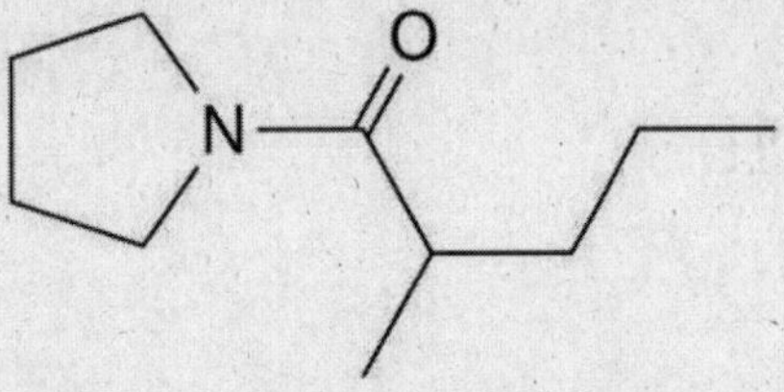

a) 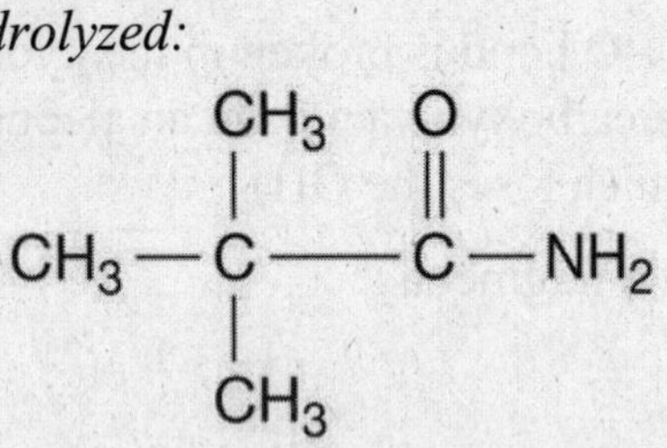 $CH_3-C(CH_3)_2-C(=O)-NH_2$

b) $CH_3-N(CH_3)-C(=O)-CH_2-CH_2-CH_2-CH_2-CH_3$

c) N, O

Amides are formed in the condensation of a carboxylic acid and an amine, so hydrolysis of an amide will form a carboxylic acid and an amine.

NOTE:

For amides, it **does** matter which C–N bond is broken in the hydrolysis–it must be the bond between the carbonyl carbon (C=O) and the nitrogen. In each case, after breaking the C–N bond, we add an OH group to the carbonyl C and an H atom to the N.

a) Breaking the bond between the carbonyl C and the N gives two fragments:

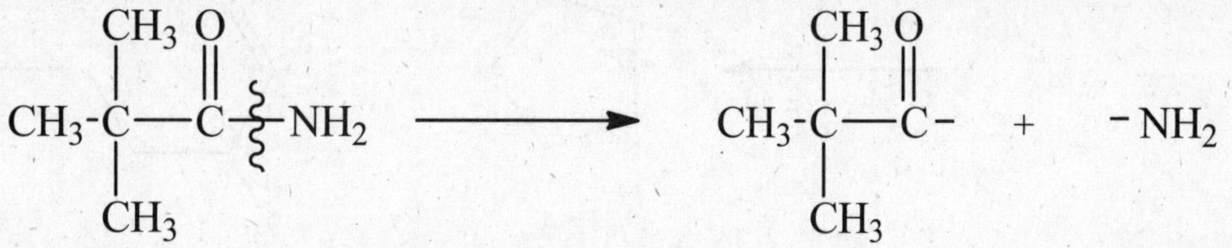

$$CH_3\text{-}C(CH_3)_2\text{-}C(=O)\text{-}NH_2 \longrightarrow CH_3\text{-}C(CH_3)_2\text{-}C(=O)\text{-} \quad + \quad \text{-}NH_2$$

Add OH to the carbonyl to form a carboxylic acid, and H to the nitrogen to form, in this case, ammonia:

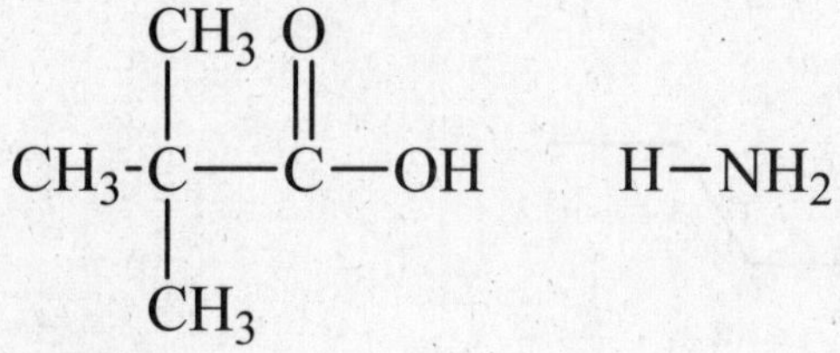

$$CH_3\text{-}C(CH_3)_2\text{-}C(=O)\text{-}OH \qquad H\text{-}NH_2$$

NOTE:

$H\text{–}NH_2$ is the same as NH_3, the familiar formula for ammonia.

b) Breaking the bond between the carbonyl C and the N gives two fragments:

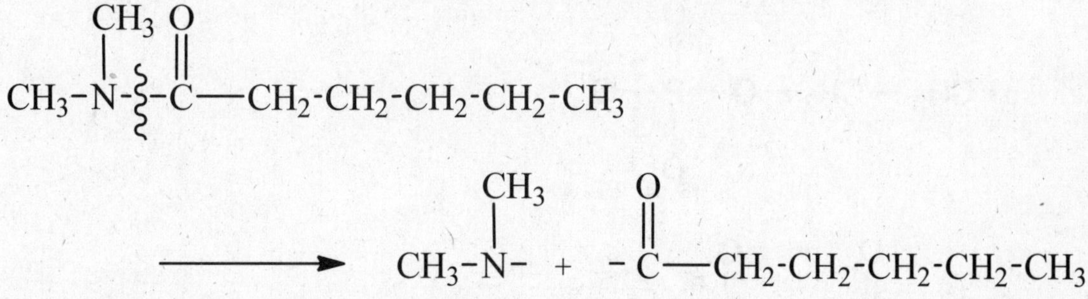

$$CH_3\text{-}N(CH_3)\text{-}C(=O)\text{-}CH_2\text{-}CH_2\text{-}CH_2\text{-}CH_2\text{-}CH_3$$

$$\longrightarrow CH_3\text{-}N(CH_3)\text{-} \quad + \quad \text{-}C(=O)\text{-}CH_2\text{-}CH_2\text{-}CH_2\text{-}CH_2\text{-}CH_3$$

Add OH to the carbonyl to form a carboxylic acid, and H to the nitrogen to form an amine:

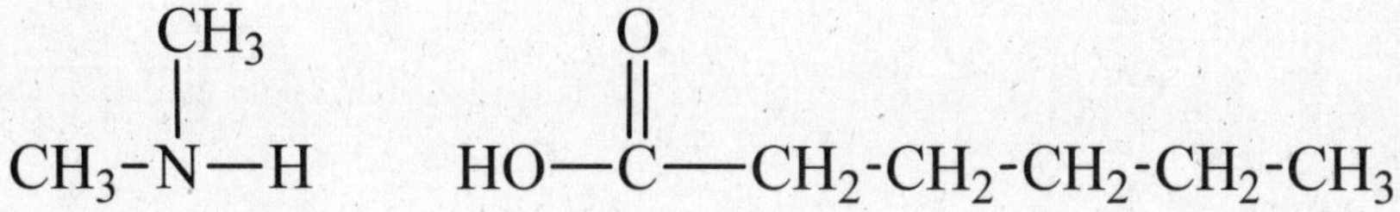

$$CH_3\text{-}N(CH_3)\text{-}H \qquad HO\text{-}C(=O)\text{-}CH_2\text{-}CH_2\text{-}CH_2\text{-}CH_2\text{-}CH_3$$

c) Breaking the bond between the carbonyl C and the N gives two fragments:

Add OH to the carbonyl to form a carboxylic acid, and H to the nitrogen to form an amine:

N–H HO

12.27 *Draw the structures of the products that are formed when each of the following compounds is hydrolyzed:*

a) $CH_3-CH_2-O-\overset{O}{\overset{\|}{P}}(-O^-)-O^-$

b) $^-O-\overset{O}{\overset{\|}{P}}(-O^-)-O-\overset{O}{\overset{\|}{C}}-$(cyclopentyl)

c) $CH_3-\overset{O}{\overset{\|}{C}}-S-CH_2-CH_3$

d) $CH_3-\underset{}{CH}(CH_3)-CH(CH_3)-CH_2-O-\overset{O}{\overset{\|}{P}}-O-\overset{O}{\overset{\|}{P}}-O^-$

See Table 12.3 for help with this one. In each case, we have to decide which bond makes the most sense to hydrolyze. The reactions shown in Table 12.3 as hydrolysis reactions are all condensation reactions in reverse.

a) The hydrolysis of a phosphoester will produce an alcohol and a phosphate ion. In this reaction, the O between the C and the P is the focus of the hydrolysis. It turns out not to matter which of its two bonds we break, but since we formed a P–O bond in Section 12.2, we'll break one here:

$$CH_3\text{-}CH_2\text{-}O\text{-}\wr\text{-}P(=O)(O^-)\text{-}O^- \longrightarrow CH_3\text{-}CH_2\text{-}O\text{-} \;+\; \text{-}P(=O)(O^-)\text{-}O^-$$

Adding an H atom to the O on the alkyl group and an OH group to the P gives us an alcohol and a phosphate ion:

$$CH_3\text{-}CH_2\text{-}OH \qquad HO\text{-}P(=O)(O^-)\text{-}O^-$$

b) The hydrolysis of a phosphoric anhydride will produce a carboxylic acid and a phosphate ion. The O between the C and the P is the focus of the hydrolysis. It turns out not to matter which of its two bonds we break, but we'll be consistent with the condensation reactions from Section 12.2; in those reactions we formed a P–O bond, so we'll break one here:

$$^-O\text{-}P(=O)(O^-)\text{-}\wr\text{-}O\text{-}C(=O)\text{-}C_5H_9 \longrightarrow {}^-O\text{-}P(=O)(O^-)\text{-} \;+\; \text{-}O\text{-}C(=O)\text{-}C_5H_9$$

Adding an H atom to the O on the carboxylate group and an OH group to the P gives us a carboxylic acid and a phosphate ion:

$$^-O\text{-}P(=O)(O^-)\text{-}OH \qquad H\text{-}O\text{-}C(=O)\text{-}C_5H_9$$

c) In the hydrolysis of a thioester, the bond between the carbonyl C atom and the S atom is broken:

$$CH_3\text{-}\overset{\overset{\displaystyle O}{\|}}{C}\text{-}\wr\text{-}S{-}CH_2{-}CH_3 \longrightarrow CH_3\text{-}\overset{\overset{\displaystyle O}{\|}}{C}\text{-} \;+\; \text{-}S{-}CH_2{-}CH_3$$

Adding an OH group to the carbonyl and an H atom to the sulfur atom gives a carboxylic acid and a thiol:

$$CH_3\text{-}\overset{\overset{\displaystyle O}{\|}}{C}{-}OH \qquad H{-}S{-}CH_2{-}CH_3$$

d) In the hydrolysis of a diphosphate, the bond between the two phosphate fragments is broken. It doesn't matter which P–O bond is broken (try it!):

$$CH_3\text{-}\underset{\underset{\displaystyle CH_3}{|}}{CH}{-}\underset{\underset{\displaystyle CH_3}{|}}{CH}{-}CH_2{-}O{-}\overset{\overset{\displaystyle O}{\|}}{\underset{\underset{\displaystyle O^-}{|}}{P}}{-}O\text{-}\wr\text{-}\overset{\overset{\displaystyle O}{\|}}{\underset{\underset{\displaystyle O^-}{|}}{P}}{-}O^-$$

$$\longrightarrow CH_3\text{-}\underset{\underset{\displaystyle CH_3}{|}}{CH}{-}\underset{\underset{\displaystyle CH_3}{|}}{CH}{-}CH_2{-}O{-}\overset{\overset{\displaystyle O}{\|}}{\underset{\underset{\displaystyle O^-}{|}}{P}}{-}O\text{-} \;+\; \text{-}\overset{\overset{\displaystyle O}{\|}}{\underset{\underset{\displaystyle O^-}{|}}{P}}{-}O^-$$

Adding an OH group to the P and an H atom to the O completes the products of the hydrolysis:

$$CH_3\text{-}\underset{\underset{\displaystyle CH_3}{|}}{CH}{-}\underset{\underset{\displaystyle CH_3}{|}}{CH}{-}CH_2{-}O{-}\overset{\overset{\displaystyle O}{\|}}{\underset{\underset{\displaystyle O^-}{|}}{P}}{-}O{-}H \qquad HO{-}\overset{\overset{\displaystyle O}{\|}}{\underset{\underset{\displaystyle O^-}{|}}{P}}{-}O^-$$

Solutions to Section 12.5 Core Problems

12.29 *Draw the structures of the organic products that are formed when the following compounds are hydrolyzed under physiological conditions:*

$$a)\; CH_3{-}\underset{\underset{\displaystyle CH_3}{|}}{CH}{-}\underset{\underset{\displaystyle CH_3}{|}}{CH}{-}CH_2{-}\overset{\overset{\displaystyle O}{\|}}{C}{-}O{-}CH_3$$

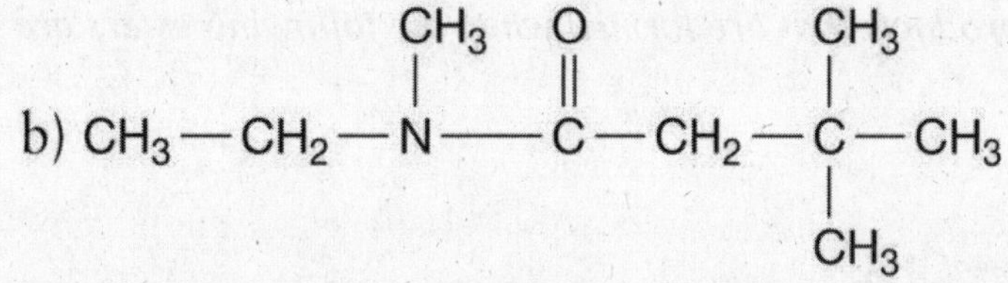

$$\text{b) } CH_3-CH_2-\underset{\displaystyle |}{\overset{\displaystyle CH_3}{\overset{|}{N}}}-\overset{\displaystyle O}{\overset{\|}{C}}-CH_2-\underset{\displaystyle CH_3}{\underset{|}{\overset{\displaystyle CH_3}{\overset{|}{C}}}}-CH_3$$

Under physiological conditions (around pH 7), carboxylic acids will actually exist in the carboxylate ion form, and amine functional groups will actually exist in the ammonium ion form. Other functional groups will appear the same as always (other functional groups are not acidic or basic enough to be ionized at pH 7.) The most straightforward thing to do is to draw the hydrolysis products as usual, then look for carboxylic acids and amines, and convert them to their conjugates.

a) Hydrolysis of an ester gives a carboxylic acid and an alcohol:

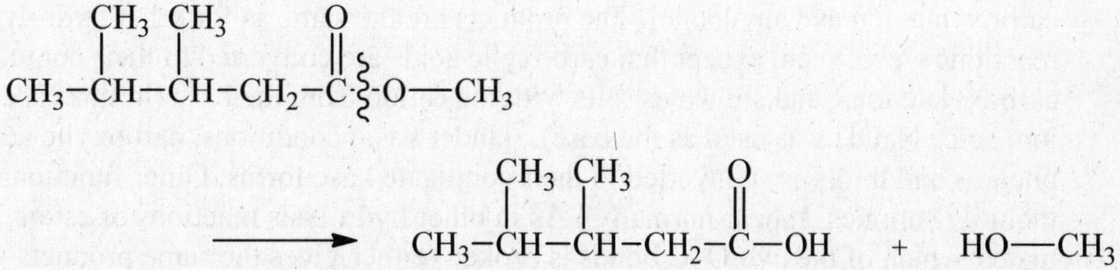

$$CH_3-\overset{\displaystyle CH_3}{\overset{|}{CH}}-\overset{\displaystyle CH_3}{\overset{|}{CH}}-CH_2-\overset{\displaystyle O}{\overset{\|}{C}}\wr O-CH_3$$

$$\longrightarrow CH_3-\overset{\displaystyle CH_3}{\overset{|}{CH}}-\overset{\displaystyle CH_3}{\overset{|}{CH}}-CH_2-\overset{\displaystyle O}{\overset{\|}{C}}-OH \quad + \quad HO-CH_3$$

At physiological pH, the carboxylic acid will appear as a carboxylate ion, but the alcohol group will be unchanged (alcohol groups are not acidic or basic):

$$CH_3-\overset{\displaystyle CH_3}{\overset{|}{CH}}-\overset{\displaystyle CH_3}{\overset{|}{CH}}-CH_2-\overset{\displaystyle O}{\overset{\|}{C}}-O^- \quad + \quad HO-CH_3$$

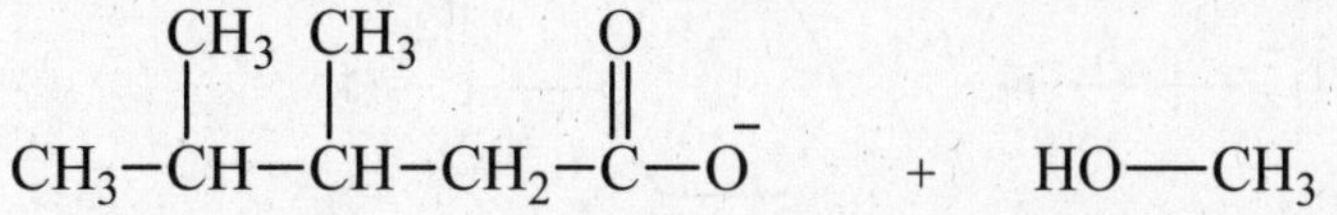

b) Hydrolysis of an amide gives an amine and a carboxylic acid. The bond that is broken in the hydrolysis must be the bond between the N atom and the carbonyl C atom:

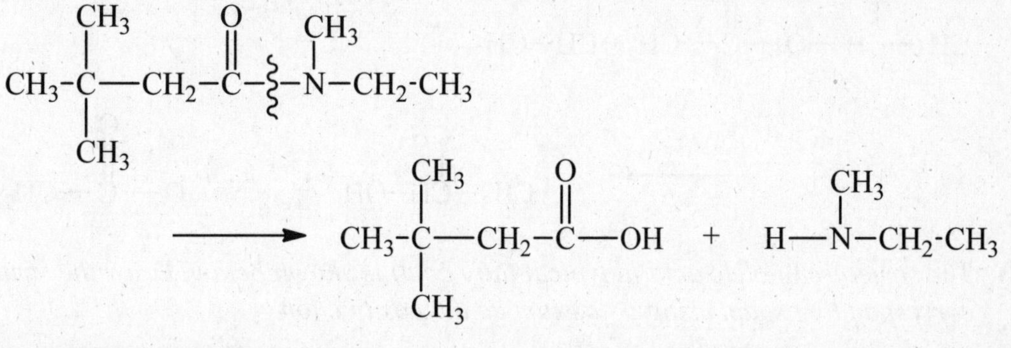

$$CH_3-\underset{\displaystyle CH_3}{\underset{|}{\overset{\displaystyle CH_3}{\overset{|}{C}}}}-CH_2-\overset{\displaystyle O}{\overset{\|}{C}}\wr\overset{\displaystyle CH_3}{\overset{|}{N}}-CH_2-CH_3$$

$$\longrightarrow CH_3-\underset{\displaystyle CH_3}{\underset{|}{\overset{\displaystyle CH_3}{\overset{|}{C}}}}-CH_2-\overset{\displaystyle O}{\overset{\|}{C}}-OH \quad + \quad H-\overset{\displaystyle CH_3}{\overset{|}{N}}-CH_2-CH_3$$

Under physiological conditions, carboxylic acids are converted to carboxylate ions, and amines are converted to ammonium ions:

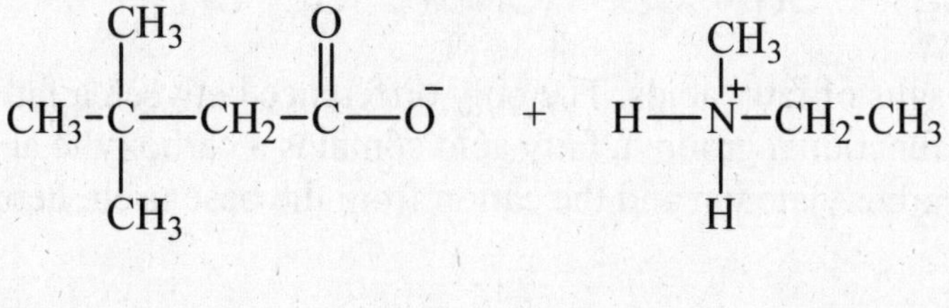

$$CH_3-\underset{\displaystyle CH_3}{\underset{|}{\overset{\displaystyle CH_3}{\overset{|}{C}}}}-CH_2-\overset{\displaystyle O}{\overset{\|}{C}}-O^- \quad + \quad H-\underset{\displaystyle H}{\underset{|}{\overset{\displaystyle CH_3}{\overset{|}{N^+}}}}-CH_2-CH_3$$

12.31 *Draw the structures of the products that are formed when the following esters are saponified using NaOH:*

a) 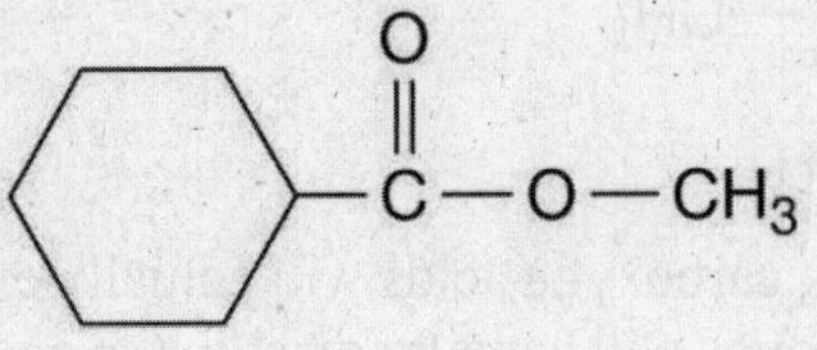 (cyclohexyl)$-\overset{O}{\overset{\|}{C}}-O-CH_3$

b) 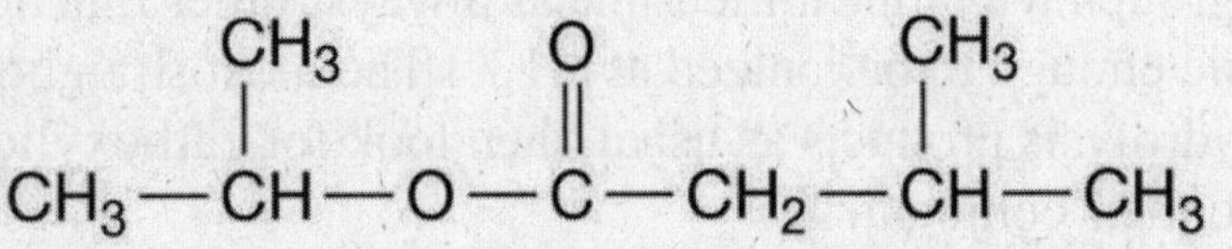 $CH_3-\underset{}{\overset{CH_3}{\overset{|}{CH}}}-O-\overset{O}{\overset{\|}{C}}-CH_2-\overset{CH_3}{\overset{|}{CH}}-CH_3$

"Saponification" is hydrolysis with a strong base; saponification of an ester gives a carboxylate ion and an alcohol. The products are the same as for other hydrolysis reactions we've seen, except that carboxylic acids are converted to their conjugate base carboxylate ions, and shown as salts with the cation from the base (in this case sodium ion, since NaOH was used as the base). (Under basic conditions, carboxylic acids, phenols and thiols are converted to their conjugate base forms. Other functional groups, including amines, appear normally.) As in other hydrolysis reactions of esters, it doesn't matter which of the two O–C bonds is broken (either gives the same products–try it!).

a)

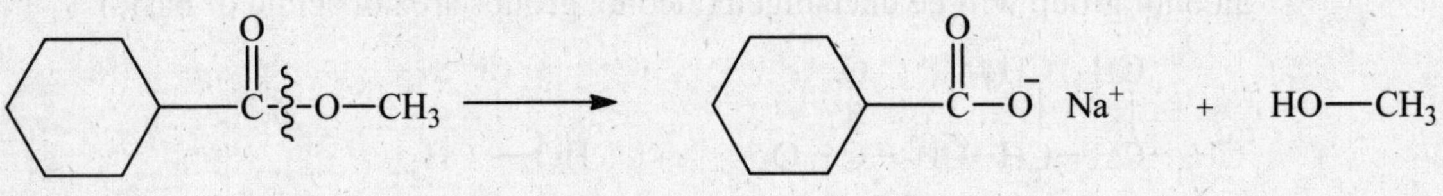

$$\text{(cyclohexyl)}-\overset{O}{\overset{\|}{C}}-O-CH_3 \longrightarrow \text{(cyclohexyl)}-\overset{O}{\overset{\|}{C}}-O^{-}\ Na^{+} + HO-CH_3$$

b)

$$CH_3-\overset{CH_3}{\overset{|}{CH}}-O-\overset{O}{\overset{\|}{C}}-CH_2-\overset{CH_3}{\overset{|}{CH}}-CH_3$$

$$\longrightarrow CH_3-\overset{CH_3}{\overset{|}{CH}}-OH + Na^{+}\ {}^{-}O-\overset{O}{\overset{\|}{C}}-CH_2-\overset{CH_3}{\overset{|}{CH}}-CH_3$$

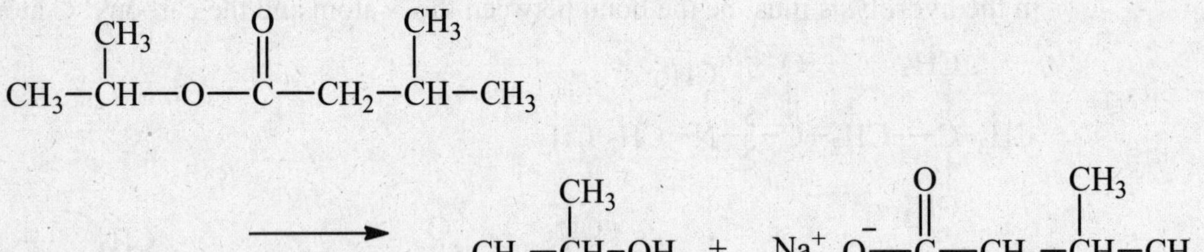

12.33 *The structure of oleic acid (a typical fatty acid) is shown below. Draw the structure of the corresponding soap, using potassium as the positive ion.*

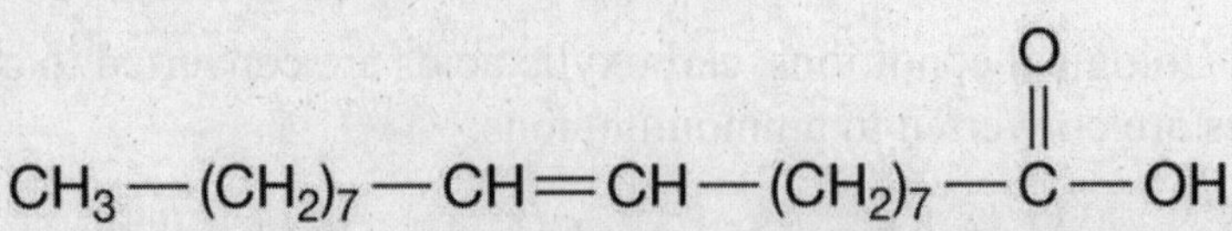

$$CH_3-(CH_2)_7-CH=CH-(CH_2)_7-\overset{O}{\overset{\|}{C}}-OH$$

"Soaps" are carboxylate salts of fatty acids. The only difference between a fatty acid and a soap is the form of the functional group: a fatty acid contains a carboxylic acid group, while a soap contains a carboxylate ion and the cation from the base used, here specified as K^+.

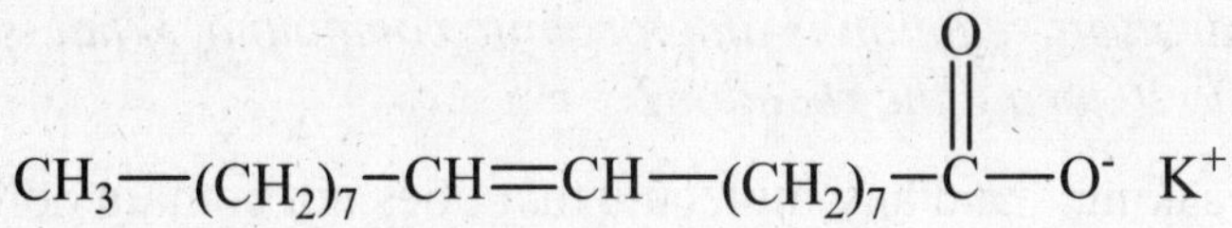

$$CH_3—(CH_2)_7—CH{=}CH—(CH_2)_7—\overset{\overset{O}{\|}}{C}—O^-\ K^+$$

Solutions to Concept Questions

* indicates more challenging problems

12.35 *Both the dehydration reaction and the condensation reaction remove water from organic molecules. How do these two reactions differ from each other?*

In a dehydration, the H and OH come from adjacent single-bonded C atoms in the same molecule, and a double bond is formed between the adjacent C atoms. In a condensation, the H and OH come from different molecules, and a single bond (usually C–O, C–N, C–S or P–O) is formed, coupling the two molecular fragments together.

12.37 *What functional group is formed when the following compounds condense?*

The best strategy here is not to try to memorize a sentence for each condensation, but to understand the reaction and know the functional groups. A quick sketch will then get you the right answer.

a) *two alcohols* Draw them:

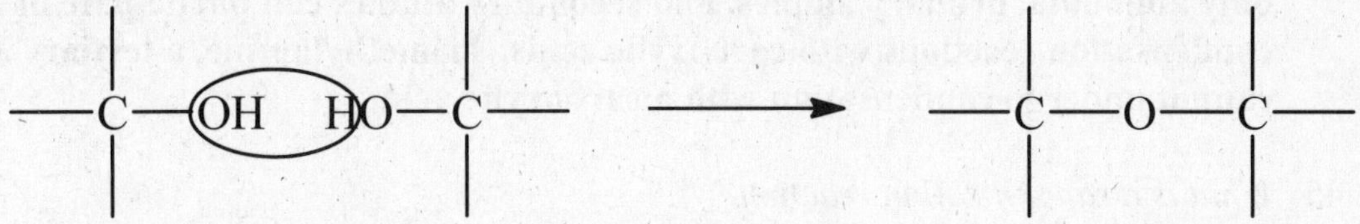

—C—OH HO—C— ⟶ —C—O—C—

alcohol + alcohol **ether**

b) *an alcohol and a carboxylic acid* Draw them:

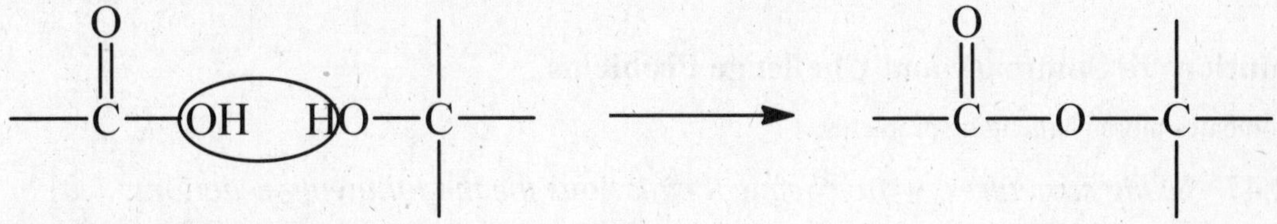

—C(=O)—OH HO—C— ⟶ —C(=O)—O—C—

carboxylic acid + alcohol **ester**

c) *a carboxylic acid and an amine* Draw them:

—C(=O)—OH H—N— ⟶ —C(=O)—N—

carboxylic acid + amine **amide**

12.39 *One of the reactants in any hydrolysis reaction is an inorganic compound. What is this compound, and what happens to it during the reaction?*

"Hydrolysis" means "water-breaking," and any molecule that does not contain both carbon and hydrogen is classified as "inorganic." **Every hydrolysis reaction involves a water molecule. In a hydrolysis, an organic molecule is broken into two fragments, and a water molecule is broken up into H and OH fragments. The H atom adds to an N or O atom in the fragment of the hydrolyzed organic molecule, and the OH group adds to a carbon atom in the other fragment of the organic molecule.**

12.41 *When an ester is hydrolyzed, the pH of the solution changes. Does the pH go up, or does it go down? Explain your answer.*

The hydrolysis of an ester produces a carboxylic acid and an alcohol. Since carboxylic acids are acidic (while esters and alcohols are neutral), the product solution will be more acidic than the reactant solution. Therefore the pH of the solution goes down when an ester is hydrolyzed.

12.43 *Explain why trimethylamine cannot condense with carboxylic acids.*

In a condensation reaction between a carboxylic acid and an amine, the OH group from the carboxylic acid and an H atom from the amine group combine to form water. If there is no H atom on the amine N atom, the condensation cannot occur, so only ammonia, primary amines, and secondary amines can participate in condensation reactions with carboxylic acids. Trimethylamine, a tertiary amine, cannot undergo condensation with a carboxylic acid.

12.45 *What is a saponification reaction?*

"Saponification" is the hydrolysis of an ester by a strong base, to produce a carboxylate salt and an alcohol. If the ester is a fat molecule, then the product of the hydrolysis is a soap. The term "saponification" comes from the Latin for soap.

Solutions to Summary and Challenge Problems

* indicates more challenging problems.

12.47 *Using structures, write chemical equations for the following reactions:*

a) *The dehydration of 2-propanol*

In a dehydration reaction, an H atom and an OH group are removed from adjacent single-bonded C atoms in the same molecule, forming a double bond. To make this easier to see, the 2-propanol molecule is shown with the OH group and an H on an adjacent carbon both pointing up, in a convenient orientation for removal:

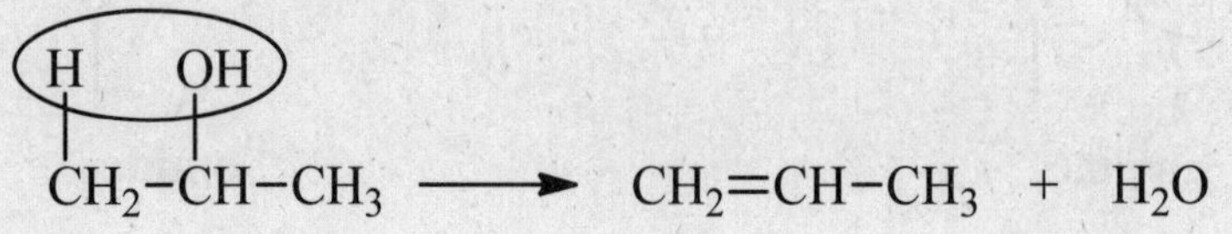

$$\underset{\text{H}}{\overset{|}{CH_2}}-\underset{\text{OH}}{\overset{|}{CH}}-CH_3 \longrightarrow CH_2=CH-CH_3 + H_2O$$

b) The condensation of two molecules of 2-propanol

In a condensation, the H and OH come from different molecules, and a C–O (as is the case here) or C–N single bond is formed, coupling the two remaining fragments together into a larger molecule. To make this easier to see, the 2-propanol molecules are shown with their OH groups facing each other, in a convenient orientation for condensation:

$$(CH_3)_2CH\text{–}OH \quad HO\text{–}CH(CH_3)_2 \longrightarrow (CH_3)_2CH\text{–}O\text{–}CH(CH_3)_2 + H_2O$$

12.49 *Malonic acid contains two carboxylic acid groups, so it can condense with two molecules of 1-propanol. Draw the structure of the organic product that will be formed in this reaction.*

$$HO\text{–}\overset{O}{\overset{\|}{C}}\text{–}CH_2\text{–}\overset{O}{\overset{\|}{C}}\text{–}OH$$

Malonic acid

In a condensation, an H atom is removed from an organic functional group on one molecule, and an OH group is removed from another. The reaction forms a water molecule and a C–O (as is the case here) or C–N single bond, coupling the two remaining molecular fragments together into a larger molecule. As always, draw the reactant molecules, line up the functional groups to face each other, and pick out the H and OH that will from the water molecules:

$$CH_3\text{-}CH_2\text{-}CH_2\text{-}OH \quad HO\text{–}\overset{O}{\overset{\|}{C}}\text{–}CH_2\text{–}\overset{O}{\overset{\|}{C}}\text{–}OH \quad HO\text{-}CH_2\text{-}CH_2\text{-}CH_3$$

$$\longrightarrow \mathbf{CH_3\cdot CH_2\cdot CH_2\text{-}O\text{–}\overset{O}{\overset{\|}{C}}\text{–}CH_2\text{–}\overset{O}{\overset{\|}{C}}\text{–}O\text{–}CH_2\cdot CH_2\cdot CH_3} + 2\,H_2O$$

*12.51 *a) Aspirin can be made by condensing salicylic acid with acetic acid to form an ester. Using this information, draw the structure of aspirin.*

$$C_6H_4(OH)\text{–}\overset{O}{\overset{\|}{C}}\text{–}OH$$

Salicylic acid

We draw the two molecules with their functional groups facing each other, and select the H and OH that will be removed to form a water molecule. The problem specifies that the product is an ester, which is the product of the condensation of a carboxylic acid and an alcohol, so we line up the acetic acid molecule (a carboxylic acid) with the phenol OH (which reacts in this case as an alcohol) on salicylic acid:

$$\text{C}_6\text{H}_4(\text{C(=O)–OH})(\text{OH}) + \text{HO–C(=O)–CH}_3 \longrightarrow \text{C}_6\text{H}_4(\text{C(=O)–OH})(\text{O–C(=O)–CH}_3) + H_2O$$

aspirin

NOTE:
We could also have lined up the acetic acid molecule with the carboxylic acid group on salicylic acid, but this would not have given us an ester as specified in the problem (the group that forms in the condensation of two carboxylic acids is called an "acid anhydride").

b) *Oil of wintergreen can be made by condensing salicylic acid with methanol to form an ester. Using this information, draw the structure of oil of wintergreen.*

We draw the two molecules with their functional groups facing each other, and select the H and OH that will be removed to form a water molecule. The problem specifies that the product is an ester, which is the product of the condensation of a carboxylic acid and an alcohol, so we line up the methanol molecule (an alcohol) with the carboxylic acid OH on salicylic acid.

$$\text{C}_6\text{H}_4(\text{C(=O)–OH})(\text{OH}) + \text{HO–CH}_3 \longrightarrow \text{C}_6\text{H}_4(\text{C(=O)–O–CH}_3)(\text{OH}) + H_2O$$

oil of wintergreen

NOTE:
We could also have lined up the methanol molecule with the phenol OH group on salicylic acid, but this would not have given us an ester as specified in the problem (the group that forms in the condensation of two alcohols is an ether).

12.53 *The following alcohol can condense with two phosphate ions. Draw the structure of the organic product of this reaction.*

$$HO–CH_2–CH_2–CH_2–OH$$

As usual, we draw the three structures, line up the –OH groups participating in the condensation, identify the H and OH to be removed, and draw the product, which in this case is a phosphoester:

$$^{-}O-\overset{\overset{O}{\|}}{\underset{\underset{O^{-}}{|}}{P}}-OH \quad HO-CH_2\text{-}CH_2\text{-}CH_2-OH \quad HO-\overset{\overset{O}{\|}}{\underset{\underset{O^{-}}{|}}{P}}-O^{-}$$

$$\longrightarrow \quad ^{-}O-\overset{\overset{O}{\|}}{\underset{\underset{O^{-}}{|}}{P}}-O-CH_2\text{-}CH_2\text{-}CH_2-O-\overset{\overset{O}{\|}}{\underset{\underset{O^{-}}{|}}{P}}-O^{-} \quad + \; 2\,H_2O$$

NOTE:
While the entire reaction is shown, the question only asks for the organic product.

12.55 *Draw the structure of the organic product that is formed when three molecules of the following compound condense to form a single molecule.*

$$HO-\underset{}{\overset{\overset{CH_3}{|}}{CH}}-CH_2-\overset{\overset{CH_3}{|}}{CH}-OH$$

As usual, we draw the three structures, line up the OH groups participating in the condensation, identify the H and OH to be removed, and draw the product, in this case an ether:

$$HO-\overset{\overset{CH_3}{|}}{CH}-CH_2\text{-}\overset{\overset{CH_3}{|}}{CH}-OH \quad HO-\overset{\overset{CH_3}{|}}{CH}-CH_2\text{-}\overset{\overset{CH_3}{|}}{CH}-OH \quad HO-\overset{\overset{CH_3}{|}}{CH}-CH_2\text{-}\overset{\overset{CH_3}{|}}{CH}-OH$$

$$\longrightarrow \quad HO-\overset{\overset{CH_3}{|}}{CH}-CH_2\text{-}\overset{\overset{CH_3}{|}}{CH}-O-\overset{\overset{CH_3}{|}}{CH}-CH_2\text{-}\overset{\overset{CH_3}{|}}{CH}-O-\overset{\overset{CH_3}{|}}{CH}-CH_2\text{-}\overset{\overset{CH_3}{|}}{CH}-OH \quad + 2\;H_2O$$

NOTE:
While the entire reaction is shown, the question only asks for the organic product.

12.57 *Asparagine is one of the naturally occurring amino acids. It can be used to make a polymer. Draw the structure of the compound that is formed when three molecules of asparagine condense. (Hint: the amide group in asparagine does not react.)*

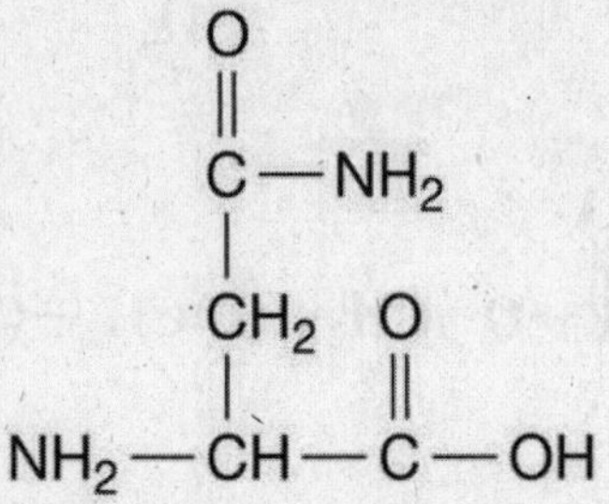

```
       O
       ||
       C—NH2
       |
       CH2 O
       |   ||
NH2—CH—C—OH
```

Asparagine

If the amide group does not react, that leaves a carboxylic acid group and an amine group to participate in the condensation reaction, which will form an amide functional group. Draw the structure three times, line up the groups participating in the condensation, identify the H and OH to be removed, and draw the product:

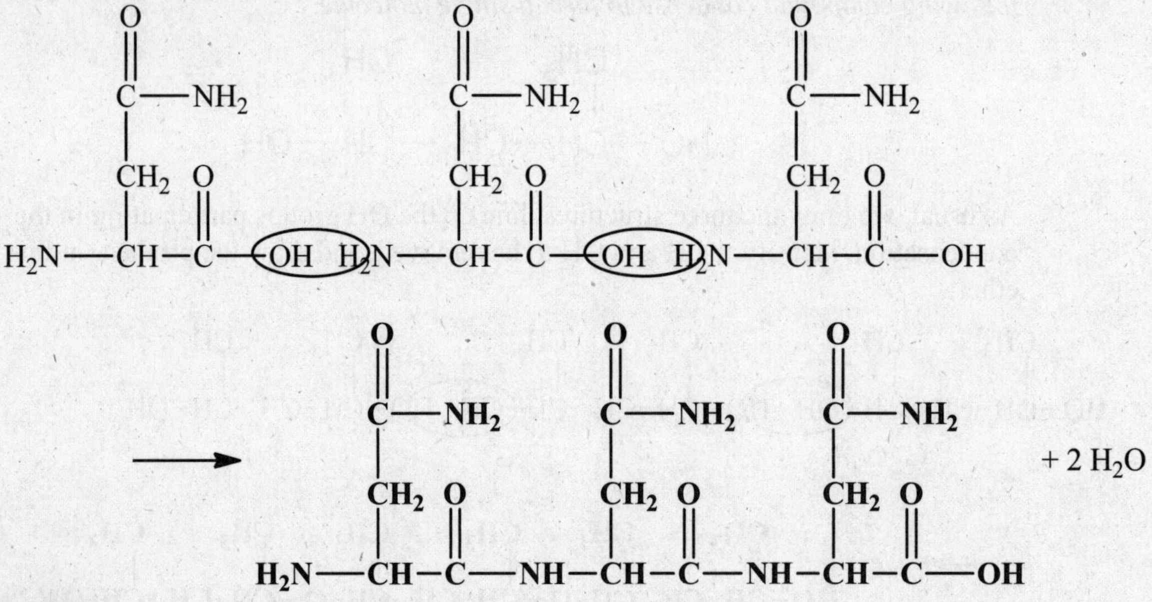

```
       O                O                O
       ||               ||               ||
       C—NH2            C—NH2            C—NH2
       |                |                |
       CH2 O            CH2 O            CH2 O
       |   ||           |   ||           |   ||
H2N—CH—C—(OH H)N—CH—C—(OH H)N—CH—C—OH
```

```
          O              O              O
          ||             ||             ||
          C—NH2          C—NH2          C—NH2
          |              |              |              + 2 H2O
  →       CH2 O          CH2 O          CH2 O
          |   ||         |   ||         |   ||
   H2N—CH—C—NH—CH—C—NH—CH—C—OH
```

NOTE:

While the entire reaction is shown, the question only asks for the structure of the compound formed.

*12.59 *The structures of the amino acids glycine and valine are shown here. You can make three different molecules by condensing two molecules of glycine with one molecule of valine. Draw the structures of these three compounds.*

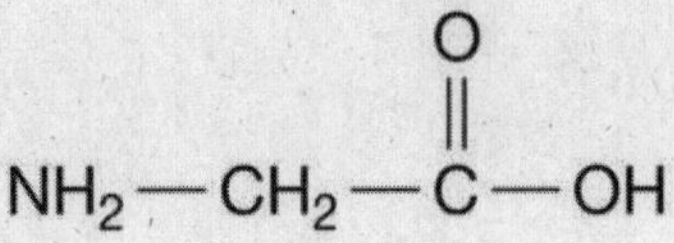

```
                                    CH3
                                    |
                O              CH3—CH  O
                ||                  |  ||
NH2—CH2—C—OH          NH2—CH—C—OH
```

Glycine 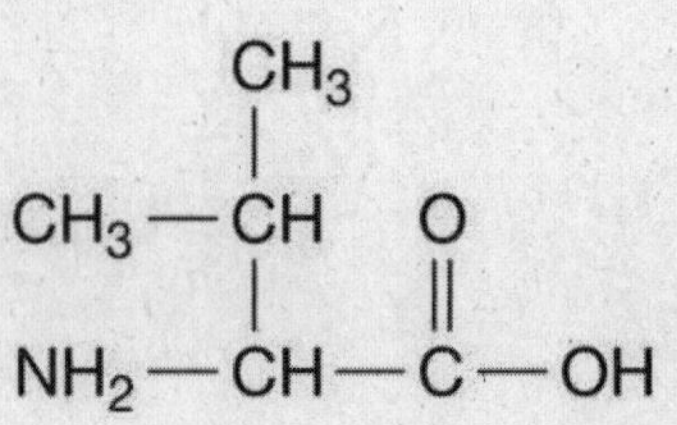 **Valine**

The different molecules depend on the order in which we connect the three molecules. Using the three-letter abbreviations for the amino acids, the options are (1) Gly-Gly-Val, (2) Gly-Val-Gly, and (3) Val-Gly-Gly. Here's the complete reaction for the first option:

$$H_2N-CH_2-\overset{\overset{O}{\|}}{C}-OH \quad H_2N-CH_2-\overset{\overset{O}{\|}}{C}-OH \quad H_2N-\underset{}{\overset{\overset{CH_3}{|}}{\overset{H_3C-CH}{|}}}{CH}-\overset{\overset{O}{\|}}{C}-OH$$

Glycine Glycine Valine

$$\longrightarrow H_2N-CH_2-\overset{\overset{O}{\|}}{C}-NH-CH_2-\overset{\overset{O}{\|}}{C}-NH-\overset{\overset{CH_3}{|}}{\overset{H_3C-CH}{|}}{CH}-\overset{\overset{O}{\|}}{C}-OH + 2\,H_2O$$

And here are the product structures for the other two combinations:

$$H_2N-CH_2-\overset{\overset{O}{\|}}{C}-NH-\overset{\overset{CH_3}{|}}{\overset{H_3C-CH}{|}}{CH}-\overset{\overset{O}{\|}}{C}-NH-CH_2-\overset{\overset{O}{\|}}{C}-OH$$

Glycine Valine Glycine

$$H_2N-\overset{\overset{CH_3}{|}}{\overset{H_3C-CH}{|}}{CH}-\overset{\overset{O}{\|}}{C}-NH-CH_2-\overset{\overset{O}{\|}}{C}-NH-CH_2-\overset{\overset{O}{\|}}{C}-OH$$

Valine Glycine Glycine

*12.61 *Two carboxylic acids can condense to form a compound called an anhydride. Using structures show the condensation reaction of two molecules of acetic acid.*

As with any other condensation, we draw the two molecules with their functional groups facing each other, identify the OH and H to be removed, and draw the product molecule by connecting the two fragments. The structure of an anhydride is like an ester with an extra carbonyl group:

$$CH_3-\overset{\overset{O}{\|}}{C}-OH \quad HO-\overset{\overset{O}{\|}}{C}-CH_3 \longrightarrow CH_3-\overset{\overset{O}{\|}}{C}-O-\overset{\overset{O}{\|}}{C}-CH_3 + H_2O$$

anhydride functional group

12.63 *Draw the structures of the products of the reactions in parts a through e of Problem 12.60 as they will actually appear at physiological pH.*

See Section 11.6, and in particular Table 11.5, for a review of the behavior of organic acids and bases at physiological pH. Under physiological conditions (around pH 7), carboxylic acids will actually exist in the carboxylate ion form, and amine functional groups will actually exist in the ammonium ion form. Organic phosphates can be shown with either two O^- groups, or one O^- and one OH (they are at equilibrium under physiological conditions, so either representation is acceptable). Other functional groups will appear the same as always (other functional groups, including phenols and thiols, are not acidic or basic enough to be ionized at pH 7.) The most straightforward thing to do is to draw the hydrolysis products as usual, then look for carboxylic acids and amines, and convert them to their conjugates.

a) Hydrolysis of an ether gives two alcohols. Alcohols are neither acidic nor basic and appear in their usual form at physiological pH:

$CH_3CH_2{\cdot}OH$ $\quad$ $HO{\cdot}CH_2CH_2CH_3$

b) Hydrolysis of an ester gives an alcohol and a carboxylic acid. At physiological pH the carboxylic acid will appear as the carboxylate ion. Alcohols are neither acidic nor basic and appear in their usual form:

$CH_3{-}CH(CH_3){-}OH$ $\quad$ $^-O{-}C(=O){-}H$

c) Hydrolysis of an ester gives an alcohol and a carboxylic acid. At physiological pH the carboxylic acid will appear as the carboxylate ion. Alcohols are neither acidic nor basic and appear in their usual form:

$CH_3{-}CH(CH_3){-}CH_2{-}CH(CH_3){-}C(=O){-}O^-$ $\quad$ $HO{-}CH_3$

d) Hydrolysis of an amide gives an amine and a carboxylic acid. This one is harder to see, so here are the unionized forms of the products first:

$CH_3{-}CH_2{-}C(=O){-}OH$ $\quad$ HN (pyrrolidine ring)

At physiological pH the carboxylic acid will appear as the carboxylate ion, while the amine will appear as the ammonium ion:

$CH_3{-}CH_2{-}C(=O){-}O^-$ $\quad$ H_2N^+ (pyrrolidinium ring)

e) Hydrolysis of an amide gives an amine and a carboxylic acid. The products in their unionized forms are:

$$CH_3-CH_2-NH_2 \qquad HO-\overset{O}{\overset{\|}{C}}-CH_2-CH_3$$

At physiological pH the carboxylic acid will appear as the carboxylate ion, while the amine will appear as the ammonium ion:

$$CH_3-CH_2-\overset{+}{N}H_3 \qquad {}^{-}O-\overset{O}{\overset{\|}{C}}-CH_2-CH_3$$

12.65 *Using structures, write the chemical equations for the hydrolysis reactions of each of the following compounds. Draw the organic products in their unionized forms.*

a) $CH_3-\overset{O}{\overset{\|}{C}}-S-CH_2-CH_3$

b) $C_6H_5-\overset{O}{\overset{\|}{C}}-O-\overset{O}{\overset{\|}{P}}(-O^-)-O^-$

"Unionized" (read as "un-ionized") means that we're ignoring any ionization of the molecules at physiological pH, and answering the question as we would have at the start of the chapter.

a) Hydrolysis of a thioester produces a carboxylic acid and a thiol. Break the C–S bond, add the OH group to the carbonyl C atom and the H atom to the S atom:

$$CH_3-\overset{O}{\overset{\|}{C}}-S-CH_2-CH_3 + H_2O \longrightarrow CH_3-\overset{O}{\overset{\|}{C}}-OH + HS-CH_2-CH_3$$

b) Hydrolysis of an organic phosphate produces a carboxylic acid and a phosphate ion. It doesn't matter whether we break the C–O bond or the O–P bond, as long as each ends up with an –OH group after the hydrolysis:

$$C_6H_5-\overset{O}{\overset{\|}{C}}-O-\overset{O}{\overset{\|}{P}}(-O^-)-O^- + H_2O \longrightarrow C_6H_5-\overset{O}{\overset{\|}{C}}-OH + HO-\overset{O}{\overset{\|}{P}}(-O^-)-O^-$$

NOTE:
The phosphate is shown as an ion; the problem specifies that organic products should be shown in their unionized forms, but phosphate ion is not organic.

12.67 *a) When the following compound is hydrolyzed, the products are three amino acids. Draw the structures of these amino acids in their unionized forms.*

Hydrolysis of amides always targets the bond between the carbonyl C atom and the N atom, producing a carboxylic acid and an amine. It may help to draw the unionized products first and then change groups to show their forms at physiological pH:

$$NH_2{-}CH(CH_3){-}C({=}O){-}NH{-}CH(CH_2{-}C({=}O){-}OH){-}C({=}O){-}NH{-}CH(CH_2{-}CH(CH_3){-}CH_3){-}C({=}O){-}OH \quad + 2\,H_2O$$

$$\longrightarrow NH_2{-}CH(CH_3){-}C({=}O){-}OH \;+\; NH_2{-}CH(CH_2{-}C({=}O){-}OH){-}C({=}O){-}OH \;+\; NH_2{-}CH(CH_2{-}CH(CH_3){-}CH_3){-}C({=}O){-}OH$$

(unionized products)

b) Draw the structures of the products in part a as they appear at physiological pH.

At physiological pH, all the carboxylic acid groups will appear as carboxylate ions, and all the amine groups will appear as ammonium ions:

$$\overset{+}{N}H_3{-}CH(CH_3){-}C({=}O){-}O^- \qquad \overset{+}{N}H_3{-}CH(CH_2{-}C({=}O){-}O^-){-}C({=}O){-}O^- \qquad \overset{+}{N}H_3{-}CH(CH_2{-}CH(CH_3){-}CH_3){-}C({=}O){-}O^-$$

Don't forget to ionize the second carboxylic acid group in the center molecule.

*12.69 *Compounds that contain an ester group within a ring of atoms are called lactones. Lactones can be hydrolyzed, but only one product is formed, rather than two. Draw the structure of the product that will be formed when the lactone below is hydrolyzed. You may draw the unionized form of the product.*

(structure: benzene ring fused to a five-membered lactone ring containing CH_2, $C{=}O$ and O)

This is just like any other hydrolysis of an ester. The only difference is that the product groups, a carboxylic acid group and an alcohol, are located in the same molecule. It may help to rotate the molecule into a position that makes the ester group a little more familiar:

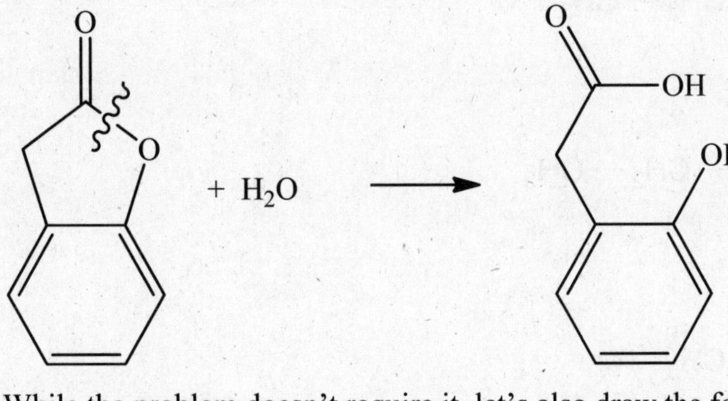

+ H_2O ⟶

While the problem doesn't require it, let's also draw the form of the molecule as it would appear at physiological pH, with the carboxylic acid group ionized to a carboxylate ion:

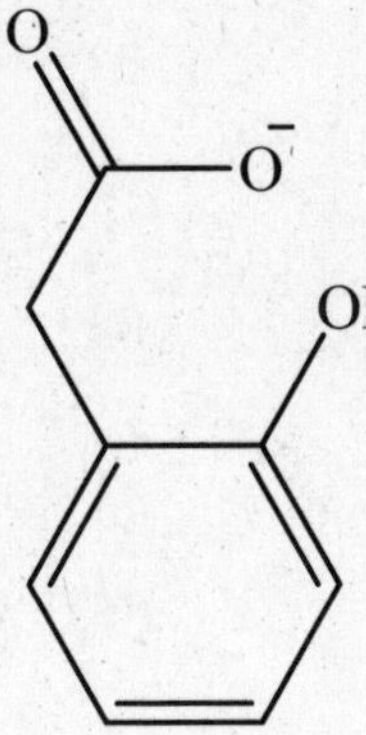

or

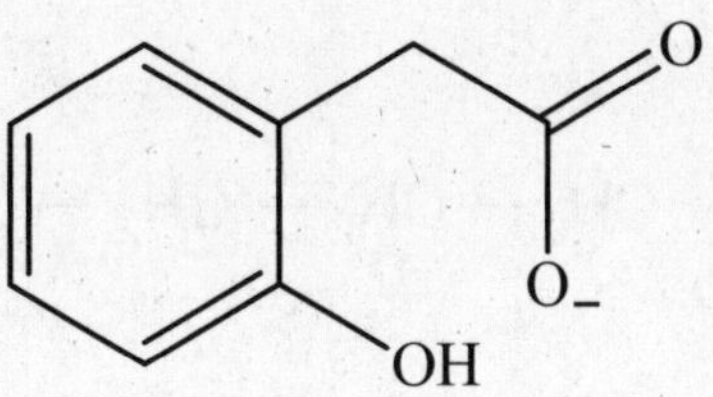

12.71 *From each of the following pairs of compounds, select the compound that has the higher solubility in water:*

a) $CH_3-CH_2-O-CH_2-CH_3$ or

$CH_3-CH_2-CH_2-CH_2-CH_3$

b) $CH_3-(CH_2)_3-O-(CH_2)_3-CH_3$ or

$CH_3-CH_2-O-CH_2-CH_3$

c) $CH_3-CH_2-CH_2-O-CH_2-CH_2-CH_3$ or

$CH_3-CH_2-O-CH_2-O-CH_2-CH_3$

$$CH_3-\overset{\overset{\displaystyle O}{\|}}{C}-O-CH_2-CH_3$$

d) 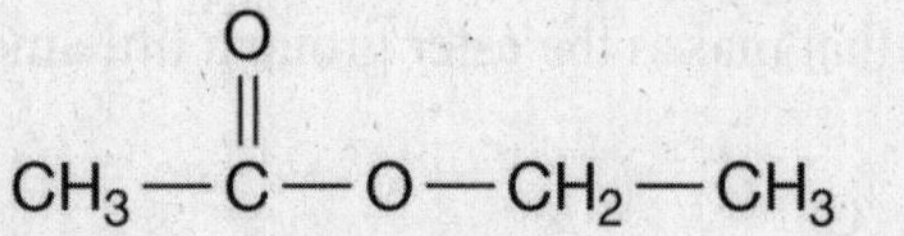 or

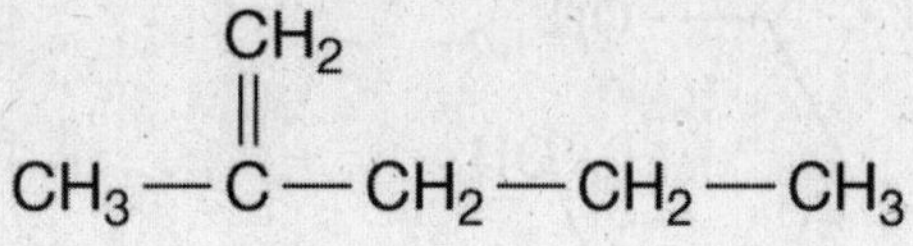

$$CH_3-\overset{\overset{\displaystyle CH_2}{\|}}{C}-CH_2-CH_2-CH_3$$

e) 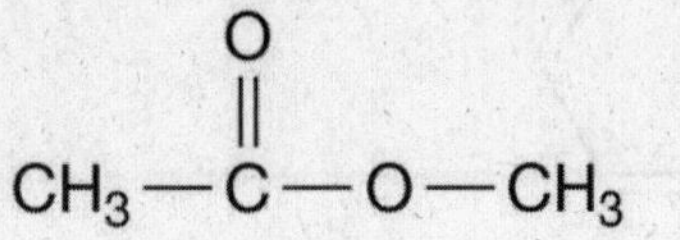 $$CH_3-\overset{\overset{\displaystyle O}{\|}}{C}-O-CH_3$$ or

$$CH_3-\overset{\overset{\displaystyle O}{\|}}{C}-O-(CH_2)_5-CH_3$$

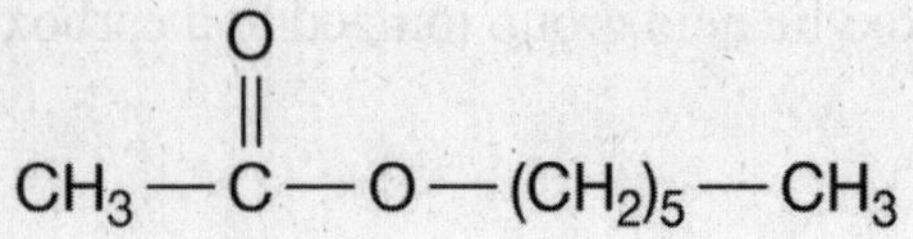

f) 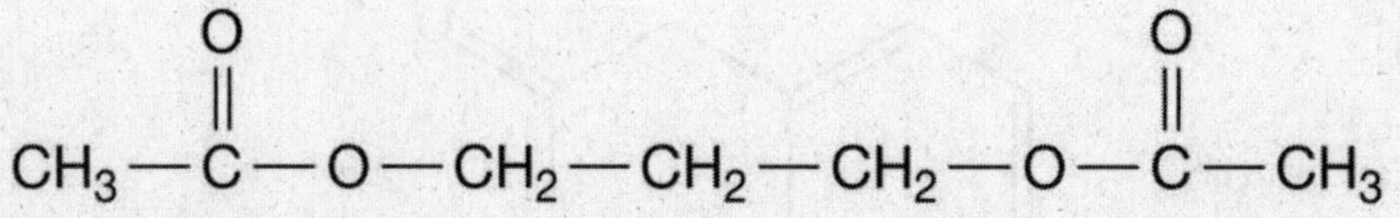 $$CH_3-\overset{\overset{\displaystyle O}{\|}}{C}-O-CH_2-CH_2-CH_2-O-\overset{\overset{\displaystyle O}{\|}}{C}-CH_3$$ or

$$CH_3-\overset{\overset{\displaystyle O}{\|}}{C}-O-CH_2-CH_2-CH_2-CH_2-CH_2-CH_3$$

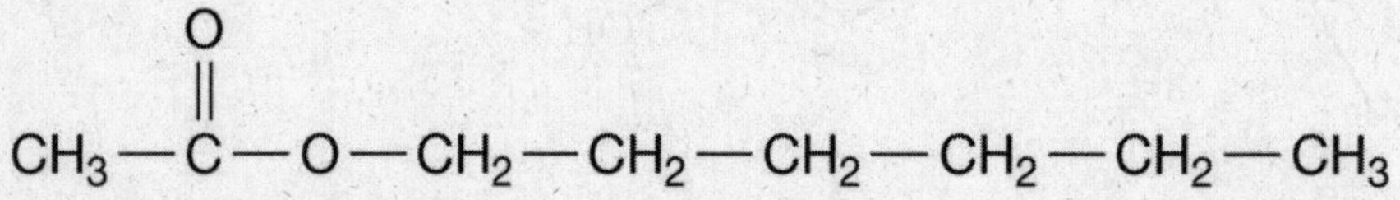

Chapter 5 review! Aqueous solubility of organic compounds depends on the balance between hydrophobic (non-hydrogen-bonding) and hydrophilic (hydrogen-bonding) areas of the molecule.

a) The ether group in diethyl ether is somewhat hydrophilic (it can be a hydrogen-bond acceptor), while pentane, a hydrocarbon, is completely hydrophobic. **Diethyl ether is more soluble in water than pentane is.**

b) Both molecules have (hydrophilic) ether functional groups, but dibutyl ether has much larger hydrophobic areas than does diethyl ether. **Diethyl ether is more soluble in water than dibutyl ether.**

c) Both molecules are about the same size, but the second molecule has two hydrophilic ether functional groups, while the first molecule, dipropyl ether, only has one. **The second molecule is more hydrophilic, so this compound is more soluble in water.**

NOTE:
It's beyond the scope of this text to come up with an IUPAC name for the second molecule!)

d) The first molecule has a hydrophilic, hydrogen-bond-accepting ester group; the second, 2-methyl-1-pentene, is completely hydrophobic (like all hydrocarbons). **The ester compound is more soluble in water than the hydrocarbon compound.**

e) Each molecule has an ester functional group, but the second molecule has much larger hydrophobic areas than the first. **The first compound is more soluble in water** because its hydrophobic area is smaller.

f) The two molecules are about the same size, but the first molecule has two ester functional groups while the second has only one. **The first compound will be more soluble in water** because it is more hydrophilic.

*12.73 *All of the compounds below are soluble in water. Which of these will produce an acidic solution, which will produce a basic solution, and which will produce a neutral solution?*

a) $CH_3-CH_2-CH_2-NH_2$

b) $CH_3-CH_2-C(=O)-NH_2$

c) $CH_3-C(=O)-O-CH_3$

d) $CH_3-C(=O)-OH$

e) CH_3-CH_2-OH

f) $CH_3-C(=O)-H$

g) 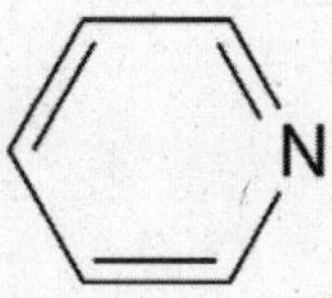 (pyridine ring with N)

h) 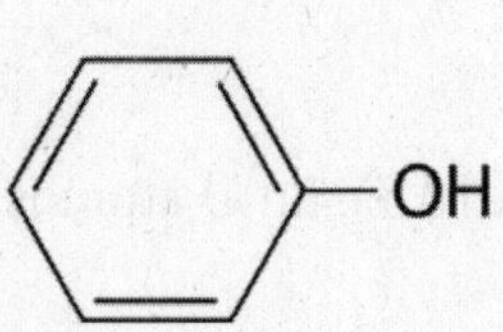 (benzene ring)—OH

Chapter 11 review! The acidic functional groups are carboxylic acids, phenols and thiols. Amines are basic. Other functional groups, including alcohols, esters and amides, are neutral, as are hydrocarbon groups.

a) Amine compound: **basic.**

b) Amide compound: **neutral.**

c) Ester compound: **neutral.**

d) Carboxylic acid compound: **acidic.**

e) Alcohol compound: **neutral.**

f) Aldehyde compound: **neutral.**

g) Amine compound: **basic.**

h) Phenol: **acidic.**

*12.75 *Complete the following sequence of reactions by drawing the structures of compounds A through D. Hint: compound D has the molecular formula* $C_4H_8O_2$.

$$CH_2=CH_2 + H_2O \longrightarrow \text{Compound A}$$

$$\text{Compound A} \xrightarrow{\text{Oxidation}} \text{Compound B}$$

$$\text{Compound B} \xrightarrow{\text{Oxidation}} \text{Compound C}$$

$$\text{Compound A} + \text{Compound C} \xrightarrow{\text{Condensation}} \text{Compound D} + H_2O$$

The first step is the hydration of an alkene (section 9.1), forming an alcohol (remember that in a hydration reaction, we add an OH group to one of the carbon atoms in a double bond and an H atom to the other, forming a C–C single bond):

OH H
| |
$CH_2{=}CH_2 + H_2O \longrightarrow CH_2{-}CH_2$
compound A

The second step is the oxidation of an alcohol to form a carbonyl group (section 10.2). Since compound A is a primary alcohol, the product is an aldehyde (remember that in the oxidation of an alcohol, two H atoms are removed, one from the alcohol O and one from the carbon to which the alcohol group is attached, forming a C=O double bond. Since we no longer need to see the H atom that was added to the carbon atom in Step 1, it has been combined with the other H atoms and shown as $-CH_3$):

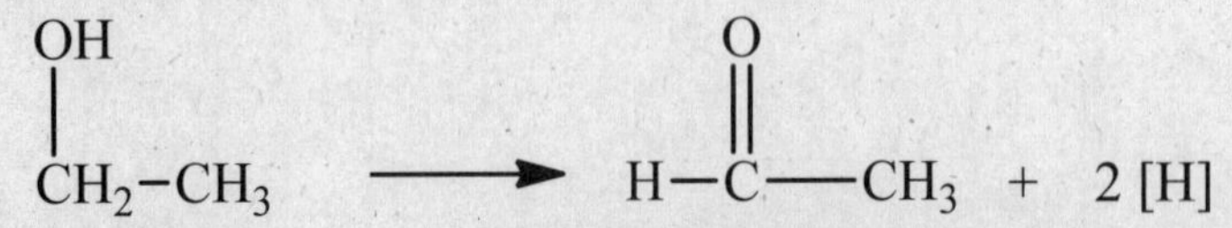

OH O
| ||
$CH_2{-}CH_3 \longrightarrow H{-}C{-}CH_3 + 2\,[H]$
compound A compound B

Oxidation of an aldehyde forms a carboxylic acid (section 10.5); an O atom is inserted between the carbonyl C atom and its attached H:

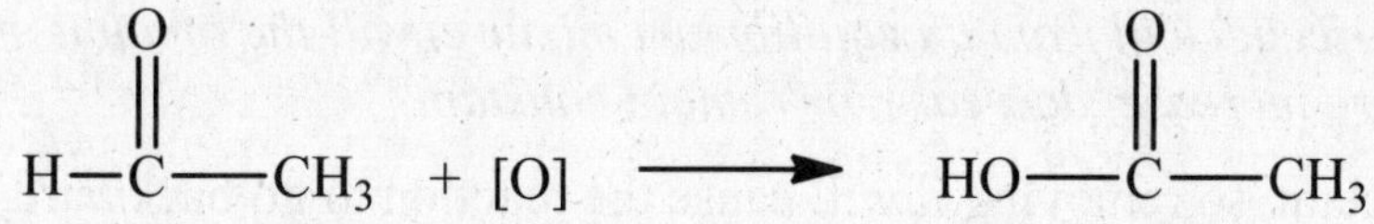

$$\mathrm{H-\overset{\overset{\displaystyle O}{\|}}{C}-CH_3 \; + [O] \longrightarrow HO-\overset{\overset{\displaystyle O}{\|}}{C}-CH_3}$$

compound B compound C

The reaction of compound C, a carboxylic acid, and compound A, an alcohol, will be a condensation reaction, forming an ester functional group. The compound A molecule has been turned around to make the condensation reaction easier to see; make sure that you can see that it's the same molecule as in Step 1:

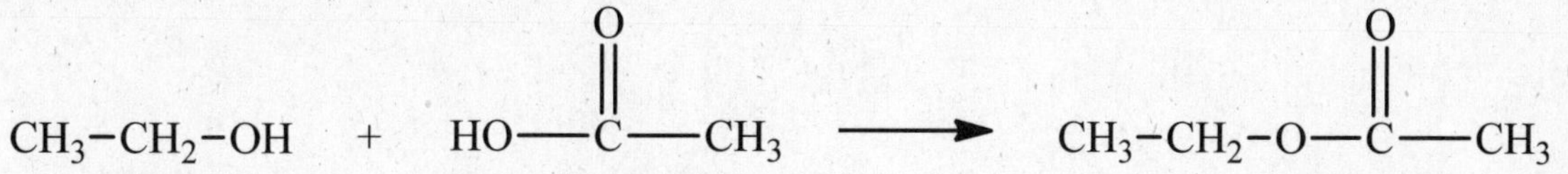

$$\mathrm{CH_3-CH_2-OH \; + \; HO-\overset{\overset{\displaystyle O}{\|}}{C}-CH_3 \longrightarrow CH_3-CH_2-O-\overset{\overset{\displaystyle O}{\|}}{C}-CH_3}$$

compound A compound C compound D

*12.77 *The following condensation reaction forms an equilibrium mixture:*

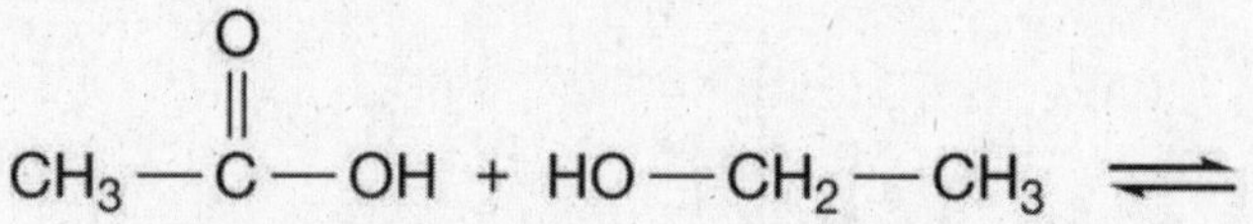

$$\mathrm{CH_3-\overset{\overset{\displaystyle O}{\|}}{C}-OH + HO-CH_2-CH_3 \rightleftharpoons}$$

Acetic acid **Ethanol**

$$\mathrm{CH_3-\overset{\overset{\displaystyle O}{\|}}{C}-O-CH_2-CH_3 + H_2O}$$

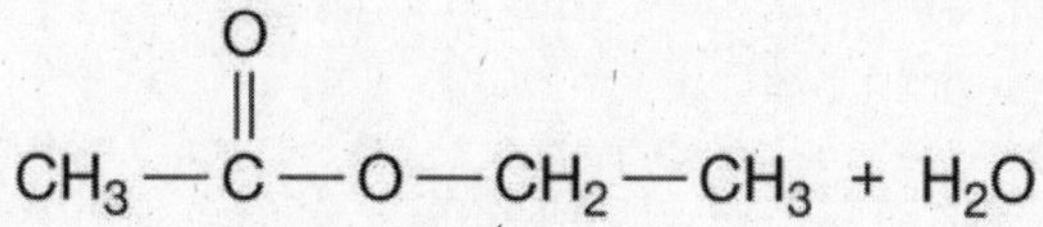

Ethyl acetate

Chapter 6 review! Remember LeChâtelier's principle: *when we disturb an equilibrium, the reaction will go in the direction that counteracts the disturbance.* Adding reactants, or removing products, will cause the reaction to go forward to restore equilibrium; removing reactants, or adding products, will cause the reaction to go backward to restore equilibrium.

a) *If some acetic acid is added to an equilibrium mixture, will the amount of ethyl acetate in the mixture increase, decrease, or remain constant?*

   Acetic acid is a reactant, so adding it will cause the reaction to go forward, **increasing the amount of ethyl acetate.**

b) *If some ethanol is added to an equilibrium mixture, will the amount of acetic acid in the mixture increase, decrease, or remain constant?*

   Ethanol is a reactant, so adding it will cause the reaction to go forward, using up some acetic acid. **The amount of acetic acid will decrease.**

c) *If you remove some acetic acid from an equilibrium mixture, will the amount of ethyl acetate in the mixture increase, decrease, or remain constant?*

Acetic acid is a reactant, so removing it will cause the reaction to go backward, **decreasing the amount of ethyl acetate.**

Chapter 13

13

Proteins

Solutions to Section 13.1 Core Problems

13.1 *The unionized form of the amino acid valine is shown here. Identify each of the following pieces of the molecule:*

a) *The amino group*

The amino group is the $-NH_2$ group on the amino acid. Remember (Chapter 11) that while "amine" refers to any nitrogen with one or more alkyl groups attached, an "amino" group is specifically a primary amine or $-NH_2$ group. (The amino acid proline is an exception to this.)

b) *The acid group*

The acid group is the carboxylic acid attached to the alpha carbon. (Aspartic acid and glutamic acid also have a second carboxylic acid group, but because it is not attached directly to the alpha carbon, it's not the acid group referred to in the "amino acid" name.)

c) *The α-carbon atom*

The alpha carbon atom connects the amino group to the acid group. In the twenty common amino acids, the alpha carbon has at least one hydrogen atom attached to it.

d) *The side chain*

The side chain is the other group of atoms (besides the acid group and the amino group) attached to the alpha carbon atom. In valine, it's an isopropyl group, but in other amino acids it can be a hydrogen atom or an alkyl group, or can include other functional groups.

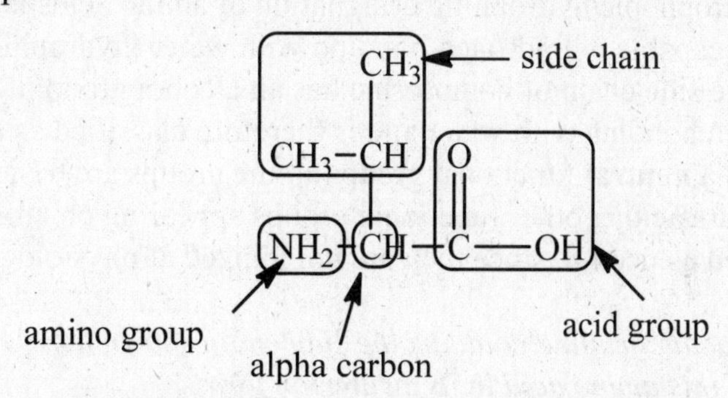

13.3 *Homoserine is an amino acid that is not used to build proteins, but it is produced when some of the normal amino acids are broken down.*

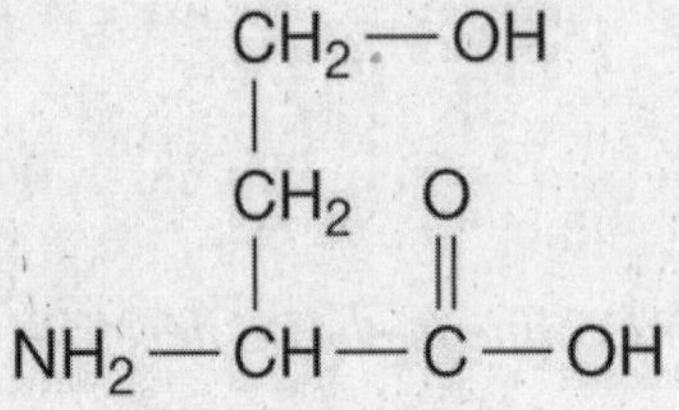

```
      CH2—OH
      |
      CH2  O
      |    ||
NH2—CH—C—OH
```

Homoserine

a) *Identify the side chain in homoserine.*

The side chain of any amino acid is the group attached to the alpha carbon atom:

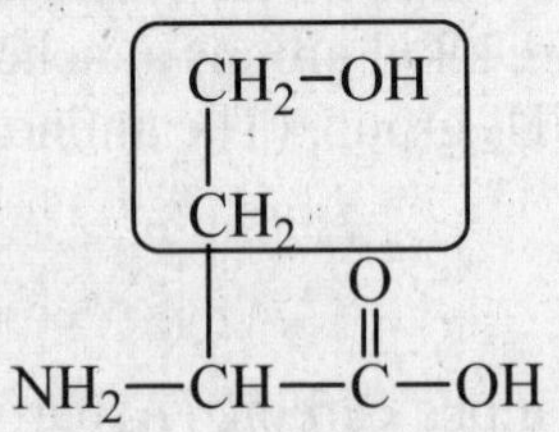

```
      CH2–OH
      |
      CH2
      |    O
      |    ||
NH2—CH—C—OH
```

b) *Draw the structure of homoserine in the zwitterion form.*

All amino acids behave the same way at physiological pH: amine groups are converted to their ammonium ion form, and carboxylic acid groups are converted to their carboxylate ion form.

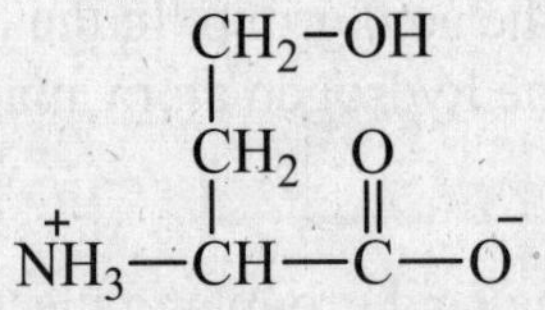

```
       CH2–OH
       |
       CH2  O
 +     |    ||    −
NH3—CH—C—O
```

c) *Is homoserine a hydrophilic or a hydrophobic amino acid? If it is hydrophilic, is it acidic, basic, or neutral?*

The hydrophobic/hydrophilic designation of amino acids is based on whether the side chain is capable of hydrogen bonding with water (hydrophilic) or not (hydrophobic). Since the side chain of homoserine has an alcohol group, it can participate in hydrogen bonding with water and is therefore classified as **hydrophilic**. The alcohol group is a **neutral** functional group (amine groups are basic and carboxylic acid groups are acidic; other functional groups appearing on amino acid side chains are classified as neutral, since they are not ionized at physiological pH.)

13.5 *The amino acid citrulline contains the side chain shown here. Draw the complete structure of this amino acid in its zwitterion form.*

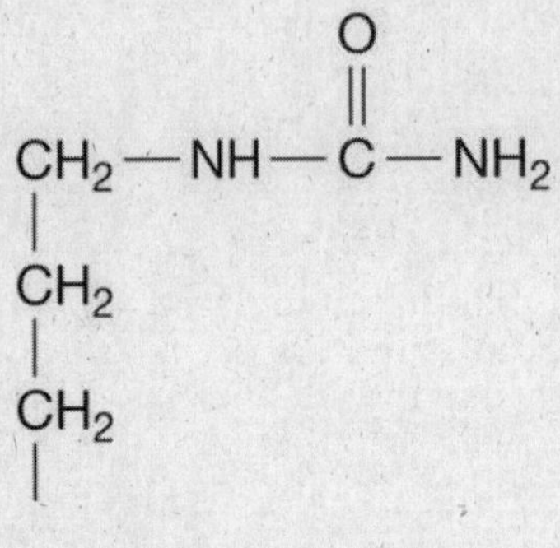

```
                O
                ||
CH2—NH—C—NH2
|
CH2
|
CH2
|
```

The side chain of citrulline (as it appears at pH 7)

All amino acids have the same basic structure of the alpha carbon and the amino and acid groups attached to it. Since the problem specifies the zwitterion form, we show the amine group converted to its ammonium ion form, and the carboxylic acid group in its carboxylate ion form:

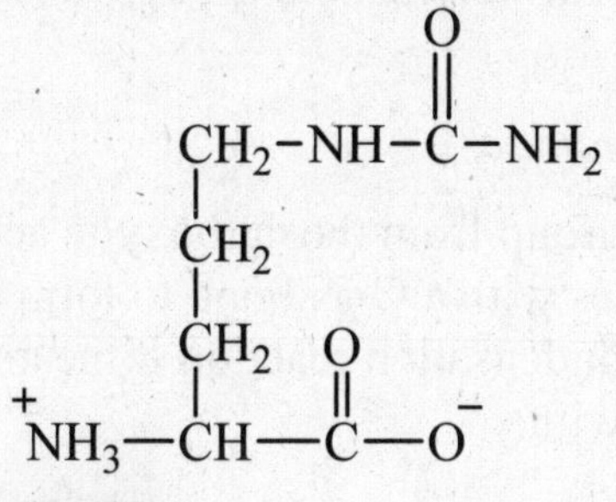

```
                     O
                     ‖
   CH2–NH–C–NH2
   |
   CH2
   |
   CH2  O
+  |    ‖    −
NH3—CH—C—O
```

NOTE:

The side chain contains an amide, not a normal amine; amides are not ionized at pH 7, and the figure actually specifies that the side chain is shown "as it appears at pH 7," so we don't change its form in the zwitterion.

13.7 *Lysine is a weak acid under physiological conditions. Explain why lysine is classified as a basic amino acid. (The structure of lysine is shown in Table 13.1.)*

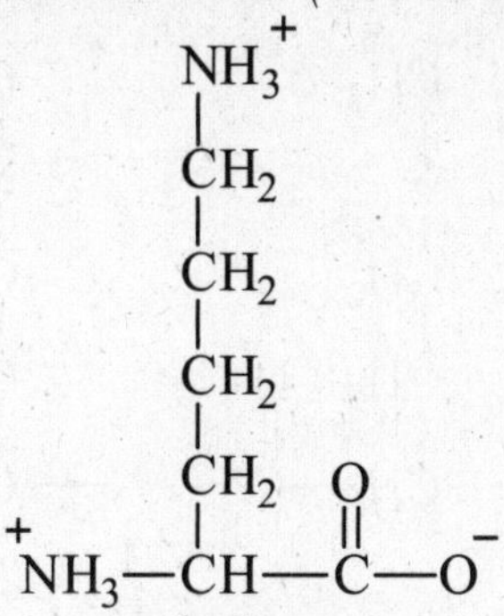

```
        +
     NH3
     |
     CH2
     |
     CH2
     |
     CH2
     |
     CH2  O
+    |    ‖    −
NH3—CH—C—O
     lysine
```

Amino acids are classified as acidic, basic or neutral according to the functional groups on their side chains, and the classifications refer to the *unionized* forms. **Lysine in its unionized form has an amine group, a basic functional group, on the side chain; at pH 7 the basic amine group is converted to the weakly acidic ammonium ion form.**

13.9 *Threonine (see Table 13.1) contains two chiral carbon atoms. Draw the structure of threonine, and circle the two chiral carbon atoms in this amino acid.*

Chapter 9 review! Any carbon atom that has four *different* atoms or groups bonded to it is a chiral carbon atom. The alpha carbon atom in an amino acid is usually chiral (glycine is an exception, since its alpha carbon atom has two H atoms attached). Threonine also has one chiral carbon atom in its side chain:

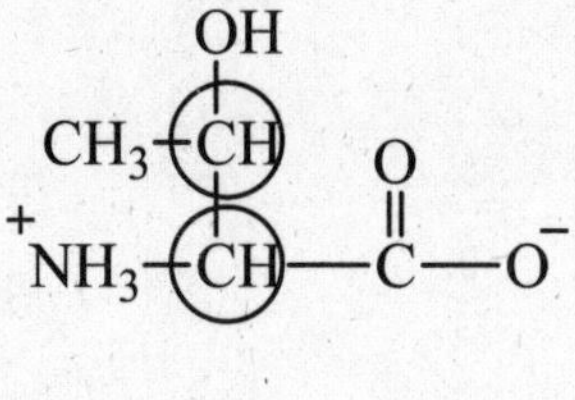

```
        OH
        |
   CH3–(CH)   O
+       |     ‖    −
NH3–(CH)—C—O
```

Solutions to Section 13.2 Core Problems

13.11 *Draw the structures of the following molecules as they appear at pH 7:*

Polypeptides are always assembled by lining up the amino acids involved and using condensation reactions between the amine groups and carboxylic acid groups to form amide (peptide) groups.

a) *The dipeptide that contains two molecules of methionine.*

Draw two molecules of methionine, take an –OH group from the carboxylic acid and an –H from the amino group, and link the molecules with a C–N bond to form a peptide group. It's easier to see the setup for the condensation reaction if the amino acids are drawn in their unionized forms to begin with:

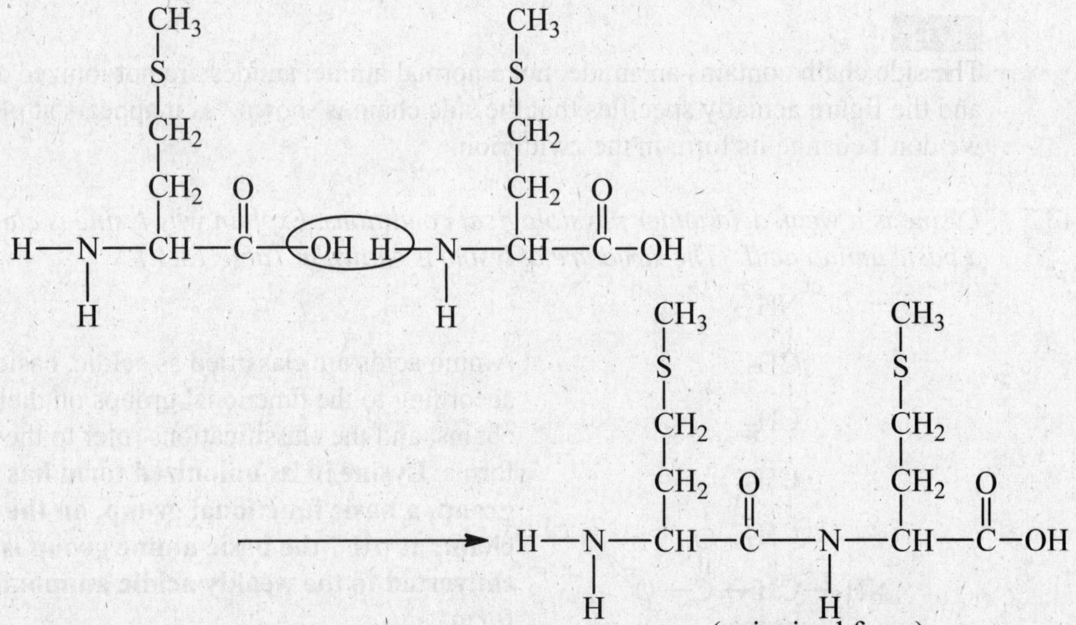

$$\mathrm{H{-}N(H){-}CH(CH_2CH_2SCH_3){-}C({=}O){-}OH \;\; H{-}N(H){-}CH(CH_2CH_2SCH_3){-}C({=}O){-}OH}$$

$$\longrightarrow \mathrm{H{-}N(H){-}CH(CH_2CH_2SCH_3){-}C({=}O){-}N(H){-}CH(CH_2CH_2SCH_3){-}C({=}O){-}OH}$$

(unionized form)

Then, since the problem specifies showing the molecule as it appears at pH 7, we can draw the zwitterion form, showing amines in their ammonium ion forms and carboxylic acids in their carboxylate ion forms:

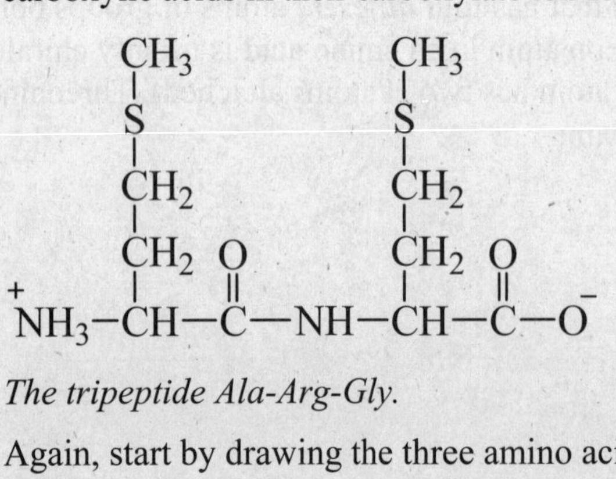

$$\mathrm{\overset{+}{N}H_3{-}CH(CH_2CH_2SCH_3){-}C({=}O){-}NH{-}CH(CH_2CH_2SCH_3){-}C({=}O){-}O^-}$$

b) *The tripeptide Ala-Arg-Gly.*

Again, start by drawing the three amino acids in their unionized forms (when you get good at seeing these, you can skip this step and go straight to attaching the carbonyl carbon on one amino acid and the amino nitrogen on the other. This step does not have to be shown, it just makes it easier to see the condensation reaction). Take the

–OH from the carboxylic acid group and an –H atom from the amino group, and link the carbon and nitrogen atoms to form peptide groups.

NOTE:
Since the side chain in Arg is not involved in the reaction, there's no benefit to showing the side chain in its unionized form, so it is left in the ionized form:

$$\underset{\text{Ala}}{H-\underset{H}{N}-\overset{CH_3}{CH}-\overset{O}{\overset{\|}{C}}-OH} \quad \underset{\text{Arg}}{H-\underset{H}{N}-\overset{\overset{+}{N}H_2=C(-NH_2)-NH-CH_2-CH_2-CH_2}{CH}-\overset{O}{\overset{\|}{C}}-OH} \quad \underset{\text{Gly}}{H-\underset{H}{N}-CH_2-\overset{O}{\overset{\|}{C}}-OH}$$

$$\longrightarrow H-\underset{H}{N}-\overset{CH_3}{CH}-\overset{O}{\overset{\|}{C}}-NH-\overset{\overset{+}{N}H_2=C(-NH_2)-NH-CH_2-CH_2-CH_2}{CH}-\overset{O}{\overset{\|}{C}}-NH-CH_2-\overset{O}{\overset{\|}{C}}-OH$$

Then we redraw the tripeptide product as a zwitterion, with the amino group in its ammonium ion form and the carboxylic acid group in its carboxylate ion form:

$$\overset{+}{N}H_3-\overset{CH_3}{CH}-\overset{O}{\overset{\|}{C}}-NH-\overset{\overset{+}{N}H_2=C(-NH_2)-NH-CH_2-CH_2-CH_2}{CH}-\overset{O}{\overset{\|}{C}}-NH-CH_2-\overset{O}{\overset{\|}{C}}-O^-$$

13.13 *The structure of a tripeptide is shown here.*

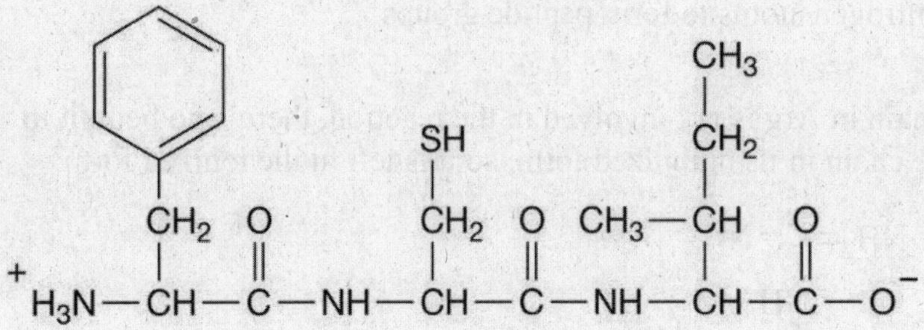

$$\overset{+}{H_3N}-\underset{\substack{|\\CH_2-C_6H_5}}{CH}-\overset{O}{\overset{\|}{C}}-NH-\underset{\substack{|\\CH_2SH}}{CH}-\overset{O}{\overset{\|}{C}}-NH-\underset{\substack{|\\CH(CH_3)CH_2CH_3}}{CH}-\overset{O}{\overset{\|}{C}}-O^-$$

a) *Circle the backbone of this tripeptide.*

The backbone of any polypeptide is the string made of the amino groups, alpha carbons, and carboxylic acids of the constituent amino acids, linked together by peptide bonds–that is, everything but the side chains:

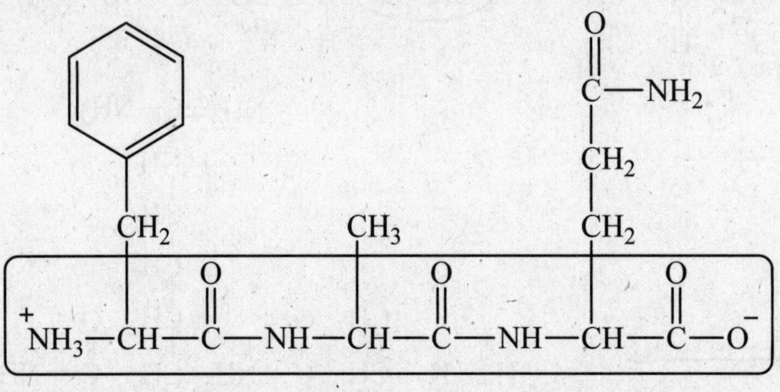

$$\overset{+}{NH_3}-\underset{\substack{|\\CH_2-C_6H_5}}{CH}-\overset{O}{\overset{\|}{C}}-NH-\underset{\substack{|\\CH_3}}{CH}-\overset{O}{\overset{\|}{C}}-NH-\underset{\substack{|\\CH_2CH_2C(=O)NH_2}}{CH}-\overset{O}{\overset{\|}{C}}-O^-$$

b) *Which amino acids were used to make this tripeptide?*

We identify the component amino acids by their side chains, using Table 13.1. **The side chain on the first amino acid is that of phenylalanine; the second is alanine; and the third is glutamine.**

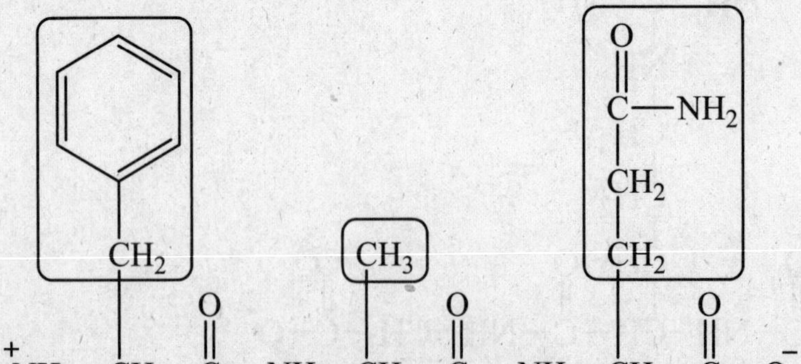

$$\overset{+}{NH_3}-\underset{\substack{|\\\boxed{CH_2-C_6H_5}}}{CH}-\overset{O}{\overset{\|}{C}}-NH-\underset{\substack{|\\\boxed{CH_3}}}{CH}-\overset{O}{\overset{\|}{C}}-NH-\underset{\substack{|\\\boxed{CH_2CH_2C(=O)NH_2}}}{CH}-\overset{O}{\overset{\|}{C}}-O^-$$

c) *What is the N-terminal amino acid in this tripeptide?*

"Terminal" means "end," and "N-terminal" means "the end with the nitrogen," the end of the polypeptide that has an ammonium ion group rather than a carboxylate group. In this case, **phenylalanine** is the N-terminal amino acid, and glutamine is the C-terminal amino acid. By convention the N-terminal group is always written on the left end of the structure.

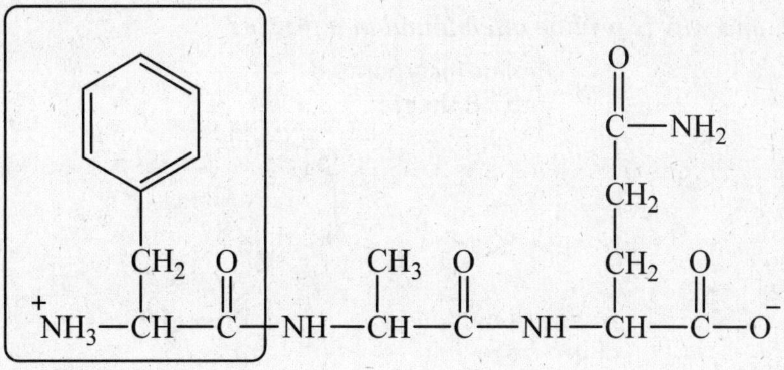

13.15 *What is the primary structure of a protein?*

The primary structure of a protein or polypeptide is simply the list of what amino acids are connected in what order.

13.17 *Describe the shape of the polypeptide backbone and the locations of the side chains in an α-helix.*

In an alpha helix, the polypeptide backbone twists into a spiral (think of a coiled spring). The side chains on each loop form hydrogen bonding interactions with the side chains on the next loops (above and below), as in Figure 13.6 in the text.

13.19 *In a peptide group, which two atoms form hydrogen bonds with other peptide groups?*

See Chapter 4 for a review of hydrogen bonding if you need to at this time. Hydrogen bond *donors* contain hydrogen atoms that are covalently bonded to O or N (these H atoms will have δ+ charges); hydrogen bond *acceptors* contain O or N atoms with δ– charges. In the peptide group, the N atom is bonded to a H atom (except in the case of proline); it is therefore a hydrogen bond donor. The N atom has a δ– charge, while the H atom bonded to it has a δ+ charge. In the carbonyl group, the O atom has a δ– charge. **The two atoms that form hydrogen bonds between peptide groups are O and H: the carbonyl O atom on one peptide group forms a hydrogen bond with the H atom on nitrogen in another peptide group.**

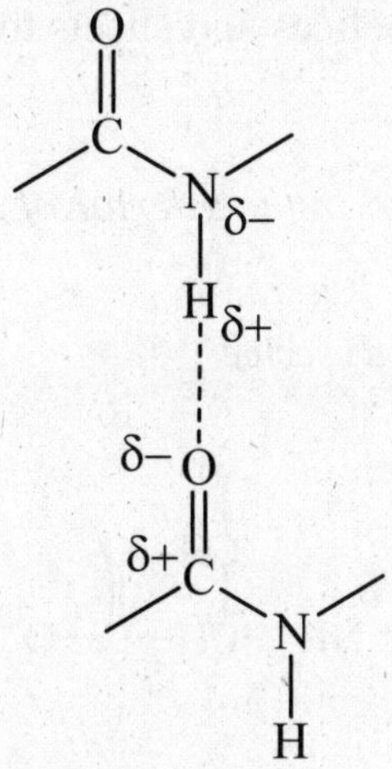

13.21 *What is a β-turn, and why is proline often found in a β-turn?*

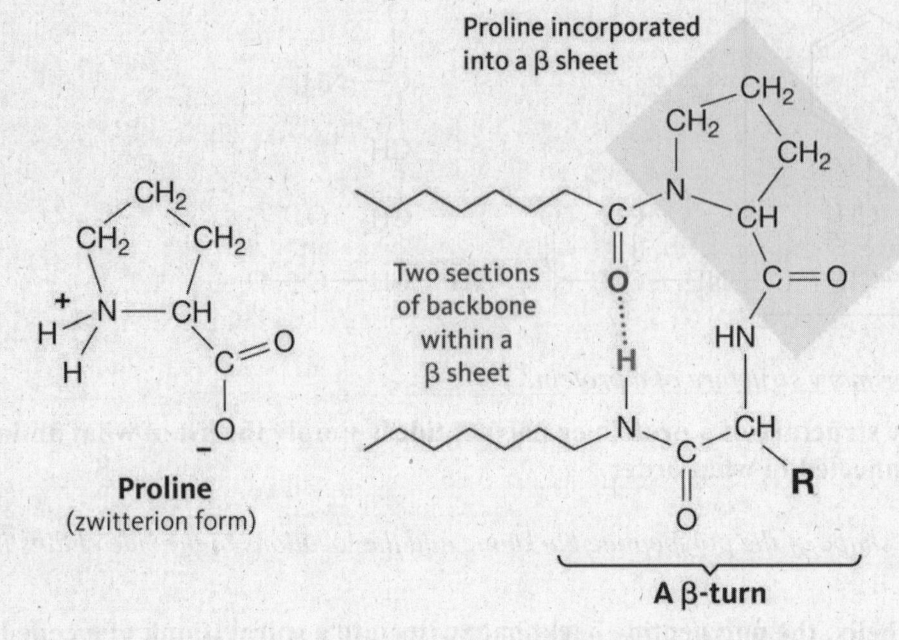

When proline is incorporated into a protein, the five-membered ring in proline keeps the protein backbone from being able to twist like a normal helix or lie flat like a normal sheet. The proline group also cannot participate in hydrogen bonding with other peptide groups, because there's no hydrogen atom on the proline nitrogen. Proline tends to produce an abrupt bend or kink in an otherwise regular secondary structure. This gives it the ability to form the U-turn of the backbone of the peptide so that it can fold back on itself to form sheets, instead of just one long line or helix.

Solutions to Section 13.3 Core Problems

13.23 *Describe the shape of a globular protein.*

The "glob-" part here is a good reminder: **globular proteins twist back on themselves to form compact globs like balls of string,** in contrast to proteins that form long fibers (fibrous proteins). Globular proteins contain a combination of helices and sheets to form a compact ball.

13.25 *Which of the following amino acids is the most likely to be found in the interior of a protein? Explain your answer.*

a) glycine *b) phenylalanine* *c) glutamic acid* *d) serine*

a) $^{+}NH_3-CH_2-C(=O)-O^-$

b) $^{+}NH_3-CH(CH_2C_6H_5)-C(=O)-O^-$

c) $^{+}NH_3-CH(CH_2-CH_2-C(=O)-O^-)-C(=O)-O^-$

d) $^{+}NH_3-CH(CH_2-OH)-C(=O)-O^-$

Hydrophobic side chains tend to concentrate on the inside of a protein, and hydrophilic side chains tend to appear on the outside. Glutamic acid and serine, with hydrophilic side chains, are likely to be found on the outside, in contact with the aqueous solution. **Glycine and phenylalanine, with hydrophobic side chains, are likely to be found in the interior of a protein, forming hydrophobic interactions with other hydrophobic side chains. Of the two, phenylalanine, with its larger side chain, is likely to form stronger hydrophobic interactions than glycine.** ("Hydrophobic interactions" are just dispersion forces, as introduced in Chapter 4.)

13.27 *Draw structures to show the two ways that the side chain of tyrosine can form a hydrogen bond with water.*

Chapter 4 review! Hydrogen bonding is extremely important through the rest of the book, so if you are having trouble with any of the hydrogen bonding questions you should go back and review the appropriate sections. A hydrogen bond is an attraction between an H atom covalently bonded to N or O (in which case the H atom has a δ+ charge), and an N or O atom with a lone pair of electrons and a δ– charge. The side chain of tyrosine has both of these, so it can form two different interactions with water molecules:

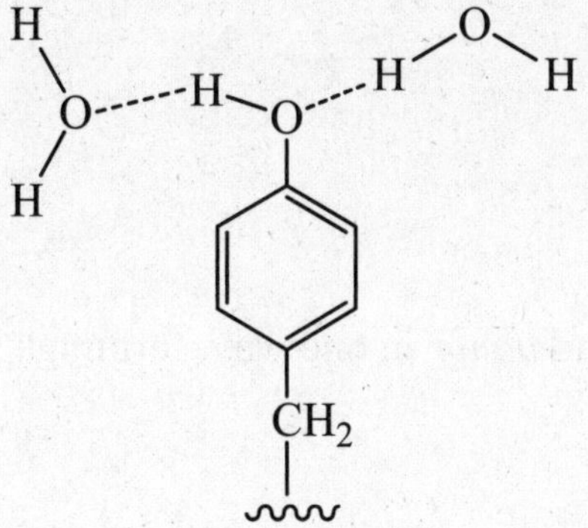

13.29 *Draw structures showing two ways to form hydrogen bonds between the side chains of serine and asparagine.*

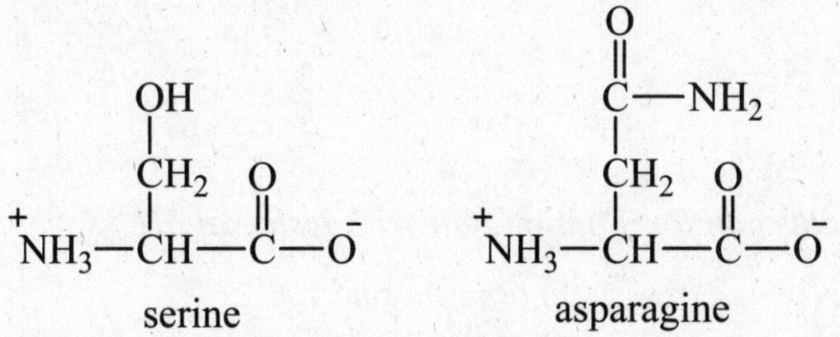

serine

asparagine

Since we're only asked to form hydrogen bonds between the side chains, let's show just the side chains for simplicity:

OH
|
CH_2
|
serine

O
‖
C—NH_2
|
CH_2
|
asparagine

A hydrogen bond is an attraction between an H atom covalently bonded to N or O (in which case the H atom has a δ+ charge), and an N or O atom with a lone pair of electrons and a δ– charge. This gives us three possible arrangements for hydrogen bonding between these two side chains (you're asked to draw two of them). Asparagine has both N and O atoms that can *accept* hydrogen bonds from serine, forming hydrogen bonding interactions 1 and 2:

(structure: serine O—H ① ··· O=C of asparagine; asparagine NH_2 ② ··· H—O of serine)

serine   asparagine   serine

However asparagine can only act as a hydrogen bond *donor* in one way, forming hydrogen bonding interaction 3:

(structure: asparagine C(=O)—N(H)—H ③ ··· O(H)—CH_2 of serine)

asparagine   serine

13.31 *Which of the following amino acids can form an ion pair with lysine at pH 7?*

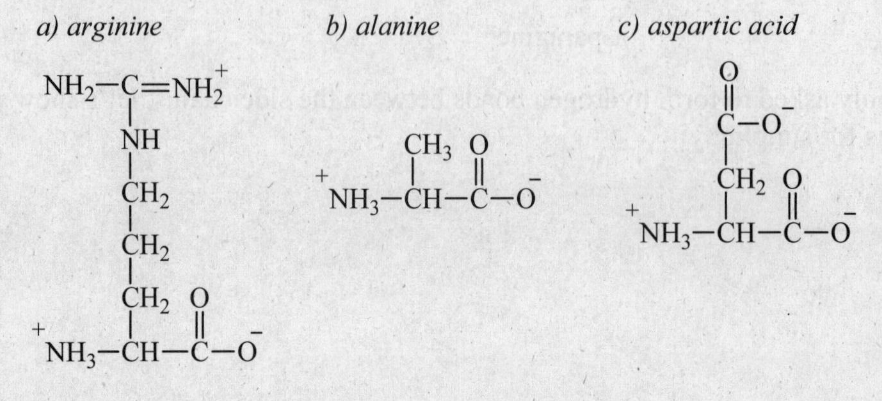

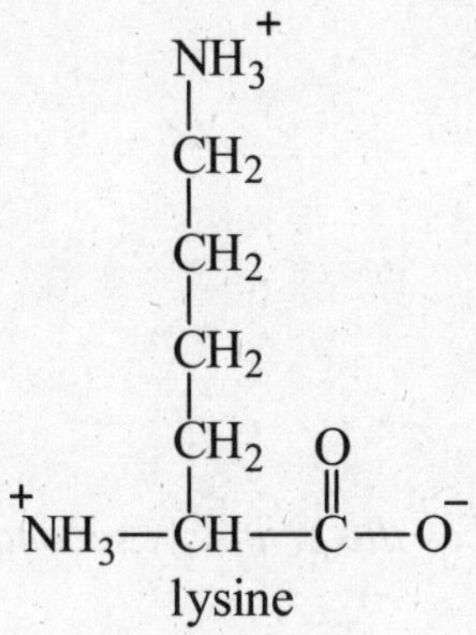

$$\begin{array}{l} NH_3^+ \\ | \\ CH_2 \\ | \\ CH_2 \\ | \\ CH_2 \\ | \\ CH_2 \quad O \\ | \qquad \| \\ \overset{+}{N}H_3—CH—C—O^- \end{array}$$

lysine

"Ion pairs" are combinations of + and – charged side chains. At pH 7, lysine has a positive charge (its amine group is converted to the ammonium ion form), so it will form an ion pair with a negatively charged side chain. Arginine, like lysine, is positively charged at pH 7 and will not form an ion pair (two positive charges will repel, not attract, each other). Alanine is a hydrophobic amino acid and its side chain is not ionized at pH 7. **Aspartic acid is the only one of these that is negatively charged at pH 7 and can form an ion pair with lysine.**

13.33 *What interactions occur between the following?*

a) *The side chains of asparagine and threonine.*

Asparagine and threonine are both hydrophilic and neutral. Their side chains will form hydrogen bonds. Several hydrogen-bonding interactions are possible, one of which is shown:

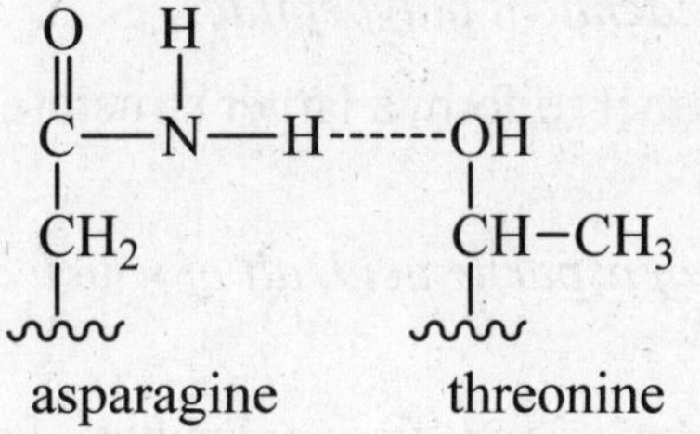

$$\begin{array}{l} O \quad H \\ \| \quad\; | \\ C—N—H\text{------}OH \\ | \qquad\qquad\quad | \\ CH_2 \qquad\quad CH-CH_3 \\ | \qquad\qquad\quad | \\ \sim\sim \qquad\qquad \sim\sim \end{array}$$

asparagine threonine

b) *The side chains of two molecules of methionine.*

Methionine is hydrophobic. The side chains of methionine molecules will form hydrophobic interactions. The side chain of methionine has no ability to form hydrogen-bonding interactions.

$$\begin{array}{l} CH_3 \\ | \\ S \\ | \\ CH_2 \\ | \\ CH_2 \\ | \\ \sim\sim \end{array}$$

methionine

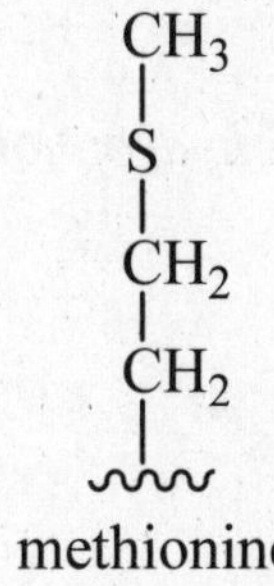

c) *A water molecule and the side chain of threonine.*

Threonine is hydrophilic. Its alcohol group will form hydrogen bonds with water molecules.

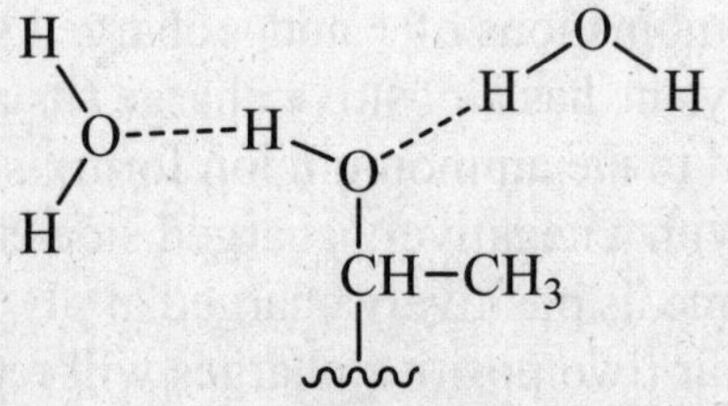

H O H
O----H O H
H
CH–CH_3

threonine

13.35 *For each of the following statements, tell whether it describes the primary, secondary, tertiary, or quaternary structure of a protein:*

See Table 13.2 for the summary of the different levels of structure in proteins. Primary structure is the backbone of the polypeptide, the list of which amino acids, in what order; secondary structure is the way the backbone twists or folds to form helices or sheets; tertiary structure is the way the helices and sheets assemble into balls, fibers or other shapes; and quaternary structure is the way that multiple polypeptides assemble together to build a larger structure.

a) Isocitrate dehydrogenase is made from eight identical polypeptides.

This describes eight proteins assembling together to form a larger structure: **quaternary structure.**

b) Bovine ribonuclease contains five molecules of aspartic acid, all of which are on the exterior of the protein.

This describes both an aspect of primary structure ("five molecules of aspartic acid") **and an aspect of tertiary structure** (the molecules are "on the exterior of the protein.") We don't know anything about the secondary structure (how the backbone is arranged into helices or sheets).

c) Bovine ribonuclease contains three sections of α-helix.

This describes the **secondary structure**.

d) The interior of citrate synthase contains a cluster made up of the hydrophobic amino acids methionine, phenylalanine, leucine, and alanine.

This is a description of the **primary structure** of the protein.

Solutions to Section 13.4 Core Problems

13.37 *What happens to a protein when it is denatured?*

Since the function of a protein depends sensitively on how it's shaped, and the shape (at least in terms of secondary, tertiary and quaternary structures) is determined by hydrogen-bonding and ion-pairing interactions, a change in the environment of a protein can have a dramatic effect on its structure and function. The change in the environment could be:

- a change in solvent (a solvent with different hydrogen-bonding abilities than water would disrupt the hydrophilic and hydrophobic interactions in the protein; a hydrophobic solvent could, in effect, turn a protein inside out, since hydrophilic side chains normally found on the exterior might flip inside the protein to find each other);

- a change in pH (the ion-pair interactions between side chains depend on the ionization of the side chains at physiological pH; at high pH, basic side chains will appear with un-ionized amine groups, while at low pH, acidic side chains will be converted back to the un-ionized carboxylic acid groups);
- a change in temperature (at higher temperatures, molecular motion is increased; proteins may essentially shake loose from their normal configurations and get locked into different shapes);
- violent agitation (with an effect similar to that of high temperatures);
- addition of an ionic salt to the solution (allowing ionic side chains to form ion pairs with the new ions, rather than with side chains of other amino acids).

The simple answer is that denaturing is a change in the structure of the protein, caused by any of these factors, that destroys its normal function.

13.39 *Why does each of the following conditions denature proteins?*

a) *Adding a solution of HCl*

The structure of a protein depends in part on ion-pair interactions between positively and negatively charged side chains (ammonium ion groups and carboxylate ion groups, respectively). At low pH, acidic side chains will be converted back to the unionized carboxylic acid groups, which cannot form ion pairs with ammonium ion groups. In addition, any repulsion interactions between like-charged carboxylate ion groups are lost, which may also have an effect on the structure of the protein.

b) *Adding a solution that contains Hg^{2+} ions*

Mercury(II) ions form strong attractions to sulfur atoms, and can break the disulfide bridges between cysteine amino acid side chains in a protein.

c) *Agitating the protein solution*

Violently shaking a protein solution can cause the hydrogen bonds and ion-pair interactions between side chains to shake loose from each other and form new interactions with different parts of the molecule, changing the structure of the protein molecule.

Solutions to Section 13.5 Core Problems

13.41 *What is an enzyme?*

An enzyme is a protein that catalyzes a reaction. This is a good time to look again at some vocabulary words: to say that something is a "protein" is a structural categorization–it's made of amino acids linked together by peptide bonds. The word "protein" doesn't say anything about the function of the molecule, only its chemical makeup. The word "catalyst," on the other hand, is a job description–it is a substance that provides a lower-energy pathway for a reaction, thereby causing the reaction to go faster. (Now is a good time to go back to Chapter 6 for a review of catalysts if you've forgotten!) The catalyst could be an element or compound, it could be ionic or covalent, it could be a salt or a protein or anything–calling it a catalyst doesn't say anything about its chemical makeup, only its function.

The word "enzyme" encompasses both: a substance must be **both** a protein (made up of amino acids, specific primary/secondary/tertiary structures, etc) **and** a catalyst (causes a reaction to go faster without being consumed in the reaction) to be classified as an enzyme. There are many proteins that aren't enzymes, and many catalysts that aren't enzymes.

13.43 *Sucrase is a protein that is produced by your intestinal tract. It catalyzes the hydrolysis of sucrose (table sugar), breaking it down into two simple sugars:*

$$\textit{sucrose} + H_2O \rightarrow \textit{glucose} + \textit{fructose}$$

Identify the substrates, the products, and the enzyme in this reaction.

A "substrate" is the reactant in a catalyzed reaction. **In this reaction, sucrose and water are the substrates. The products are glucose and fructose** (the word "product" has its usual meaning). The "enzyme" is the protein catalyst in the reaction; we're told in the problem that sucrase is a protein and a catalyst, so **sucrase is the enzyme in the reaction.**

13.45 *Define the following terms:*

a) active site

The active site is the specially-shaped cavity in the enzyme that substrate molecules fit into. It's a docking station for the reactant molecules.

b) enzyme–substrate complex

The enzyme-substrate complex is the enzyme with a substrate molecule properly docked in its active site. While there are (usually) no covalent bonds between the enzyme and substrate, the substrate molecule locks fairly securely into place through some combination of dispersion forces, ion pairing, and hydrogen bonding interactions. With the substrate molecule(s) properly in place in the active site, the enzyme can catalyze a reaction.

13.47 *Many enzymes catalyze reactions that involve phosphate ions. Several of these enzymes contain a magnesium ion at the active site. Explain why the magnesium ion helps the enzyme to bind to phosphate.*

At physiological pH, phosphate ions have a –1 or –2 charge (remember that two forms of the phosphate ion exist in equilibrium at physiological pH). **Phosphate ions are negatively charged and will be attracted to positively charged ions such as magnesium ion (Mg^{2+}). Having a magnesium ion at the active site will help attract the phosphate ions into place.**

13.49 *How does the presence of an enzyme affect the activation energy of a reaction?*

Chapter 6 review! A catalyst speeds up a reaction by providing an alternate pathway with a lower activation energy. **An enzyme lowers the activation energy of a reaction.**

13.51 *The following graph shows the relationship between activity and pH for two enzymes, papain and catalase. At what pH does each enzyme show the highest activity?*

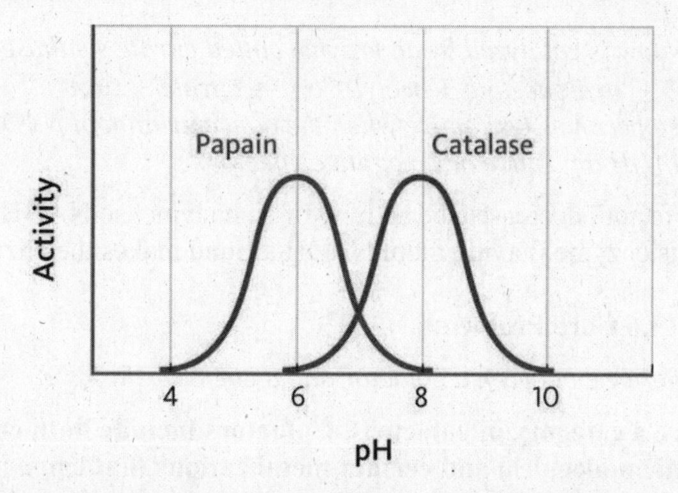

Activity is indicated by the height of the curve for each enzyme. **The highest point on the activity curve for papain is around pH 6, and the highest point on the activity curve for catalase is around pH 8.**

13.53 *Does the activity of a typical enzyme decrease, increase, or remain the same as the temperature rises from 10°C to 30°C? Why is this?*

Human body temperature is 37°C, so enzymes in humans would be expected to have high activity near that temperature. In practice the activity of an enzyme usually increases with temperature right up until the point that the protein structure begins to denature, usually around 50 °C. **As the temperature rises from 10°C to 30°C, the enzyme is getting closer and closer to its optimal temperature range, so the activity of the enzyme would increase.** The activity is expected to be low at 10°C, increase as the temperature increases to 50°C, then decrease again if the temperature continues to rise beyond that point, eventually denaturing if the temperature climbs too high.

13.55 *Complete the following sentences with the correct terms:*

a) *A molecule that is similar to the substrate and that blocks the active site of an enzyme is called a/an* **competitive inhibitor**.

"Inhibitors" are substances that block reactions or block the effect of a catalyst; "competitive," in this context, means that the inhibitor molecule competes with the proper substrate for the active site in an enzyme, and to compete well it would have to be structurally similar to the substrate (or it wouldn't fit properly into the active site of the enzyme.)

NOTE:

These definitions are consistent with the common use of these terms.

b) *A molecule that binds to an enzyme outside the active site and that makes the enzyme more active is called a/an* **positive effector**.

"Effectors" change the activity of enzymes without binding directly into their active sites; a "negative effector" slows down the catalytic reaction by making the enzyme bind the substrate less effectively, while a "positive effector" makes the reaction faster by increasing the enzyme's ability to bind substrate molecules.

13.57 *The reaction below is catalyzed by an enzyme called citrate synthase.*

oxaloacetate + acetyl-CoA → citrate + CoA

Citrate synthase becomes less active when the concentration of NADH in the cell increases. Is NADH a positive or a negative effector?

A "negative effector" decreases the activity of an enzyme, so **NADH is a negative effector** for this enzyme (having more NADH around makes the enzyme less active.)

Solutions to Section 13.6 Core Problems

13.59 *What is the difference between a cofactor and a coenzyme?*

Coenzymes are a category of cofactors. Cofactors include both coenzymes (cofactors that are organic molecules) and certain metal cations that can act as cofactors.

13.61 *FAD is a coenzyme. Is it also a cofactor? Explain.*

Coenzymes are a category of cofactor, so FAD is both.

13.63 *How do our bodies obtain metallic cofactors?*

All metallic cofactors must be a part of the diet. Since the metal ions are specific elements, there is no way for them to be manufactured by the body; they must be present in our foods. (Remember from Chapter 2 that elements cannot be made or changed by chemical reactions.)

13.65 *What vitamin do our bodies need in order to make NAD^+?*

Niacin is the B vitamin that our bodies use to make NAD^+.

Solutions to Section 13.7 Core Problems

13.67 *Some people cannot make an enzyme called phenylalanine 4-monooxygenase, which catalyzes the first step in the sequence of reactions that breaks down the amino acid phenylalanine. This disorder is called phenylketonuria (PKU), and is treated by restricting the amount of phenylalanine in the diet. Explain why PKU cannot be treated with a pill that contains the needed enzyme.*

Remember that all enzymes are proteins; the very acidic conditions in the stomach will denature proteins that aren't adapted to low pH, destroying their ability to act as enzymes, and the digestive enzymes in the gastrointestinal tract will digest proteins of almost any kind, so protein drugs can't usually be administered in pill form.

13.69 *We need all 20 amino acids to make proteins. Why are only some of the amino acids classified as "essential"?*

The essential amino acids are the ones our bodies cannot make from other nutrients (specifically, other amino acids)**, so they must be present in sources of dietary protein.**

13.71 *Describe the two parts of the nutritional protein requirement.*

The essential amino acids must be present in the proteins we eat, because our bodies cannot manufacture these amino acids. However, in the case of the other amino acids that can be manufactured by the body, the nitrogen in those amino acids still has to come from protein–it's the only significant source of nitrogen in the diet (fats and carbohydrates do not contain nitrogen). So in addition to having enough of the essential amino acids, the diet must include enough total protein to supply the body with nitrogen for the manufacture of non-essential amino acids and any other nitrogen-containing compounds.

13.73 *Explain why a person who is undernourished can consume adequate amounts of both essential and non-essential amino acids, yet still exhibit symptoms of protein deficiency.*

Protein is not the body's preferred fuel. If a person's diet doesn't contain enough fats or carbohydrates to supply energy needs, the body can use protein as a source of fuel, but that may leave the person without enough protein for other purposes.

13.75 *What kinds of organisms can make amino acids using inorganic sources of nitrogen such as ammonium ions and nitrate ions?*

Plants can make amino acids using these sources of nitrogen, but animals must consume their nitrogen in the form of amino acids.

13.77 *What is nitrogen fixation, and why is it important to life on earth?*

The nitrogen that makes up about 80% of the Earth's atmosphere is in the form of N_2, which is extremely unreactive and cannot be used chemically by plants or animals. Certain bacteria are capable of using N_2 to make ammonium ions, which can then be converted by plants into amino acids, which in turn can be used by animals to make other amino acids and proteins. **The bacterial process of turning atmospheric N_2 into the NH_4^+ ions that are biologically useful is called nitrogen fixation, and without it neither plants nor animals could make compounds containing nitrogen.**

13.79 *What chemical compound do our bodies excrete when we need to get rid of excess nitrogen?*

Mammals excrete excess nitrogen in the form of urea, $(NH_2)_2C{=}O$. Birds and most reptiles excrete excess nitrogen as uric acid, and fish excrete it as ammonium ions.

Solutions to Concept Questions

* indicates more challenging problems.

13.81 *What is an amino acid, and why are amino acids important to all organisms?*

Although any compound that contains an amino (–NH_2) group and a carboxylic acid (–COOH) group is an amino acid, this term generally refers to a substance that contains an amine group and a carboxylic acid group separated by one carbon atom, called the "alpha carbon." Amino acids are the building blocks of proteins, a group of compounds that are essential to the function of all life on Earth.

$$\underset{\text{amino group}}{\boxed{NH_2}}-\underset{\text{alpha carbon}}{\boxed{CH}}-\underset{\text{acid group}}{\boxed{\overset{O}{\overset{\|}{C}}-OH}}$$

13.83 *Valine is a polar molecule, because it contains two strongly hydrophilic functional groups (the amino group and the acid group). Why then is valine always classified as a hydrophobic amino acid?*

$$NH_2-\underset{\boxed{CH_3-\underset{|}{\overset{CH_3}{\overset{|}{CH}}}}}{CH}-\overset{O}{\overset{\|}{C}}-OH$$

Since all amino acids share the common structure of the amine group, the alpha carbon, and the carboxylic acid group, different amino acids are classified by their side chains. The side chain of valine is a hydrocarbon group, which is hydrophobic, so valine is classified as a hydrophobic amino acid.

13.85 *Draw the structure of a peptide group, and show which two atoms participate in hydrogen bonds.*

donor group: $N^{\delta-}-H^{\delta+}$

acceptor: $^{\delta-}O=C^{\delta+}$

The hydrogen atom on the N in one peptide group forms a hydrogen bond with the carbonyl O atom in another peptide group.

13.87 *What are the two common secondary structures in proteins, and how do they differ from one another?*

Primary structure is the backbone of the polypeptide, the list of which amino acids, in what order; secondary structure is the way the backbone twists or folds to form helices or sheets; tertiary structure is the way the helices and sheets assemble into balls, fibers or other shapes; and quaternary structure is the way that multiple polypeptides assemble together to build a larger structure. See Figures 13.5 and 13.6 for pictures of **the two common secondary structures, alpha helices and beta sheets. In an alpha helix, the polypeptide backbone twists into a spiral** (think of a coiled spring). **The side chains on each loop form hydrogen bonding interactions with the side chains on the next loops**

above and below, as in Figure 13.6 in the text. **In a beta sheet, the polypeptide backbone forms wavy flat lines that u-turn back on themselves repeatedly to form sheets. The side chains on each line form hydrogen bonding interactions with those on the next line.**

13.89 *Why do amino acids that have hydrocarbon side chains generally appear in the interior of a globular protein?*

Globular proteins are shaped like tangled balls of string. Proteins in organisms are commonly in aqueous environments. The strong interactions between water molecules and hydrophilic side chains cause proteins to fold and twist up in such a way that the hydrophobic side chains end up on the inside of the protein, where they can attract each other through dispersion force interactions, while the hydrophilic side chains are on the outside of the protein, in contact with the water molecules.

13.91 *Explain why membrane proteins contain an unusually high proportion of hydrophobic amino acids.*

Membranes are made of hydrophobic materials, which keep the aqueous solutions inside and outside the membrane separated from each other. Any portion of a protein that interacts with the membrane will be hydrophobic as well.

13.93 *What is the difference between the secondary structure and the tertiary structure of a protein?*

Secondary structure is the way the backbone (primary structure) twists or folds to form helices or sheets; tertiary structure is the way the helices and sheets (secondary structures) assemble into globular, fibrous or other shapes.

13.95 *Your saliva contains an enzyme that breaks down starch. If you mix your saliva with a little ethanol, the enzyme becomes inactive. Explain.*

Enzymes are proteins. Since the function of a protein depends sensitively on how it's shaped, and the shape (at least in terms of secondary, tertiary and quaternary structures) is determined by hydrogen-bonding and ion-pairing interactions, a change in the environment of a protein can have a dramatic effect on its structure and function. A solvent with different hydrogen-bonding abilities than water would disrupt the hydrophilic and hydrophobic interactions in the protein, causing it to change shape and deactivating (denaturing) the protein. Ethanol is more hydrophobic than water, so it causes the enzyme to rearrange itself, and the protein becomes inactive.

13.97 *Describe the three steps that happen when an enzyme catalyzes a reaction.*

(1) The substrate (reactant) molecules dock into the active site (cavity) of the enzyme, forming an enzyme-substrate complex.

(2) The interactions in the enzyme-substrate complex cause shifts in the bond strengths and polarities of the substrate molecules, allowing them to react to form product molecules.

(3) The product molecules are released from the active site, leaving the enzyme free to take in new substrate molecules and repeat the reaction.

13.99 *What is the activity of an enzyme, and what is the typical range of enzyme activities?*

The "activity" of an enzyme is a measure of how quickly it catalyses reactions, usually reported in terms of the number of reactions per unit time. The fastest enzymes catalyze up to a million reactions per second, while the slowest catalyze only about one reaction per second, but the typical range is 10 to 1000 reactions per second.

13.101 *Two samples of an enzyme are dissolved in water. Sample #1 is heated to 100°C, and sample #2 is cooled to 0°C. At these temperatures, the enzyme has no detectable activity. Next, both samples are brought to 37°C. At this temperature, sample #2 is active, but sample #1 is not. Explain this difference.*

The inactivity of the enzyme at 0°C (sample #2) is due to the very slow molecular motion at this temperature; the enzyme hasn't been damaged, it just can't get substrate molecules into the active site and products back out again. When it warms up again to its optimal temperature range, its activity is back to normal. However, the inactivity of the enzyme at 100°C (sample #1) is attributable to the protein having been denatured by the high temperature; most proteins denature beginning around 50°C. Denaturing is an irreversible process in which the protein rearranges into a different shape from the active form. When the protein cools back down again, it still stays in the new rearranged form, and therefore has irreversibly lost its normal activity.

13.103 *What is the difference between a competitive inhibitor and a negative effector? Which of these usually has a chemical structure that resembles the substrate?*

Competitive inhibitors occupy the active site of an enzyme directly, so they usually have chemical structures that are similar to the substrate the enzyme normally acts on (they have to *compete* with the substrate for a position in the active site.) Negative effectors interact with the enzyme at some location besides the active site, changing the shape of the enzyme enough to decrease its ability to attract substrate molecules into the active site.

13.105 *What is the difference between an essential amino acid and a nonessential amino acid?*

While there are many amino acids that are required for normal function of the body, many of them can be manufactured by the body from other amino acids. For these non-essential amino acids, any dietary source of amino acids will do. The essential amino acids are the eight (or ten) amino acids that cannot be manufactured from other amino acids in the diet; they must be obtained in complete form from dietary protein sources. See Table 13.4.

13.107 *Humans can use the nitrogen atom in leucine to make the amino group in serine, but we cannot use the nitrogen atom in serine to make leucine. Why is this?*

Leucine is an essential amino acid, so it must be obtained intact from dietary sources. Serine is a nonessential amino acid; our bodies can manufacture it from other nutrients in the diet. The nitrogen atom must be supplied from an amino acid; it could come from any dietary amino acid, including leucine.

Solutions to Summary and Challenge Problems

* indicates more challenging problems.

13.109 *Gamma-carboxyglutamic acid is used to build certain proteins that play a role in blood clotting.*

```
      O          O
      ||         ||
HO—C—CH—C—OH
             |
           CH2  O
             |      ||
NH2—CH—C—OH
```

Gamma-Carboxyglutamic acid

a) *Identify the side chain in this amino acid, and classify the amino acid as hydrophobic or hydrophilic. If it is hydrophilic, classify it as acidic, basic or neutral.*

The side chain of any amino acid is the group attached to the alpha carbon. The side chain of this amino acid contains two carboxylic acid groups; these groups are hydrophilic and acidic, so this amino acid is classified as **hydrophilic and acidic.**

```
      O          O
      ||         ||
HO—C—CH—C—OH
             |
           CH2
             |      O
             |      ||
NH2—CH—C—OH
```

b) *Draw the structure of this amino acid as it will appear at pH 7.*

At pH 7, amine groups are converted to the ammonium ion form, and carboxylic acid groups are converted to the carboxylate ion form:

```
     O          O
     ||         ||
⁻O—C—CH—C—O⁻
            |
          CH2  O
 +         |      ||
NH3—CH—C—O⁻
```

c) *Are there any chiral carbon atoms in this molecule? If so, which ones are they?*

Chapter 9 review! Any carbon atom that has four *different* atoms or groups bonded to it is a chiral carbon atom. **The alpha carbon atom in this molecule, as in most amino acids, is chiral** (the alpha carbon atom in glycine is an exception, since its alpha carbon atom has two H atoms attached). This amino acid has no other chiral carbon atoms:

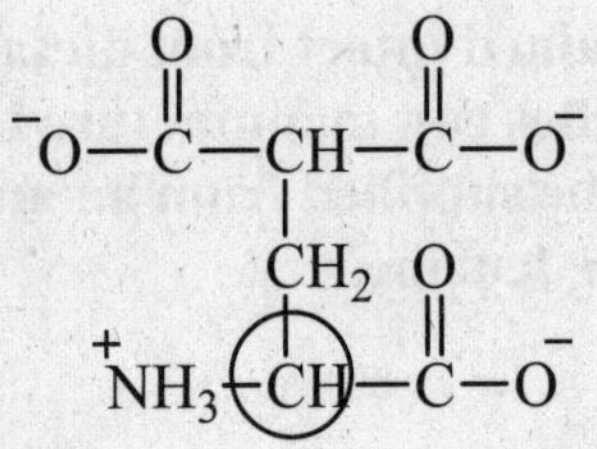

$$
\begin{array}{c}
{}^{-}O-\overset{\overset{\displaystyle O}{\|}}{C}-CH-\overset{\overset{\displaystyle O}{\|}}{C}-O^{-} \\
| \\
CH_2\quad \overset{\displaystyle O}{\|} \\
\overset{+}{N}H_3-\underset{\bigcirc}{CH}-C-O^{-}
\end{array}
$$

NOTE:
The top center carbon has two identical carboxylate groups attached.

*13.111 *List all of the possible tripeptides that can be made using arginine, isoleucine, and valine (one molecule of each). (One possible tripeptide is Arg-Ile-Val.)*

This is a question about the primary structure of a polypeptide: which amino acids, in what order. Remember that because each amino acid has an amino end and an acid end, reversing the order gives a different molecule (the N-terminal amino acid in one chain will be C-terminal in the other), and list all possible orders:
Arg-Ile-Val (given in problem)
Arg-Val-Ile
Val-Arg-Ile
Val-Ile-Arg
Ile-Arg-Val
Ile-Val-Arg

13.113 *When the following compound is hydrolyzed, the products are four amino acids. Three of the four are common amino acids, but the other is not.*

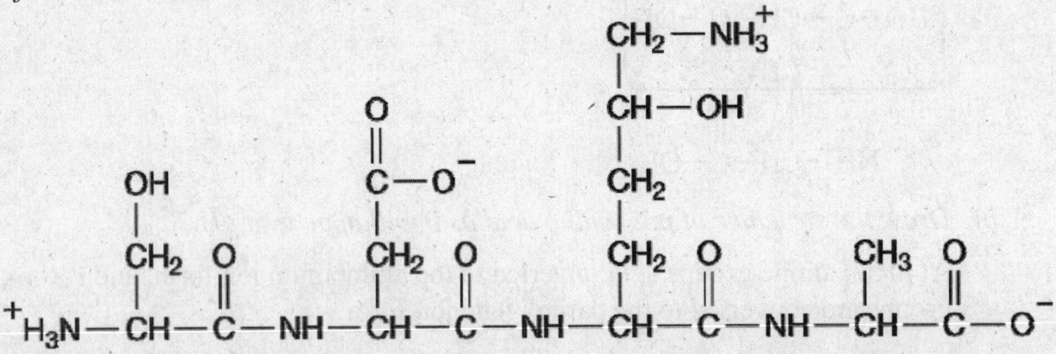

$$
\begin{array}{l}
\qquad\qquad\qquad\qquad\qquad\qquad\qquad\qquad CH_2-NH_3^{+} \\
\qquad\qquad\qquad\qquad\qquad\qquad\qquad\qquad | \\
\qquad\qquad\qquad\qquad O \qquad\qquad\qquad CH-OH \\
\qquad\qquad\qquad\qquad \| \qquad\qquad\qquad\quad | \\
\quad OH \qquad\qquad\qquad C-O^{-} \qquad\qquad CH_2 \\
\quad | \qquad\qquad\qquad\qquad | \qquad\qquad\qquad\quad | \\
\quad CH_2 \quad O \qquad\quad CH_2 \quad O \qquad\quad CH_2 \quad O \qquad\quad CH_3 \quad O \\
\quad | \qquad \| \qquad\qquad | \qquad \| \qquad\qquad | \qquad \| \qquad\qquad | \qquad \| \\
{}^{+}H_3N-CH-C-NH-CH-C-NH-CH-C-NH-CH-C-O^{-}
\end{array}
$$

a) *Circle the peptide groups in this molecule. How many peptide groups are there?*

Peptide groups are amide functional groups (Ch 12) formed in the condensation of amino acids. **Linking four amino acids forms three peptide groups:**


```
                                              +
                                    CH2-NH3
                                    |
                      O             CH-OH
                      ||     -      |
       OH             C--O          CH2
       |              |             |
       CH2  O         CH2  O        CH2  O         CH3  O
 +     |    ||        |    ||       |    ||        |    ||   -
 NH3--CH---C---NH---CH---C---NH---CH---C---NH---CH---C--O
```

b) *Draw the structures of the four amino acids that are formed when this compound is hydrolyzed. (Draw the structures as they appear at pH 7.)*

You can actually do this by writing out each hydrolysis reaction step, but you don't need to; remember that amino acids all have a common structure of the amino group, the alpha carbon, and the carboxylic acid group, and only the side chains are different. If you can identify the four side chains in the polypeptide, you can draw the four amino acids without thinking specifically about the hydrolysis reaction. Remember to show amine groups in their ammonium ion forms, and carboxylic acid groups in their carboxylate ion forms:

```
                                                           +
                                                  CH2-NH3
                                                  |
                            O                     CH-OH
                            ||    -               |
   OH                       C--O                  CH2
   |                        |                     |
   CH2  O                   CH2  O                CH2  O                CH3  O
 + |    ||   -          +   |    ||   -       +   |    ||   -       +   |    ||   -
NH3-CH--C--O           NH3-CH---C--O         NH3-CH---C--O          NH3-CH---C--O
   serine                aspartic acid          (uncommon)              alanine
```

c) *Identify the three common amino acids, using the information in Table 13.1.*

Amino acids are identified by their side chains. These were identified and labeled in the answer to part b) above.

d) *Classify the fourth amino acid as hydrophobic or hydrophilic. If it is hydrophilic, classify it as acidic, basic or polar neutral.*

```
          +
CH2-NH3              CH2-NH2
|                    |
CH-OH                CH-OH
|                    |
CH2                  CH2
|                    |
CH2                  CH2
|                    |
~~~                  ~~~
form at pH 7         unionized form
```

The side chain of the fourth amino acid contains both an alcohol group and an ammonium ion group (in its form at pH 7). These are both hydrophilic groups, so the amino acid is **hydrophilic**. Remember that amino acids are classified as acidic, neutral or basic according to their *unionized* forms; in its unionized form, this amino acid would contain a (basic) amine group in its side chain, so it is classified as a **basic** amino acid.

e) What is the C-terminal amino acid in this molecule?

The C-terminal amino acid is the amino acid on the end with the carboxylate group (and by convention is written on the right of the peptide chain), in this case **alanine.**

f) What is the N-terminal amino acid in this molecule?

The N-terminal amino acid is the amino acid on the end with the amino group (and by convention is written on the left of the peptide chain), in this case **serine**.

*13.115 *The rare amino acid that follows, called selenocysteine, is found in a few microorganisms. Two molecules of this amino acid can combine to form a diselenide bridge, analogous to the disulfide bridge that is formed by cysteine. Draw the structure of the diselenide bridge that is formed by two selenocysteine molecules.*

$$^{+}H_3N-CH(CH_2SeH)-C(=O)-O^{-}$$

Though this problem is marked as a challenge problem, it's not that bad. By analogy to a "disulfide," which contains a –S–S– group connecting two molecules, a "diselenide" will contain a –Se–Se– group connecting two molecules. We can write the reaction by analogy to the oxidation reaction that takes place between two thiol groups (Section 10.4), starting by rotating the two molecules to put the side chains facing each other so they can react:

$$^{+}NH_3-CH(COO^{-})-CH_2-SeH \;+\; HSe-CH_2-CH(COO^{-})-NH_3^{+}$$

$$\longrightarrow\; ^{+}NH_3-CH(COO^{-})-CH_2-Se-Se-CH_2-CH(COO^{-})-NH_3^{+} \;+\; 2\,[H]$$

To make it easier to see what's going on, here's the same reaction, showing only the parts that actually react (the side chains of the amino acids):

$$\text{—}CH_2\text{-}SeH \quad + \quad HSe\text{—}CH_2\text{—} \longrightarrow \text{—}CH_2\text{—}Se\text{—}Se\text{—}CH_2\text{—}$$

Remember that the problem only asks you for the diselenide bridge, not the reaction or even the complete product:

$$\mathbf{\text{—}CH_2\text{—}Se\text{—}Se\text{—}CH_2\text{—}}$$

13.117 *The following sequence of amino acids occurs in a globular protein, and forms an α-helix. Would you expect this section of the polypeptide to be located in the interior or on the exterior of the protein?*

Leu–Ser–Phe–Ala–Ala–Ala–Met–Asn–Gly–Leu–Ala

Only serine (Ser) and asparagine (Asn) are hydrophilic; the other 9 amino acids are hydrophobic. Since hydrophobic amino acids tend to fold to the inside of globular proteins, leaving hydrophilic groups in contact with the aqueous solution outside the protein, **this sequence is likely to be located on the interior of the protein.**

13.119 *Draw an energy diagram that compares the activation energy of a reaction with and without an enzyme.*

Chapter 6 review! Enzymes are catalysts; replace the word "enzyme" with the word "catalyst" and this is a Chapter 6 question. Catalysts lower the activation energy of the reaction without changing the energies of the reactants and products. Here's a sample energy diagram for a reaction that is slightly endothermic:

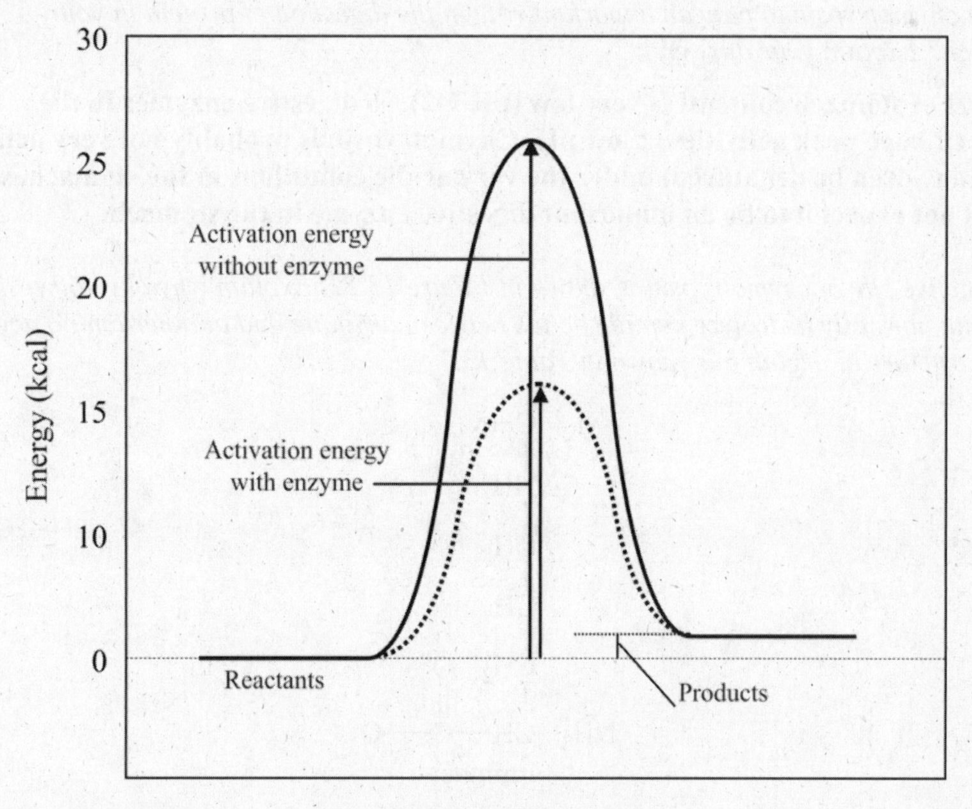

13.121 *The enzyme propionyl-CoA carboxylase contains a molecule of lysine covalently bonded to a molecule of biotin. The structure of biotin is shown here, and the structure of lysine is shown in Table 13.1.*

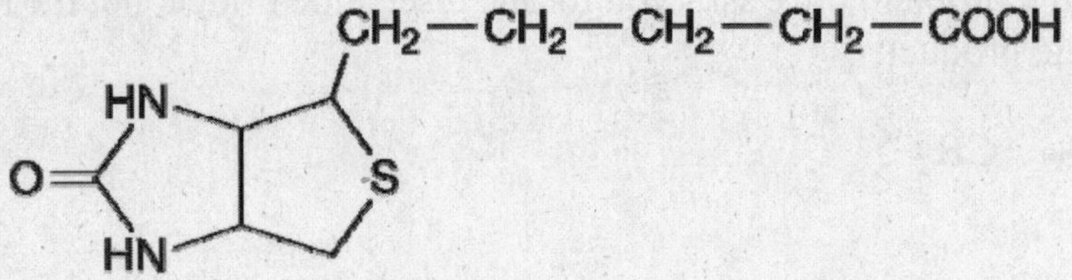

$CH_2-CH_2-CH_2-CH_2-COOH$

HN

O= S **Biotin**

HN

a) *Is lysine a cofactor for this enzyme? Is it a coenzyme?*

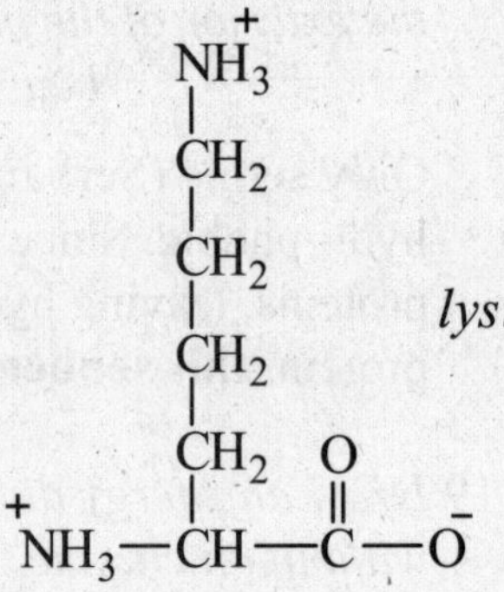

A cofactor is a molecule or ion that is required for the function of an enzyme, but is not, itself, an amino acid. Coenzymes are a subset of cofactors, so amino acids cannot be cofactors or coenzymes. **Lysine is an amino acid, so it cannot be a cofactor or a coenzyme.**

$$\overset{+}{NH_3}-CH_2-CH_2-CH_2-CH_2-CH(\overset{+}{NH_3})-C(=O)-O^-$$

lysine

b) *Is biotin a cofactor for this enzyme? Is it a coenzyme?*

Biotin is a cofactor: it is a required part of the enzyme but is not an amino acid. The two groups of cofactors are coenzymes (cofactors that are organic molecules) and metal cations. **As an organic cofactor, biotin is also a coenzyme.**

13.123 *The activity of chymotrypsin reaches a peak around pH 7 to 8. Based on this, would you expect chymotrypsin to play an important role in the digestion of protein in your stomach? Explain your answer.*

The pH of stomach contents is very low (pH 1-2), so digestive enzymes in the stomach have peak activities at low pH. Chymotrypsin is probably not very active (and may even be denatured) under the very acidic conditions in the stomach, so we would not expect it to be an important digestive enzyme in the stomach.

*13.125 *The active site of chymotrypsin is shown in Figure 13.21. Explain why chymotrypsin does not normally hydrolyze peptide bonds between arginine and another amino acid. (The structure of arginine is shown in Table 13.1.)*

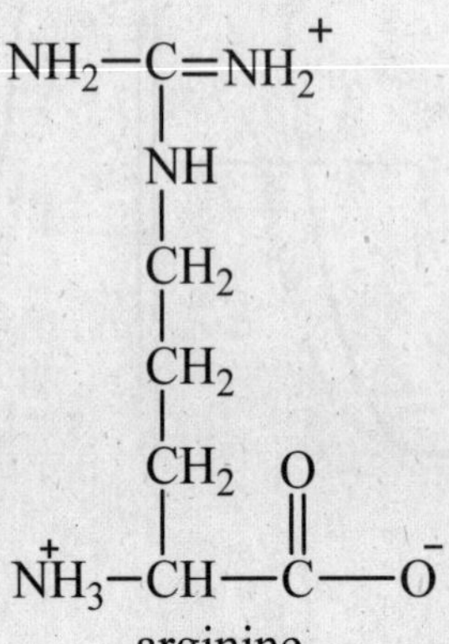

$$NH_2-C(=\overset{+}{NH_2})-NH-CH_2-CH_2-CH_2-CH(\overset{+}{NH_3})-C(=O)-O^-$$

arginine

The active site in chymotrypsin contains a nonpolar pocket that can accommodate an aromatic side chain, but does not fit other amino acids. As a result, chymotrypsin can only hydrolyze a protein at a peptide group that is next to an aromatic side chain. Arginine has a long, hydrophilic side chain, not a hydrophobic, ring-shaped aromatic side chain, so arginine does not fit into the active site in chymotrypsin. Therefore chymotrypsin does not normally hydrolyze peptide bonds between arginine and another amino acid.

*13.127 *Lysozyme is an enzyme that destroys the outer walls of some types of bacteria. It is found in egg white (and many other biological fluids). The pH–activity curve of lysozyme peaks around pH 5, but the pH of egg white is normally around 8 to 9. What does this suggest about the effect of bacterial contamination on the pH of egg white?*

Bacterial contamination probably makes the egg white solution become more acidic. It makes sense that if bacterial contamination would typically cause the pH of egg white to become more acidic, the egg white would contain an enzyme that would become active when the egg white becomes acidic, and destroy the bacteria. That way, the egg would be protected against bacterial contamination.

13.129 *Atorvastatin (Lipitor) is widely used to lower cholesterol levels. This medication is a competitive inhibitor of HMG-CoA reductase, one of the enzymes that your body uses to build cholesterol. Would you expect the structure of atorvastatin to resemble the structure of the substrate in the reaction that is catalyzed by HMG-CoA reductase?*

Yes, we would expect the atorvastatin molecule to resemble the substrate in the reaction that is normally catalyzed by this enzyme. Competitive inhibitors have to compete with substrate molecules for the active site of an enzyme, so they usually have structures that closely resemble those of the normal substrate molecules.

13.131 *NAD^+ and FAD are both coenzymes. Intracellular fluid contains a significant concentration of NAD^+, but it has no FAD. Based on this information, which of these coenzymes is permanently bonded to its enzyme?*

Coenzymes are required for enzymes to function properly. **If the fluid doesn't contain FAD, then the FAD must be permanently bonded to its enzyme (the enzyme molecules must be bringing their own FAD).** The presence of NAD^+ in the intracellular fluid indicates that NAD^+ is not permanently bonded to its enzymes but instead moves in and out of the active site of the enzymes, just as other substrate molecules do.

13.133 *All amino acids contain the elements hydrogen, carbon, nitrogen, and oxygen. In human nutrition, which of these elements can only be supplied by other amino acids?*

Hydrogen, carbon and oxygen can all be obtained from other nutrients, but nitrogen is only supplied in the form of amino acids in the diet.

13.135 *What is a complete protein? What types of foods supply complete proteins?*

Complete proteins are proteins that contain significant amounts of all 8 essential amino acids. Most plant sources lack one or more of the essential amino acids, but animal protein sources (meat, fish, dairy, insects) supply them all. These animal proteins are therefore complete proteins.

Chapter 14

Carbohydrates and Lipids

Solutions to Section 14.1 Core Problems

14.1 *Tagatose is a monosaccharide that has been proposed as a "low-calorie sweetener". One molecule of tagatose contains six carbon atoms. What is the molecular formula of tagatose?*

Monosaccharides contain C, H, and O in a fixed ratio of 1:2:1 (CH_2O), so the molecular formula is $\mathbf{C_6H_{12}O_6}$.

14.3 *Classify each of the following monosaccharides as a triose, a tetrose, a pentose, or a hexose.*

a)

b)

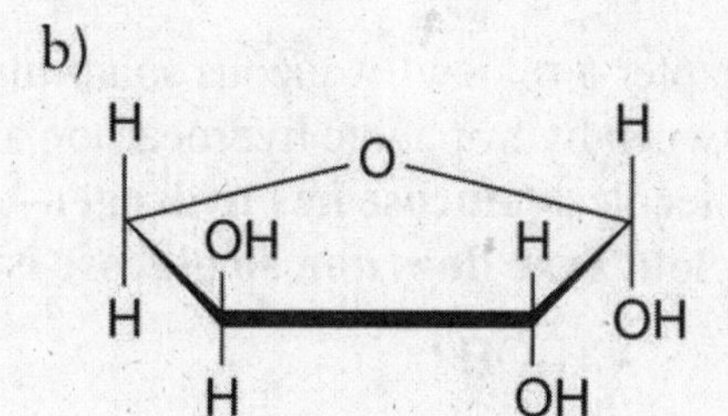

Monosaccharides are classified according to the number of carbon atoms in the molecule. The prefixes (tri = 3, tetr = 4, pent = 5, hex = 6) are the same ones we learned in Chapter 3 for naming covalent compounds.

a) Six carbon atoms (five in the ring and one below): **hexose**

b) Four carbon atoms (all in the ring): **tetrose**

14.5 *Find and circle carbon 4 in each of the monosaccharides in Problem 14.3.*

The numbering system for the Haworth projection of a monosaccharide begins at the right side of the molecule, starting with the carbon atom that is not part of the ring (if any) or with the rightmost carbon in the ring, then numbering the carbon atoms sequentially clockwise.

a)

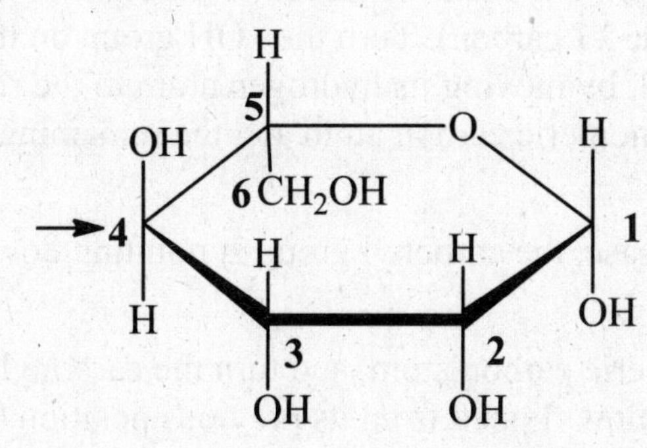

b)

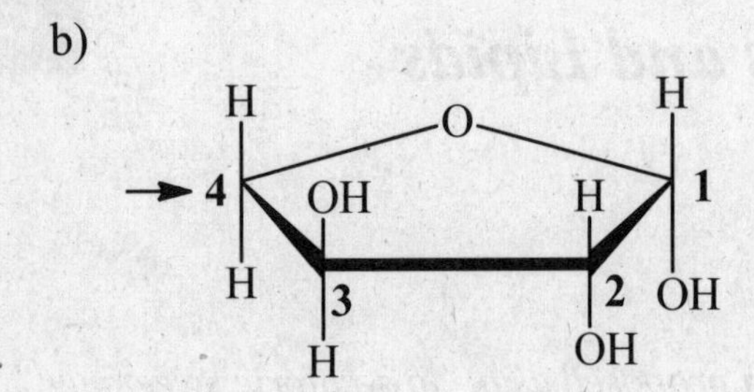

14.7 *Which of the five common monosaccharides contain a six-membered ring?*

See Figure 14.3. **Galactose, mannose and glucose contain 6-membered rings.** Remember that the trios, tetrose, etc. designations depend on the total number of carbon atoms (including any carbon atoms outside the ring), not the number of atoms in the ring.

14.9 *Cyclohexane and glucose both contain a six-membered ring. Which compound should have the higher solubility in water, and why?*

Chapter 5 review! Aqueous solubility of organic compounds depends on the balance between hydrophobic hydrocarbon areas and hydrophilic functional groups in the molecules. **Glucose has hydrogen-bonding alcohol functional groups, while cyclohexane does not, so glucose is much more soluble in water than cyclohexane is.**

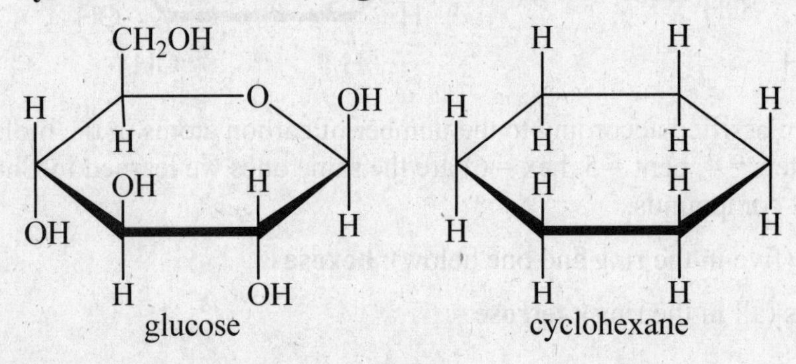

Solutions to Section 14.2 Core Problems

14.11 *Draw structures to show that when the ring in ribose opens during mutarotation, the open-chain product contains an aldehyde group.*

In a mutarotation:

(1) Break the bond between the ring O atom and the next carbon to the right of it, the *anomeric carbon* (in this molecule, the #1 carbon). Turn the –OH group on the anomeric carbon atom into a carbonyl, by moving its hydrogen atom to the ring O and forming a double bond between the anomeric carbon atom and the remaining oxygen atom.

(2) Flip the carbonyl group over (in this case, the carbonyl group is pointing down, so we flip it up.)

(3) Reattach the ring O atom to the anomeric carbon atom, and turn the carbonyl back into an alcohol. The alcohol group is now flipped from its previous position (in this case, it was below the ring and is now above.)

Step 1 (forming the carbonyl) is the important one here: **the carbonyl group is part of an aldehyde functional group, because the carbonyl C atom is also bonded to an H atom.**

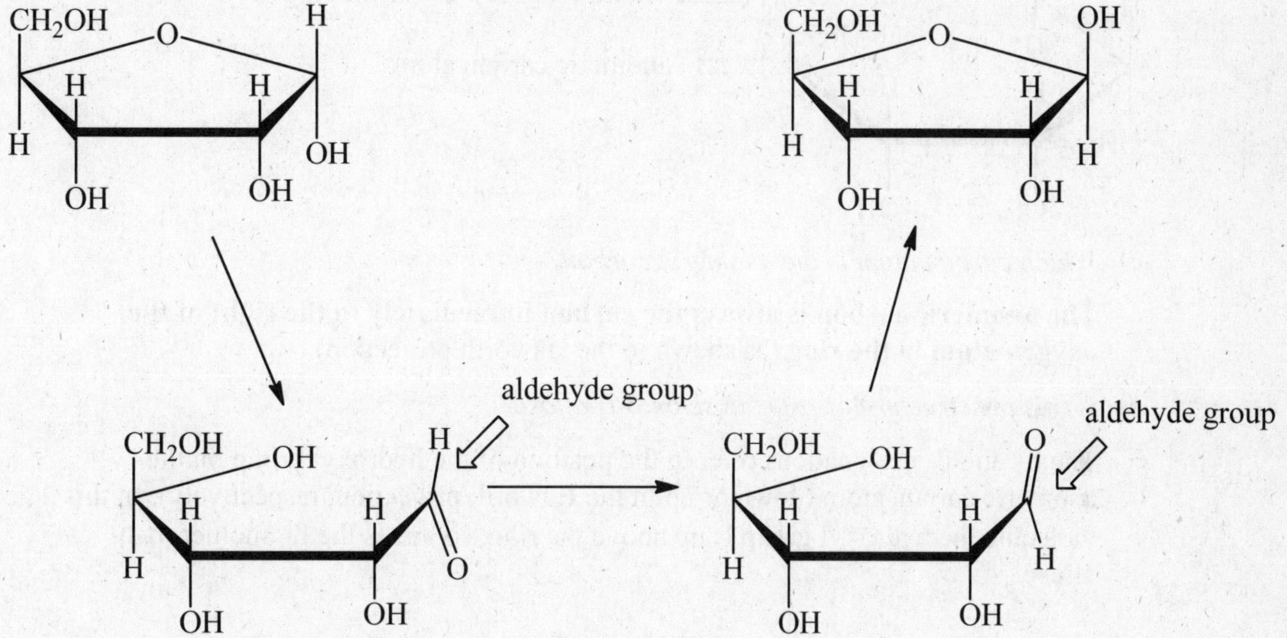

aldehyde group

aldehyde group

14.13 *The following monosaccharide is called β-D-ribulose.*

a) Is β-D-ribulose a reducing sugar? Explain why or why not.

b) Is β-D-ribulose an aldose, or is it a ketose?

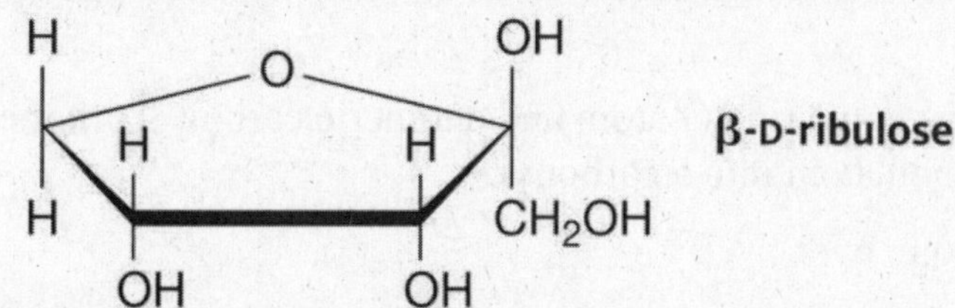

In Step 1 of the mutarotation, β-D-ribulose forms a ketone group, so it is a **ketose** (part b). Reducing sugars are aldoses (or sugars that convert to aldoses in Benedict's reagent), so **β-D-ribulose is not a reducing sugar** (part a).

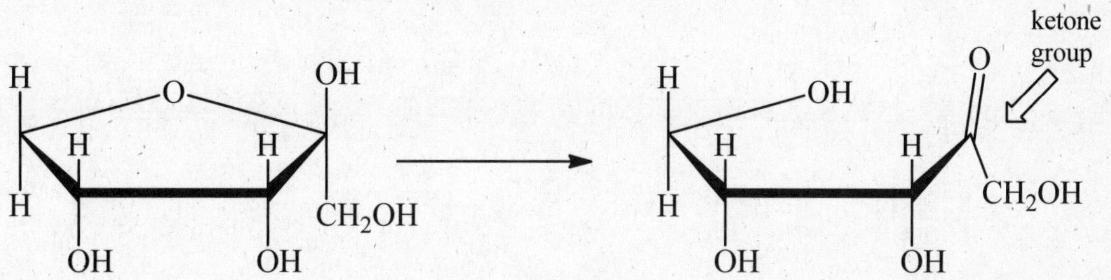

ketone group

14.15 *The structure of one of the anomers of D-allose is shown here:*

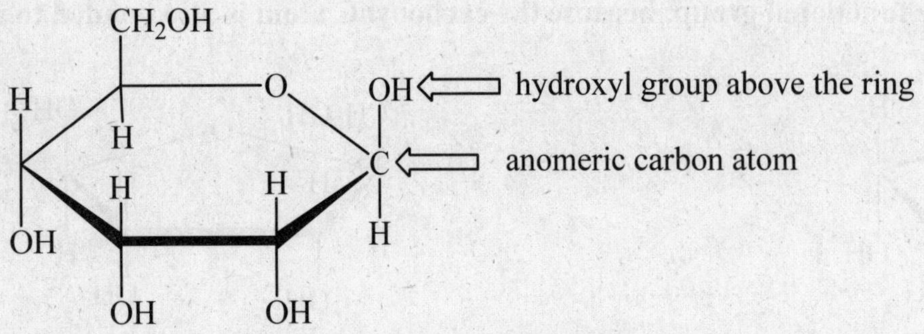

CH_2OH

OH ⟸ hydroxyl group above the ring

C ⟸ anomeric carbon atom

a) *Which carbon atom is the anomeric carbon?*

The anomeric carbon is always the carbon immediately to the right of the oxygen atom in the ring (as shown in the Haworth projection).

b) *Is this the structure of α-D-allose or β-D-allose?*

The α- and β- designations refer to the position of the hydroxyl group on the anomeric carbon atom (down or up in the Haworth projection, respectively). In this molecule the hydroxyl group is up above the ring, so this is the β- anomer: **β-D-allose.**

14.17 *Draw structures to show how α-D-allose can change into β-D-allose. (The structure of one of the anomers of D-allose is given in Problem 14.15.)*

The α- and β- forms of a sugar switch back and forth by mutarotation. Since we identified the structure in 14.15 as β-D-allose, we first need to draw α-D-allose, with the –OH group below the anomeric carbon atom. Then we carry out the steps of a mutarotation:

(1) Break the bond between the ring O atom and anomeric carbon. Turn the –OH group on the anomeric carbon atom into a carbonyl.

(2) Flip the carbonyl group over.

(3) Reattach the ring O atom to the anomeric carbon atom, and turn the carbonyl back into an alcohol.

(For a more detailed review of mutarotation, see Problem 14.11.)

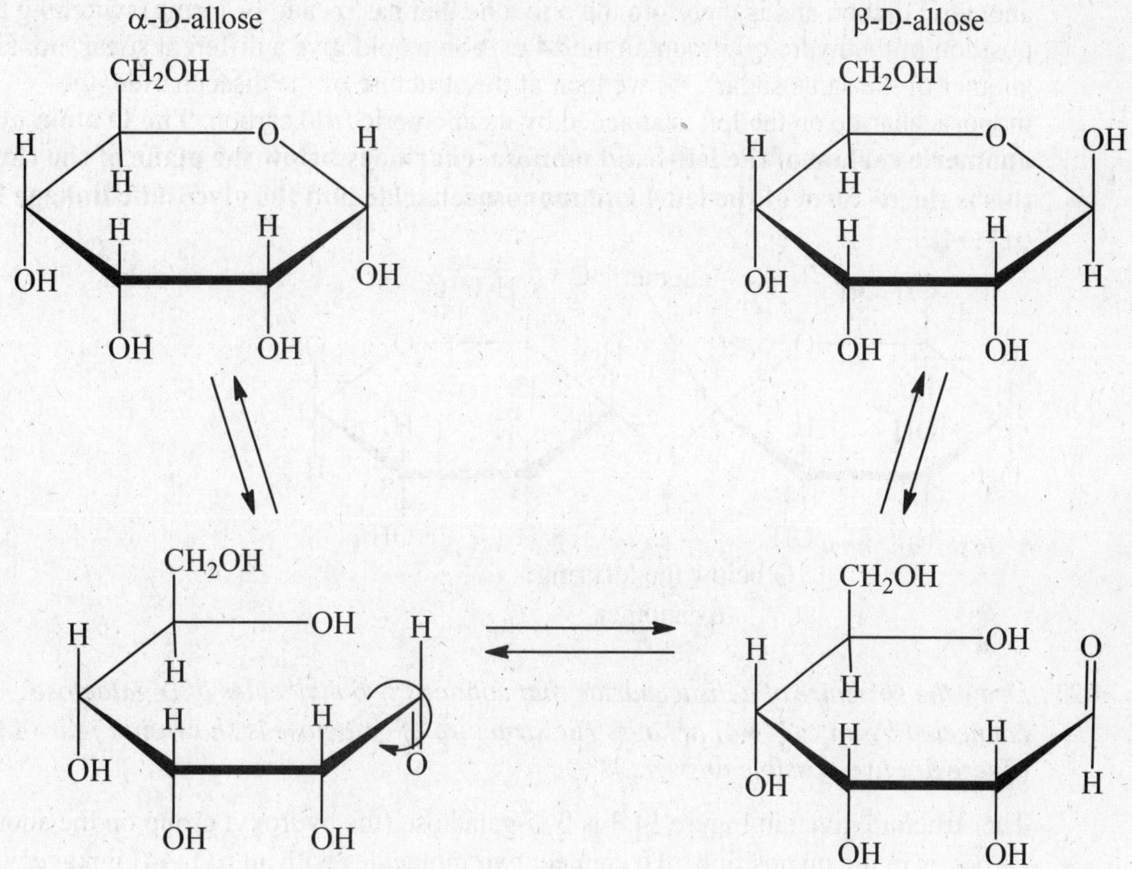

α-D-allose

β-D-allose

14.19 *Which of the following molecules is the enantiomer of α-D-fructose?*
a) β-D-fructose **b) α-L-fructose** *c) β-L-fructose*

Enantiomeric pairs of sugar molecules are labeled as D and L forms. Any other change made to the structure (α- to β-, change in the position of any other atoms to give a different sugar, etc) of a sugar does not result in the formation of an enantiomer. **The enantiomer of α-D-fructose is α-L-fructose.**

Solutions to Section 14.3 Core Problems

14.21 *Tell whether the glycosidic linkage in the molecule below is α(1→4) or β(1→4), and explain how you can tell:*

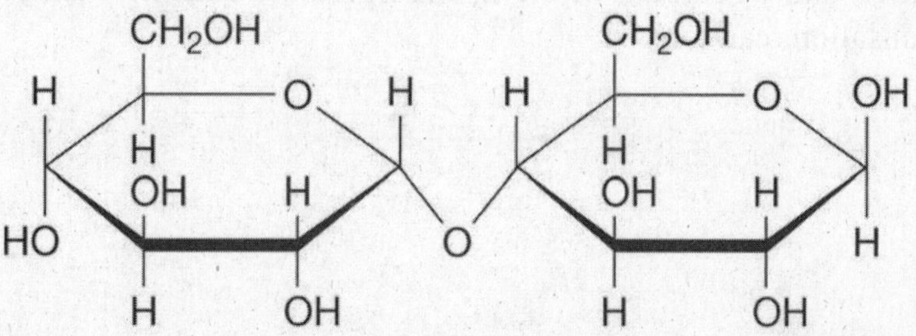

Let's review this naming system: the (1→4) part tells us that the condensation reaction that produces the glycosidic linkage takes place between the hydroxyl groups on the #1 carbon atom on one sugar and the #4 carbon atom on the other. The #1 sugar is the anomeric carbon and is therefore the only one that has α- and β- forms (switching the position of the hydroxyl group on the #4 carbon would give a different sugar, not an anomer of the same sugar). As we look at the structure of the disaccharide, the monosaccharide on the left is attached by its anomeric (#1) carbon. **The O atom on the anomeric carbon of the left-hand monosaccharide is below the plane of the ring, so this is the α- form of the left-hand monosaccharide and the glycosidic linkage is α(1→4).**

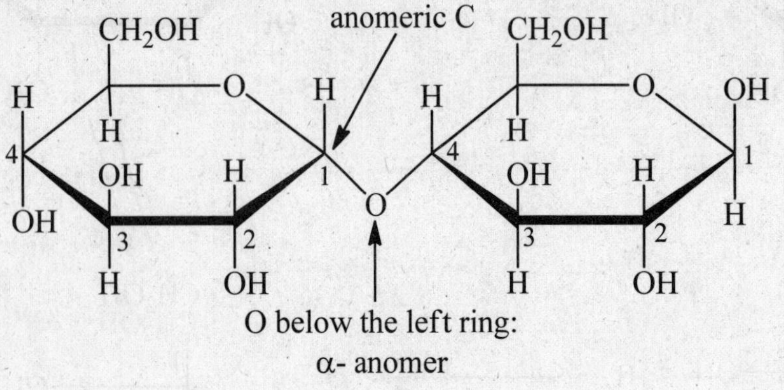

anomeric C

O below the left ring: α- anomer

14.23 *Draw the structure of a disaccharide that contains two molecules of D-galactose connected by an α(1→4) linkage. The structure of galactose is shown in Figure 14.3. (There are two possible answers.)*

The structure given in Figure 14.3 is β-D-galactose (the hydroxyl group on the anomeric carbon is in the up position). To connect two molecules with an α(1→4) linkage, we'll have to draw one of the two molecules as α-D-galactose:

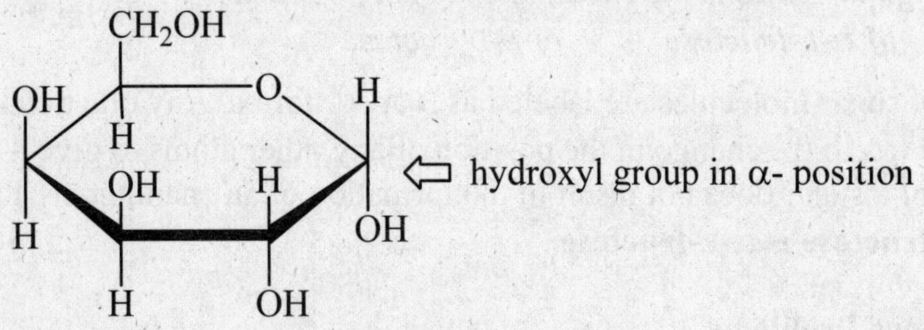

⇦ hydroxyl group in α- position

We'll use this α-D-galactose molecule in a condensation with another galactose molecule, in this case the original β-D-galactose. We line up the two molecules so that the hydroxyl groups on the #1 and #4 carbons of the two molecules are facing each other, in position for the condensation reaction:

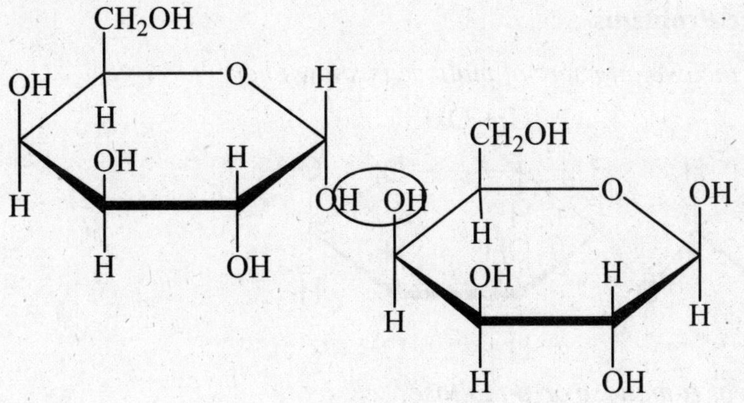

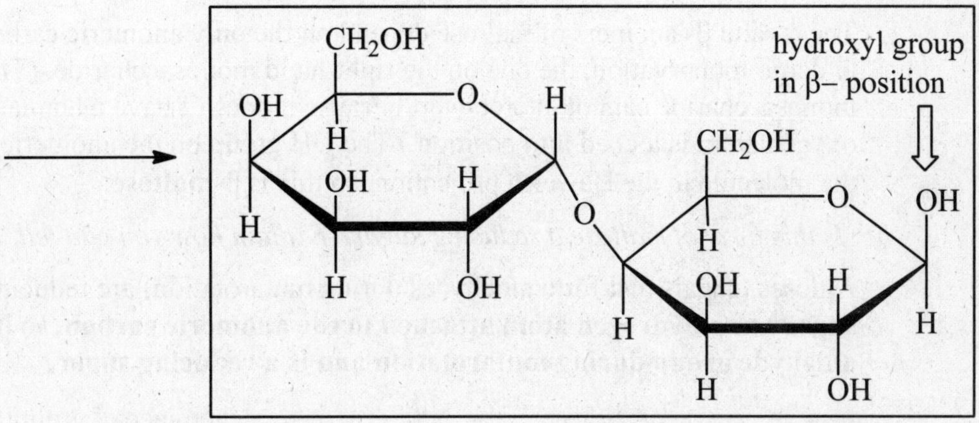

CH$_2$OH
OH
O
H
H
OH
H
CH$_2$OH
H
OH OH
O
OH
H
H
OH
H
OH
H
H
H
H
OH

CH$_2$OH
hydroxyl group
in β– position
OH
O
H
H
CH$_2$OH
OH
H
O
OH
H
O
H
H
OH
H
H
OH
H
H
H
OH

The left-hand molecule has to be the α- form, because the problem specifies the α(1→4) linkage. The second molecule could be in the α- form, however, and we've only shown it in the β- form, so the other possible answer option is to have both molecules in the α-form:

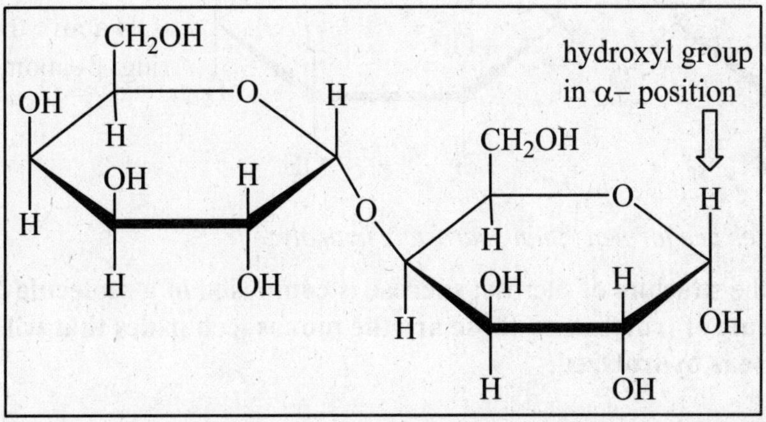

CH$_2$OH
hydroxyl group
in α– position
OH
O
H
H
CH$_2$OH
OH
H
O
H
H
O
H
H
OH
OH
H
H
OH
H
H
OH

Solutions to Section 14.4 Core Problems

14.25 *The structure of one of the two anomers of maltose is shown here.*

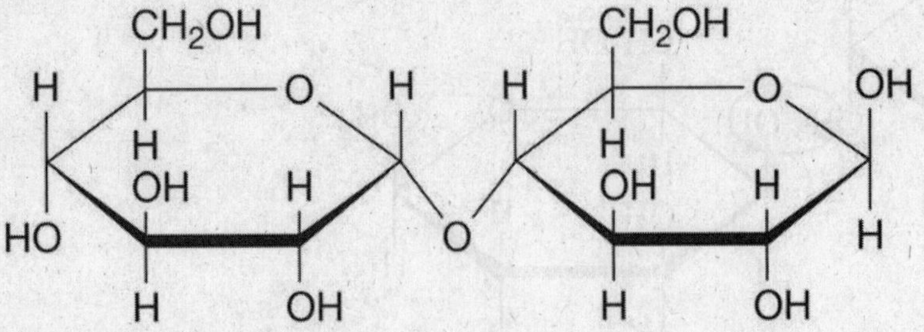

a) *Is this the structure of α-maltose or β-maltose?*

The α- and β- anomers of maltose depend on the only anomeric carbon that can still undergo mutarotation, the one on the right-hand monosaccharide. (The left-hand monosaccharide cannot interconvert because it doesn't have a hemiacetal group–the oxygen atom is locked into position.) The OH group on the anomeric carbon is above the molecule in the Haworth projection, so this is **β-maltose.**

b) *Is this form of maltose a reducing sugar? Explain how you can tell.*

Aldoses (sugars that form aldehydes during mutarotation) are reducing sugars. **This sugar has a hydrogen atom attached to the anomeric carbon, so it forms an aldehyde group during mutarotation and is a reducing sugar.**

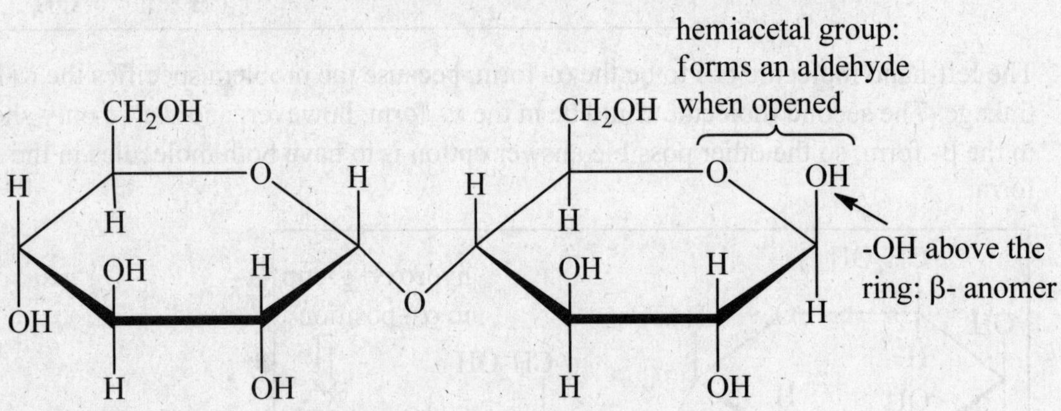

14.27 *What monosaccharides are formed when sucrose is hydrolyzed?*

Figure 14.13 shows the structure of sucrose; **sucrose is composed of a molecule of glucose and a molecule of fructose, so those are the monosaccharides that will be formed when sucrose is hydrolyzed.**

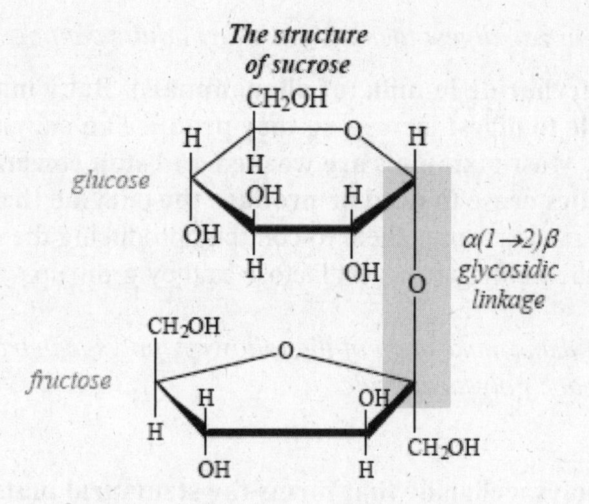

14.29 *a) Does sucrose have two anomeric forms? Explain.*

It does not. The structure of sucrose is shown in Figure 14.13. Anomeric forms interconvert through mutarotation reactions, which require a hemiacetal group at the anomeric carbon (the carbon atom to the right of the oxygen atom in the ring must have an –OH group on it for the forms to switch back and forth.) **Because the two monosaccharides are attached to each other through their anomeric carbons, neither of the two anomeric carbon can undergo mutarotation, so there is only one possible form of sucrose.** Sucrose does not contain a hemiacetal group, so it cannot undergo mutarotation.

See the answer to 14.11 for a review of mutarotation if you need to.

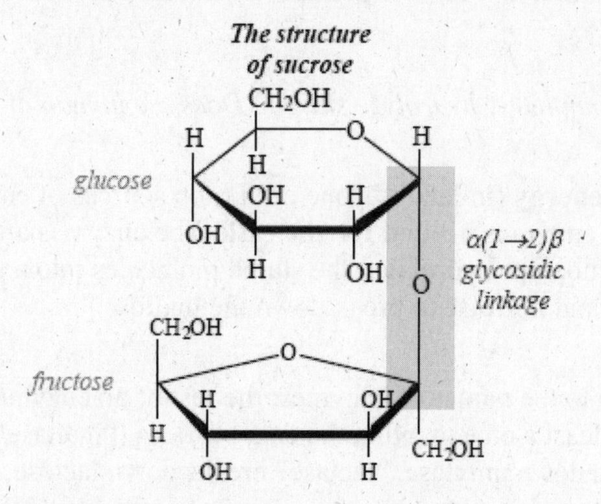

b) Is sucrose a reducing sugar? Explain.

Reducing sugars form aldehyde groups when the hemiacetal group breaks open during mutarotation. **Sucrose does not contain a hemiacetal group, so it is not a reducing sugar.**

14.31 *Virtually all children can digest lactose, but many adults cannot. Why is this?*

Lactose is the disaccharide in milk (of all mammals). Baby mammals live on milk, and need to be able to digest lactose, so they produce an enzyme that can digest the lactose molecules. Most mammals are weaned and stop consuming milk at some point, so their bodies cease to need or produce the enzyme that digests lactose. A mutation in some humans causes them to continue producing the enzyme to adulthood rather than losing their ability to digest lactose as they grow up.

14.33 *What types of organisms make each of the following polysaccharides, and what is the biological role of each polysaccharide?*

a) chitin

Chitin is the polysaccharide that forms the structural material in the exoskeletons of insects and other organisms with exoskeletons.

b) amylose

Amylose is an energy-storage polysaccharide produced by plants.

14.35 *What monosaccharide is used to build starch and cellulose?*

Both starch and cellulose are polysaccharides of **glucose.**

14.37 *What types of glycosidic bonds are present in each of polysaccharides in Problem 14.33?*

Chitin, like cellulose, is a structural material and contains $\beta(1\rightarrow4)$ glycosidic bonds. Amylose is an energy-storage material and is digestible by animals; it contains $\alpha(1\rightarrow4)$ glycosidic bonds.

14.39 *What enzymes are required to completely hydrolyze starch? Does the human digestive system make these enzymes?*

Humans can digest starch for energy (in fact, it's one of our top sources of energy in the diet), **so we do produce the enzymes needed for the task.** The enzymes are **amylase** and a **debranching enzyme** to chop up (hydrolyze) the starch molecules into a mixture of glucose and maltose molecules, and **maltase** to break down the maltose.

NOTE:

The –ase ending that is common to the names of enzymes; the rest of an enzyme's name is usually a description of, or at least a clue to, what the enzyme does ("maltase" breaks down maltose, "amylase" breaks down amylase, "lactase" breaks down lactose, "alcohol dehydrogenase" dehydrogenates alcohols, etc.)

Solutions to Section 14.5 Core Problems

14.41 *Classify each of the following fatty acids as saturated, monounsaturated, or polyunsaturated.*

a) $CH_3-(CH_2)_{12}-\overset{\overset{\displaystyle O}{\|}}{C}-OH$

b) $CH_3-(CH_2)_4-CH{=}CH-CH_2-CH{=}CH-(CH_2)_7-\overset{\|}{C}-OH$

If an organic molecule contains even one C=C or C≡C, it is referred to as "unsaturated." However, a double or triple bond between atoms of other elements, or between a carbon atom and an atom of another element, does not count. If there is exactly one C=C, the molecule is "monounsaturated;" if there are two or more C=C, the molecule is "polyunsaturated."

a) **Saturated.** This molecule has only C–C single bonds (the C=O bond in the carbonyl is irrelevant to the saturated/unsaturated designation).

NOTE:
Every C atom in the chain has two H atoms attached, so there's no way for there to be a double bond between C atoms in the chain.

b) **Polyunsaturated**. The chain contains two C=C double bonds.

14.43 *a) Predict whether the melting points of each of the fatty acids in Problem 14.41 will be higher or lower than room temperature.*

Saturated fatty acids, like those in bacon grease, are generally solid at room temperature (and would therefore have melting points higher than room temperature), while unsaturated fatty acids, like those in vegetable oil, are generally liquid at room temperature (and would therefore have melting points lower than room temperature). **We would therefore expect the melting point of substance a) to be higher than room temperature, and the melting point of substance b) to be lower than room temperature.**

b) Predict whether each of these fatty acids will be a solid or a liquid at room temperature.

We would expect substance a), a saturated fatty acid, to be solid at room temperature, and substance b), an unsaturated fatty acid, to be liquid at room temperature.

14.45 *Draw the structures of the triglycerides that contain the following fatty acids:*

"Triglycerides" contain three fatty acid molecules attached through condensation reactions to a glycerol backbone (the "tri" in triglyceride specifies three fatty acids.)

$$\begin{array}{l} HO{-}CH_2 \\ \quad\;\; | \\ HO{-}CH \quad \text{glycerol} \\ \quad\;\; | \\ HO{-}CH_2 \end{array}$$

a) three molecules of myristic acid

Myristic acid is a saturated fatty acid with 14 carbon atoms:

$$CH_3{-}(CH_2)_{12}{-}\overset{\overset{\large O}{\|}}{C}{-}OH$$

Attaching three of these fatty acids to a glycerol molecule gives us the following triglyceride:

$$\begin{array}{l} \quad\quad\quad\quad\quad\;\; O \\ \quad\quad\quad\quad\quad\;\; \| \\ CH_3-(CH_2)_{12}-C-O-CH_2 \\ \quad\quad\quad\quad\quad\;\; O \quad\;\; | \\ \quad\quad\quad\quad\quad\;\; \| \quad\;\;\; | \\ CH_3-(CH_2)_{12}-C-O-CH \\ \quad\quad\quad\quad\quad\;\; O \quad\;\; | \\ \quad\quad\quad\quad\quad\;\; \| \quad\;\;\; | \\ CH_3-(CH_2)_{12}-C-O-CH_2 \end{array}$$

b) *two molecules of lauric acid and one molecule of linolenic acid (there is more than one possible answer for this part)*

lauric acid

$$CH_3-(CH_2)_{10}-\overset{\overset{\large O}{\|}}{C}-OH$$

linolenic acid

$$CH_3\text{-}CH_2\text{-}CH=CH-CH_2-CH=CH-CH_2-CH=CH-(CH_2)_7-\overset{\overset{\large O}{\|}}{C}-OH$$

We attach two lauric acid molecules and one linolenic acid molecule to a glycerol backbone:

$$\begin{array}{r} CH_3-(CH_2)_{10}-\overset{\overset{O}{\|}}{C}-O-CH_2 \\ | \;\;\; \\ CH_3\text{-}CH_2\text{-}CH=CH-CH_2-CH=CH-CH_2-CH=CH-(CH_2)_7-\overset{\overset{O}{\|}}{C}-O-CH \;\; \\ | \;\;\; \\ CH_3-(CH_2)_{10}-\overset{\overset{O}{\|}}{C}-O-CH_2 \end{array}$$

The linolenic acid could also be on the top or bottom carbon of the glycerol, so there are two possible structures for this triglyceride (make sure you can draw the other one, and see that it doesn't matter whether you put the linolenic acid on the top or bottom, it's the same molecule either way.)

14.47 *Rank the following triglycerides from highest to lowest melting point.*
triglyceride #1: contains 3 molecules of stearic acid
triglyceride #2: contains 3 molecules of oleic acid
triglyceride #3: contains 2 molecules of oleic acid and 1 molecule of stearic acid

Remember saturated fats tend to have high melting points, while unsaturated fats tend to have low melting points. Stearic acid is saturated and oleic acid is unsaturated, so the more stearic acid in a fat, the higher its melting point, while the more oleic acid a fat contains, the lower its melting point will be. **Triglyceride #1, a completely saturated fat, will have the highest melting point; triglyceride #3, with a mix of saturated and unsaturated fatty acids, will be in the middle; and triglyceride #2, with three unsaturated fatty acids, will have the lowest melting point of the three.**

14.49 *The essential fatty acid linolenic acid is sometimes called alpha-linolenic acid. Certain plant oils contain an isomer of this compound, called gamma-linolenic acid. The structure of gamma-linolenic acid is shown below. Classify this fatty acid using the omega system.*

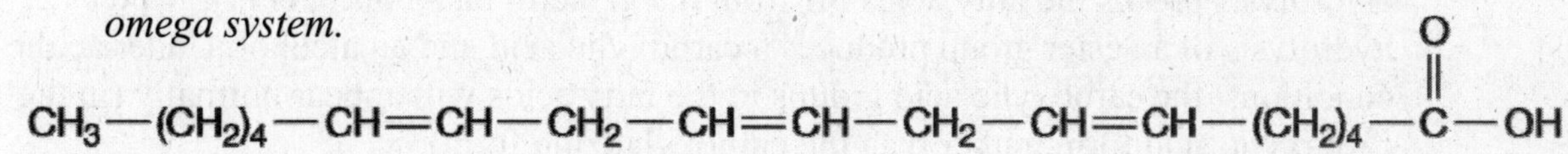

$$CH_3-(CH_2)_4-CH=CH-CH_2-CH=CH-CH_2-CH=CH-(CH_2)_4-\overset{\overset{\large O}{\|}}{C}-OH$$

For purposes of classifying fatty acids according to the omega system, the only part of the molecule we care about is the end of the molecule away from the carboxylic acid group. We count from the last carbon in the chain (the omega carbon, the one farthest from the carboxylic acid) to the first double bond we reach. We can ignore the rest of the molecule. In this molecule, the first double bond we reach is the #6 bond in the molecule, so this is an **omega-6 fatty acid**.

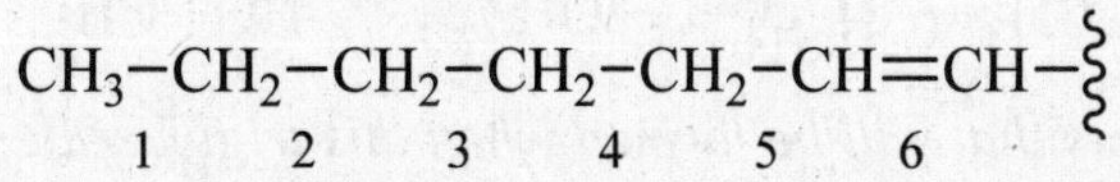

$$\underset{1}{CH_3}-\underset{2}{CH_2}-\underset{3}{CH_2}-\underset{4}{CH_2}-\underset{5}{CH_2}-\underset{6}{CH}=CH-\wr$$

Solutions to Section 14.6 Core Problems

14.51 a) *Draw the structure of the product that is formed when the following triglyceride is completely hydrogenated.*

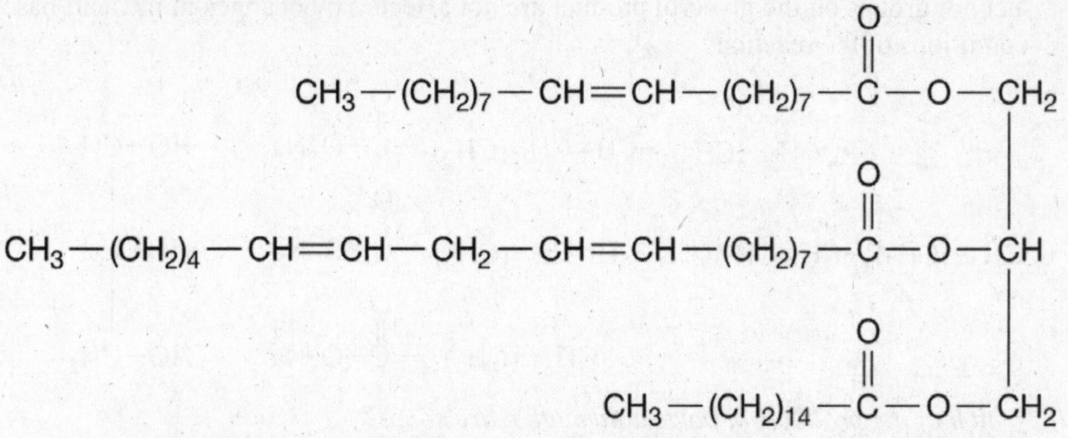

$$CH_3-(CH_2)_7-CH=CH-(CH_2)_7-\overset{\overset{\large O}{\|}}{C}-O-CH_2$$
$$CH_3-(CH_2)_4-CH=CH-CH_2-CH=CH-(CH_2)_7-\overset{\overset{\large O}{\|}}{C}-O-CH$$
$$CH_3-(CH_2)_{14}-\overset{\overset{\large O}{\|}}{C}-O-CH_2$$

Hydrogenation turns all double and triple bonds between carbon atoms into single bonds (converts alkenes and alkynes to alkanes). All of the carbon atoms in the interior of the chain will end up as $-CH_2-$ groups, so you can count up the total number of carbon atoms in the chain and combine them in parentheses:

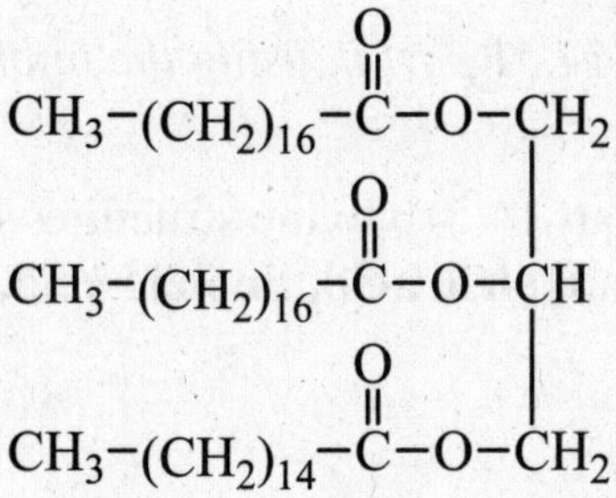

$$CH_3-(CH_2)_{16}-\overset{\overset{\large O}{\|}}{C}-O-CH_2$$
$$CH_3-(CH_2)_{16}-\overset{\overset{\large O}{\|}}{C}-O-CH$$
$$CH_3-(CH_2)_{14}-\overset{\overset{\large O}{\|}}{C}-O-CH_2$$

You can also draw out the condensed formula or line formula if you prefer (not shown).

b) *Draw the structures of the products that will be formed when this triglyceride is hydrolyzed using a 2 M solution of H_2SO_4.*

Hydrolysis breaks the fatty acids off from the glycerol backbone. As in Chapter 12, hydrolysis of an ester group produces a carboxylic acid and an alcohol. Under acidic conditions, the carboxylic acid groups in the fatty acids will appear normally (in the carboxylic acid form, rather than the carboxylate ion form):

$$CH_3{-}(CH_2)_7{-}CH{=}CH{-}(CH_2)_7{-}\overset{\overset{\large O}{\|}}{C}{-}OH$$

$$CH_3{-}(CH_2)_4{-}CH{=}CH{-}CH_2{-}CH{=}CH{-}(CH_2)_7{-}\overset{\overset{\large O}{\|}}{C}{-}OH$$

$$CH_3{-}(CH_2)_{14}{-}\overset{\overset{\large O}{\|}}{C}{-}OH$$

$$\begin{array}{l} HO{-}CH_2 \\ \quad\;\; | \\ HO{-}CH \\ \quad\;\; | \\ HO{-}CH_2 \end{array}$$

c) *Draw the structures of the products that will be formed when this triglyceride is hydrolyzed using a 2 M solution of NaOH.*

The products are the same as in the hydrolysis in part b), except that under basic conditions, the carboxylic acid groups in the fatty acid products will be converted to their carboxylate ions and will form salts with the sodium ions from the NaOH. The alcohol groups on the glycerol product are not affected by changes in the acid/base conditions of the reaction:

$$CH_3{-}(CH_2)_7{-}CH{=}CH{-}(CH_2)_7{-}\overset{\overset{\large O}{\|}}{C}{-}O^{-}\,Na^{+}$$

$$CH_3{-}(CH_2)_4{-}CH{=}CH{-}CH_2{-}CH{=}CH{-}(CH_2)_7{-}\overset{\overset{\large O}{\|}}{C}{-}O^{-}\,Na^{+}$$

$$CH_3{-}(CH_2)_{14}{-}\overset{\overset{\large O}{\|}}{C}{-}O^{-}\,Na^{+}$$

$$\begin{array}{l} HO{-}CH_2 \\ \quad\;\; | \\ HO{-}CH \\ \quad\;\; | \\ HO{-}CH_2 \end{array}$$

d) *Which of the products of parts a through c are soaps?*

Soaps are ionic salts formed by combining the carboxylate forms of fatty acid molecules with cations. **The three fatty acid salts in part c) are soaps**; the glycerol molecules in b) and c), the fatty acids in b), and the saturated fat in a) are not soaps.

14.53 *Identify the fatty acids that are formed in Problem 14.51 part b, using the information in Table 14.5.*

Compare the structures of the three fatty acids in Part 14.51b to the structures in the table. The three fatty acids, in order from top to bottom, are **oleic acid, linoleic acid, and palmitic acid.**

14.55 *What is a trans fatty acid, and under what circumstances are trans fatty acids formed?*

***Trans* fatty acids are unsaturated fatty acids in which the alkene functional group is *trans* rather than *cis*. Naturally occurring unsaturated fatty acids almost always contain *cis* double bonds, but during the process of hydrogenating fats (to turn liquid unsaturated oils into solid saturated fats) some of the double bonds, upon coming into contact with the hydrogenation catalyst, are not hydrogenated but instead flip to the *trans* form.**

14.57 *What products are formed when a triglyceride is hydrolyzed in the digestive tract?*

Digestive enzymes (lipases–the "-ase" indicates an enzyme, while the "lip-" refers to lipids or fats) remove two of the fatty acids from a triglyceride molecule, **forming a monoglyceride and two fatty acid molecules. The fatty acids are converted to the carboxylate ion form in the digestive tract.**

14.59 *What is the function of bile salts in fat digestion?*

Fats, being hydrophobic, tend to clump together in the aqueous environment of the digestive tract and form large globs that can't be efficiently attacked by digestive enzymes. Bile salts (which act like detergents) help to break the large globs of fat into smaller globs for more efficient and rapid digestion. (Remember from Chapter 6 that the rate of a reaction is affected by the area of contact between the reactants–if the fat is in smaller droplets, the digestion reactions can happen more rapidly.)

Solutions to Section 14.7 Core Problems

14.61 *Which of the building blocks of a glycerophospholipid are strongly hydrophilic?*

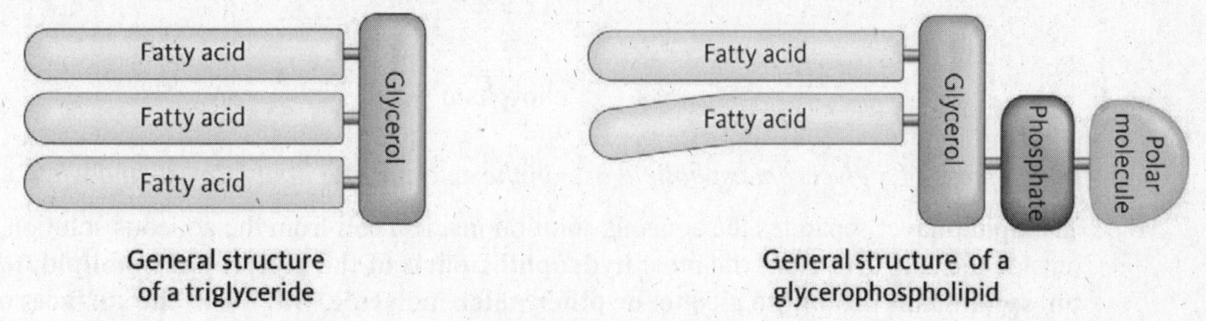

"Glycero-" means the molecule contains glycerol; "phospho-" means it contains phosphate; and "lipid" indicates a fatty molecule. **The phosphate ion is strongly hydrophilic**, but glycerophospholipids also contain another polar molecule, usually an amino alcohol attached through its side chain; **the amino alcohol is also strongly hydrophilic**, as its amine and alcohol groups are exposed.

14.63 *Draw the structure of the glycerophospholipid that contains two molecules of stearic acid and a molecule of serine.*

$$CH_3-(CH_2)_{16}-COOH \qquad HO-CH_2-CH(COO^-)-NH_3^+$$

Stearic acid **Serine**

The "glycerophospho" part of the word tells us we need a glycerol molecule and a phosphate group. We look in Table 14.5 for the structure of stearic acid and attach two stearic acid molecules to the glycerol backbone, then find the structure of serine in Table 13.1 and attach it to the phosphate group (remember that serine is acting here as an amino alcohol and attaching through its alcohol group):

$$\begin{array}{l} CH_3-(CH_2)_{16}-C(=O)-O-CH_2 \\ CH_3-(CH_2)_{16}-C(=O)-O-CH \\ \qquad\qquad\qquad\qquad\qquad\quad CH_2-O-P(=O)(O^-)-O-CH_2-CH(COO^-)-NH_3^+ \end{array}$$

stearic acid; glycerol; serine; phosphate

14.65 *Which parts of a glycerophospholipid are on the exterior of a lipid bilayer, and why?*

The lipid bilayer separates the aqueous solution inside a cell from the aqueous solution outside the cell. Therefore **the most hydrophilic parts of the glycerophospholipid, the phosphate and the amino alcohol or other polar molecule, will be on the surfaces of the lipid bilayer**, while the hydrophobic parts of the molecule, the fatty acid chains, will make up the interior of the lipid bilayer.

14.67 *CO_2 can cross a lipid bilayer, but HCO_3^- cannot. Why is this?*

The interior of the lipid bilayer is nonpolar and strongly hydrophobic. CO_2 is also nonpolar, and as such is soluble in the nonpolar interior region. Ions, on the other hand, are much more strongly attracted to the water in the aqueous solutions in contact with the exterior surfaces of the lipid bilayer, and will not dissolve in the nonpolar interior of the membrane (this has less to do with their inability to dissolve in the nonpolar phase than with a strong attraction for the aqueous phase).

14.69 *What is the function of transport proteins, and why do cells need transport proteins?*

The basic purpose of the phospholipid bilayer is to separate the solution inside the membrane from the solution outside. However, cells routinely need to move materials in and out across the membrane (in a selective way, of course). **Transport proteins allow selective passage of molecules or ions that cannot pass freely through the phospholipid bilayer.**

14.71 *Cells can absorb the amino acid valine from the surrounding fluid when the concentration of valine is higher inside the cell than outside.*

Remember back to Chapter 5: things tend to spread out more or less evenly given the chance, so a solute like valine would normally be expected to spread (dialyze or diffuse) through a membrane from areas of higher concentration to areas of lower concentration. This problem is stating the opposite: that even when the concentration of valine inside is higher than the outside, the cell can still bring in even more valine. We now classify the normal Chapter 5 tendency of solute concentrations to even out as "passive transport" (because it happens without the cell having to do anything special) and the opposite process–increasing rather than decreasing differences in concentration between the sides of a membrane–as "active transport" (because something in the cell has to make it happen, it can't just let it work itself out.)

a) *Is this an example of active or passive transport?*

Since the valine is moving in the opposite direction from that predicted by Chapter 5, this is an example of **active transport**.

b) *Does this type of transport require energy?*

Passive transport happens on its own, but **active transport requires an input of energy.**

c) *Does this type of transport produce a concentration gradient?*

Concentration gradients are differences in the concentration of a solute across a membrane. Passive transport (normal dialysis or diffusion) evens out the differences in concentration of a solute inside and outside of the cell, but active transport maintains or increases these differences in concentration, **producing a concentration gradient.**

14.73 *Why are concentration gradients important in living organisms?*

Producing a concentration gradient requires an input of energy, and as such is a way of storing potential energy. Cells can then use the stored energy to carry out other processes.

Solutions to Section 14.8 Core Problems

14.75 *What is the function of cholesterol in the human body?*

Cholesterol in cell membranes allows the membranes to remain flexible over a wide range of temperatures, and the substance serves as the starting material for all other steroids the body produces.

14.77 *Which type of steroid hormone has the following function?*

a) *Regulation of the concentration of sodium ions in the blood.*

Mineralocorticoids are responsible for regulating the concentrations of Na^+ and K^+ concentration in blood plasma. The root "mineral-" here is a clue that we're dealing with inorganic ions.

b) *Regulation of the development of the female reproductive system.*

Estrogens are the group of hormones responsible for regulating maturation of the reproductive system in females.

14.79 *A chemist carries out a reaction that makes a small change to the structure of a steroid hormone. Is this change likely to have a significant impact on the physiological effect of the hormone, or will it probably have only a minor effect? Explain.*

It's likely to have a significant impact on the effect of the hormone. Even small changes to the structures of hormone molecules can cause dramatic and unpredictable differences in their functions.

14.81 *Why do steroids require specialized molecules to carry them throughout the body?*

Steroids are large hydrocarbon molecules with (generally) only a few hydrophilic functional groups, so they tend to be quite insoluble in water but soluble in fats. To move through the body, steroids have to be carried in the blood, which is an aqueous mixture. Since steroids are not soluble in blood, they have to attach to specialized protein molecules that can carry them in the blood and other aqueous body fluids.

14.83 *Which two types of lipoproteins are the primary carriers of cholesterol in the blood? How do their roles differ?*

Low density lipoproteins (LDLs) and high density lipoproteins (HDLs) are the main cholesterol transport systems in blood. LDL carries stored cholesterol from the liver to other tissues. HDL carries cholesterol back to the liver.

Solutions to Section 14.9 Core Problems

14.85 *Many complex molecules in our bodies require glucose as a building block. List two ways in which our bodies can obtain the glucose they need to build these molecules.*

Our bodies are capable of (1) manufacturing glucose from other nutrients (amino acids, glycerol, etc), but generally our bodies (2) obtain glucose from dietary carbohydrates.

14.87 a) *What types of fatty acids are classified as essential, and why?*

Our bodies can make saturated and monounsaturated fatty acids from carbohydrates or amino acids, but we cannot make polyunsaturated fatty acids, some of which are needed for the body to function properly. As a result, **linoleic acid and linolenic acid, which contain two and three alkene groups respectively, are dietary**

requirements for all humans and all other mammals and are called essential fatty acids.

b) *How do our bodies obtain these essential fatty acids?*

Our diet must include fats that contain these fatty acids.

14.89 a) *What are the starting materials in photosynthesis, and what are the products?*

Photosynthesis is the process by which plants turn water and carbon dioxide (the starting materials) into glucose and O_2 (g) (the products).

b) *What is the source of the energy that is needed in photosynthesis?*

Photosynthesis is a highly endothermic process; **the energy is provided in the form of sunlight.** There are clues to this in the name: "photo" has to do with light, "synthesis" means "putting together."

Solutions to Concept Questions

* indicates more challenging problems.

14.91 *your body uses α-D-glucose to build glycogen. Which of the following could our bodies use as a source of the α-D-glucose? Explain your answers.*

a) β-D-glucose *b) α-L-glucose* *c) lactose*

a) It's easy to convert back and forth between α- and β- forms of a sugar, **so a) β-D-glucose is fine as a source of α-D-glucose.**

b) There is no way that our bodies can convert between enantiomers of sugars, so **α-L-glucose cannot be used as a source of α-D-glucose.**

c) The structure of lactose is shown in Figure 14.13. Lactose is made from β-D-glucose and β-D-galactose, so **hydrolysis of lactose is one possible source of glucose for the body**, at least for a person who can digest lactose (see Problem 14.31).

14.93 *Rank the following solvents based on how well glucose should dissolve in them. Start with the best solvent.*

acetone *water* *3-pentanone*

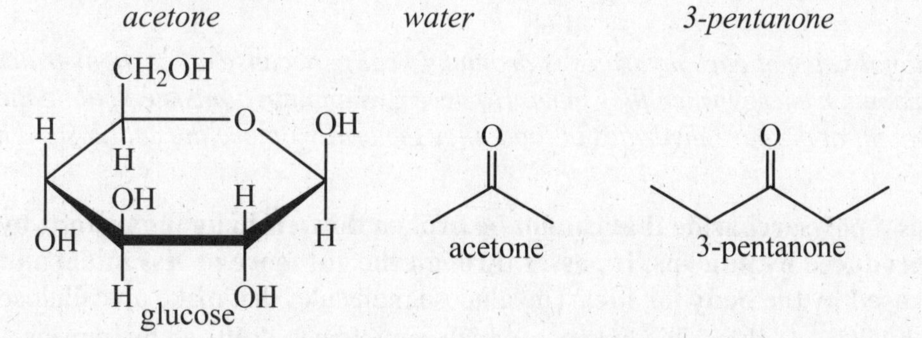

Chapter 5 review! Glucose, with an alcohol group on every carbon atom, is very hydrophilic and highly soluble in water. The more hydrophilic the solvent, the more soluble glucose is expected to be, and it's expected to be relatively insoluble in hydrophobic solvents (the glucose molecules would rather stay together and form hydrogen bonds than disperse into a hydrophobic solvent, where dispersion forces are the only source of attraction between molecules.) 3-pentanone is the least hydrophilic/most hydrophobic of the three solvents listed, so the solubility of glucose in 3-pentanone is expected to be the lowest of the three. Acetone, with smaller alkyl groups, is less hydrophobic than 3-pentanone, so glucose will be more soluble in acetone than in 3-pentanone.

In order, then:
water ← **most hydrophilic, best ability to dissolve glucose, best solvent**
acetone
3-pentanone ← **most hydrophobic, least ability to dissolve glucose, poorest solvent**

14.95 *Why can't your body use the L forms of carbohydrates?*

The enzymes that process carbohydrates are chiral, just as the carbohydrates themselves are, with active sites that are specific to the shapes of the D forms of the carbohydrates. The L forms don't fit in the active sites of the enzymes (just as your right foot doesn't fit properly in your left shoe.)

14.97 *Lactase (the enzyme that breaks down lactose) can also hydrolyze the glycosidic bond in cellobiose. However, lactase cannot break down maltose or sucrose. Based on the structures of these disaccharides, explain why this is reasonable.*

The structures of all these disaccharides can be found in Sections 14.3 and 14.4. Lactose and cellobiose both have β(1→4) glycosidic linkages, so it is not surprising that the same enzyme could hydrolyze both. The glycosidic linkages in maltose and sucrose are α(1→4) and α(1→2)β, respectively, so the glycosidic linkages in these disaccharides won't fit properly into the active site of the enzyme for hydrolysis.

14.99 *If you form a glycosidic bond that links carbon 1 of glucose molecule to carbon 1 of a second glucose molecule, will the product be a reducing sugar? Why or why not?*

It will not. The #1 carbon atom is the anomeric carbon, the carbon atom that's involved in the ring opening to form the hemiacetal. If both anomeric carbons are involved in the glycosidic linkage, there is no hemiacetal group, and the disaccharide cannot be a reducing sugar. See Section 14.6.

14.101 *The nutritional value of carbohydrates is around 4 kcal/g, because all carbohydrates produce this much energy when they burn. However, when nutritionists calculate the Calorie content of the carbohydrates in food, they generally ignore the cellulose. Why is this?*

Cellulose is a polysaccharide that cannot be broken down into its sugar units by enzymes produced by humans. It passes through the gut more or less intact and cannot be used by the body for fuel. The glucose molecules that make up cellulose are connected by β(1→4) glycosidic linkages, which cannot be hydrolyzed by humans.

14.103 *Why are the sodium salts of fatty acids more soluble than the original fatty acids?*

The carboxylate ion group is much more hydrophilic than the corresponding carboxylic acid group.

14.105 a) *Why is the melting point of a saturated fatty acid higher than the melting point of an unsaturated fatty acid that contains the same number of carbon atoms?*

Saturated fatty acid molecules have more linear shapes and stack together more easily than do *cis*-unsaturated fatty acids, which tend to have bent shapes because of the double bonds. One of the main differences between solids and liquids is orderly stacking of the particles. The poor ability of *cis*-unsaturated fatty acids to stack neatly means the molecules don't attract each other as strongly, and it takes less energy to overcome the dispersion force attractions between the molecules in the solid phase and melt the sample.

b) *The melting points of trans fatty acids are considerably higher than the melting points of cis fatty acids that contain the same number of carbon atoms. Why is this?*

The overall shape of *trans* fatty acids is much closer to that of saturated fats than it is to the shape of *cis* fatty acids. *Trans* double bonds don't cause a kink in the hydrocarbon tail, but rather allow it to remain straight as in saturated fatty acids.

O
OH
saturated fatty acid

O
OH
trans-fatty acid

O
OH
cis-fatty acid

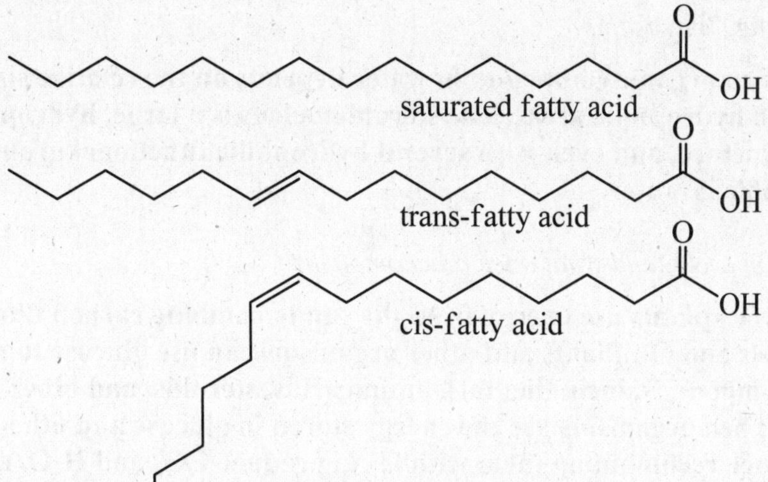

14.107 *Glycerophospholipids form lipid bilayers when they are mixed with water, but they do not do so when they are mixed with ethanol. Why is this?*

The phosphate groups on the phospholipids are strongly hydrophilic (attracted to water), but the fatty tails are hydrophobic. The lipid bilayers form in water because of the tendency of the hydrophobic fatty acid tails to form a separate hydrophobic phase away from the water molecules. Ethanol has an alkyl group that can interact with the hydrophobic tails, which drastically reduces their tendency to aggregate together and exclude solvent.

14.109 *Why can't ions cross a lipid bilayer?*

The interior of the lipid bilayer is nonpolar and strongly hydrophobic. Ions are strongly attracted to the water in the aqueous solutions in contact with the exterior surfaces of the lipid bilayer, and will not dissolve in the nonpolar interior of the membrane (this has less to do with an inability to dissolve in the nonpolar phase than with a strong attraction for the aqueous phase).

14.111 *Cells can only move glucose across the cell membrane when there is much higher concentration of sodium ions outside the cell than there is inside the cell. Explain how these are connected.*

The sodium ion concentration gradient is a form of potential energy for the cell; the cell can take the tendency of the sodium ion to move in one direction, releasing energy, and use that energy to carry out other processes that require energy, such as moving glucose across the membrane.

14.113 *All of the steroid hormones contain hydrophilic functional groups, yet they have very low solubilities in water. Why is this?*

The solubility of an organic compound in water depends on the relative sizes of its hydrophobic and hydrophilic areas. The steroid nucleus is a large, hydrophobic hydrocarbon structure, and even with several hydrophilic functional groups it tends to remain insoluble in water.

14.115 *What is the carbon cycle, and what does it accomplish?*

In the carbon cycle, plants use energy from the sun to combine carbon dioxide and water into glucose and O_2. Plants and other organisms can use glucose to make other organic compounds, including fats, amino acids, steroids, and other carbohydrates. Then organisms use the energy stored in glucose and other organic compounds as fuel, recombining them with O_2 to produce CO_2 and H_2O, and the cycle repeats. The carbon cycle takes energy from sunlight, and turns it into a form of stored energy that can be used by all the other organisms in the world, the ones that can't carry out photosynthesis (everything but plants).

Solutions to Summary and Challenge Problems

14.117 *The following two compounds are made from glucose. Which of these compounds (if any) can be oxidized by the Benedict's reagent? Explain your answer.*

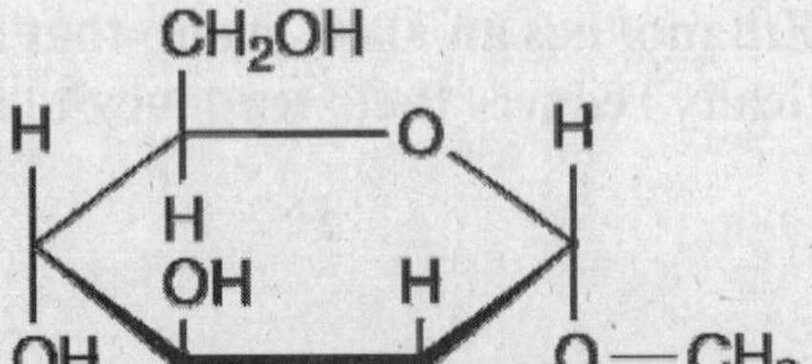

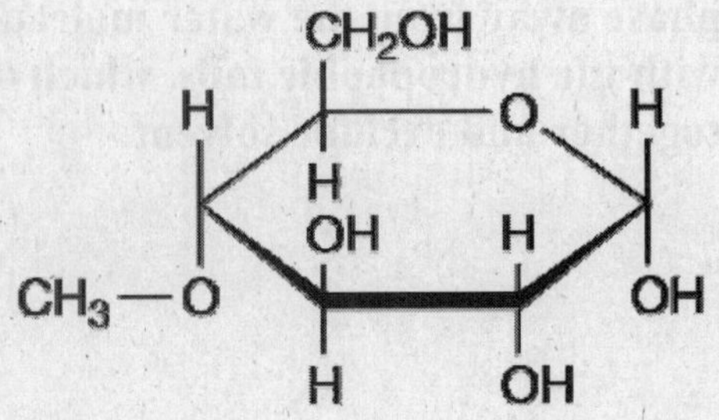

Benedict's reagent is used to detect reducing sugars, those that form aldehydes when the hemiacetal group opens. The first molecule does not have a hemiacetal group (there's no hydroxyl group on the anomeric carbon–the oxygen has a methyl group instead of a hydrogen), so it will not react with Benedict's reagent. The second molecule can form an aldehyde when the hemiacetal group opens, so it will react with Benedict's reagent:

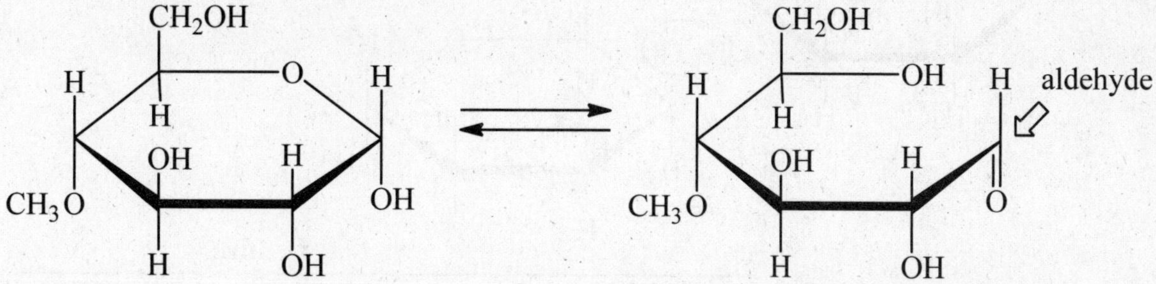

14.119 *The monosaccharide below is called D-talose. The hydrogen and hydroxyl groups on the anomeric carbon atom have been omitted.*

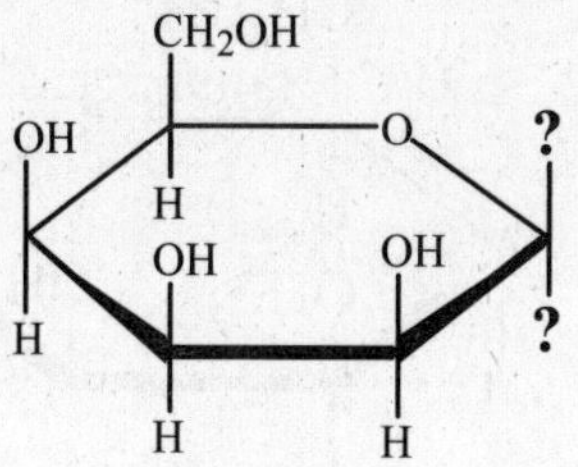

D-talose

a) *Draw the structure of α-D-talose by adding H and OH to this structure.*

To draw the α- form of the monosaccharide, we need to show the –OH group on the anomeric carbon pointing down, below the ring:

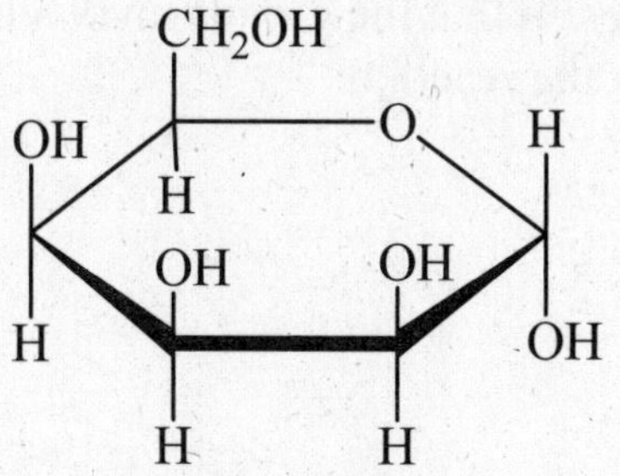

α-D-talose

b) *Draw the structure of the disaccharide that is formed when two molecules of α-D-talose form an α(1→4) glycosidic linkage.*

In the α anomer, the –OH group on the anomeric carbon is below the ring. We're only asked for the product, but it will be easier to see if we set up the condensation reaction between the two molecules. First, we orient the two molecules of α-D-talose so that the –OH on the #1 carbon in one molecule is facing the –OH on the #4 carbon atom in the other, in position for a condensation reaction. Then we identify the H and

OH that are to be removed, and write the product molecule with its glycosidic (ether) linkage:

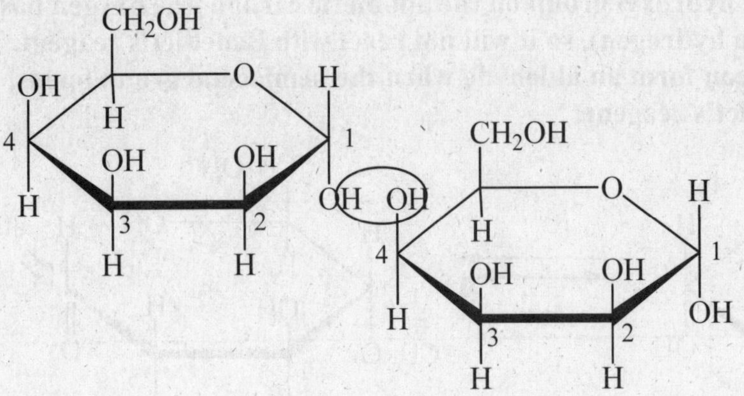

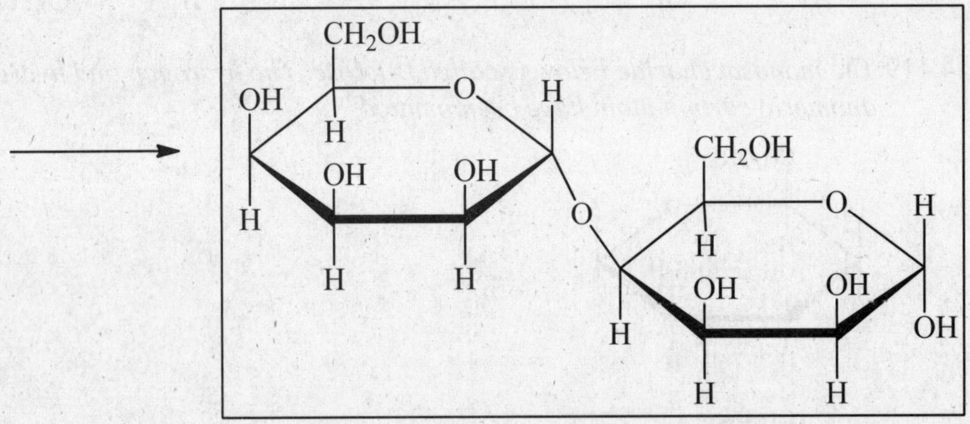

14.121 *Draw the structure of the molecule that will be formed when three molecules of β-D-glucose are bonded together by β(1→4) glycosidic linkages.*

In the β anomer, the –OH group on the anomeric carbon is above the ring. As in any other condensation reaction, we first orient the molecules so that the groups involved in the condensation are facing each other, in position for the reaction:

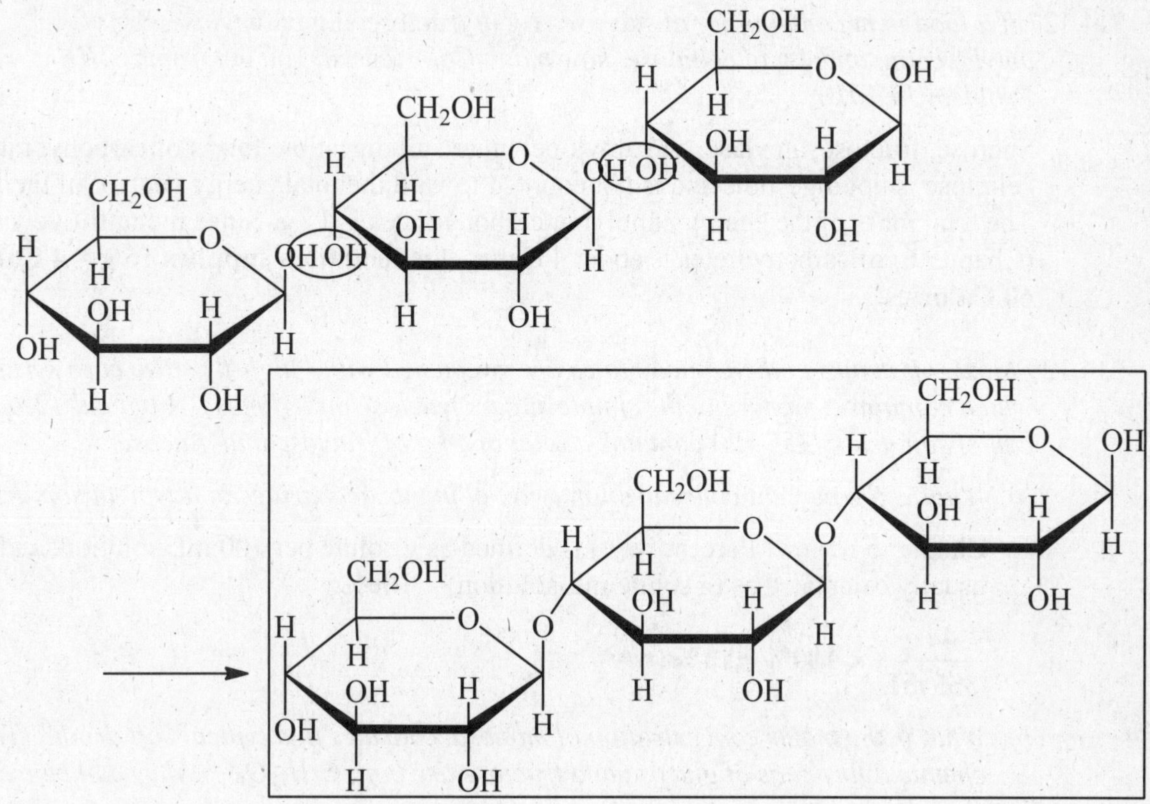

*14.123 *Glucose has the molecular formula $C_6H_{12}O_6$. What is the molecular formula of the compound that contains four molecules of glucose connected to one another by α(1→4) glycosidic linkages?*

Linking 4 molecules requires forming 3 glycosidic linkages, and each time we form a glycosidic linkage through a condensation reaction we produce a water molecule (H_2O). **So our strategy is to multiply the molecular formula by 4, to get $C_{24}H_{48}O_{24}$, then subtract three water molecules (6 H and 3 O) to get $C_{24}H_{42}O_{21}$.**

NOTE:

It doesn't actually matter whether the glycosidic linkages are α(1→4) or some other type; any glycosidic linkage is formed by a condensation reaction, so the type of linkage has no effect on the molecular formula, only on the shape of the structure.

14.125 *How does the structure of chitin differ from that of cellulose, and how are their structures similar to each other?*

The structures of chitin and cellulose both contain long chains of monosaccharides linked by β(1→4) glycosidic linkages. However, while cellulose is composed of long chains of glucose molecules, chitin is made up of long chains of a modified sugar called N-acetylglucosamine.

*14.127 *If a food item contains 3 g of sucrose, 1 g of fructose, 3 g of amylose, 8 g of amylopectin, and 2 g of cellulose, how many Calories will you get from it? (Hint: refer to Problem 14.101.)*

Sucrose, fructose, amylase, and amylopectin are all digestible forms of carbohydrates, but cellulose is not digestible and is not counted toward the total energy content of the food. The total mass of the energy-supplying carbohydrates is 15 g. Since the nutritive value (Chapter 6) of carbohydrates is about 4 Cal/g, **this food item supplies 15 g × 4 Cal/g = 60 Calories.**

*14.129 *Most soft drinks in the United States are sweetened with "high-fructose corn syrup," which contains a mixture of the simple sugars fructose and glucose. A typical 12-ounce can of soft drink (355 mL) contains a total of 44 g of fructose and glucose.*

a) *What is the percent concentration (w/v) of monosaccharides in a typical soft drink?*

Chapter 5 review! Percent (w/v) is defined as g solute per 100 mL solution, and usually calculated as (g solute/mL solution) × 100%:

$$\frac{\mathbf{44\,g}}{\mathbf{355\,mL}} \times \mathbf{100\% = 12\,\%\,(w/v)}$$

b) *What is the molar concentration of monosaccharides in a typical soft drink? (The chemical formulas of glucose and fructose are both $C_6H_{12}O_6$, so they can be treated as if they were the same compound.)*

"Molar concentration" means molarity, moles solute/L solution. First we need to determine the number of moles of sugar in 44 g, and to do that we need to add up the molar mass of the formula $C_6H_{12}O_6$, 180.2 g/mol. The volume of the soda solution in liters is 0.355 L (Chapter 1). Then,

$$\frac{\mathbf{44\,g}}{\mathbf{0.355\,L}} \times \frac{\mathbf{1\,mol}}{\mathbf{180.2\,g}} = \mathbf{0.69\,M}$$

c) *Is a soft drink isotonic, hypertonic, or hypotonic? (Recall that the isotonic concentration is around 0.28 M.)*

The soft drink solution is strongly hypertonic, with a concentration of sugar alone (not counting salts, coloring agents, other flavor compounds, etc) between two and three times the total solute concentration in blood plasma.

14.131 *A student draws the following line structure for palmitoleic acid:*

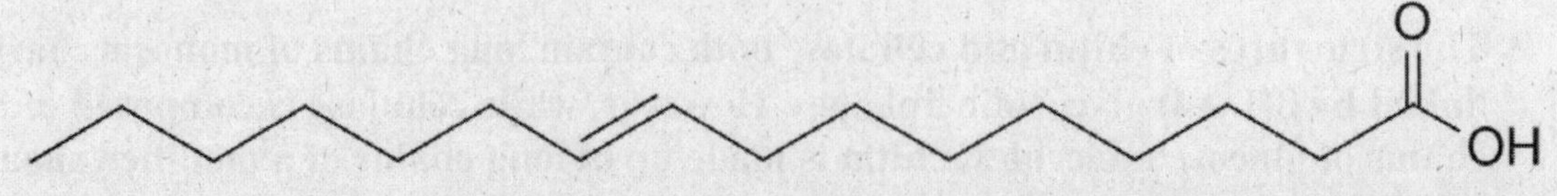

Why is this structure potentially misleading? (Hint: what does this structure imply about the double bond?) Draw it in a fashion that would not be misleading.

As the line structure is drawn, the double bond is shown as a *trans* double bond. The condensed structure would look like this:

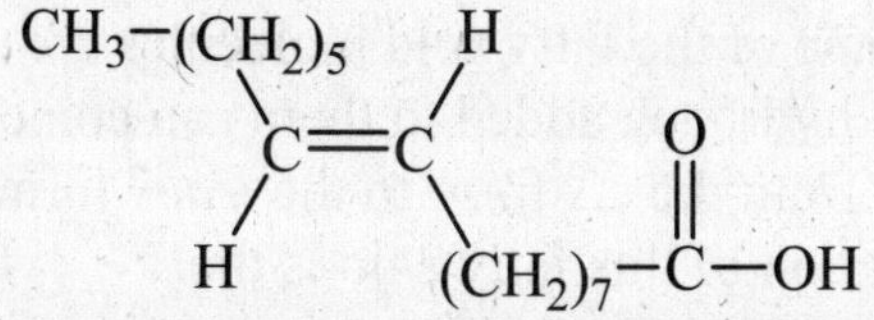

$$CH_3{-}(CH_2)_5{-}CH{=}CH{-}(CH_2)_7{-}\overset{O}{\overset{\|}{C}}{-}OH$$

(trans form: $CH_3-(CH_2)_5$ and H above; H and $(CH_2)_7-C(=O)-OH$ below)

Naturally-occurring unsaturated fatty acids contain *cis* double bonds. There are several possible ways to redraw the structure showing the double bond in its *cis* form:

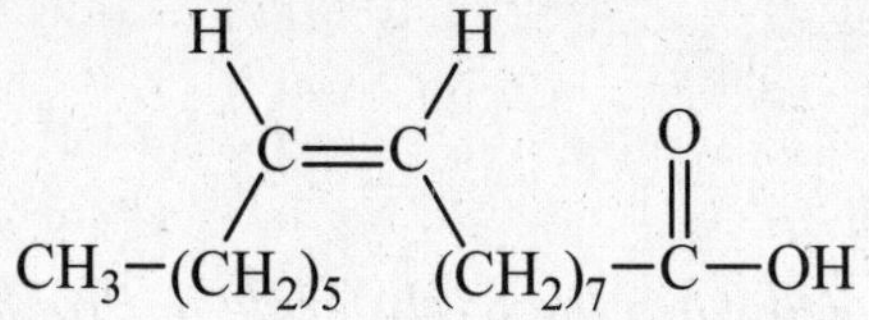

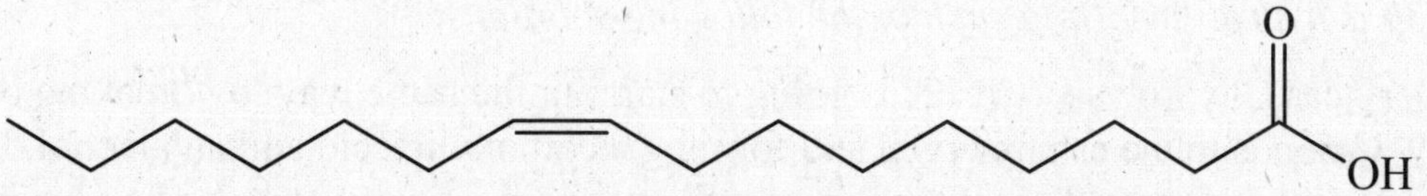

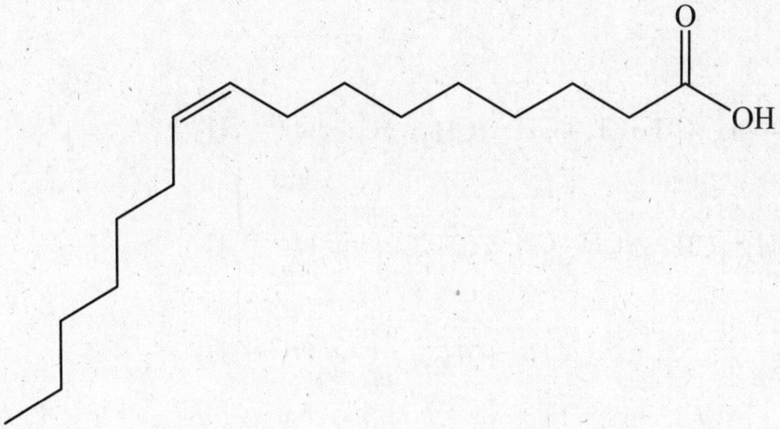

H H
C=C
$CH_3-(CH_2)_5$ $(CH_2)_7-\overset{O}{\overset{\|}{C}}-OH$

O
OH

O
OH

14.133 *When a sample of the following fatty acid is partially hydrogenated, two products are formed. Draw the structures of these products. (Hint: One contains an alkene group and the other does not.)*

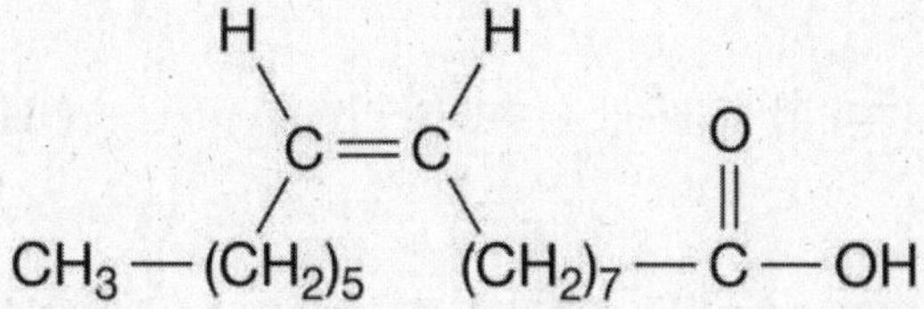

H H
C=C
$CH_3-(CH_2)_5$ $(CH_2)_7-\overset{O}{\overset{\|}{C}}-OH$

The hydrogenation process turns alkenes into alkanes and unsaturated fats into saturated fats. When the molecule is hydrogenated, the following molecule is produced:

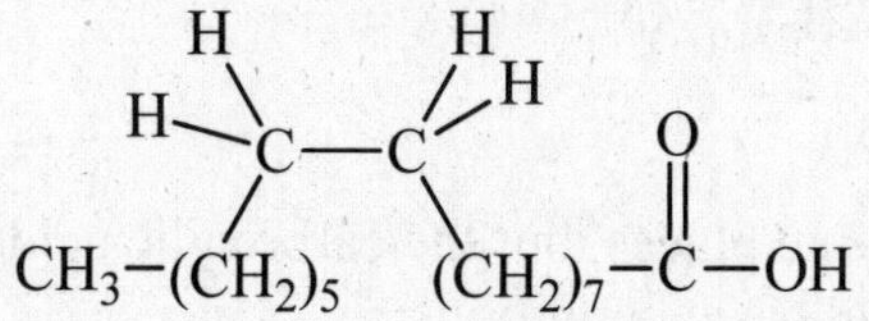

H H
H–C–C–H
$CH_3-(CH_2)_5$ $(CH_2)_7-\overset{O}{\overset{\|}{C}}-OH$

This saturated fat molecule can also be written as $CH_3{-}(CH_2)_{14}{-}COOH$.

In "partial hydrogenation" of a sample of fat, some of the fatty acid molecules are hydrogenated, but even the ones that don't have hydrogen added to them can come into contact with the hydrogenation catalyst and flip from the *cis* form to the *trans* form. Therefore the other product of partial hydrogenation of this fatty acid is the corresponding *trans* fatty acid:

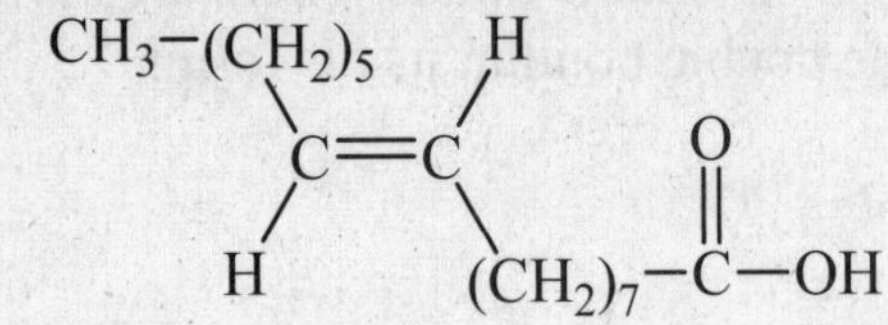

$$CH_3-(CH_2)_5-CH=CH-(CH_2)_7-\overset{O}{\overset{\|}{C}}-OH \quad (trans;\ H \text{ atoms on opposite sides})$$

14.135 *Draw the structures of the products that are formed when the triglyceride in Problem 14.134 is hydrolyzed using each of the following solutions:*

In every case, hydrolysis (Ch 12) is going to start out the same way: by breaking one of the O–C bonds to the ester oxygen and forming a carboxylic acid and an alcohol. Doing this three times for the three ester groups will give us three fatty acids and a glycerol molecule:

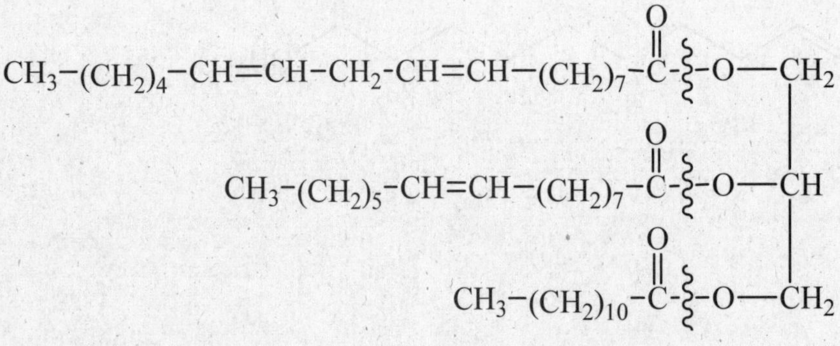

$$CH_3-(CH_2)_4-CH=CH-CH_2-CH=CH-(CH_2)_7-\overset{O}{\overset{\|}{C}}\text{-}\wr\text{-}O-CH_2$$
$$CH_3-(CH_2)_5-CH=CH-(CH_2)_7-\overset{O}{\overset{\|}{C}}\text{-}\wr\text{-}O-CH$$
$$CH_3-(CH_2)_{10}-\overset{O}{\overset{\|}{C}}\text{-}\wr\text{-}O-CH_2$$

a) 2 M H_2SO_4

In acidic conditions, the carboxylic acid groups appear normally (in their carboxylic acid form). Alcohols are not affected by the pH of the solution, so the glycerol also appears normally:

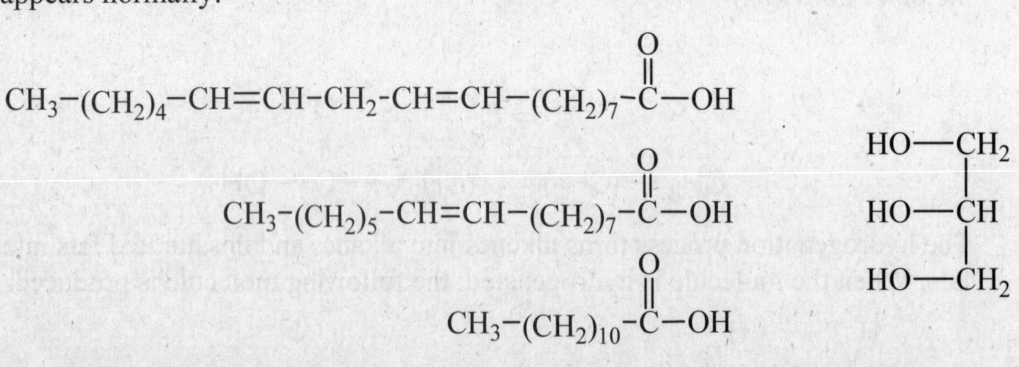

$$CH_3-(CH_2)_4-CH=CH-CH_2-CH=CH-(CH_2)_7-\overset{O}{\overset{\|}{C}}-OH$$
$$CH_3-(CH_2)_5-CH=CH-(CH_2)_7-\overset{O}{\overset{\|}{C}}-OH$$
$$CH_3-(CH_2)_{10}-\overset{O}{\overset{\|}{C}}-OH$$
$$HO-CH_2 \quad HO-CH \quad HO-CH_2$$

b) 2 M NaOH

The hydrolysis breaks the molecule in the same places, but the carboxylic acid groups are converted to their carboxylate salts by the basic conditions:

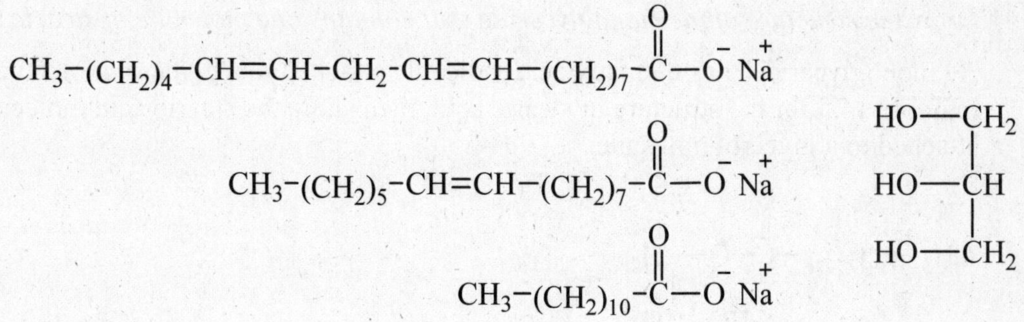

$$CH_3-(CH_2)_4-CH=CH-CH_2-CH=CH-(CH_2)_7-\overset{O}{\overset{\|}{C}}-O^- \ Na^+$$

$$CH_3-(CH_2)_5-CH=CH-(CH_2)_7-\overset{O}{\overset{\|}{C}}-O^- \ Na^+$$

$$CH_3-(CH_2)_{10}-\overset{O}{\overset{\|}{C}}-O^- \ Na^+$$

$$\begin{array}{l} HO-CH_2 \\ \quad\;\; | \\ HO-CH \\ \quad\;\; | \\ HO-CH_2 \end{array}$$

c) 2 M KOH

The only difference between NaOH and KOH is that the cation in the salt will be K^+ instead of Na^+:

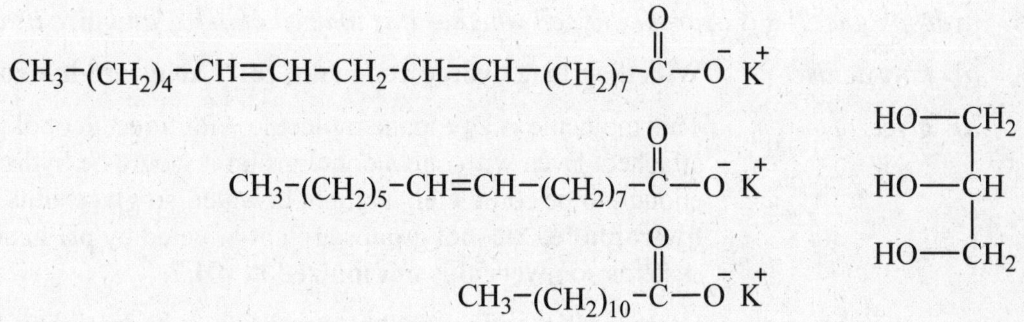

$$CH_3-(CH_2)_4-CH=CH-CH_2-CH=CH-(CH_2)_7-\overset{O}{\overset{\|}{C}}-O^- \ K^+$$

$$CH_3-(CH_2)_5-CH=CH-(CH_2)_7-\overset{O}{\overset{\|}{C}}-O^- \ K^+$$

$$CH_3-(CH_2)_{10}-\overset{O}{\overset{\|}{C}}-O^- \ K^+$$

$$\begin{array}{l} HO-CH_2 \\ \quad\;\; | \\ HO-CH \\ \quad\;\; | \\ HO-CH_2 \end{array}$$

14.137 *A sample of triglycerides from the liver of a fish contains 23% saturated fatty acids, 55% monounsaturated fatty acids, and 22% polyunsaturated fatty acids. The sample melts at 4°C.*

a) Is this mixture a solid or a liquid at room temperature? How can you tell?

Chapter 4 review–you don't have to know anything about saturated and unsaturated fats to answer this question. We're told the sample melts at 4°C, so it's solid below that temperature and liquid above. Room temperature is 20–25°C, so **the mixture is liquid at room temperature.**

b) Is this a fat or an oil? Explain your answer.

Chemically it's a fat (a triglyceride of fatty acids) but in common terminology it would be called an oil, because it is liquid at room temperature.

14.139 *If linoleic acid is completely hydrogenated, what fatty acid will be formed? Draw the structure and give the name of the product.*

$CH_3–(CH_2)_4–CH=CH–CH_2–CH=CH–(CH_2)_7–COOH$ linoleic acid

Complete hydrogenation turns both double bonds into single bonds:

$$CH_3–(CH_2)_4– CH_2– CH_2–CH_2– CH_2– CH_2– (CH_2)_7–COOH$$

We then rewrite the condensed formula, combining all $–CH_2–$ groups together in parentheses, and look at Table 14.5 to identify the product:

$CH_3–(CH_2)_{16}–COOH$ stearic acid

14.141 *Draw the structure of the monoglyceride that contains one molecule of stearic acid.*

A "monoglyceride" has one fatty acid molecule attached to a glycerol molecule. We look in Table 14.5 for the structure of stearic acid, then show the stearic acid molecule attached to a glycerol molecule:

$$CH_3-(CH_2)_{16}-\overset{\overset{\displaystyle O}{\|}}{C}-O-CH_2$$
$$HO-CH$$
$$HO-CH_2$$

14.143 *Tell whether each of the following components of a glycerophospholipid is hydrophilic or hydrophobic. If it is hydrophilic, tell whether it is ionized at pH 7, and give its charge.*

a) fatty acids With their long hydrocarbon tails, **fatty acids are hydrophobic.**

b) glycerol This molecule is a propane molecule with three alcohol groups attached. Even with one alcohol group it would be hydrophilic enough to be completely soluble in water, so **glycerol is strongly hydrophilic.** Alcohol groups are not affected by pH in aqueous systems so **glycerol is not ionized at pH 7.**

c) amino alcohol **Hydrophilic:** amino alcohols have a short hydrocarbon chain with at least two strongly hydrophilic functional groups, an amine group and an alcohol group, attached. Alcohol groups are not ionized at physiological pH, but amine groups are converted to their ammonium ion forms, so **the amino alcohol is ionized, with a +1 charge, at pH 7.**

d) phosphate As we saw in Chapter 11, **phosphate ions are ionized at pH 7, with two ionic forms, the –1 form and the –2 form, in equilibrium.**

14.145 *Which of the following can pass through a lipid bilayer (without the assistance of a transport protein)?*

The interior of the lipid bilayer is highly hydrophobic. Nonpolar molecules that are soluble in hydrocarbon solvents can pass through by simple diffusion, without the assistance of a transport protein, but ions and highly polar, hydrophilic molecules cannot.

a) $H_2PO_4^-$ *b) asparagine (an amino acid)* *c) acetaldehyde*

d) CO_2 *e)* $CH_3–NH_3^+$ *f) sucrose*

d) CO_2 **(nonpolar) and c) acetaldehyde (polar but not strongly hydrophilic) can pass through the lipid bilayer,** but a) $H_2PO_4^-$, b) asparagine and e) $CH_3–NH_3^+$ are charged under physiological pH conditions, and f) sucrose is too strongly hydrophilic to leave the aqueous solutions inside and outside the cell and pass through the lipid bilayer.

*14.147 *Which would you expect to have the higher solubility in water, aldosterone or testosterone? Explain your answer. (The structures of these steroids are given in Table 14.10.)*

The aqueous solubility of steroids, like that of other organic compounds, depends on the balance between hydrophobic and hydrophilic areas of the molecules. The steroid nucleus is hydrophobic; aldosterone has two alcohol groups (which are highly hydrophilic), along with two ketone groups and an aldehyde group (all moderately hydrophilic). Testosterone has only one alcohol group and one ketone group. **Aldosterone has more hydrophilic groups than testosterone and is expected to be more soluble in water.** (Neither compound is very soluble, however; most steroids aren't.)

*14.149 *Androstenedione ("andro") has been used by athletes to increase muscle mass. In the body, it is converted into testosterone. What type of reaction is this? (Hint: look at the structure of testosterone.)*

The only difference between the two molecules is the conversion of a ketone group in Androstenedione to an alcohol group in testosterone. This conversion requires adding two H atoms to a carbonyl; the reaction type is **reduction of a carbonyl to form an alcohol** as in Chapter 12.

androstenedione

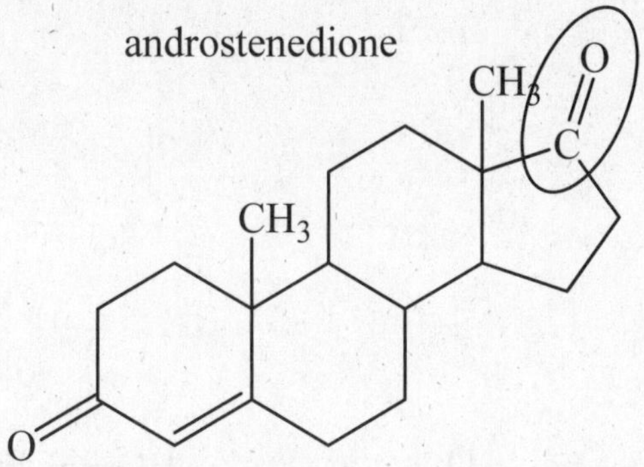

CH_3 O C CH_3 O

testosterone

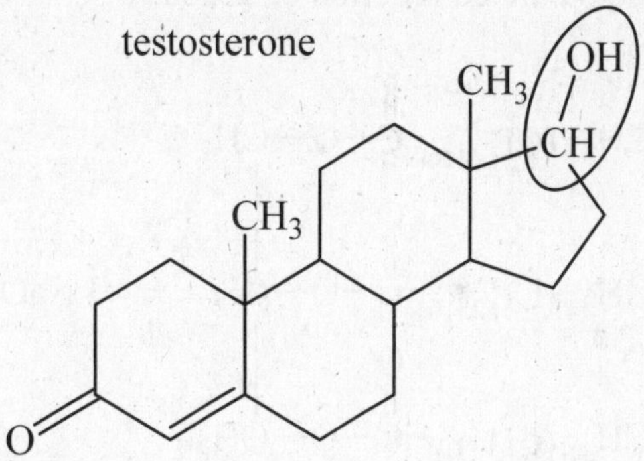

CH_3 OH CH CH_3 O

14.151 *What types of proteins transport each of the following types of lipids in the blood?*

a) *triglycerides*	The **chylomicrons** are the primary transport mechanisms for triglycerides.
b) *cholesterol*	The **LDLs and HDLs** are the main cholesterol transport system.
c) *steroid hormones*	The bulk of the estradiol and testosterone is carried by a protein called ***sex hormone binding globulin* (SHBG).** Most of the cortisol, aldosterone, and progesterone is carried by ***corticosteroid binding globulin* (CBG).**

14.153 a) *Classify DHA using the omega system. (The structure of DHA is shown on page 617.)*

$CH_3-CH_2-CH=CH-CH_2-CH=CH-CH_2-CH=CH-CH_2-CH=CH-CH_2-CH=CH-CH_2-CH=CH-(CH_2)_2-COOH$

For purposes of classifying fatty acids according to the omega system, we count from the last carbon in the chain (the omega carbon, the one farthest from the carboxylic acid group) to the first double bond we reach. We can ignore the rest of the molecule.

In this molecule, the first double bond we reach is the #3 bond in the molecule, so this is an **omega-3 fatty acid**.

b) *What is the molecular formula of DHA?*

Chapter 2 review! Add up all the atoms in the molecule: $\mathbf{C_{22}H_{32}O_2}$.

c) *Write a balanced chemical equation for the hydrogenation of DHA, using H_2 as the source of hydrogen.*

Since DHA has six double bonds, we'll need six molecules of H_2 (12 total atoms) to carry out a complete hydrogenation. One molecule of H_2 is needed to saturate each double bond. All 12 atoms will be added to the molecule and therefore to the molecular formula:

$$\mathbf{C_{22}H_{32}O_2 + 6\ H_2 \rightarrow C_{22}H_{44}O_2}$$

*14.155 *A triglyceride contains only stearic acid. If you react 10.0 grams of this triglyceride with enough NaOH to completely saponify the triglyceride, what mass of soap will you form?*

The structure of stearic acid is given in Table 14.5. We draw the triglyceride and write the balanced reaction equation:

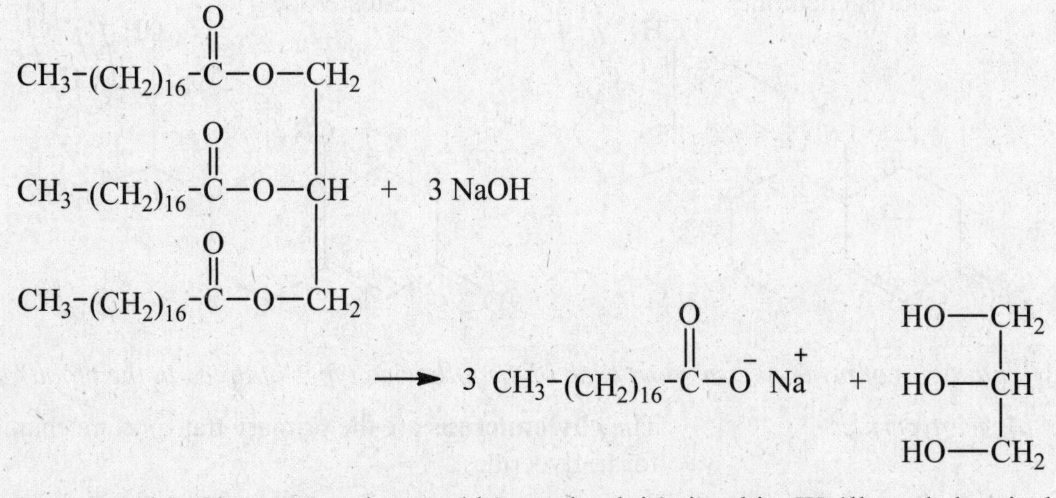

$$\begin{array}{l} CH_3-(CH_2)_{16}-\overset{O}{\overset{\|}{C}}-O-CH_2 \\ CH_3-(CH_2)_{16}-\overset{O}{\overset{\|}{C}}-O-CH \quad + \quad 3\ NaOH \\ CH_3-(CH_2)_{16}-\overset{O}{\overset{\|}{C}}-O-CH_2 \end{array}$$

$$\longrightarrow 3\ CH_3-(CH_2)_{16}-\overset{O}{\overset{\|}{C}}-\overset{-}{O}\ \overset{+}{Na} \quad + \quad \begin{array}{l} HO-CH_2 \\ HO-CH \\ HO-CH_2 \end{array}$$

From here it's Chapter 6 review, and it's not hard, it's just big. We'll need chemical formulas of the triglyceride and the fatty acid salt (the *soap*) so that we can calculate their molar masses. The chemical formula of the triglyceride is $C_{57}H_{110}O_6$, and its molar mass is 891.5 g/mol. The chemical formula of the soap is $C_{18}H_{35}O_2Na$ and its molar mass is 306.5 g/mol. For every one molecule of the triglyceride that reacts, three molecules of soap are formed, so the mass relationship between the triglyceride and the soap is

$$891.5\text{ g triglyceride} = 3(306.5)\text{ g soap} = 919.5\text{ g soap}$$

Starting with 10.0 g of the triglyceride,

$$\mathbf{10.0\ g\ triglyceride \times \frac{919.5\ g\ soap}{891.5\ g\ triglyceride} = 10.3\ g\ soap}$$

*14.157 *Fatty acids react with $Ca(OH)_2$ to form an insoluble white solid that is one of the components of "soap scum." Write a chemical equation that shows this reaction, using myristic acid as the fatty acid.*

Fatty acids reacts with OH^- to form carboxylate ion with –1 charge. Since $Ca(OH)_2$ has two hydroxide ions, it will react with two molecules of the fatty acid. (The formula of myristic acid is in Table 14.5.)

$$\mathbf{2\ CH_3\text{–}(CH_2)_{12}\text{–}COOH + Ca(OH)_2 \rightarrow [CH_3\text{–}(CH_2)_{12}\text{–}COO^-]_2Ca^{2+} + 2\ H_2O}$$

*14.159 *The structure of aldosterone is shown in Table 14.10.*

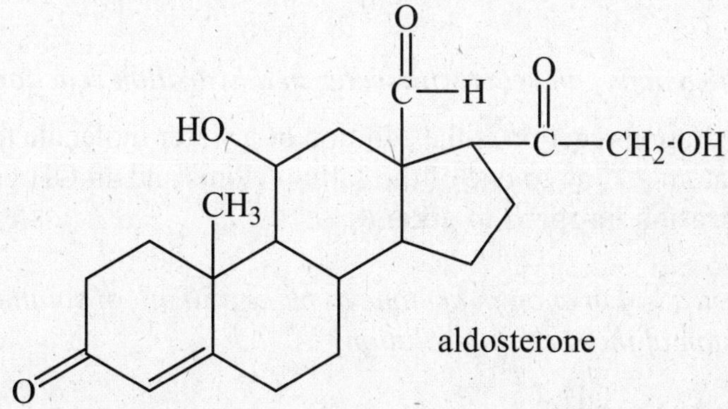

aldosterone

a) *Circle and name all of the functional groups in aldosterone.*

This question is Chapter 8 review, with the exception of the ketone groups that were introduced in Ch 10.

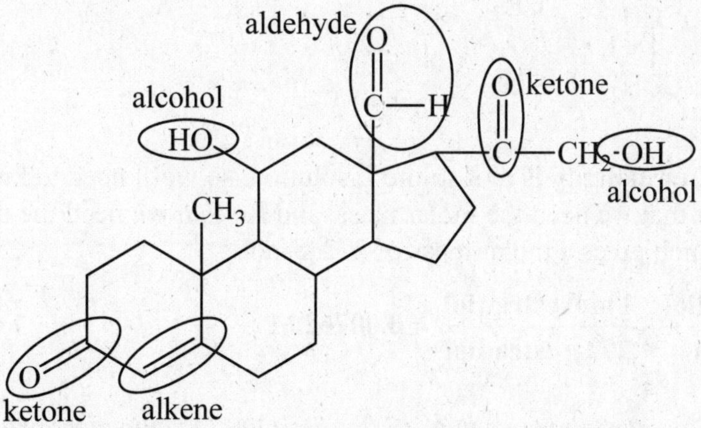

aldehyde
ketone
alcohol
alcohol
ketone
alkene

b) *Which of these functional groups can be oxidized?*

Chapter 10 review: oxidation involves an increase in the number of bonds to oxygen atoms or the removal of hydrogen atoms. **Primary and secondary alcohols can be oxidized to form carbonyls**; the primary alcohol group on the right of the molecule can be oxidized to form an aldehyde group, and the secondary alcohol at the top left can be oxidized to form a ketone. **Aldehydes can also be oxidized**; oxidation of the aldehyde group will form a carboxylic acid.

c) *Which of these functional groups can be reduced?*

Reduction of organic compounds usually involves the addition of hydrogen atoms. Carbonyl groups can be reduced to form alcohols, so **the aldehyde group and the two ketone groups can be reduced (forming primary and secondary alcohols, respectively)**. Hydrogenation of alkenes is a reduction reaction, so **the alkene group can also be reduced to an alkane group** (forming a saturated molecule).

d) *Which of these functional groups can react with acetic acid to form an ester?*

Chapter 12 review: esters are formed in the condensation reaction between a carboxylic acid and an alcohol. **Either alcohol group can react with acetic acid to form an ester.**

e) *Which of these functional groups can react with water in a hydration reaction?*

Chapter 9 review: the hydration reaction is the addition of a water molecule to a C=C double bond, with an H atom adding to one of the carbon atoms and an OH group adding to the other. **Hydration happens to alkenes.**

*14.161 *If you dissolve 0.725 g of estradiol in enough acetone to make 350 mL of solution, what will be the molar concentration of the resulting solution?*

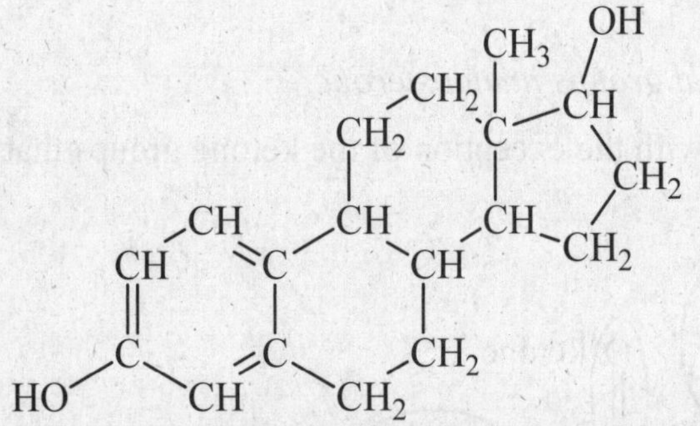

```
                              OH
                    CH3       /
             CH2    |      CH
       CH2  /       C  ---/
        |           |      \
        |           |       CH2
   CH   CH         CH      /
CH   ==C     CH  /    CH2
||     |      |
C      C      CH2
HO   CH   CH2
```

estradiol

"Molar concentration" or molarity is mol solute/L solution, so we'll need to know the moles of estradiol. For that we need the molar mass, and for *that* we need the molecular formula, $C_{18}H_{24}O_2$, which gives a molar mass of 272 g/mol.

$$\frac{\mathbf{0.725\ g\ estradiol}}{\mathbf{0.350\ L\ soln}} \times \frac{\mathbf{1\ mol\ estradiol}}{\mathbf{272\ g\ estradiol}} = \mathbf{0.00762\ M}$$

*14.163 *Glycoproteins contain a carbohydrate bonded to a protein. In some glycoproteins, the side chain of the amino acid serine condenses with the hydroxyl group on the carbon 1 of mannose. Draw the structure of the product of this condensation.*

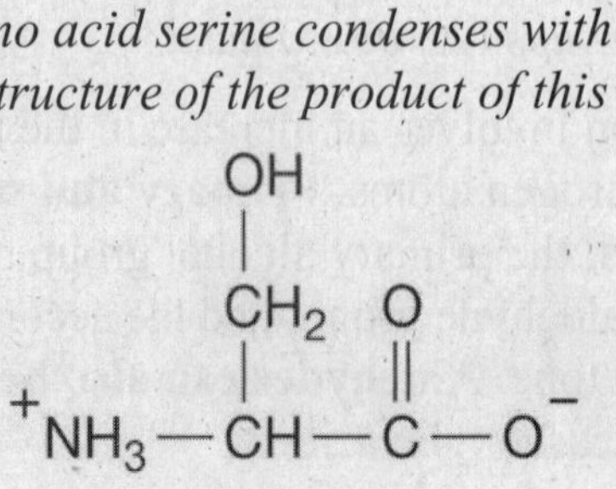

```
          OH
          |
          CH2  O
          |    ||
 +NH3 —  CH — C — O-
```

Serine

The structure of β-mannose is given in Figure 14.3. The problem specifies that the condensation reaction is between the *–OH on the #1 (anomeric) carbon* of mannose and the *side chain* of the serine, so just as we've done before, we draw the two molecules with the involved alcohol groups facing each other, in position for reaction:

$+ H_2O$

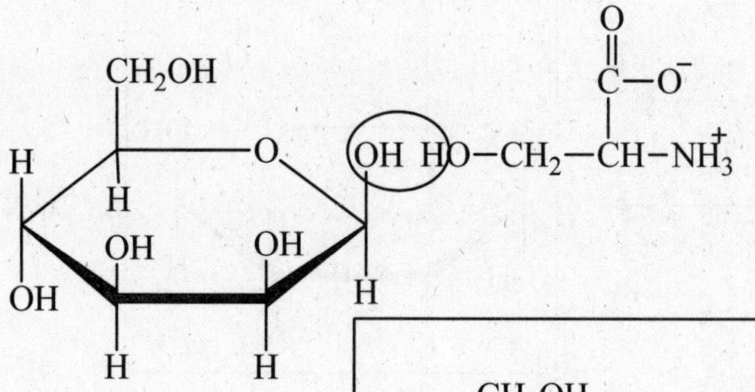

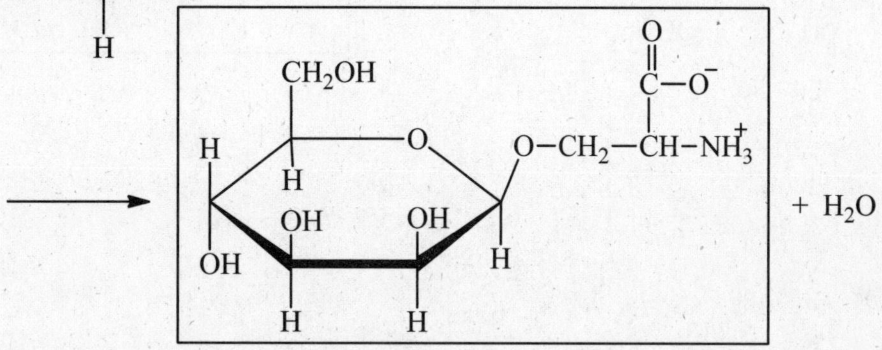

We could also have used α-mannose in the same reaction to make the α- form of the product.

*14.165 *When our cells absorb a molecule of glucose, they condense the glucose with a phosphate ion to form glucose-6-phosphate. The condensation reaction involves the alcohol group on carbon 6 of glucose. Draw the structure of the product of this condensation.*

The numbering of monosaccharides begins at the right side of the molecule, starting with the carbon atom that is not part of the ring (if any) or with the rightmost carbon in the ring and counting clockwise around the ring:

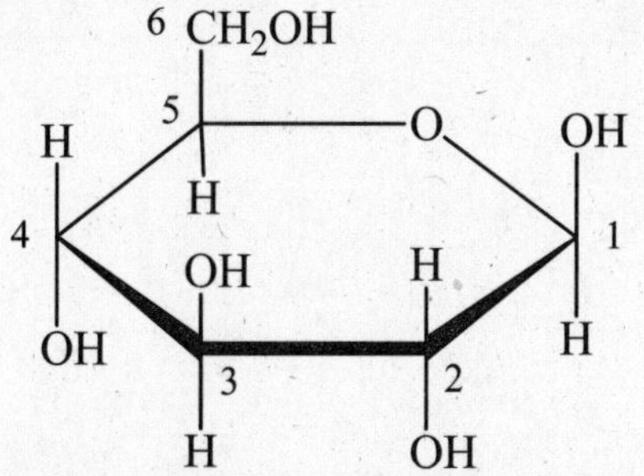

As always in condensation reactions, we draw the two molecules with the condensation groups facing each other, in position for reaction:

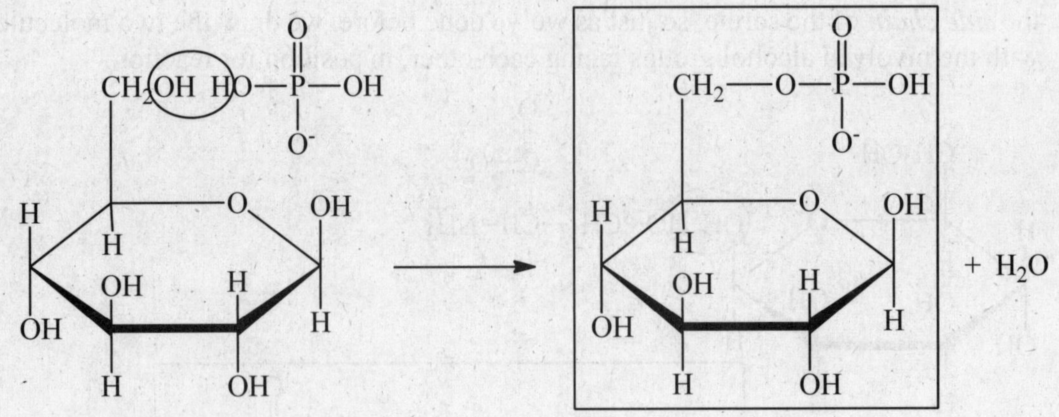

O
CH$_2$OH HO—P—OH
O$^-$

H O OH
H
OH H
OH H
H OH

→

O
CH$_2$—O—P—OH
O$^-$

H O OH
H
OH H
OH H
H OH

+ H_2O

Chapter 15

15

Metabolism: The Chemical Web of Life

Solutions to Section 15.1 Core Problems

15.1 *What are the chemical building blocks of ATP?*

ATP is "adenosine triphosphate"– the name gives you a list of the building blocks, **an adenosine nucleoside (adenine and ribose) and three phosphate groups.**

15.3 *Write a chemical equation for the reaction that occurs when your body stores energy in the form of ATP. (You may use abbreviations such as ATP and ADP.)*

$$\mathbf{ADP + P_i + energy \rightarrow ATP + H_2O}$$

The reaction that creates ATP is a condensation reaction between ADP and a phosphate ion, abbreviated P_i for "inorganic phosphate." Like all condensation reactions (Chapter 12), it produces a water molecule. (The reaction is endothermic; this is easy to remember if you know that *using*–hydrolyzing–ATP is how the body produces energy.) It helps to know the words that ADP (adenosine *di*phosphate) and ATP (adenosine *tri*phosphate) stand for; it makes sense that to turn a diphosphate into a triphosphate you'd have to add a phosphate.

15.5 *Your body is constantly building proteins from amino acids. This process requires energy, which is supplied by ATP. Is this an anabolic process, or is it a catabolic process?*

"Building" reactions, those that produce more complex molecules from simpler ones, are **anabolic** and require energy.

15.7 *Both of the following reactions are exothermic. Our bodies use one of them to supply the energy to make a molecule of ATP. Which one? Explain your answer.*

phosphoarginine + H_2O → *arginine* + P_i + *7.7 kcal*
glucose-1-phosphate + H_2O → *glucose* + P_i + *5.0 kcal*

Making a ATP requires at least 7.3 kcal of energy per mole of ATP, **so only the first reaction can be used to make ATP.**

15.9 *Both of the following reactions require energy, which is supplied by ATP. However, one of these reactions only produces a low concentration of product, while the other goes nearly to completion. Which is which? Explain your answer.*

glucose + P_i + *5.0 kcal* → *glucose-1-phosphate* + H_2O
creatine + P_i + *10.3 kcal* → *creatine phosphate* + H_2O

The hydrolysis of ATP produces 7.3 kcal/mol, so only reactions that (like the first reaction above) require that much energy or less will go to completion. The second reaction requires more than 7.3 kcal, so it forms an equilibrium mixture.

15.11 *Our bodies can use fructose as an energy source. The first step of this pathway is the condensation of fructose with phosphate ion.*

$$\text{fructose} + P_i + \text{energy} \rightarrow \text{fructose-1-phosphate} + H_2O$$

ATP and ADP are also involved in this reaction.

a) *Do our bodies break down ATP when we make fructose-1-phosphate, or do we make ATP? How can you tell?*

The reaction is endothermic (energy is on the reactants side of the equation), so we must break down ATP to supply the energy for this step.

b) *Write a chemical equation for the formation of fructose-1-phosphate that includes ATP and ADP.*

We can do this by combining the given equation with the equation that breaks down ATP:

$$\text{fructose} + P_i \rightarrow \text{fructose-1-phosphate} + H_2O$$
$$+ \text{ATP} + H_2O \rightarrow \text{ADP} + P_i$$
$$\text{fructose} + \text{ATP} + P_i + H_2O \rightarrow \text{fructose-1 phosphate} + \text{ADP} + H_2O + P_i$$

Removing species that appear on both sides of the equation (P_i and H_2O) gives us:

fructose + ATP → fructose-1 phosphate + ADP

A few things to notice: (1) The energy on the reactants side in the first equation is supplied by the energy on the products side in the second equation. (2) ATP is a source of phosphate, the P_i, in the first equation is supplied by the breakdown of the ATP in the second equation. (3) Since one reaction is a condensation and the other is a hydrolysis, the water molecule never shows up in the overall equation; the H and OH groups are exchanged directly between the reactant molecules.

c) *Is the reaction you wrote in part "b" an activation reaction? Why or why not?*

The reaction uses energy from ATP to make a high-energy organic phosphate (the definition of an activation reaction), so it is an activation reaction.

Solutions to Section 15.2 Core Problems

15.13 *What two compounds do our bodies use to remove hydrogen atoms from organic molecules when we burn these molecules to obtain energy?*

Chapter 10 review! **NAD^+ and FAD remove H atoms from organic molecules, forming NADH + H^+ or $FADH_2$.**

15.15 *Ethanol (C_2H_6O) contains six hydrogen atoms. However, when our bodies burn ethanol to obtain energy, we form twelve hydrogen atoms for each ethanol molecule we break down. Where did the other hydrogen atoms come from?*

All of the C atoms in the reactant end up in CO_2 molecules, and the O atoms in the CO_2 product molecules come mostly from water molecules. **The 6 extra H atoms come from the 3 water molecules that supply the oxygen:**

$$C_2H_6O + 3\ H_2O \rightarrow 2\ CO_2 + 12\ [H]$$

15.17 *Brain cells normally burn glucose as their only fuel. However, during a prolonged fast, brain cells can also burn 3-hydroxybutyric acid. Write a chemical equation showing how this compound is broken down to CO_2 and hydrogen atoms. Use [H] to represent hydrogen atoms that are removed by the redox coenzymes.*

$$CH_3-\underset{}{\overset{OH}{\overset{|}{CH}}}-CH_2-\overset{O}{\overset{\|}{C}}-OH$$

3-Hydroxybutyric acid

First, convert the formula for 3-hydroxybutyric acid to a molecular formula, then write the skeleton equation:

$$C_4H_8O_3 + ?\ H_2O \rightarrow ?\ CO_2 + ?\ [H]$$

Then follow the steps:

Step 1: Balance C, by writing the coefficient for CO_2 (in this case, 4 CO_2).
Step 2: Balance O, by writing the coefficient for H_2O (in this case, 5 H_2O).
Step 3: Balance H, by writing the coefficient for [H] (in this reaction, 18).

$$\mathbf{C_4H_8O_3 + 5\ H_2O \rightarrow 4\ CO_2 + 18\ [H]}$$

15.19 *When a brain cell burns a molecule of 3-hydroxybutyric acid to obtain energy, it transfers four hydrogen atoms to FAD. The rest of the hydrogen atoms are removed by NAD^+. Using this information and your answer to Problem 15.17, write a balanced chemical equation for the breakdown of 3-hydroxybutyric acid that includes all of the redox coenzymes.*

Adding four H atoms to FAD will require two molecules of FAD:

$$4\ [H] + 2\ FAD \rightarrow 2\ FADH_2$$

This leaves 14 H atoms to react with NAD^+, requiring seven molecules of NAD^+. Each NAD^+ produces one NADH and one H^+:

$$4\ [H] + 7\ NAD^+ \rightarrow 7\ NADH + 7\ H^+$$

Adding these reactions to the reaction we wrote in 15.17:

$$C_4H_8O_3 + 5\ H_2O \rightarrow 4\ CO_2 + 18\ [H]$$

gives the overall equation:

$$\mathbf{C_4H_8O_3 + 5\ H_2O + 2\ FAD + 7\ NAD^+ \rightarrow 4\ CO_2 + 2\ FADH_2 + 7\ NADH + 7\ H^+}$$

Solutions to Section 15.3 Core Problems

15.21 *Describe the function of the mitochondria in a cell.*

The matrix of a mitochondrion contains the enzymes that break down all of the amino acids and fats, and most of the carbohydrates, used as fuel by the organism to produce energy. The mitochondria are the "energy factories" of the cell, converting the energy of oxidation reactions into the chemical energy of ATP.

15.23 *What is the function of the electron transport chain?*

See Figure 15.7. **The electron transport chain does three things:**

(1) It recycles the used redox coenzymes NADH or $FADH_2$ back into NAD^+ and FAD for reuse. (This step produces electrons.)
(2) It uses the H^+ ions and electrons from step 1 and combines them with O_2 from the bloodstream to form H_2O molecules (and produce energy).
(3) It stores the energy from step, by using the energy to push H^+ ions across the inner membrane into the intermembrane space, creating a concentration and charge gradient.

15.25 *Describe how the mitochondrion makes an H^+ concentration gradient.*

A concentration gradient is a difference in the concentration of a solute in different regions (across a membrane in this case). **The energy produced by the conversion of O_2 and H^+ is used to push other H^+ ions from the matrix, across the inner membrane, and into the space between the inner and outer membranes (intermembrane space). Therefore the H^+ ion concentration is greater in the intermembrane space, outside the inner membrane, than inside it: a concentration gradient.**

15.27 a) *When the electron transport chain oxidizes a molecule of NADH, what does it form?*

The electron transport chain turns NADH (the reduced form of the coenzyme) back into NAD^+ (the oxidized, active form of the coenzyme), producing an H^+ ion and two electrons.

b) *How many hydrogen ions are transferred through the inner membrane as a molecule of NADH is oxidized?*

The H^+ ion and the electrons from the oxidation of NADH are combined with oxygen from O_2 to form H_2O, producing energy. **The energy produced by these reactions is enough to transport ten H^+ ions across the membrane for every one molecule of NADH.**

c) *How many molecules of ATP (on average) can the mitochondrion make using the energy from the oxidation of one molecule of NADH?*

Making one molecule of ATP from ADP requires the equivalent energy of three H^+ ions, plus the energy of another H^+ ion to release the ATP to the cytosol for a total of 4 H^+. Since each NADH provides the energy to transport 10 H^+ through the inner membrane, **the oxidation of one molecule of NADH is enough energy to make 2.5 molecules of ATP from ADP ($10 \div 4 = 2.5$).** (Of course the reaction never really produces a half molecule of ATP–this is the average energy equivalent.)

d) *If a mitochondrion oxidizes 20 molecules of NADH, how many molecules of ATP can it make?*

One NADH is enough energy for 2.5 ATP, so **if a mitochondrion oxidizes 20 molecules of NADH, it can make 50 molecules of ATP from ADP:**

20 NADH × 2.5 ATP/NADH = 50 ATP

15.29 *A membrane separates solution A from solution B. Solution A contains 0.100 M Na^+ and 0.099 M Cl^-. Solution B contains 0.200 M Na^+ and 0.201 M Cl^-.*

a) *Is there a concentration gradient across this membrane? Why or why not?*

A concentration gradient is a difference in the concentration of a solute in different regions. **Both ions have greater concentrations in solution B than in solution A, so there is a concentration gradient across this membrane.**

b) *Is there a charge gradient across this membrane? Why or why not?*

Solution A has slightly more positive charge than negative charge (net +), while solution B has slightly more negative charge than positive charge (net –), so there is a charge gradient across this membrane.

15.31 *Explain how H^+ transport by the electron transport chain produces a concentration gradient across the inner mitochondrial membrane.*

The electron transport chain pushes H^+ ions from the interior (matrix) of the mitochondrion, through the inner membrane and into the space between the membranes, creating a greater concentration of H^+ ions in the intermembrane space than inside the membrane.

15.33 *If a mitochondrion oxidizes two molecules of NADH and two molecules of $FADH_2$, what is the total number of molecules of ATP that the mitochondrion will make available to the cell?*

Two molecules of NADH	×	10 H^+/NADH	=	20 H^+ ions transported
Two molecules of $FADH_2$	×	6 $H^+/FADH_2$	=	12 H^+ ions transported
Total				32 H^+ ions transported

Producing one molecule of ATP and releasing it to the cytosol requires the equivalent energy of four H^+ ions, so **this process would produce a total of eight molecules of ATP for the cell to use:**

$$\mathbf{32\ H^+ \times 1\ ATP/4\ H^+ = 8\ ATP}$$

Alternatively, we can use the relationships between the redox coenzymes and ATP directly (skipping over the H^+ ion):

Two molecules of NADH	×	2.5 ATP/NADH	=	5 ATP
Two molecules of $FADH_2$	×	1.5 ATP/$NADH_2$	=	3 ATP
Total			=	8 ATP

Solutions to Section 15.4 Core Problems

15.35 *What is an activation reaction, and how is ATP involved in this type of reaction?*

Activation reactions use ATP to form high-energy organic phosphates. The conversion of ATP to ADP provides both the inorganic phosphate group and the energy to create the high-energy molecules.

15.37 *What organic compound is broken down during glycolysis, and what product is formed when this molecule is broken down?*

"Glyco-" often has to do with glucose (as in "glycogen," Chapter 14) and "-lysis" is breaking (as in "hydrolysis," Ch 12, and "hemolysis," Ch 5), so **"glycolysis" means "breaking down glucose." At the end of the ten-step glycolysis process, the breakdown produces pyruvate ions.**

15.39 *Explain why lactic acid fermentation is an anaerobic pathway.*

Lactic acid fermentation does not require an input of oxygen; there is no net oxidation of the reactants, so the process is called "anaerobic." The process does not convert NAD^+ (an oxidizing agent) to NADH; the absence of redox coenzymes from a metabolic pathway is a clue that it is anaerobic.

15.41 *The last reaction in lactic acid fermentation is:*

$$\text{pyruvate} + NADH + H^+ \rightarrow \text{lactate} + NAD^+$$

What is the source of the NADH for this reaction?

The NADH for this reaction is made in one of the earlier reactions of glycolysis. Since the overall process of lactic acid fermentation shows neither a net use nor production of NAD^+, the NADH for the last step must have been produced in an earlier step.

Solutions to Section 15.5 Core Problems

15.43 *In the oxidative decarboxylation of pyruvate, what happens to the three carbon atoms in the pyruvate ion?*

The carboxylate carbon atom is converted to carbon dioxide (as in the Chapter 11 introduction of decarboxylation), while **the fragment containing the other two carbon atoms combines with Coenzyme A (a thiol) to produce acetyl-CoA (a thioester),** the starting material for the citric acid cycle.

15.45 *The citric acid cycle oxidizes an organic molecule, breaking part of it down into carbon dioxide. What is this organic molecule?*

The starting material in the citric acid cycle is acetyl-CoA.

15.47 *How many decarboxylation reactions are there in the citric acid cycle? Is this reasonable, given that the citric acid cycle breaks down the acetyl group of acetyl-CoA?*

There are two decarboxylation reactions (part of step 3 and step 4). This makes sense, because the acetyl group of acetyl-CoA has two carbon atoms that need to be converted to CO_2.

15.49 *What is the purpose of converting citrate into isocitrate during the second step of the citric acid cycle?*

The next (third) step in the citric acid cycle is the oxidation of an alcohol group to form a carbonyl. The conversion in the second step turns a tertiary alcohol (citrate), which cannot be oxidized to form a carbonyl, into a secondary alcohol (isocitrate), which can be oxidized to form a carbonyl.

15.51 *What is the total ATP yield when one molecule of acetyl-CoA is broken down to CO_2? Include the ATP that is formed by oxidative phosphorylation.*

The citric acid cycle produces one molecule of ATP directly by oxidative phosphorylation, plus three molecules of NADH (at 2.5 ATP per NADH, enough for 7.5 ATP) and one molecule of $FADH_2$ (which produces energy for 1.5 ATP). **The total yield of ATP is 7.5 + 1.5 + 1 = 10 ATP.**

15.53 *In some cells, the oxidation of glucose produces 32 molecules of ATP. How many of these are made by the electron transport chain?*

The electron transport chain is the process that converts NADH and $FADH_2$ back into NAD^+ and FAD. In the oxidation of one molecule of glucose, the oxidation of 10 NADH and 2 $FADH_2$ produce 25 ATP and 3 ATP, respectively, so **28 molecules of ATP are made by the electron transport chain.**

Solutions to Section 15.6 Core Problems

15.55 *Stearic acid has the chemical formula $C_{18}H_{36}O_2$. Write a balanced equation showing how stearic acid can be broken down into CO_2 and hydrogen atoms. Use [H] to represent hydrogen atoms that are removed by NAD^+ and FAD.*

First, we write the skeleton equation:

$$C_{18}H_{36}O_2 + ?\ H_2O \rightarrow ?\ CO_2 + ?\ [H]$$

Then follow the steps:

Step 1: Balance C, by writing the coefficient for CO_2 (in this case, 18 CO_2).
Step 2: Balance O, by writing the coefficient for H_2O (in this case, 34 H_2O).
Step 3: Balance H, by writing the coefficient for [H] (in this reaction, 36 + 2(34) = 104 [H]).

$$\mathbf{C_{18}H_{36}O_2 + 34\ H_2O \rightarrow 18\ CO_2 + 104\ [H]}$$

15.57 *What reaction must occur before any fatty acid can be oxidized?*

The first step is an activation reaction, in which the fatty acid carboxylate ion combines with coenzyme A to form fatty-acyl coenzyme A.

15.59 *Write chemical equations for the reactions in one cycle of beta oxidation, starting with the fatty-acyl CoA shown here.*

$$CH_3-(CH_2)_{10}-CH_2-CH_2-\overset{O}{\overset{\|}{C}}-S-CoA$$

The steps in the beta oxidation are:

(1) Dehydrogenation (Chapter 10) by FAD (Section 10.6), removing hydrogen atoms from the alpha and beta carbons to produce a C=C double bond:

$$CH_3-(CH_2)_{10}-\overset{H}{\overset{|}{C}H}-\overset{H}{\overset{|}{C}H}-\overset{O}{\overset{\|}{C}}-S-CoA \; + \; FAD \longrightarrow CH_3-(CH_2)_{10}-CH{=}CH-\overset{O}{\overset{\|}{C}}-S-CoA \; + \; FADH_2$$

(2) Hydration (Chapter 9) of the double bond, with the alcohol group being placed on the beta carbon (the carbon atom farther from the carbonyl group):

$$CH_3-(CH_2)_{10}-CH{=}CH-\overset{O}{\overset{\|}{C}}-S-CoA \; + \; H_2O \longrightarrow CH_3-(CH_2)_{10}-\overset{OH}{\overset{|}{C}H}-\overset{H}{\overset{|}{C}H}-\overset{O}{\overset{\|}{C}}-S-CoA$$

(3) Oxidation (Section 10.2) of the alcohol group to form a new carbonyl:

$$CH_3-(CH_2)_{10}-\overset{OH}{\overset{|}{C}H}-CH_2-\overset{O}{\overset{\|}{C}}-S-CoA \; + \; NAD^+ \longrightarrow CH_3-(CH_2)_{10}-\overset{O}{\overset{\|}{C}}-CH_2-\overset{O}{\overset{\|}{C}}-S-CoA \; + \; NADH \; + \; H^+$$

(4) Breaking the carbon-carbon bond to the right of the ketone group, and adding coenzyme A (Section 11.2):

$$CH_3-(CH_2)_{10}-\overset{O}{\overset{\|}{C}}-\wr-CH_2-\overset{O}{\overset{\|}{C}}-S-CoA \; + \; HS\text{-}CoA \longrightarrow CH_3-(CH_2)_{10}-\overset{O}{\overset{\|}{C}}-S-CoA \; + \; CH_3-\overset{O}{\overset{\|}{C}}-S-CoA$$

NOTE:
The remaining fatty acid fragment is left in activated form, ready to start the next cycle. We'd have to redraw it to show the alpha and beta carbons explicitly:

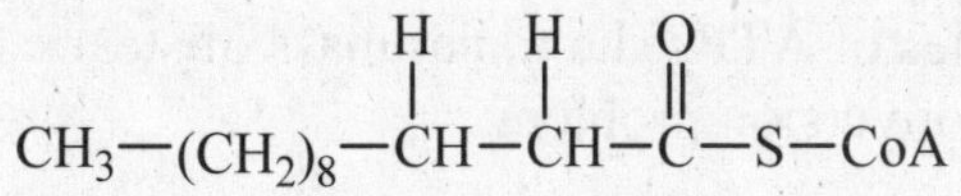

$$CH_3-(CH_2)_8-\underset{}{\overset{H}{\overset{|}{C}}}H-\overset{H}{\overset{|}{C}}H-\overset{O}{\overset{||}{C}}-S-CoA$$

15.61 *a) How many cycles of beta oxidation are required to break down one molecule of stearic acid?*

Beta oxidation removes two carbon atoms at a time in the form of acetyl-CoA. Stearic acid has a total of 18 carbon atoms, and **will require 8 cycles of beta oxidation to break down** (the 8th cycle will leave a 2-carbon fragment.)

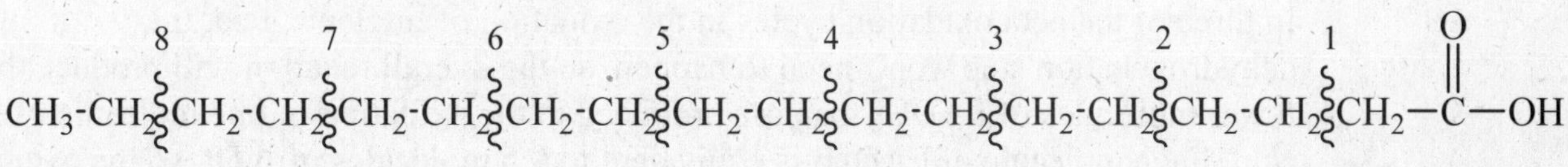

$$CH_3\text{-}CH_2 \overset{8}{\wr} CH_2\text{-}CH_2 \overset{7}{\wr} CH_2\text{-}CH_2 \overset{6}{\wr} CH_2\text{-}CH_2 \overset{5}{\wr} CH_2\text{-}CH_2 \overset{4}{\wr} CH_2\text{-}CH_2 \overset{3}{\wr} CH_2\text{-}CH_2 \overset{2}{\wr} CH_2\text{-}CH_2 \overset{1}{\wr} CH_2-\overset{O}{\overset{||}{C}}-OH$$

b) How many molecules of acetyl-CoA will be formed when one molecule of stearic acid is broken down?

Each two-carbon fragment forms one molecule of acetyl-CoA, so an 18-carbon fatty acid will form **9 molecules of acetyl-CoA.**

c) What is the overall ATP yield when one molecule of stearic acid is broken down to CO_2?

The initial activation of the molecule consumes the energy of 2 ATP.
Each beta oxidation cycle produces:

one NADH (energy equivalent to:	2.5 ATP)
+ one $FADH_2$ (energy equivalent to:	1.5 ATP)
for a total of:	4 ATP produced per beta oxidation cycle.

Each citric acid cycle produces:

one ATP directly:	1 ATP
three NADH (energy equivalent to:	7.5 ATP)
+ one $FADH_2$ (energy equivalent to:	1.5 ATP)
for a total of:	10 ATP produced per citric acid cycle.

(Write those numbers down once you've worked them out, you'll need them again in later problems: 4 ATP per beta oxidation, 10 ATP per citric acid cycle.)

In the oxidation of stearic acid, there are 8 beta oxidation cycles ($8 \times 4 = 32$ ATP), 9 molecules of acetyl-CoA requiring 9 citric acid cycles ($9 \times 10 = 90$ ATP). Subtracting the initial 2 ATP used in activation gives:
-2 (activation) + 32 (from beta-oxidation) + 90 (from citric acid cycle) = 120 ATP
The overall ATP yield when one molecule of stearic acid is broken down to CO_2 is 120 molecules.

d) *Calculate the number of moles of ATP your body will form when it burns 100 g of stearic acid. The molecular formula of stearic acid is $C_{18}H_{36}O_2$.*

If one molecule of stearic acid produces 120 molecules of ATP, then one mole of stearic acid produces 120 moles of ATP. The molar mass of stearic acid is 284.47 g/mol. From here it's a unit conversion problem:

$$\mathbf{100\,g\,C_{18}H_{36}O_2 \times \frac{1\,mol}{284.47\,g} \times \frac{120\,mol\,ATP}{1\,mol\,C_{18}H_{36}O_2} = 42.2\,mol\,ATP}$$

15.63 *Linolenic acid contains the same number of carbon atoms as stearic acid (18), but it contains three double bonds. How many molecules of ATP can our body form when we oxidize one molecule of linolenic acid?*

In three of the beta oxidation cycles in the oxidation of linolenic acid, the dehydrogenation step won't need to happen, so the overall reaction will produce three fewer molecules of $FADH_2$ than are produced in the oxidation of one molecule of stearic acid. Each molecule of $FADH_2$ is equivalent to 1.5 molecules of ATP, so the overall reaction will produce 4.5 fewer molecules of ATP than in the oxidation of stearic acid, or **115.5 molecules of ATP**.

NOTE:
It's not really producing a half molecule of ATP, it's an energy equivalence.)

Solutions to Section 15.7 Core Problems

15.65 *What is a transamination reaction?*

In a transamination reaction, the amino group of an amino acid is moved to a different organic molecule, usually α-ketoglutaric acid.

15.67 *The first reaction in the breakdown of the amino acid isoleucine is a transamination. Draw the structure of the product that is formed when isoleucine is transaminated. (The structure of isoleucine is shown in Table 13.1.)*

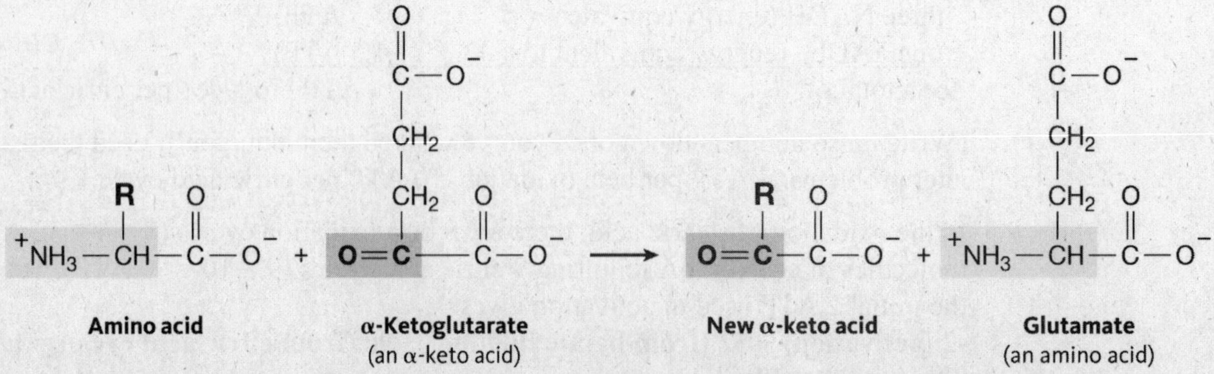

$$^{+}NH_3{-}CH(R){-}COO^- + O{=}C(CH_2CH_2COO^-){-}COO^- \longrightarrow O{=}C(R){-}COO^- + {}^{+}NH_3{-}CH(CH_2CH_2COO^-){-}COO^-$$

Amino acid | α-Ketoglutarate (an α-keto acid) | New α-keto acid | Glutamate (an amino acid)

See Figure 15.19 above. In the transamination of isoleucine, the amino group in the reactant isoleucine molecule is converted to a ketone in the product molecule. The complete reaction is shown below:

$$CH_3CH_2CH(CH_3)CH(\overset{+}{N}H_3)COO^- + {}^-OOCCH_2CH_2C(=O)COO^- \longrightarrow CH_3CH_2CH(CH_3)C(=O)COO^- + {}^-OOCCH_2CH_2CH(\overset{+}{N}H_3)COO^-$$

15.69 *Most nitrogen atoms are removed from amino acids in the form of a specific ion. What is that ion?*

When our bodies break down amino acids, we convert virtually all of the nitrogen atoms into **ammonium ions, NH_4^+**.

15.71 *How many molecules of ATP must our bodies break down in order to dispose of one ammonium ion?*

Our bodies must break down four molecules of ATP for every molecule of urea we make. Since we make urea from two ammonium ions, **the urea cycle consumes two molecules of ATP to dispose of one ammonium ion.**

15.73 *When your body breaks down a molecule of proline, it produces eight molecules of NADH, two molecules of $FADH_2$, two molecules of ATP, five molecules of CO_2, and one NH_4^+ ion. Calculate the total yield of ATP from the breakdown of a molecule of proline. Be sure to consider the urea cycle.*

Remember one NADH is the energy equivalent of 2.5 ATP, and one $FADH_2$ is the energy equivalent of 1.5 ATP. It takes the equivalent of 2 ATP to dispose of one ammonium ion. Adding it all up:

8 NADH	×	2.5 ATP/ NADH	=	20 ATP
2 $FADH_2$	×	1.5 ATP/$FADH_2$	=	3 ATP
2 ATP			=	2 ATP
1 NH_4^+			=	–2 ATP
		total	=	**23 ATP**

Solutions to Section 15.8 Core Problems

15.75 *What monosaccharide is formed in gluconeogenesis, and what organic molecules are used to build it?*

"Gluco" = glucose, "neo" = new, "genesis" = making. **Gluconeogenesis makes one new molecule of glucose from two pyruvate ions.**

15.77 *Which involves a larger number of molecules of ATP, glycolysis or gluconeogenesis?*

Gluconeogenesis involves a larger number of molecules of ATP. Making a new molecule of glucose (gluconeogenesis) breaks down six ATP, while using a molecule of glucose (glycolysis) makes two ATP. This is consistent with the general rule that anabolic pathways consume more ATP than is made in the corresponding catabolic pathway.

15.79 a) *How many molecules of ATP are required to activate a fatty acid when our bodies use it to build a triglyceride?*

It takes a total of **seven molecules of ATP** to make one molecule of triglyceride: six to activate the fatty acids (two for each fatty acid) and one to activate the glycerol.

b) *What compound is formed when we activate a fatty acid?*

We activate fatty acids by converting them into **fatty acyl-CoA**, just as we do when we break them down to obtain energy.

15.81 *How many molecules of ATP are needed to add one molecule of glucose to glycogen?*

Glycogen is a polysaccharide (Chapter 14). For every molecule of glucose that is added to a polysaccharide, our bodies must break down **two molecules of ATP**.

15.83 *What is a futile cycle?*

A futile cycle is the simultaneous production and breakdown of the same type of molecule: a catabolic pathway and the corresponding anabolic pathway happening at the same time within a cell. Thus the only net result of a futile cycle is the breakdown of ATP (since the anabolic pathway consumes more ATP than is made in the corresponding catabolic pathway).

15.85 *The liver can both build and break down glycogen, using the paired reactions below. Glycogen phosphorylase breaks down glycogen into glucose-1-phosphate, and glycogen synthase is one of the two enzymes that build glycogen from glucose-1-phosphate. The activity of glycogen phosphorylase increases when AMP is present. What effect (if any) would you expect AMP to have on glycogen synthase? Explain your answer.*

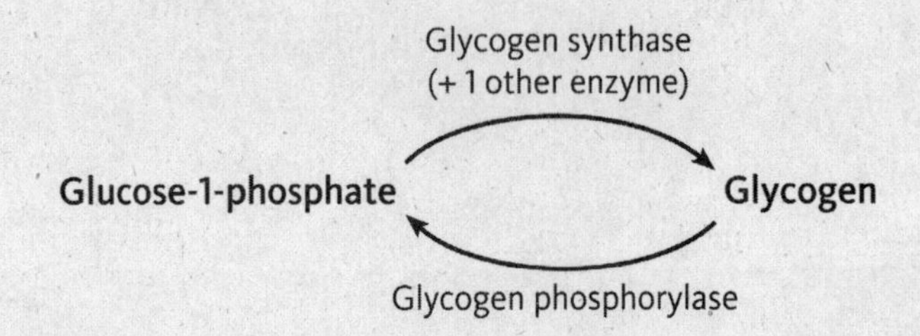

A pair of enzymes that are involved in opposing (anabolic and catabolic) processes will usually respond in opposite ways to AMP (and to citrate ion as well). **If AMP increases the activity of glycogen phosphorylase, then we would expect AMP to inhibit the effect (decrease the activity) of glycogen synthase.** This would help the cell to avoid futile cycles.

Solutions to Concept Questions

* indicates more challenging problems.

15.87 *Describe the role of the ATP cycle in metabolism.*

The ATP cycle allows cells to use food as an energy source for other cellular functions that require energy. ATP is synthesized from precursors in cellular respiration and functions as a store of chemical energy. This ATP can then be broken down, supplying energy to other cellular processes. Cells use ATP to carry out processes that require energy (anabolic processes), and make new ATP through catabolic breakdown of carbohydrate, fat or protein molecules.

15.89 *Explain why each of the following is considered to be a high-energy molecule.*

High-energy molecules are starting materials in exothermic (energy-releasing) reactions in cells. In general, high energy molecules contain weak bonds that are easy to break, while stronger bonds are formed in the formation of the product molecules. Reactions in which the bonds being broken are weaker than the bonds being formed are exothermic.

a) *ATP* — **Hydrolysis of ATP to form ADP is an exothermic reaction, in which weak bonds are broken and stronger bonds are formed** (Figure 15.2).

b) *NADH* — **The transfer of hydrogen atoms from NADH to oxygen atoms in O_2, producing water molecules, is an exothermic reaction.**

c) *acetyl-CoA* — **The oxidation of acetyl-CoA through the citric acid cycle is an exothermic reaction.**

15.91 *Describe how the mitochondria use a concentration gradient to link the oxidation of NADH and $FADH_2$ to the formation of ATP.*

The oxidation of NADH and $FADH_2$ produces H^+ ions. These ions are transported across the inner membrane of the mitochondrion (active transport), producing a higher H^+ ion concentration in the space between the membranes than in the matrix inside the inner membrane. The tendency of normal, passive transport to cause H^+ ions to flow the opposite direction is a source of energy that can be used to power the production of ATP.

15.93 *What metabolic pathways are required in order to oxidize a molecule of glucose to CO_2? In what parts of a cell do these pathways occur?*

Glycolysis breaks down glucose into pyruvate ions; this process occurs in the cytosol of the cell and does not involve the mitochondria. The pyruvate ions undergo **oxidative decarboxylation** in the mitochondria by HS-CoA to produce acetyl-CoA. The molecules of acetyl-CoA are then broken down into CO_2 through the **citric acid cycle**, which also occurs in the **mitochondria.**

15.95 *During glycolysis, four separate molecules of ADP are converted into ATP. However, the ATP yield from glycolysis is always given as 2 molecules of ATP, instead of four. Explain.*

The glycolysis pathway includes two steps that are activation reactions, which require the energy produced by converting ATP to ADP. Therefore the net ATP yield is 2 ATP (4 produced – 2 used = 2).

15.97 *What metabolic pathways are required to oxidize a fatty acid to CO_2? In what parts of a cell do these pathways occur?*

The first step is an **activation reaction**, in which the fatty acid carboxylate ion combines with HS-CoA (coenzyme A) to form a fatty-acyl CoA. This molecule then undergoes cycles of **beta oxidation** until it is completely broken down into two-carbon fragments of acetyl-CoA. The molecules of acetyl-CoA are then broken down into CO_2 through the **citric acid cycle. All of these processes take place in the mitochondria.**

15.99 *Explain why your body obtains more ATP from a saturated fatty acid than it does from unsaturated fatty acid with the same number of carbon atoms.*

A beta oxidation cycle begins with a dehydrogenation step, which converts a saturated fatty acid to an unsaturated fatty acid. This reaction produces one molecule of $FADH_2$, the energy equivalent of 1.5 molecules of ATP. In an unsaturated hydrocarbon, one or more of the beta oxidation cycles will not have a dehydrogenation step, so the overall reaction will produce fewer molecules of $FADH_2$ and thus fewer molecules of ATP than are produced in the oxidation of a saturated fatty acid of the same size. (See Section 15.6.)

15.101 *What is the function of the urea cycle, and why is it important in the metabolism of amino acids?*

In the metabolism of amino acids, the nitrogen from the amino acids is converted into NH_4^+ ions, which are toxic to body tissues. The urea cycle combines the ammonium ions with hydrogen carbonate ions (HCO_3^-) to form urea, a relatively safe compound that can be stored by the body until it can be excreted.

15.103 *When your body oxidizes the amino acid valine to obtain energy, it converts valine into succinyl-CoA. Explain how this pathway allows your body to reduce the number of enzymes it must make. (Hint: Review the reactions of the citric acid cycle.)*

Every different molecule the body has to break down requires a new enzyme, unless the molecule can be broken down into something the body already has a pathway to deal with. Converting valine into succinyl-CoA is useful because succinyl-CoA is one of the molecules in the citric acid cycle, so the body already has enzymes that can turn succinyl-CoA into succinate ion (and subsequently into CO_2 through the rest of the steps in the cycle.)

15.105 *Why is it important that cells not carry out both a catabolic pathway and the corresponding anabolic pathway at the same time?*

The purpose of catabolism is to break down large molecules into smaller ones to produce energy, and the purpose of anabolism is to build larger molecules from smaller ones. **If a cell carries out both pathways at the same time–building and breaking down the same molecule–there is no net production of large molecules, but because anabolic cycles use more energy than the corresponding catabolic cycles produce, there is a net breakdown of ATP. It's called a "futile cycle" because it wastes ATP without producing anything useful for the cell.**

15.107 *Under what circumstances does your liver carry out gluconeogenesis at a significant rate?*

The activities of the enzymes that control key steps of gluconeogenesis are favored by high concentrations of citrate ion and low concentrations of AMP in the cells. Under the opposite conditions, glycolysis is favored.

Solutions to Summary and Challenge Problems

15.109 *Your body can make cholesterol, using acetyl groups from acetyl-CoA.*

a) *Is this an anabolic pathway, or is it a catabolic pathway?*

This process turns small molecules (two-carbon fragments in acetyl-CoA) into large molecules (cholesterol, with its large steroid nucleus). **It is an anabolic pathway.**

b) *Does this pathway make ATP, or does it consume ATP?*

Anabolic pathways always require a net input of energy, which means they will always consume ATP.

c) *Does this pathway make ADP, or does it consume ADP?*

When ATP is consumed, it is turned into ADP, so this pathway makes ADP.

c) *What are some possible sources of the acetyl-CoA for this pathway?*

Acetyl-CoA is produced in the catabolism of fatty acids, saccharides, and amino acids, so the acetyl-CoA used to make cholesterol could come from any of those food sources.

15.111 *Your body can interconvert the amino acids glutamine and glutamate (the pH 7 form of glutamic acid), using the two reactions shown below. One of these reactions is always combined with the hydrolysis of ATP. Which one is it, and how can you tell?*

$$\text{glutamate} + NH_4^+ + 3.4\ \text{kcal} \rightarrow \text{glutamine} + H_2O$$
$$\text{glutamine} + H_2O \rightarrow \text{glutamate} + NH_4^+ + 3.4\ \text{kcal}$$

Hydrolysis of ATP produces energy; it's the conversion of ATP into ADP. Reactions that require energy are teamed up with hydrolysis of ATP to supply the energy. **The first reaction requires an input of energy, so it will be combined with the hydrolysis of ATP.**

15.113 *Each of the following nutrients can be oxidized to produce energy in the form of ATP.*
glucose lactose stearic acid (a fatty acid) isoleucine (an amino acid)

Figure 15.20 is a good place to look for the review of the relationships between nutrients and processes.

For each of the following pathways, tell which of those four nutrients requires this pathway in its catabolism:

a) *glycolysis*

Glycolysis is the catabolic pathway that breaks down **glucose**. Also, **lactose** is a disaccharide made up of two glucose molecules (Section 14.4), so after an initial hydrolysis step separates the glucose molecules, they'll need glycolysis as well.

b) *transamination*

Isoleucine: transamination is the first step in the catabolism of amino acids. Notice the "-amin-" part of the word—amino acids are the nutrients that contain amine groups (carbohydrates and fats do not contain amines).

c) *citric acid cycle*

Saccharides, fats, and amino acids are all broken down into acetyl-CoA, which all goes into the citric acid cycle. **All of these nutrients will end up in the citric acid cycle.**

d) *oxidative deamination*

Isoleucine. There's that "-amin-" again; amino acids require oxidative deamination.

e) *beta oxidation*

Stearic acid: beta oxidation is the catabolic pathway for fats.

f) *urea cycle*

Isoleucine. Urea contains nitrogen; if you can just remember that, you don't have to memorize the structure or formula of it to link it to amino acids, the only nutrients that supply nitrogen.

g) *oxidative phosphorylation*

Oxidative phophorylation is the sequence of reactions that harnesses energy from NADH and $FADH_2$ to make ATP. Since catabolism of all of these nutrients produces NADH and $FADH_2$, **they all require oxidative phosphorylation.**

15.115 *Several amino acids are converted into succinic acid ($H_2C_4H_4O_4$) when they are broken down. The succinic acid is then oxidized to CO_2. Write balanced equations for each of the following reactions:*

a) *The combustion of succinic acid.*

Remember that combustion reactions (Chapters 6 and 8) are always reactions with O_2. Carbon in the reactants is converted to CO_2 and hydrogen in the reactants is converted to H_2O. We write a general reaction and then balance it. (We run into an even/odd problem with O, as often happens, but it's easily solved by doubling the number of molecules/moles of succinic acid.)

$$2\ H_2C_4H_4O_4 + 7\ O_2 \rightarrow 8\ CO_2 + 6\ H_2O$$

b) *The oxidation of succinic acid to CO_2 and hydrogen atoms (You may represent the hydrogen atoms as [H]).*

First, we write the skeleton equation:

$$H_2C_4H_4O_4 + ?\ H_2O \rightarrow ?\ CO_2 + ?\ [H]$$

Then follow the steps:
Step 1: Balance C, by writing the coefficient for CO_2 (in this case, 4 CO_2).
Step 2: Balance O, by writing the coefficient for H_2O (in this case, 4 H_2O).
Step 3: Balance H, by writing the coefficient for [H] (in this reaction, 6 + 8 = 14 [H]).

$$H_2C_4H_4O_4 + 4\ H_2O \rightarrow 4\ CO_2 + 14\ [H]$$

c) *The reaction of succinic acid with NAD^+ and FAD, to form CO_2, NADH and $FADH_2$ (This reaction requires one molecule of FAD and more than one molecule of NAD^+).*

NAD^+ and FAD take up the hydrogen atoms represented as [H] in the reaction in part (b). NAD^+ becomes NADH + H^+, consuming 2 [H] per NAD^+, and FAD becomes $FADH_2$, consuming 2 [H] per FAD. If the reaction has only one FAD, that leaves 12 [H] to be taken up by 6 NAD:

$$H_2C_4H_4O_4 + 4\ H_2O + FAD + 6\ NAD^+ \rightarrow 4\ CO_2 + FADH_2 + 6\ NADH + 6\ H^+$$

15.117 *List all of the starting materials and products in glycolysis, including redox coenzymes.*

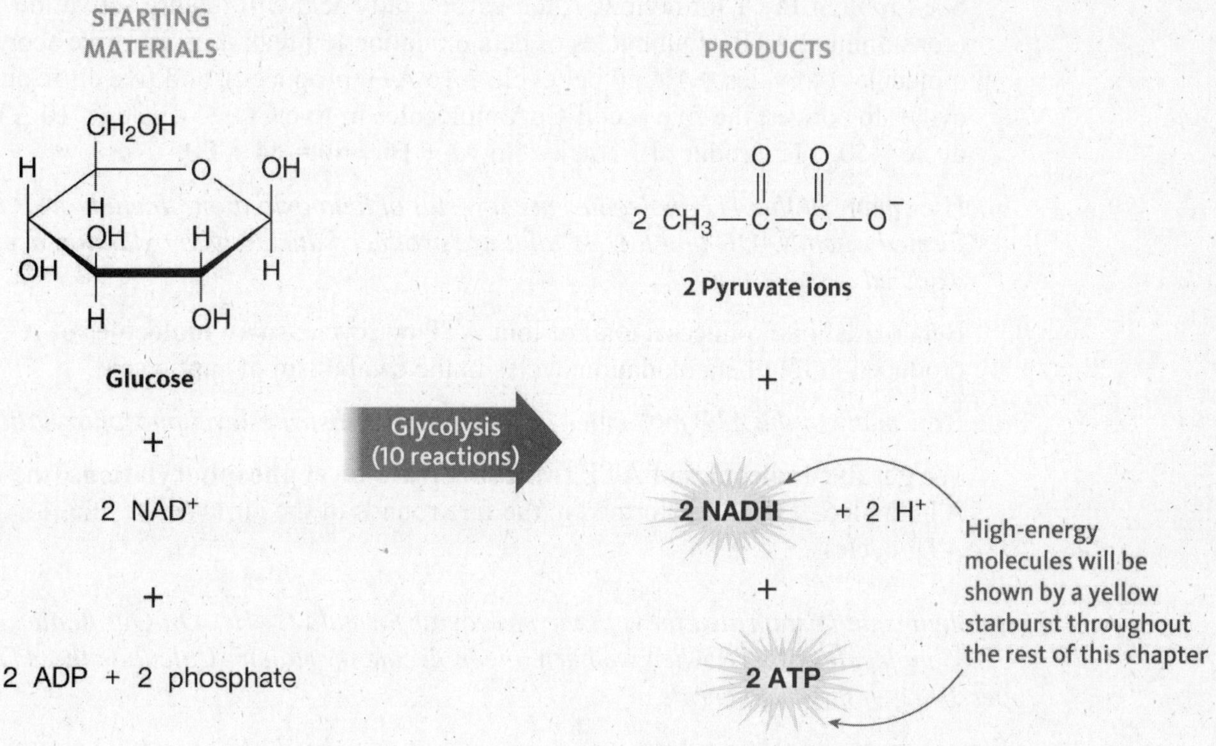

STARTING MATERIALS

Glucose + $2\ NAD^+$ + 2 ADP + 2 phosphate

Glycolysis (10 reactions)

PRODUCTS

$2\ CH_3{-}C(=O){-}C(=O){-}O^-$ (2 Pyruvate ions) + 2 NADH + 2 H^+ + 2 ATP

High-energy molecules will be shown by a yellow starburst throughout the rest of this chapter

See Figure 15.11 on previous page. **The reaction combines a molecule of glucose ($C_6H_{12}O_6$) with 2 NAD^+, 2 ADP, and 2 phosphate ions, to form 2 pyruvate ions ($CH_3COCO_2^-$), 2 NADH, 2 H^+, and 2 ATP.**

15.119 *What metabolic pathway is involved when milk spoils? What chemical compound is responsible for the unpleasant flavor of the spoiled milk?*

Lactic acid fermentation produces lactate ions (the conjugate base of lactic acid) in muscle cells and lactic acid in the spoilage of milk. **Lactic acid** is responsible for the characteristic flavor of spoiled milk.

15.121 *Some organisms can convert ethanol into acetyl-CoA using the following sequence of reactions:*

$$\text{ethanol} + NAD^+ \rightarrow \text{acetaldehyde} + NADH + H^+$$
$$\text{acetaldehyde} + NAD^+ + \text{CoA} \rightarrow \text{acetyl-CoA} + NADH + H^+$$

The acetyl-CoA is then broken down to CO_2 by the citric acid cycle. What is the total ATP yield when one molecule of ethanol is oxidized to CO_2?

Each molecule of acetyl-CoA that goes through the citric acid cycle yields the energy equivalent of 10 ATP (we calculated this in problem 15.61c).The reactions given here produce a total of 2 NADH, the energy equivalent of 5 ATP (2.5 ATP/NADH). **The total ATP yield is 15 ATP.**

15.123 *Capric acid is a ten-carbon saturated fatty acid.*

a) *Calculate the ATP yield when one molecule of capric acid is broken down into CO_2.*

See Problem 15.61 for review. A ten-carbon fatty acid will require activation (consuming 2 ATP), four cycles of beta oxidation to break it up into five acetyl-CoA molecules (4 cycles × 4 ATP per cycle = 16 ATP produced), and five citric acid cycles to convert the five acetyl-CoA molecules to five CO_2 (5 cycles × 10 ATP per cycle = 50 ATP produced). Total yield: **-2 + 16 + 50 = 64 ATP.**

b) *How many of the ATP molecules are a result of beta oxidation? Include ATP that is formed when NADH and $FADH_2$ that are produced during beta oxidation are oxidized.*

Beta oxidation produces a total of four ATP per cycle, so **16 molecules of ATP** are produced in four beta oxidation cycles in the catabolism of capric acid.

c) *How many of the ATP molecules are formed by substrate-level phosphorylations?*

We get five molecules of ATP from substrate-level phosphorylations (the five ATP molecules that are formed in the five rounds of the citric acid cycle, 1 ATP/cycle).

15.125 *Triolein is an unsaturated fat with the molecular formula $C_{57}H_{104}O_6$. Our bodies obtain 379.5 molecules of ATP when we burn one molecule of triolein. Calculate the ATP yield per 100 g of this triglyceride.*

We are given the relationship:

$$379.5 \text{ molecules ATP} = 1 \text{ molecule } C_{57}H_{104}O_6$$

Therefore, we can also say that

$$379.5 \text{ moles ATP} = 1 \text{ mole } C_{57}H_{104}O_6$$

To find the ATP yield per 100 g of triolein, $C_{57}H_{104}O_6$, we'll need the molar mass of triolein, 885.4 g/mol. From here it's a conversion problem:

$$\mathbf{100\,g\,C_{57}H_{104}O_6 \times \frac{1\,mol}{885.4\,g} \times \frac{379.5\,mol\,ATP}{1\,mol\,C_{18}H_{36}O_2} = 42.9\,mol\,ATP}$$

15.127 *When the amino acid lysine is broken down, it undergoes a transamination reaction involving the amino group in the side chain. Draw the structure of the product of this reaction. (The structure of lysine is shown in Table 13.1.)*

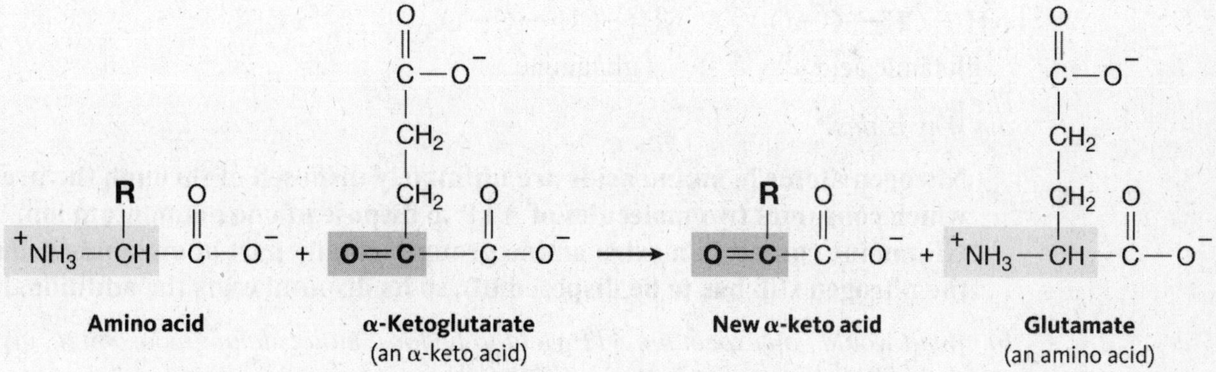

$$^{+}NH_3-CH(R)-C(=O)-O^- \; + \; O=C(-CH_2-CH_2-C(=O)-O^-)-C(=O)-O^- \longrightarrow O=C(R)-C(=O)-O^- \; + \; ^{+}NH_3-CH(-CH_2-CH_2-C(=O)-O^-)-C(=O)-O^-$$

Amino acid + **α-Ketoglutarate** (an α-keto acid) → **New α-keto acid** + **Glutamate** (an amino acid)

See Figure 15.19 above. In the transamination of lysine, the amino group in the side chain of the reactant lysine molecule is converted to a ketone in the product. The complete reaction is shown below:

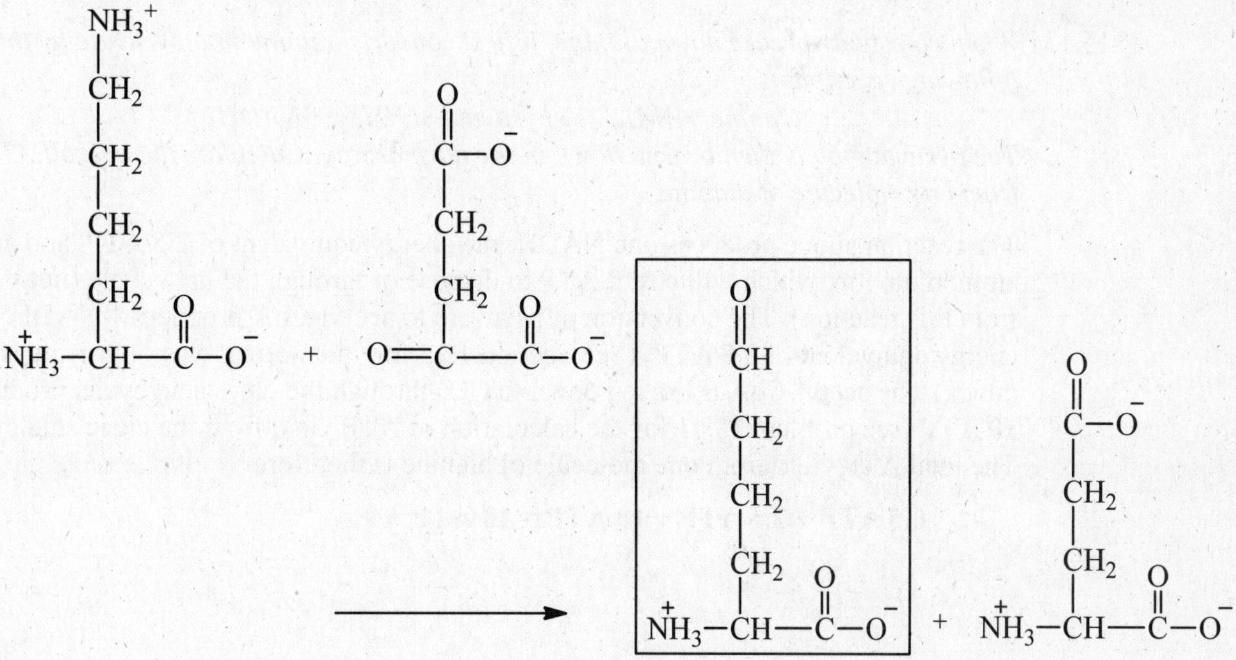

$$\overset{+}{N}H_3-CH(-CH_2-CH_2-CH_2-CH_2-NH_3^{+})-C(=O)-O^- \; + \; O=C(-CH_2-CH_2-C(=O)-O^-)-C(=O)-O^-$$

$$\longrightarrow \boxed{\overset{+}{N}H_3-CH(-CH_2-CH_2-CH_2-CH=O)-C(=O)-O^-} \; + \; \overset{+}{N}H_3-CH(-CH_2-CH_2-C(=O)-O^-)-C(=O)-O^-$$

15.129 *Glutamic acid and glutamine have the same carbon skeleton. However, your body obtains 20.5 ATP from one molecule of glutamic acid, while it only obtains 18.5 ATP from a molecule of glutamine.*

The structures of glutamic acid and glutamine are found in Table 13.1:

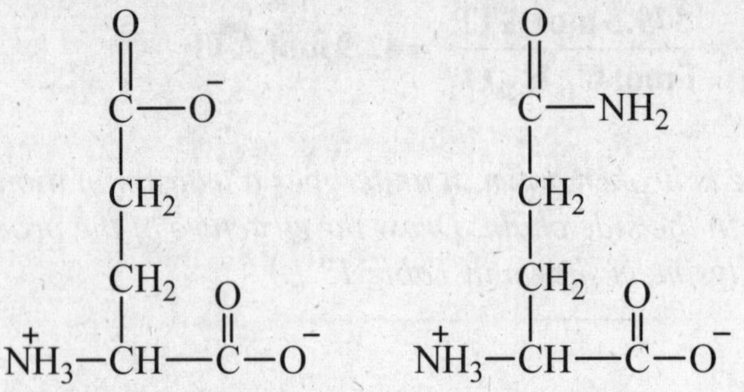

$$\underset{\text{glutamic acid}}{\overset{+}{N}H_3-\underset{\underset{\underset{\underset{C(=O)-O^-}{|}}{CH_2}}{|}}{\overset{|}{CH}}-C(=O)-O^-}$$

(Glutamic acid: $O=C(-O^-)-CH_2-CH_2-CH(\overset{+}{N}H_3)-C(=O)-O^-$)

(Glutamine: $O=C(-NH_2)-CH_2-CH_2-CH(\overset{+}{N}H_3)-C(=O)-O^-$)

glutamic acid glutamine

a) *Why is this?*

Nitrogen atoms in amino acids are ultimately disposed of through the urea cycle, which consumes two molecules of ATP to dispose of one ammonium ion. Glutamine contains an extra amino group (actually part of an amide group, but the nitrogen still has to be disposed of), so its disposal costs the additional 2 ATP.

b) *What would you expect the ATP yield to be for a molecule of glucosamine, given that the ATP yield for glucose is 32 ATP? (Glucosamine is identical to glucose except for having an amino group on carbon #2 instead of a hydroxyl group.)*

Since it will cost two ATP to dispose of one nitrogen, we'd expect glucosamine to have an ATP yield of 32 – 2 = **30 ATP**.

15.131 *When your body breaks down alanine, it first converts alanine into pyruvate by the following reaction:*

$$alanine + NAD^+ \rightarrow pyruvate + NH_4^+ + NADH$$

The pyruvate ion is then broken down in the normal way. Calculate the overall ATP yield from one molecule of alanine.

The reaction above produces one NADH, the energy equivalent of 2.5 ATP, and an ammonium ion, which will cost 2 ATP to dispose of through the urea cycle (**net 0.5 ATP** from this reaction). The conversion of pyruvate to acetyl-CoA produces 1 NADH, the energy equivalent of **2.5 ATP**. (See Section 15.5 for "the normal way" pyruvate is broken down.) The acetyl-CoA is broken down to CO_2 through the citric acid cycle, producing **10 ATP** (see problem 15.61 for the calculation of ATP yield from the citric acid cycle). The total ATP yield from one molecule of alanine is therefore

0.5 ATP + 2.5 ATP + 10 ATP = 13 ATP

15.133 *When your body breaks down alanine, it first converts the alanine into pyruvate. Based on this, can your body use alanine to make glucose? If so, how?*

Yes, the body can use alanine to make glucose. Pyruvate is the starting material in gluconeogenesis, the process of making glucose, so any molecule whose catabolism forms pyruvate can be used to make glucose.

*15.135 *It takes 44 molecules of ATP to make a molecule of palmitic acid from eight molecules of acetyl-CoA. How many molecules of ATP will be wasted in the following futile cycle?*

$$\text{palmitic acid} \rightarrow 8 \text{ acetyl-CoA} \rightarrow \text{palmitic acid}$$

Breaking down one molecule of palmitic acid into eight molecules of acetyl-CoA will take seven cycles of beta oxidation (7 cycles × 4 ATP per cycle = 28 ATP), and will cost 2 ATP in activation, for a total production of 26 ATP. If it costs 44 ATP to build the molecule again, the net waste of ATP in this futile cycle is **44 – 26 = 18 ATP**.

*15.137 *The concentration of AMP in a cell is normally quite low, but it increases when the concentration of ATP drops. Use this fact to explain why it makes sense for AMP to be a positive effector of phosphofructokinase (PFK), rather than a negative effector.*

PFK is an enzyme involved in glycolysis, which produces ATP. When the ATP concentration in a cell is low, the cell needs to carry out glycolysis to produce more ATP. If low concentrations of ATP cause high concentrations of AMP, then AMP will be a positive effector for enzymes that produce more ATP, including PFK.

You can also think of it from the opposite direction: low ATP causes high AMP (as stated in the problem), so AMP should be a negative effector for anything that uses ATP (which would decrease the concentration of ATP even further) and a positive effector for anything that produces ATP.

15.139 *The following graph shows the rate of gluconeogenesis for a person who is engaging in an exercise session.*

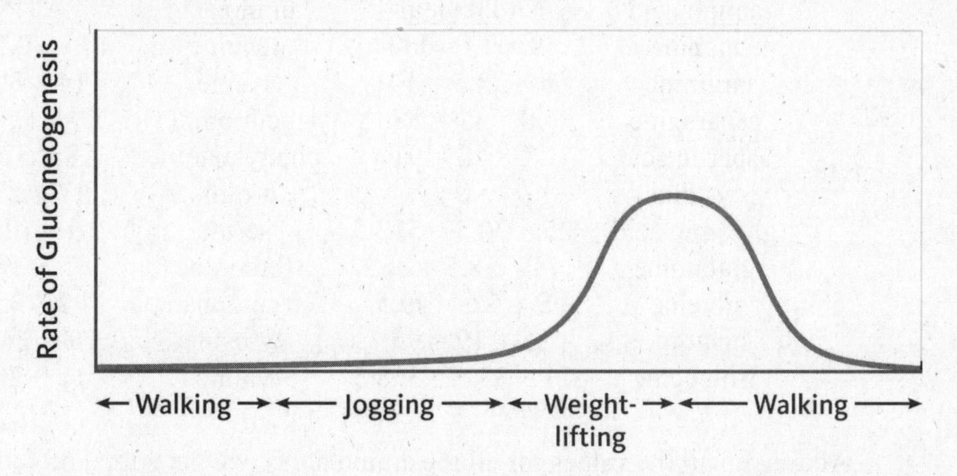

a) *Why does the rate of gluconeogenesis increase sharply during the weight-lifting portion of the session?*

See Section 15.8. **Active muscles turn glucose into lactate ions (the form of lactic acid at physiological pH) through the lactic acid fermentation process. The liver**

carries out gluconeogenesis to rebuild lactate ions into glucose, and return the glucose to the bloodstream.

b) *Why does the rate remain high for a few minutes after the weight-lifting session is completed?*

It takes some time for all the lactate ions produced in the muscles to get into the bloodstream and travel to the liver, and some time for the liver to catch up and return the glucose concentration to its normal level.

*15.141 *The main protein in cow's milk is called casein. The amino acid composition of casein is shown in the following table.*

amino acid	*number*	*amino acid*	*number*
alanine	*9*	*leucine*	*17*
arginine	*6*	*lysine*	*14*
asparagine	*8*	*methionine*	*5*
aspartic acid	*7*	*phenylalanine*	*8*
cysteine	*0*	*proline*	*17*
glutamic acid	*25*	*serine*	*16*
glutamine	*14*	*threonine*	*5*
glycine	*9*	*tryptophan*	*2*
histidine	*5*	*tyrosine*	*10*
isoleucine	*11*	*valine*	*11*

a) *Using this information and the ATP values in Table 15.7, calculate the overall ATP yield from one molecule of casein.*

This isn't exactly hard, it just takes a while. You can calculate the values by hand and then add them up, or you may want to make a spreadsheet to do the math if you know how.

amino acid	ATP yield	amino acid	ATP yield
alanine	$9 \times 13 = 117$	leucine	$17 \times 32.5 = 552.5$
arginine	$6 \times 23.5 = 141$	lysine	$14 \times 30 = 420$
asparagine	$8 \times 11 = 88$	methionine	$5 \times 21.5 = 107.5$
aspartic acid	$7 \times 13 = 91$	phenylalanine	$8 \times 32 = 256$
cysteine	0	proline	$17 \times 23 = 391$
glutamic acid	$25 \times 20.5 = 512.5$	serine	$16 \times 10.5 = 168$
glutamine	$14 \times 18.5 = 259$	threonine	$5 \times 18 = 90$
glycine	$9 \times 5.5 = 49.5$	tryptophan	$2 \times 35 = 70$
histidine	$5 \times 19 = 95$	tyrosine	$10 \times 34.5 = 345$
isoleucine	$11 \times 33.5 = 368.5$	valine	$11 \times 26 = 286$

Adding up all the values for all the amino acids, we get a total of **4407.5 molecules of ATP** from one molecule of this protein.

NOTE:

For part (c) that this also means 4407.5 moles of ATP per mole of protein; it's the same relationship.

b) *Calculate the average ATP yield per amino acid in this protein (divide the overall ATP yield by the number of amino acids in the protein).*

There are a total of **199 amino acids** in casein. The average ATP yield is

4407.5 ÷ 199 = 22.15 ATP per amino acid.

c) *The formula weight of this protein is 22,950 amu. Calculate the ATP yield per 100 g for this protein.*

22950 amu is also 22950 g/mol.

$$\mathbf{100\ g\ casein \times \frac{1\ mol}{22950\ g} \times \frac{4407.5\ mol\ ATP}{1\ mol\ casein} = 19.2\ mol\ ATP}$$

d) *How does the ATP yield you obtained in part c compare with the values for starch and triglycerides in Table 15.8?*

According to Table 15.8, the average ATP yield for starches is 19.8 moles per 100 g of starch, and the average yield for various fats is around 40. **The ATP yield for this protein, 19.2 moles ATP/100 g protein, is in the same range as starch, but less than half the yield from fats.**

NOTE:

This shouldn't be a surprise. Remember from Chapter 6 that the nutritive values of carbohydrates and proteins were about the same, 4 kcal/g, while the nutritive value for fats was more than twice as high, 9 kcal/g. In Chapter 15 we're working with the nutritive values in different units (moles of ATP instead of kcal) but the relationship is the same: starch and protein have about the same nutritive value, while fats are a little over twice as high as either starch or protein.

*15.143 *The following reaction is at equilibrium in red blood cells:*

$$pyruvate + NADH + H^+ \rightarrow lactate + NAD^+$$

a) *If some of the lactate ions pass out of the cell, what will happen to the concentration of pyruvate ions? Explain your answer.*

Chapter 6 review! When a reaction at equilibrium is disturbed, it will shift in the direction that partially counteracts the disturbance. **If some lactate ions leave the cell, the reaction will shift forward and convert reactants to products to partially replace lactate ions, which will cause the concentration of pyruvate to decrease.**

b) *If the concentration of NADH in the cell increases due to some other reaction, what will happen to the concentration of pyruvate ions? Explain your answer.*

If the concentration of NADH increases, the reaction will go forward to use it up, which will also cause the concentration of pyruvate to decrease.

c) *If the pH of the cell goes up, what will happen to the concentration of lactate ions in the cell? Explain your answer.*

An increase in pH is a decrease in the concentration of H^+ ions (Ch 7). **A decrease in H^+ will cause the reaction to go backward, converting products back to reactants and increasing the concentration of pyruvate.**

*15.145 *Which (if any) of the carbon atoms in each of the following ions are chiral?*
Chapter 9 review! Chiral carbon atoms have four single bonds to four different atoms or groups of atoms. Begin by drawing the structure of each ion.

a) *lactate* **One chiral carbon atom**:

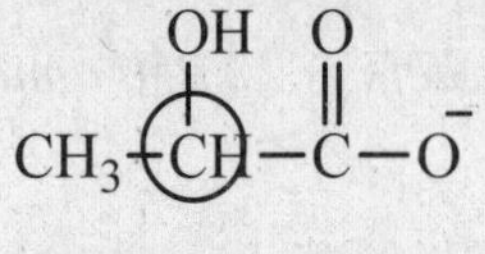

$$CH_3-\underset{\text{(circled)}}{\overset{OH}{\overset{|}{CH}}}-\overset{O}{\overset{\|}{C}}-O^-$$

b) *pyruvate* **No chiral carbon atoms**. The only carbon atom with four single bonds is attached to three hydrogen atoms:

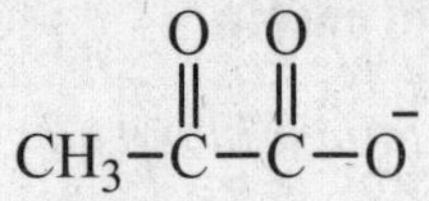

$$CH_3-\overset{O}{\overset{\|}{C}}-\overset{O}{\overset{\|}{C}}-O^-$$

C *citrate* **No chiral carbon atoms.** Notice that the groups to the left and right of the center carbon atom are the same:

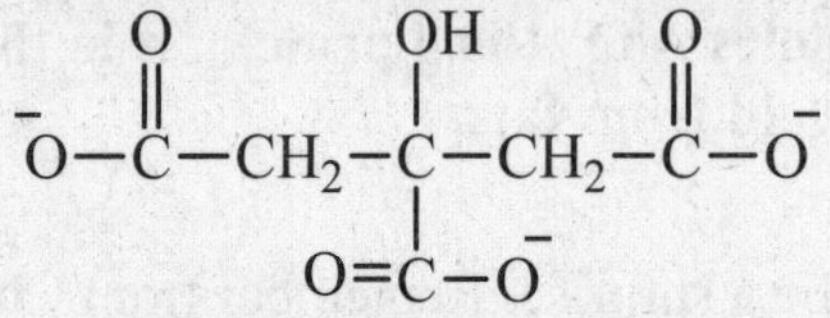

$$^-O-\overset{O}{\overset{\|}{C}}-CH_2-\underset{O=C-O^-}{\underset{|}{\overset{OH}{\overset{|}{C}}}}-CH_2-\overset{O}{\overset{\|}{C}}-O^-$$

d) *isocitrate* **Two chiral carbon atoms:**

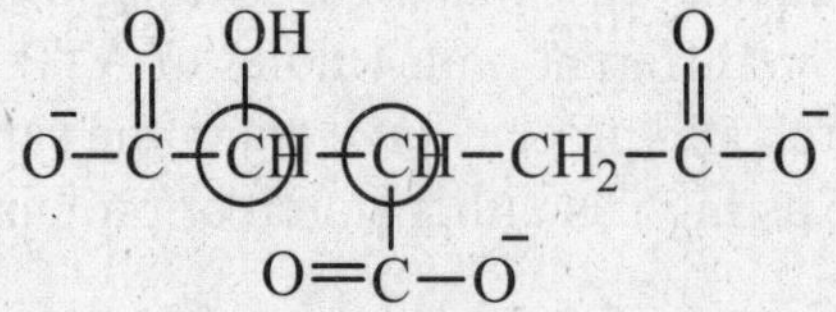

$$^-O-\overset{O}{\overset{\|}{C}}-\overset{OH}{\overset{|}{CH}}-\underset{O=C-O^-}{\underset{|}{CH}}-CH_2-\overset{O}{\overset{\|}{C}}-O^-$$

*15.147 *Your body breaks down the amino acid valine using a lengthy sequence of reactions. The first six reactions of this pathway are shown here. For each step, tell what type of reaction it is (hydration, oxidation, hydrolysis, etc.) and list any additional reactants and products (such as NAD⁺, coenzyme A, ATP, etc.).*

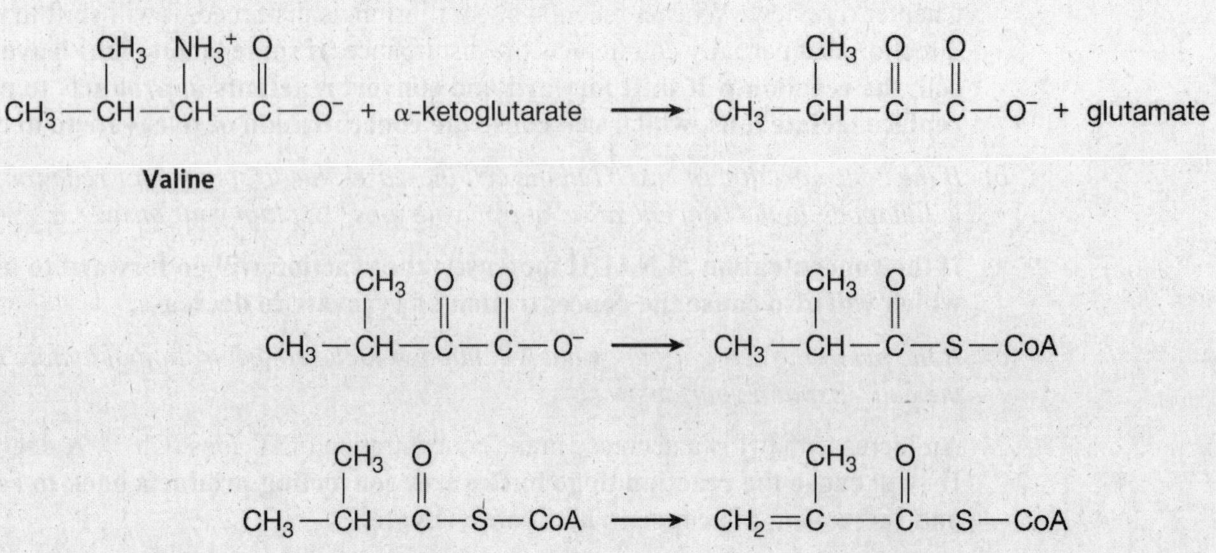

$$\underset{\text{Valine}}{CH_3-\overset{CH_3}{\overset{|}{CH}}-\overset{NH_3^+}{\overset{|}{CH}}-\overset{O}{\overset{\|}{C}}-O^-} + \alpha\text{-ketoglutarate} \longrightarrow CH_3-\overset{CH_3}{\overset{|}{CH}}-\overset{O}{\overset{\|}{C}}-\overset{O}{\overset{\|}{C}}-O^- + \text{glutamate}$$

$$CH_3-\overset{CH_3}{\overset{|}{CH}}-\overset{O}{\overset{\|}{C}}-\overset{O}{\overset{\|}{C}}-O^- \longrightarrow CH_3-\overset{CH_3}{\overset{|}{CH}}-\overset{O}{\overset{\|}{C}}-S-CoA$$

$$CH_3-\overset{CH_3}{\overset{|}{CH}}-\overset{O}{\overset{\|}{C}}-S-CoA \longrightarrow CH_2=\overset{CH_3}{\overset{|}{C}}-\overset{O}{\overset{\|}{C}}-S-CoA$$

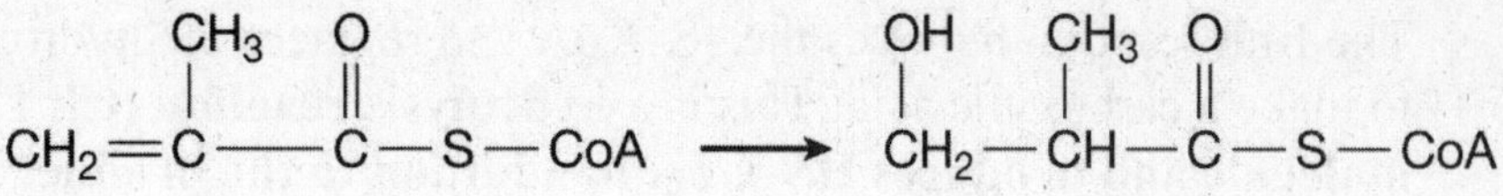

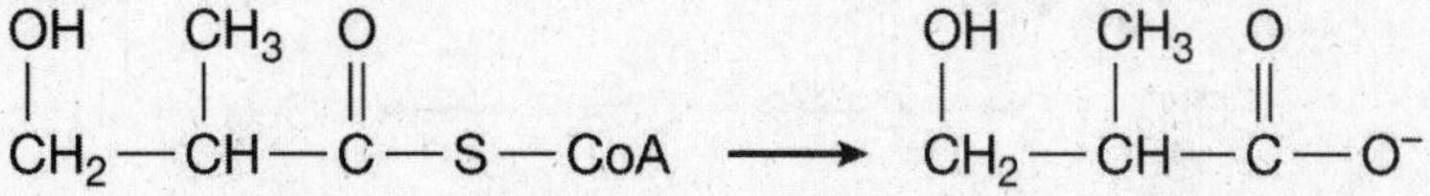

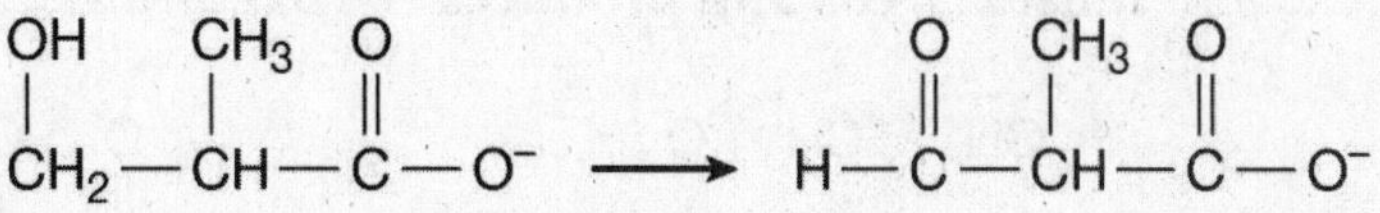

$$CH_2{=}C(CH_3){-}C({=}O){-}S{-}CoA \longrightarrow CH_2(OH){-}CH(CH_3){-}C({=}O){-}S{-}CoA$$

$$CH_2(OH){-}CH(CH_3){-}C({=}O){-}S{-}CoA \longrightarrow CH_2(OH){-}CH(CH_3){-}C({=}O){-}O^-$$

$$CH_2(OH){-}CH(CH_3){-}C({=}O){-}O^- \longrightarrow H{-}C({=}O){-}CH(CH_3){-}C({=}O){-}O^-$$

In each reaction, look carefully at the reactants and products to identify what's actually happening in the reaction.

The first reaction is a **transamination** (Figure 15.19): it converts the amino group in valine to a carbonyl. No redox coenzymes or other reactants are needed.

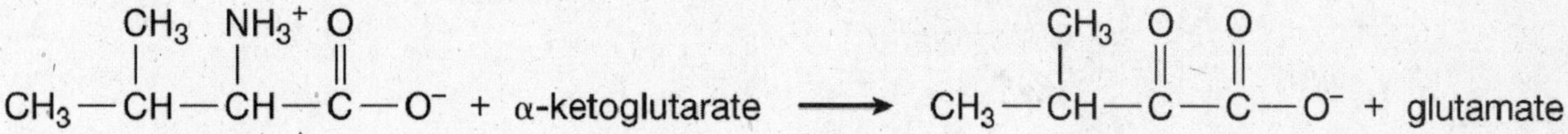

$$CH_3{-}CH(CH_3){-}CH(NH_3^+){-}C({=}O){-}O^- + \alpha\text{-ketoglutarate} \longrightarrow CH_3{-}CH(CH_3){-}C({=}O){-}C({=}O){-}O^- + \text{glutamate}$$

Valine

In the second reaction, the carboxylic acid group is removed and replaced with coenzyme-A. This reaction is an **oxidative decarboxylation** (Section 11.2). **The other reactants are HS-CoA and NAD^+, and the products are CO_2, NADH, and H^+.**

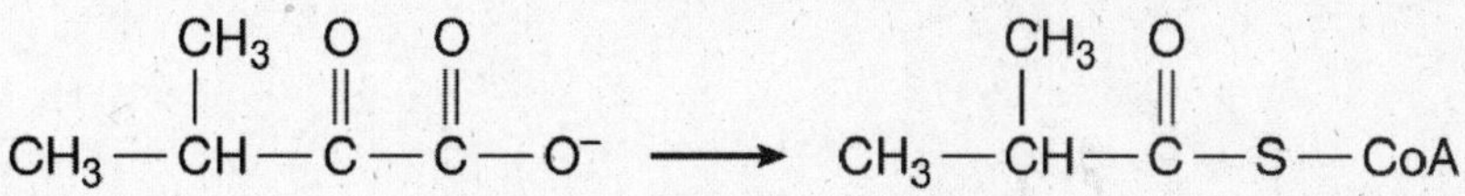

$$CH_3{-}CH(CH_3){-}C({=}O){-}C({=}O){-}O^- \longrightarrow CH_3{-}CH(CH_3){-}C({=}O){-}S{-}CoA$$

In the third reaction, two hydrogen atoms are removed to convert a saturated alkane group into an alkene. This is a **dehydrogenation** (Ch 10), and **requires a molecule of FAD, producing $FADH_2$.**

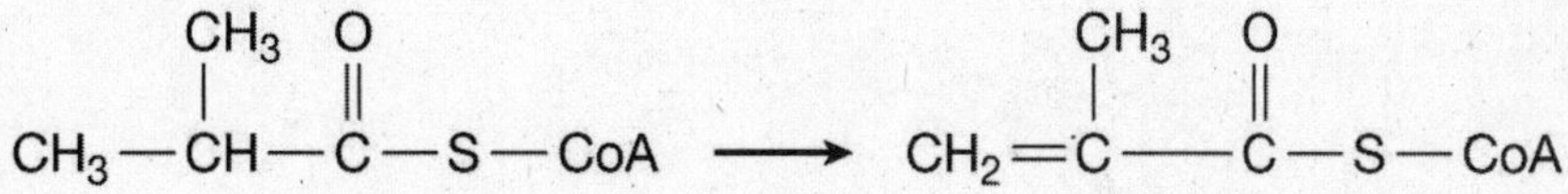

$$CH_3{-}CH(CH_3){-}C({=}O){-}S{-}CoA \longrightarrow CH_2{=}C(CH_3){-}C({=}O){-}S{-}CoA$$

The fourth reaction is a **hydration reaction** (Ch 9), which adds a water molecule to the double bond to produce an alcohol. This reaction, therefore, **requires water as a reactant**.

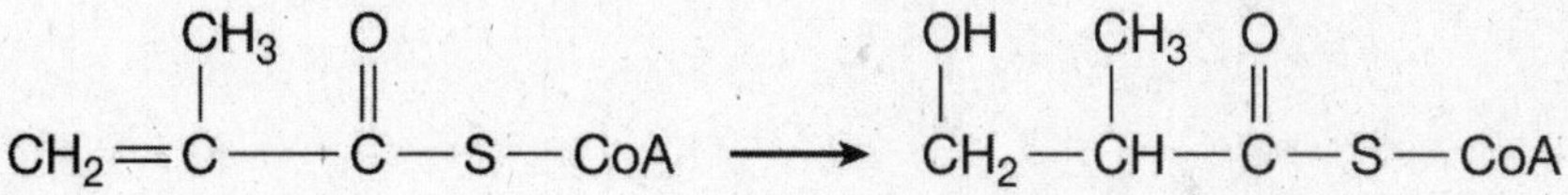

$$CH_2{=}C(CH_3){-}C({=}O){-}S{-}CoA \longrightarrow CH_2(OH){-}CH(CH_3){-}C({=}O){-}S{-}CoA$$

The fifth reaction removes the –S–CoA and replaces the hydroxyl group on the carbonyl to make a carboxylic acid. This is a **hydrolysis reaction (Ch 12); it requires a water molecule and produces HS–CoA in addition to the product shown:**

$$\underset{\text{OH}}{\underset{|}{\text{CH}_2}}-\underset{\text{CH}_3}{\underset{|}{\text{CH}}}-\overset{\text{O}}{\overset{\|}{\text{C}}}-\text{S}-\text{CoA} \longrightarrow \underset{\text{OH}}{\underset{|}{\text{CH}_2}}-\underset{\text{CH}_3}{\underset{|}{\text{CH}}}-\overset{\text{O}}{\overset{\|}{\text{C}}}-\text{O}^-$$

The last reaction shows the conversion of an alcohol group to a carbonyl, an **oxidation** reaction. **It uses NAD^+ and produces NADH and H^+.**

$$\underset{\text{OH}}{\underset{|}{\text{CH}_2}}-\underset{\text{CH}_3}{\underset{|}{\text{CH}}}-\overset{\text{O}}{\overset{\|}{\text{C}}}-\text{O}^- \longrightarrow \text{H}-\overset{\text{O}}{\overset{\|}{\text{C}}}-\underset{\text{CH}_3}{\underset{|}{\text{CH}}}-\overset{\text{O}}{\overset{\|}{\text{C}}}-\text{O}^-$$

Chapter 16

16

Nuclear Chemistry

Solutions to Section 16.1 Core Problems

16.1 *What do we mean when we say that an isotope is stable?*

Nuclei of stable isotopes do not break apart to form different nuclei. They are not radioactive.

16.3 *Which of the following is a nuclear reaction, and which is a chemical reaction?*

In a chemical reaction, the same elements are present in the products and reactants. If any element in a reaction turns into a different element, the reaction is a nuclear reaction. (See Chapter 2 for a review of the term "element" and the makeup of atoms if you need to.)

a) $CuBr_2 \rightarrow Cu + Br_2$ — **This is a chemical reaction.** The same elements, copper and bromine, are present before and after the reaction. Only electrons are rearranged; the nuclei stay the same.

b) $Ra \rightarrow Rn + He$ — **This is a nuclear reaction.** The elements in the reactants (radium) and products (radon and helium) of the reaction are different. Nuclei have been changed.

16.5 *How many protons and neutrons are there in each of the following atoms?* Chapter 2 review!

a) *An atom of* $^{81}_{34}\text{Se}$ — The bottom number is the atomic number, **the number of protons, 34.** The top number is the mass number of this isotope, protons + neutrons. The number of neutrons is therefore 81 - 34 = **47 neutrons.**

b) *An atom of copper-67* — "67" is the mass number of this isotope of copper, Cu. We look to the periodic table to find **the number of protons, 29**. Then the number of neutrons is 67 - 29 = **38 neutrons.**

16.7 *Write the nuclear symbol for each of the following atoms:*

The nuclear symbol is always in the form

$$^{\text{mass number}}_{\text{atomic number}}\text{X}$$

where X is the element symbol. We may have to figure out any of the three, depending on what we are given in the problem; the periodic table connects the atomic number and element symbol, and the mass number is the sum of the number of protons (which is the atomic number) and the number of neutrons.

a) *An atom that has an atomic number of 17 and a mass number of 37*

Look in the periodic table to find that element 17 is chlorine, Cl.

$^{37}_{17}Cl$

b) *An atom that has 26 protons and 28 neutrons*

Element number 26 is iron, Fe. The mass number is 26 + 28 = 54.

$^{54}_{26}Fe$

c) *An atom of cobalt-60*

This notation gives the name of the element with the mass number of the particular isotope (60 in this problem). Cobalt has atomic number 27.

$^{60}_{27}Co$

Solutions to Section 16.2 Core Problems

16.9 *Write the nuclear symbol for each of the following particles:*

a) *proton* protons have a mass of 1 amu and, by definition, an atomic number of 1:

$^{1}_{1}p$

b) *electron* electrons have a mass that is only a tiny fraction of an amu; the mass of an electron doesn't count toward the mass number of an isotope, so the mass number of an electron is given as 0. (Do remember, though, that electrons actually have mass—it's just very small compared to the masses of protons and neutrons.) Since protons have an atomic number of 1, and electrons and protons have opposite charge, electrons are given an "atomic number" of −1:

$^{0}_{-1}e$

16.11 *Classify each of the following nuclear reactions as an alpha decay, a beta decay, or a positron emission:*

These reactions are identified by the particles they produce. An alpha particle is a helium-4 nucleus; a beta particle is an electron; and a positron is a particle with the same mass as an electron but opposite (+1) charge. Note: since, in this case, each complete reaction is shown, it doesn't matter what other nuclei are present in the reaction; you only need to identify the small particle produced (alpha particle, beta particle or positron).

a) $^{200}_{83}\text{Bi} \rightarrow\ ^{200}_{82}\text{Pb} +\ ^{0}_{+1}\text{e}$

This reaction produces a positron (positive electron): **positron emission.**

b) $^{209}_{83}\text{Bi} \rightarrow\ ^{205}_{81}\text{Tl} +\ ^{4}_{2}\text{He}$

This reaction produces a helium-4 nucleus: **alpha decay.**

c) $^{214}_{83}\text{Bi} \rightarrow\ ^{214}_{84}\text{Po} +\ ^{0}_{-1}\text{e}$

This reaction produces an electron: **beta decay.**

16.13 *Write the nuclear equation for each of the following reactions:*

Nuclear reactions are balanced by adding up the mass numbers and the atomic numbers of all species on each side of the equation. The sums of these numbers in the reactants must equal their sums in the products. Note that "decay" and "emission" both mean that the named particle is a product.

a) *The alpha decay of* $^{209}_{84}\text{Po}$

Start by setting up the equation, putting the nuclear symbol of Po-209 as the reactant and the nuclear symbol for an alpha particle as the product.

$^{209}_{84}\text{Po} \rightarrow\ ? +\ ^{4}_{2}\text{He}$

Then figure out the mass number and atomic number for the unknown particle, to make the reaction balance. The other product will need a mass number of 196 (because 209 = 205 + 4) and an atomic number of 82 (because 84 = 82 + 2). Look in the periodic table to find that element number 82 is Pb:

$\mathbf{^{209}_{84}Po \rightarrow\ ^{205}_{82}Pb +\ ^{4}_{2}He}$

b) *The beta decay of oxygen-18*

Set up the equation with the nuclear symbol of oxygen-18 as the reactant and the nuclear symbol for a beta particle as the product. Then the other product must have mass number = 18 (18 = 18+ 0) and atomic number = 9 (8 = 9 + (−1)). Last, look up element 9 on the periodic table (F, fluorine):

$\mathbf{^{18}_{8}O \rightarrow\ ^{18}_{9}F +\ ^{0}_{-1}}e$

c) *The positron emission of oxygen-15*

Set up the equation with the nuclear symbol of oxygen-15 as the reactant and the nuclear symbol for a positron as the product. Then the other product must have mass number = 15 (15 = 15 + 0) and atomic number = 7 (8 = 7 + 1). Last, look up element 7 on the periodic table (N, nitrogen):

$\mathbf{^{15}_{8}O \rightarrow\ ^{15}_{7}N +\ ^{0}_{+1}}e$

16.15 *The following nuclear reaction is used to make one of the transuranium elements. Complete this nuclear equation.*

$${}^{209}_{98}Cf + {}^{15}_{7}N \rightarrow ? + 4\,{}^{1}_{0}n$$

The total mass number on the left (reactants) side is 209 + 15 = 224. The four neutrons on the right have a total mass of 4, so the other product needs to have a mass number of 220 (209 + 15 = 220 + 4). The sum of the atomic numbers on the left is 105, and since neutrons have atomic number = 0, the other product gets the entire 105. Element 105 is Db:

$$\mathbf{{}^{209}_{98}Cf + {}^{15}_{7}N \rightarrow {}^{220}_{105}Db + 4\,{}^{1}_{0}n}$$

16.17 *Iodine-131 is an important radioisotope in medicine. It is produced from naturally occurring tellurium-130 by the following sequence of reactions:*

$${}^{130}_{52}Te + {}^{1}_{0}n \rightarrow {}^{131}_{52}Te$$

$${}^{131}_{52}Te \rightarrow {}^{131}_{53}I + {}^{0}_{-1}e$$

Is iodine-131 made in a nuclear reactor, or is it made in a particle accelerator? Explain your reasoning.

Neutrons are produced by a nuclear reactor; since this reaction begins with a neutron bombardment, ^{131}I is probably produced in a nuclear reactor.

Solutions to Section 16.3 Core Problems

16.19 *Classify each of the following as particles or electromagnetic radiation.*

No way around this but to memorize it. Alphas, betas, positrons, neutrons are particles. Gamma rays, x-rays, and most anything with the word "light" (ultraviolet light, visible light, infrared light) or "wave" (radio wave, microwave) are electromagnetic radiation.

a) X-rays **electromagnetic radiation**

b alpha radiation **particles**

c) visible light **electromagnetic radiation**

16.21 *In the following nuclear reaction, the products weigh 0.46 mg less than the reactant. How much energy does this reaction produce when 133 g of xenon-133 reacts?*

$${}^{133}_{54}Xe \rightarrow {}^{133}_{55}Cs + {}^{0}_{-1}e$$

This is a problem we could actually have solved in Chapter 1, without knowing anything about nuclear reactions, if we had been given the appropriate conversion factors. If the reaction converts 0.46 mg of matter into energy, we can use the relationship

1 gram of mass lost = 2.15 x 10^{10} kcal

to calculate the energy produced, and round to 2 significant figures. Remember to convert mg to g first to make the units cancel properly:

$$\mathbf{0.46\ mg \times \frac{1\,g}{1000\ mg} \times \frac{2.15\times10^{10}\ kcal}{1\,g} = 9.9\times10^{6}\ kcal}$$

16.23 *In the reaction in Problem 16.21, the energy of the beta particles is 8,000,000 kcal.*

a) *What other type of radiation must this reaction produce? (Hint: Compare the energy of the beta particles to your answer to Problem 16.21.)*

Of a total of 9.9×10^6 kcal of energy converted from mass in the reaction, the beta particles only account for 8×10^6 kcal. The rest must be electromagnetic radiation (gamma rays) accompanying the beta decay.

b) *What is the energy of the radiation you identified in part a?*

The energy that remains to be accounted for is the difference between the total energy and the energy of the beta particles:

$$9.9 \times 10^6 \text{ kcal} - 8 \times 10^6 \text{ kcal} = 2 \times 10^6 \text{ kcal}.$$

This is the energy of the gamma radiation.

16.25 *Metastable krypton-81 decays by emitting gamma radiation:*

$$^{81m}_{36}Kr \rightarrow {}^{81}_{36}Kr + 4{,}400{,}000 \text{ kcal}$$

a) *How much mass is converted into energy in this reaction?*

In this problem we're given the energy and asked to calculate the mass (note that this is exactly the opposite of the process in Problem 16.21, where we were given mass and asked to calculate energy).

$$\mathbf{4.4 \times 10^6 \text{ kcal} \times \frac{1 \text{ g}}{2.15 \times 10^{10} \text{ kcal}} = 2.0 \times 10^{-4} \text{ g} = 0.20 \text{ mg}}$$

NOTE:

Our answer is rounded to 2 sig figs.

b) *The exact mass of 1 mol of ^{81}Kr is 80.916590 g. How much does 1 mol of ^{81m}Kr weigh?*

The ^{81m}Kr must weigh more than the ^{81}Kr if energy is produced in this reaction.
80.916590 g + 0.00020 g = **80.91679 g**

The mass in a), 0.00020 g, is rounded to 2 sig figs (5 decimal places), so our calculated mass of ^{81m}Kr can also be expressed to 5 decimal places.

16.27 *Which of the following are examples of ionizing radiation?*

"Ionizing radiation" is radiation with enough energy to remove an electron from a molecule or atom, turning it into a cation. Alpha, beta, and gamma radiation from nuclear reactions, X-rays, and high-energy ultraviolet light are ionizing radiation. Other forms of radiation do not have enough energy to remove electrons from molecules. Refer to Figure 16.3.

a) *gamma radiation* — **Yes, this is ionizing radiation.**

b) *infrared radiation* — **No, this is not ionizing radiation.** Infrared radiation does not have as much energy as even visible light.

c) *alpha particles* — **Yes, this is ionizing radiation.**

16.29 *How does ionizing radiation produce radicals?*

Electrons in molecules are usually paired, whether in core electron shells, in bonds, or in lone pairs. If ionizing radiation causes one electron to be knocked out of a molecule, the remaining electron is left unpaired. Atoms, molecules or ions with unpaired electrons are called radicals.

Solutions to Section 16.4 Core Problems

16.31 *Why do people who work with sources of ionizing radiation often wear a badge that contains a piece of photographic film?*

The workers wear badges to monitor their exposure to ionizing radiation. Ionizing radiation exposes photographic film, just as visible light does; but unlike visible light, which can be stopped easily by covering the film with a piece of cardboard, some forms of ionizing radiation easily penetrate the cardboard. When the film is developed, any darkening of the film indicates that the worker has been exposed to ionizing radiation.

16.33 *Which of the following traditional units is used to measure the effect of ionizing radiation on the human body?*

a) *the curie* — The curie is used to measure the amount of radiation produced by a sample, regardless of whether any of it is absorbed by tissues.

b) *the rad* — While this unit does measure a radiation dose, it does not account for the differences between types of radiation.

c) *the rem* — **This is the unit used to quantify the effect of a radiation dose on a human body.** It takes into account both the amount of ionizing radiation and the differences in damage caused by different types of radiation.

16.35 *The activity of 1 μg of ^{32}P is 41 mCi. Convert this activity into the following units:*
a) Ci *b) μCi* *c) MBq* *d) Bq* *e) GBq*

Chapter 1 review! Given appropriate conversion factors we can convert any unit into any other unit. See Table 16.5 for some useful conversion factors.

a) since the prefix milli is 1/1000 of the base unit, 1000 mCi = 1 Ci:

$$\mathbf{41\ mCi \times \frac{1\ Ci}{1000\ mCi} = 0.041\ Ci}$$

You can also realize that, as in Chapter 1, you'll have to move the decimal 3 spaces to the left for the conversion of any unit mX to X.

b) 1000 mCi = 1 Ci and 10^6 μCi = 1 Ci:

$$\mathbf{41\,mCi \times \frac{1\,Ci}{1000\,mCi} \times \frac{10^6\ \mu Ci}{1\,Ci} = 41\,000\,Ci = 4.1 \times 10^4\ Ci}$$

You can also realize that, as in Chapter 1, you'll have to move the decimal 3 spaces to the right for the conversion of any unit mX to μX.

c) From Table 16.5, 1 mCi = 37 MBq:

$$\textbf{41 mCi} \times \frac{\textbf{37 MBq}}{\textbf{1 mCi}} = \textbf{1.5} \times \textbf{10}^3 \textbf{ MBq (rounded to 2 sig fig)}$$

d) Use the answer from c). Since M = 10^6 of the base unit, 10^6 Bq = 1 MBq:

$$\textbf{1500 MBq} \times \frac{\textbf{10}^6 \textbf{ Bq}}{\textbf{1 MBq}} = \textbf{1.5} \times \textbf{10}^9 \textbf{ Bq}$$

e) G = 10^9 base units, so 1 GBq = 10^9 Bq:

$$\textbf{1.5} \times \textbf{10}^9 \textbf{ Bq} \times \frac{\textbf{1 GBq}}{\textbf{10}^9 \textbf{ Bq}} = \textbf{1.5 GBq}$$

Or, realize that we can simply replace the "$\times 10^9$" in the answer from d) with "G."

16.37 *A patient absorbs 30 mrem during radiation therapy. Express this equivalent dose in each of the following units:*
a) rem *b) Sv* *c) mSv*

a) 1000 mrem = 1 rem:

$$\textbf{30 mrem} \times \frac{\textbf{1 rem}}{\textbf{1000 mrem}} = \textbf{0.030 rem (to 2 sig fig)}$$

or recognize that you need to move the decimal 3 places to the left.

b) from Table 16.5, 100 rem = 1 Sv:

$$\textbf{0.030 rem} \times \frac{\textbf{1 Sv}}{\textbf{100 rem}} = \textbf{3.0} \times \textbf{10}^{-4} \textbf{ Sv}$$

c) 1000 mSv = 1 Sv:

$$\textbf{3.0} \times \textbf{10}^{-4} \textbf{ Sv} \times \frac{\textbf{1000 mSv}}{\textbf{1 Sv}} = \textbf{0.30 mSv}$$

16.39 *Two patients absorbed equal amounts of energy from nuclear radiation. Patient 1 was exposed to gamma radiation and received a dose of 15 mrem. Patient 2 was exposed to a proton beam with a W_R of 3. What equivalent dose did patient 2 receive?*

The rem (or in this case mrem) is a unit of "equivalent dose," which takes into account the amount of damage that is caused by the particular type of radiation to which the person was exposed. For the amount of energy absorbed, ignoring the type of radiation, we use the rad. The relationship is:

$$\text{rads} \times W_R = \text{rems}$$

From this, we can figure out the amount of energy absorbed by patient #1: gamma radiation has a W_R value of 1, so the 15 mrem exposure the patient received was also 15 mrad. Since both patients received equal amounts of energy, patient #2 must also have received 15 mrad, or an equivalent dose of:

$$\text{rads} \times W_R = 15 \text{ mrad} \times 3 = \textbf{45 mrems}$$

Solutions to Section 16.5 Core Problems

16.41 *What is background radiation?*

Background radiation is ionizing radiation that is produced by naturally occurring materials.

16.43 *In areas that have high concentrations of uranium in the soil, homeowners are told that they should seal their basement floors with a piece of non-porous plastic. Why is this?*

A plastic liner can help stop radon from diffusing into the basement. Radon is produced by the decay of uranium in soils and is a major cause of lung cancer.

16.45 *Which will produce a higher equivalent dose of radiation if it is taken internally: 10 μCi of ^{14}C (a beta emitter) or 10 μCi of ^{123}I (a gamma emitter)? Explain your answer.*

The ^{14}C will produce a higher equivalent dose, because most of the energy is absorbed by body tissue; the equivalent number of gamma photons does less damage, because most of the radiation escapes the body without being absorbed. Gamma photons have a much greater penetrating ability than beta particles and are not absorbed easily by body tissues.

16.47 *What would be an appropriate type of shielding for a sample of each of the following radioisotopes?*

a) phosphorus-32 (a beta emitter)

Beta particles are blocked by **wood, plastic, or thick cloth**. Light materials (those made of low-atomic-weight elements like C, H, N, O) are preferred, because heavier elements may produce X-rays when they absorb beta particles, so lead shielding is not the best choice.

b) technetium-99m (a gamma emitter)

Gamma rays require **thick concrete or lead** barriers to absorb the photons.

16.49 *A worker standing 10 cm away from a gamma emitter received 50 mrem of radiation in one hour. How much radiation would this worker have absorbed if he had stood 1 m away from the radioisotope?*

Exposure to radiation decreases as the square of the distance to the source increases. A distance of 1 m is ten times as far from the source as 10 cm, so the worker would have absorbed $1/10^2$ or 1/100 as many particles:
50 mrem ÷ 100 = **0.5 mrem**

Solutions to Section 16.6 Core Problems

16.51 *The half-life of phosphorus-32 is 14 days. If the activity of a sample of phosphorus-32 is 300 MBq on July 1st, what will the activity of the sample be on August 12?*

The activity decreases by half every 14 days. So:

Date	Number of half-lives	Activity of sample
July 1	0	300 MBq
July 15	1	150 MBq
July 29	2	75 MBq
Aug 12	3	**38 MBq**

16.53 *The half-life of ^{18}F is 110 minutes. A nuclear medicine clinic receives a 20-μCi sample of a compound containing ^{18}F at 1:00 p.m.. If the ^{18}F was produced at 9:20 a.m., what was its activity when it was produced?*

One solution is to work backward from 1:00 p.m. to 9:20 a.m. (Since we're working backward, we double the activity each time.) We'll make an entry for every 110 minutes (an hour and 50 minutes each time):

Time	Activity of sample
1:00 PM	20 μCi
11:10 AM	40 μCi
9:20 AM	**80 μCi**

So, as it turns out, only two half-lives have elapsed since the sample was prepared at 9:20 a.m.

Another approach is to calculate the total number of minutes between 9:20 and 1:00: it's a span of 3 hours and 40 minutes, or $3 \times 60 + 40 = 220$ minutes, or two half-lives. Therefore the original activity of the sample must have been 4 times as high as it was at 1:00 p.m.:

$$\mathbf{20\ \mu Ci \times 4 = 80\ \mu Ci}$$

16.55 *A sample of radioactive ^{13}N has an activity of 40 μCi at 10:00 a.m.. By 10:30 a.m., the activity of the sample has decreased to 5 μCi. What is the half-life of ^{13}N?*

Let's count up the number of half-lives it takes to get from 40 μCi to 5 μCi:

Number of half-lives	Activity of sample
0	40 μCi
1	20 μCi
2	10 μCi
3	5 μCi

Since the period from 10:00 to 10:30 is three half-lives, the half-life must be 30 minutes ÷ 3 = **10 minutes.**

16.57 *A sample of iodine-131 (half-life = 8 days) has an activity of 10 mCi. Approximately how many days will it take for the activity of this sample to drop to 0.1 mCi?*

Again, we can set up a table to follow the decay:

Days	Number of half-lives	Activity of sample (mCi)
0	0	10
8	1	5
16	2	2.5
24	3	1.25
32	4	0.625
40	5	0.313
48	6	0.156
56	7	0.0781

Sometime between the last two entries, the activity is 0.1 mCi. It will take **between 48 and 56 days (between 6 and 7 half-lives)** to reach an activity of 0.1 mCi.

16.59 *Two samples of wood are taken from an archeological site. Sample A has a ^{14}C activity of 0.19 Bq per gram of carbon, while sample B has a ^{14}C activity of 0.17 Bq per gram of carbon.*

Carbon-14 in living samples is constantly replaced and is maintained at a constant level, but the activity of the isotope begins to drop as soon as the organism dies and stops taking in new carbon.

a) *Which of these samples is older? Explain your answer.*

Sample B is older. The more recent a sample is, the more ^{14}C it contains, and the greater the activity of the sample. Since B has a lower activity, more of its ^{14}C has decayed, and therefore the sample must be older.

b) *Could either of these samples be 6000 years old, given that the activity of live wood is 0.22 Bq per gram of carbon?*

Carbon-14 has a half-life of 5730 years, so in 6000 years the activity of a sample would be expected to drop by a little more than half. If live wood has an activity of 0.22 Bq per gram of carbon, a sample 6000 years old should have an activity of less than 0.11 Bq per gram of carbon. These samples have much higher activity than that, and are therefore much less than 6000 years old.

Solutions to Section 16.7 Core Problems

16.61 *Why aren't alpha and beta radiation used in diagnostic imaging?*

Remember what we learned earlier about penetrating power. **Because alpha and beta radiation are almost entirely absorbed by body tissues, they don't make it through the tissue to a detector, so no image can be obtained.** (It's like trying to take a photo with the lens cover on.) A significant portion of the radiation must actually be able to pass through the tissues and be detected, to produce an image.

16.63 *Radioactive strontium (Sr) is a beta emitter and is used to relieve the pain of bone cancer, because it is absorbed by bone tissue and kills the nerve cells around the bone. Both ^{89}Sr and ^{90}Sr produce beta radiation. The half-lives of these two radioisotopes are 51 days and 29 years, respectively. Which isotope is a better choice for radiation therapy, and why?*

For radiation therapy, a short-lived, intense dose is desired, not a long-term slow dose. In order to kill diseased cells, a high dose of radiation is needed, but that dose would cause a lot of damage to healthy cells over the long term. For radiation therapy, **short half-life isotopes that decay quickly and are gone from the body in days or weeks are preferable to long-lived isotopes that will last in the body for years. Therefore, ^{89}Sr, with a 51-day half-life, is a better choice for radiation therapy than ^{90}Sr, with its 29-year half-life.**

16.65 *Explain why FDG that contains ^{18}F is useful for detecting cancer tissues.*

Cancerous cells grow and divide rapidly, taking up larger amounts of nutrients than healthy cells. Therefore cancerous cells will incorporate the FDG faster than healthy cells. As the cancer tissue accumulates a high concentration of ^{18}F atoms, the locations of the cells can be detected by the gamma emission of the ^{18}F as it decays.

16.67 *Patient A has a brain scan using ^{99m}Tc and receives 700 mrem of radiation. Patient B is given radiotherapy for a brain tumor and receives 5000 mrem of radiation. Why is patient B given a much larger radiation dose than patient A?*

Medical imaging applications use as little radiation as possible, to get the needed image while causing the least amount of damage to cells. When radiation is used to kill cancerous cells, the dose of radiation is much higher, because the point is actually to cause damage.

16.69 *Iodine-131 is a beta-emitter and is used to treat thyroid disorders. Why is radioactive iodine used, rather than a radioactive isotope of some other element?*

The thyroid gland absorbs iodine at a much higher rate than other tissues do, so radioactive iodine is very specific to thyroid imaging and treatments.

Solutions to Section 16.8 Core Problems

16.71 *What is a fission reaction?*

A fission reaction is a nuclear reaction in which a large nucleus splits into two or more similarly-sized smaller nuclei. Most fission reactions are stimulated by a neutron hitting the large nucleus.

16.73 *Balance the following fission reaction:*

$${}^{239}_{94}Pu + {}^{1}_{0}n \rightarrow {}^{104}_{42}Mo + ? + 2\,{}^{1}_{0}n$$

Fission reactions are balanced like any other nuclear reaction equation: the total of the mass numbers must be equal on both sides of the equation, and the total of the atomic numbers must also be equal on both sides of the equation. In this reaction, the total of the mass numbers on the left is 239 + 1 = 240, and the total for the known species on the right is 104 + 2(1) = 106, leaving a mass of 134 for the unknown species. The total atomic number on the left is 94 + 0 = 94, and the total for the known species on the right is 42 + 0 = 42, leaving 52 for the unknown species; element number 52 is Te. The balanced nuclear equation, therefore, is:

$$^{239}_{94}\mathbf{Pu} + {}^{1}_{0}\mathbf{n} \rightarrow {}^{104}_{42}\mathbf{Mo} + {}^{134}_{52}\mathbf{Te} + 2\,{}^{1}_{0}\mathbf{n}$$

16.75 *A large piece of pure uranium-235 can explode extremely violently. Why is this?*

If a piece of uranium-235 is large enough, then neutrons produced by naturally-decaying uranium nuclei will be captured by other nuclei in the sample, which will stimulate those nuclei to undergo fission. More neutrons will be produced in the fission reactions and a chain reaction will begin: the neutrons formed in the fission of one nucleus stimulate the fission of several more, until all the nuclei in the entire sample are undergoing fission in a short period of time, producing vast amounts of energy.

16.77 *Uranium can burn, reacting with oxygen to produce several chemical compounds. Which will produce more energy, the combustion of 1 kg of uranium or the nuclear decay of 1 kg of uranium? Why?*

Nuclear reactions always produce much more energy than any chemical reaction, if similar amounts of fuels are compared.

16.79 *In very old stars, carbon-12 atoms fuse to form magnesium-24. Write a balanced nuclear equation for this fusion reaction.*

Since the atomic number and isotope mass of magnesium-24 are twice the values for carbon-12, it takes two atoms of carbon-12 to form one atom of magnesium-24. Note, again, that the sums of the mass numbers and atomic numbers on either side of the equation are equal.

$$^{12}_{6}\mathbf{C} + {}^{12}_{6}\mathbf{C} \rightarrow {}^{24}_{12}\mathbf{Mg}$$

16.81 *What are the advantages of fusion reactions over fission reactions as a source of energy?*

Fusion reactions produce much more energy per gram of reactant than fission reactions; they use non-radioactive starting materials which are more abundant, and the products are less radioactive than those of fission reactions, alleviating waste disposal problems.

Solutions to Concept Questions

* indicates more challenging problems.

16.83 *What are the two rules for balancing a nuclear equation?*

The mass numbers must add up to the same total in the reactants and products, and the atomic numbers must add up to the same total in the reactants and products.

16.85 *Why don't chemists write the mass numbers of the atoms when they write balanced equations for ordinary chemical reactions?*

First, because ordinary chemical reactions are independent of the isotopes of the elements involved; chemical reactions depend on the chemical identities (the number of protons) of the elements involved, and not the number of neutrons. Second, because normal chemical reactions are carried out using large samples (that is, on the order of mg, g or kg), rather than individual atoms or molecules, the samples typically contain a mixture of many naturally-occurring isotopes of each element.

16.87 *Nuclear reactions can produce radiation in the form of particles or electromagnetic radiation. How do these differ from one another?*

The particles produced in nuclear reactions—alpha particles, beta particles, neutrons, etc—are actually small objects, made of matter. They have mass, take up space, and can be captured by other atoms or molecules (a beta particle, once captured by an atom, behaves like any other electron; an alpha particle, once it picks up two electrons from other atoms, behaves like any other helium atom). Photons of electromagnetic radiation, in contrast, are packets of energy. Photons don't have a rest mass, they don't take up space, and if they are captured by an atom they are completely converted into other forms of energy and no longer exist as photons (having transferred their energy into the atom in the form of heat, electron excitation, etc.).

16.89 *A friend is concerned about using her microwave oven, because she has heard people refer to microwave cooking as "nuking your food" and she is afraid that some sort of nuclear radiation is involved. How would you respond to your friend's concerns?*

Microwave "radiation" is electromagnetic radiation, like visible light, infrared light and radio waves. There's no nuclear reaction going on in a microwave oven, and it doesn't make your food (or your kitchen) radioactive, any more than being under fluorescent lights or sunlight would.

16.91 *What is an equivalent dose, and what units are used to express it?*

"Equivalent dose" is a way to measure radiation exposure that takes into account both the actual total energy of the radiation absorbed, and the relative amount of damage done by the particular type of radiation. Equivalent dose is measured in rem (or related units such as mrem). For the amount of energy absorbed, ignoring the type of radiation, we use the rad. The relationship is:

$$\text{rads} \times W_R = \text{rems}$$

where W_R is a factor related to the damage caused by the particular type of radiation.

16.93 *The half-life of cobalt-60 is around 5 years, and the half-life of cobalt-61 is 1 hour and 40 minutes. Which would have a higher activity, 10 mg of ^{60}Co or 10 mg of ^{59}Co? Explain your answer.*

Activity is a measure of how many nuclei are decaying in a particular period of time. The faster an isotope decays, the greater the activity of a sample. A short half-life means rapid decay, so a sample of cobalt-61 (with its very short half-life and rapid decay) will have a much greater activity than a similar-size sample of cobalt-60.

16.95 *A sheet of Plexiglas blocks all beta particles. Why is a thin sheet of lead sometimes added to the Plexiglas when shielding a worker from beta radiation?*

When a beta particle hits an atom of a heavy element, it can cause an X-ray to be emitted. The X-rays can be blocked by a thin sheet of lead, shielding the worker from both the beta radiation and the X-rays.

16.97 *It takes 8 days for half of a 1 g sample of iodine-131 to decay.*

a) *How long does it take for half of a 10 g sample of iodine-131 to decay? Explain.*

Half-life is the same for the decay of a given isotope, regardless of the size of the sample. A 1-g sample, a 10-g sample, a 1000-kg sample—in any of these, half of the particles will decay in 8 days.

b) *Does it take 8 days for the other half of the 1 g sample of ^{131}I to decay? Why or why not?*

No. It takes 8 days for *half* of the other half to decay. After the first 8 days, the sample of ^{131}I is a 0.5-g sample. After the second 8-day period, the sample is down to 0.25 g. Then it takes 8 days for that to decay to a 0.125-g sample, and so on.

16.99 *Positrons do not have much penetrating power, so they cannot be detected outside the body if they are produced internally. Why then are positron emitters useful in nuclear medicine?*

When positrons interact with normal matter, they undergo particle annihilation, a process in which a positron and an electron are converted completely into two gamma photons. While the positrons themselves do not escape the body tissues to be detected outside, the gamma rays that are produced do escape the body quite easily and can be detected.

16.101 *Explain why technetium-99m is never used in radiation therapy.*

This isotope is one of the few that emits only a gamma photon. Its decay does not produce alpha, beta or other particle radiation. Since gamma rays leave the body easily, without much interaction with tissues, they cannot be used to target disease cells. The point of radiation therapy is to damage and kill diseased or cancerous cells, and for this purpose, particle (alpha, beta, etc) radiation is much more effective.

16.103 *A technique called magnetic resonance imaging (MRI) uses radio waves to produce an image of soft tissues. Explain why MRI poses less of a health hazard to a patient than X-rays or CT scans.*

MRI doesn't use any ionizing radiation. Radio waves don't carry enough energy to ionize atoms or molecules in tissues.

16.105 *Life on earth would not exist were it not for a fusion reaction. Explain.*

Life on earth is dependent on the sun for energy. The sun, like other stars, is powered by the fusion of hydrogen and helium nuclei into heavier nuclei, releasing large amounts of energy.

Solutions to Summary and Challenge Problems

* indicates more challenging problems.

16.107 *Use the periodic table to answer the following questions.*

a) *How many neutrons are there in an atom of cobalt-58?*

We look in the periodic table to find that cobalt is element number 27, so it has 27 protons. 58 is the mass number, neutrons + protons, so the number of neutrons is **58 – 27 = 31.**

b) *How many protons are there in an atom of indium-111?*

If you have the name of the element, you can use the periodic table to find out how many protons it has—the mass number is irrelevant. Every atom of the element indium, In, has **49 protons** (indium is element number 49).

c) *Write the nuclear symbol for an atom that contains 30 protons and 36 neutrons.*

The element with 30 protons (atomic number 30) is zinc, Zn, and the mass number of this isotope is 30 + 36 = 66 (mass number = protons + neutrons): $\mathbf{{}^{66}_{30}Zn}$

16.109 *Chemists will often omit the atomic number when they write the symbol for a particular isotope. (For example, they may write ^{14}C, rather than $^{14}_{6}C$.) Why is this acceptable?*

The atomic number and the element symbol contain the same information, and the atomic number is the same regardless of the isotope. If you have the periodic table, you can always find the atomic number for a given element.

16.111 *Radon-221 can undergo two different types of radioactive decay. In a typical sample of radon-221 atoms, 22% of radon-221 atoms decay to francium-221, while the other 78% decay to polonium-217. Classify each of these two reactions as an alpha decay, a beta decay, or a positron emission.*

One approach here is to write the equation for each reaction and see what particles are needed to balance the equations. Francium is element 87; for the decay to francium-221, the setup is:

$$^{221}_{86}Rn \rightarrow ? + {}^{221}_{87}Fr$$

The remaining particle has a mass number of 0 and an atomic number (or, in fact, charge) of −1, a beta particle, so **the decay to francium-221 is a beta decay**:

$$^{221}_{86}Rn \rightarrow {}^{0}_{-1}e + {}^{221}_{87}Fr$$

Polonium is element 84, so for the second reaction, the setup is:

$$^{221}_{86}Rn \rightarrow ? + {}^{217}_{84}Po$$

The particle needed to balance this reaction has a mass number of 4 and an atomic number of 2, an alpha particle, so **the decay to polonium-217 is an alpha decay**:

$$^{221}_{86}Rn \rightarrow {}^{4}_{2}He + {}^{217}_{84}Po$$

16.113 *The following sequence of reactions is used to make isotopes of two transuranium elements. Complete the nuclear equations for these reactions.*

$$^{240}_{94}Pu + {}^{1}_{0}n \rightarrow \text{atom 1}$$

$$\text{atom 1} \rightarrow \text{atom 2} + {}^{0}_{-1}e$$

$$\text{atom 2} \rightarrow \text{atom 3} + {}^{4}_{2}He$$

In the first reaction, the product (atom 1) has mass number 241 and atomic number 94 (the atomic number does not change—this reaction forms another isotope of plutonium):

$$\mathbf{^{240}_{94}Pu + {}^{1}_{0}n \rightarrow {}^{241}_{94}Pu}$$

The second reaction, then, is the beta decay of plutonium-241: $^{241}_{94}Pu \rightarrow \text{atom 2} + {}^{0}_{-1}e$
To complete this equation, atom 2's mass number is 241, and its atomic number is 95 (element Am, americium):

$$\mathbf{^{241}_{94}Pu \rightarrow {}^{241}_{95}Am + {}^{0}_{-1}e}$$

The third reaction, then, is the alpha decay of americium-241: $^{241}_{95}Am \rightarrow \text{atom 3} + {}^{4}_{2}He$

The mass number of atom 3 is 237 (241 = 237 + 4) and its atomic number is 93 (95 = 93 + 2). Element 93 is Np, neptunium:

$$\mathbf{^{241}_{95}Am \rightarrow {}^{237}_{93}Np + {}^{4}_{2}He}$$

Written together, the three completed equations are:

$$\mathbf{^{240}_{94}Pu + {}^{1}_{0}n \rightarrow {}^{241}_{94}Pu}$$

$$\mathbf{^{241}_{94}Pu \rightarrow {}^{241}_{95}Am + {}^{0}_{-1}e}$$

$$\mathbf{^{241}_{95}Am \rightarrow {}^{237}_{93}Np + {}^{4}_{2}He}$$

*16.115 *Use the nuclear equation given here to answer the following questions:*

$${}^{67}_{31}Ga + {}^{0}_{-1}e \rightarrow {}^{67}_{30}Zn + 23{,}000{,}000\ kcal$$

a) *How much energy is produced when 1.00 g of ${}^{67}Ga$ turns into ${}^{67}Zn$?*

The energy of the reaction above is for one mole, or 67 grams, of ${}^{67}Ga$. We can use the relationships

$$\frac{67\ g\ {}^{67}Ga}{2.3\times10^{7}\ kcal} \quad or \quad \frac{2.3\times10^{7}\ kcal}{67\ g\ {}^{67}Ga}$$

to calculate the energy produced by one gram of ${}^{67}Ga$:

$$1.00\ g\ {}^{67}Ga \times \frac{2.3\times10^{7}\ kcal}{67\ g\ {}^{67}Ga} = 3.4\times10^{5}\ kcal$$

b) *How much mass is lost when 1.00 g of ${}^{67}Ga$ turns into ${}^{67}Zn$?*

The relationship between energy produced in a nuclear reaction and mass lost is:

1.00 gram of mass lost = 2.15×10^{10} kcal

The energy produced by the reaction of 1.00 gram of ${}^{67}Ga$ is 3.4×10^{5} kcal. The amount of mass that must be converted to produce this much energy is:

$$3.4\times10^{5}\ kcal \times \frac{1.00\ g}{2.15\times10^{10}\ kcal} = 1.6\times10^{-5}\ g$$

When 1.00 gram of ${}^{67}Ga$ turns into ${}^{67}Zn$, 1.6×10^{-5} g, or 0.016 mg, of mass is converted to energy.

16.117 *Janice works in a nuclear medicine clinic and must handle small amounts of radioactive material. List three ways in which Janice can reduce her exposure to ionizing radiation (without looking for a different line of work).*

She can use appropriate shielding (using different materials to shield herself, depending on the type of radiation involved) when she must handle radioactive materials; keep her distance from the radioactive material as much as possible when she is not handling them; and limit the amount of time she spends near radioactive sources.

16.119 *Cobalt-60 emits both beta and gamma radiation. Many disposable medical supplies, such as wound dressings and disposable syringes, are exposed to cobalt-60 after they are packaged. Explain why this is done.*

The radiation from the cobalt-60 kills any pathogenic bacteria that might have contaminated the items before they were packaged or during the packaging process. It's a method of sterilizing the supplies.

16.121 *Sodium-24 emits both beta and gamma radiation. What would be appropriate shielding for a radiologist who is using this isotope to study electrolyte transport?*

Beta particles are absorbed easily by a piece of wood or Plexiglas, but even a thick layer of lead does not completely block gamma radiation, so the radiologist should limit exposure as much as possible to keep the effective dose to a safe level. This may be accomplished in part by using remote controlled devices for handling the material.

16.123 *Technetium-99m and antimony-129 both produce gamma radiation, and they have similar half-lives (6 hours versus 4.4 hours). Why is ^{99m}Tc preferred for diagnostic imaging? The nuclear reaction for each isotope is shown here:*

$$^{99m}_{43}Tc \rightarrow\ ^{99}_{43}Tc + \gamma$$

$$^{129}_{51}Sb \rightarrow\ ^{129}_{52}Te +\ ^{0}_{-1}e + \gamma$$

The type of radiation that is most useful for diagnostic imaging is gamma radiation, because most gamma rays produced will leave the body and can be detected by the imaging instrument. Beta (and alpha) particles cannot be used for diagnostic imaging, because the particles are completely absorbed by tissues and do not leave the body. Furthermore, one goal in diagnostic imaging is to expose the body to as little radiation as possible—to use just enough radiation to get the image and no more. The decay of Technetium-99m produces only a gamma photon, making it a better choice for diagnostic imaging; the beta radiation produced in the decay of antimony-129 is useless for imaging and potentially harmful to the surrounding cells.

*16.125 *Calcium in the diet is rapidly absorbed by bone tissues and remains in the body for extended periods, while the potassium in your diet is normally excreted within a few weeks. Based on this, explain why radioactive isotopes of strontium (Sr) are considered to be much more dangerous than radioactive isotopes of rubidium (Rb). (Hint: Find these four elements on the periodic table.)*

As we learned in Chapter 2, elements in the same group tend to have similar chemical behavior. We would expect strontium and calcium ions to have similar chemistry, and to occupy similar tissues in the body. Therefore, strontium ions would accumulate in the body and be retained for long periods. Rubidium, on the other hand, would have chemistry similar to that of potassium; it will be excreted from the body rather than accumulating in tissues.

*16.127 *Earth is estimated to be around 5 billion years old. When Earth was formed, it contained both ^{238}U (half-life = 4.5 billion years) and ^{237}Np (half-life = 2.1 million years). Our planet still contains a substantial amount of ^{238}U, but it does not contain any detectable ^{237}Np. Explain this difference.*

Almost half of the original ^{238}U present at the formation of the earth is still around, while the ^{237}Np has all decayed. Five billion years is about 2400 times the half-life of ^{237}Np, which means that the amount of this isotope now present is $1/(2^{2400})$ the original amount—none at all, for practical purposes.

16.129 *You have 1 mg samples of ^{131}I and ^{123}I. The activity of the ^{131}I is 73 Ci, while the activity of the ^{123}I is 1090 Ci. Which isotope has the longer half-life? Explain your answer.*

For samples containing similar numbers of atoms, the sample with the longer half-life decays more slowly and will have a smaller activity. The activity of the ^{131}I sample is much less than that of the ^{123}I sample, indicating that it is decaying more slowly and therefore has a longer half-life.

16.131 *Earth is approximately 5 billion years old. Using this fact, explain why rocks on earth contain potassium-40 (half-life = 1.3 billion years), but no potassium-42 (half-life = 12 hours).*

5 billion years is only about 4 times the half-life of potassium-40, so about 6% of the potassium-40 that was present at the formation of the earth is still there:

Time (years)	Number of half-lives	% of original ^{40}K
0	0	100
1.3 billion	1	50
2.6 billion	2	25
3.9 billion	3	12.5
5.2 billion	4	6.25

However, any potassium-42 that was present at the formation of the earth would have decayed to below detectable limits long ago (5 billion years is about 3.5×10^{12} times the half-life of ^{42}K; it would have taken only a few weeks for the amount of ^{42}K to drop to negligible levels).

16.133 *A patient is given an injection of a solution that contains 600 MBq of ^{99m}Tc. The volume of the solution is 15 mL.*

a) *What is the concentration of ^{99m}Tc in this solution, in MBq/mL?*

Don't over think this one, just look at the units: MBq/mL.

$$\mathbf{\frac{600\ MBq}{15\ mL} = 40\ MBq/mL}$$

b) *If 2.5 mL of this solution is mixed with 12.5 mL of water, what will be the concentration of ^{99m}Tc in the new solution, in MBq/mL?*

From chapter 5, the dilution equation is $C_1V_1 = C_2V_2$.

$C_1 = 40$ MBq/mL

$V_1 = 2.5$ mL

$V_2 = 2.5 + 12.5 = 15$ mL (assuming volumes are additive, which is generally true when for dilute aqueous solutions).

$$\mathbf{(40\ MBq/mL)\ (2.5\ mL) = C_2 \times\ (15\ mL)}$$

$$\mathbf{C_2 = 6.7\ MBq/mL}$$

*16.135 *How many neutrons are produced when uranium-235 absorbs a neutron and fissions into bromine-85 and lanthanum-148?*

One approach is to write out the equation and see what is needed to balance it. "Absorbs a neutron" means that there is one neutron on the reactants side; "fissions into" means that bromine-85 and lanthanum-148 are two of the products.

$$^{235}_{92}U + {}^{1}_{0}n \rightarrow {}^{85}_{35}Br + {}^{148}_{57}La + ?\,{}^{1}_{0}n$$

The sum of the mass numbers must be balanced: a total of 236 on the left side of the equation, with the two known nuclei accounting for 85 + 148 = 233 on the right side of the equation, meaning there must also be **3 neutrons** to complete the balancing of the mass numbers. (Note that the atomic numbers are already balanced.)

Chapter 17

Nucleic Acids, Protein Synthesis, and Heredity

Solutions to Section 17.1 Core Problems

17.1 *What are the three components of a nucleotide?*

The three parts of a nucleotide are **a phosphate group, a five-carbon sugar** (ribose or deoxyribose), and one of five **organic bases** (adenine, guanine, thymine, cytosine or uracil).

17.3 *In a nucleotide, which carbon atom in the sugar is attached to the base?*

The base in a nucleotide is attached to the 1' carbon of the sugar.

17.5 *Which of the five bases occurs in DNA, but not in RNA?*

Thymine is only found in DNA, while uracil is only found in RNA. The other three bases are common to both. (If your instructor wants you to know this, there's no way around memorizing it.)

17.7 *What is the sugar in RNA?*

Remember there's a clue in the names: RNA is "ribonucleic acid" and contains **ribose**, while DNA is "deoxyribonucleic acid" and contains deoxyribose. (Also notice that the "deoxy," as you might expect from the parts of the word, means an oxygen atom has been removed: ribose has an –OH group on the 2' carbon, while deoxyribose has a hydrogen atom in the same position.)1

17.9 *Name the nucleotide below, using the standard abbreviation.*

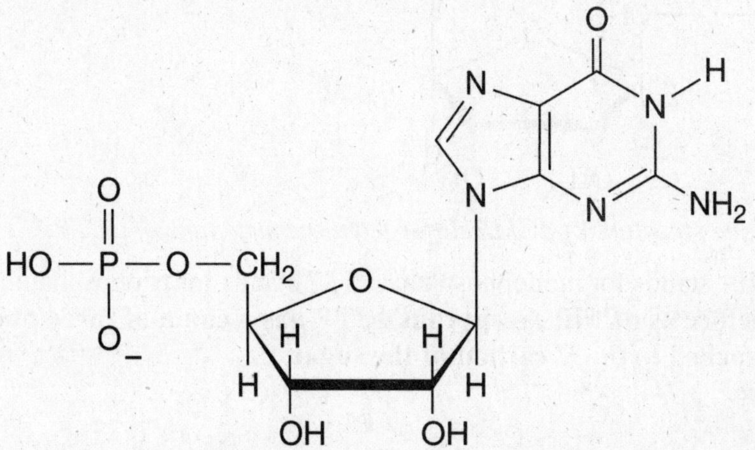

The sugar in this nucleotide is ribose (there's an –OH group on the 2' carbon), so we don't need the d at the beginning of the name. The base is guanine (Table 17.1), abbreviated G, and there is one phosphate group (monophosphate), so the abbreviation is **GMP**.

17.11 *a) Draw the structure of dCMP.*

The "d" at the beginning of dCMP tells us that the sugar is deoxyribose, and the "C" tells us that the base is cytosine. "MP" stands for monophosphate, only one phosphate group attached.

b) How does the structure of dCMP differ from the structure of CMP?

The lack of the initial letter "d" means that the sugar will be ribose (with an –OH group on the 2' carbon) instead of deoxyribose (with an –H on the 2' carbon). **CMP has the same structure as dCMP, everywhere except on the 2' carbon of the sugar, which has a hydroxyl group pointing down.**

c) How does the structure of dCMP differ from the structure of dCTP?

Where "MP" stands for monophosphate, "TP" stands for triphosphate. **dCTP has the same structure as dCMP except that dCTP has a chain of three phosphate groups attached to the 5' carbon of the sugar**:

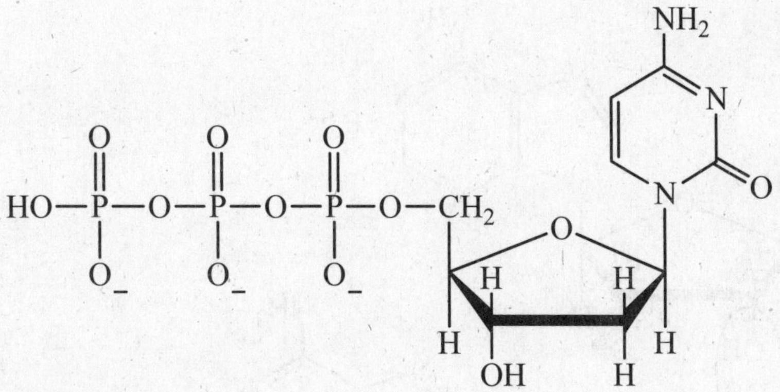

Solutions to Section 17.2 Core Problems

17.13 *In the DNA backbone, does the phosphodiester group connect two sugars to each other, two bases to one another, or a sugar to a base?*

The phosphodiester group connects the sugars of two nucleotides to each other. Remember the phrase "sugar-phosphate chain" for the backbone of the nucleic acid chain. The bases are always hanging out in space from the sugar, never part of the chain. (Think of a charm bracelet, with the sugars and phosphate groups as the links of the bracelet, and the bases as the charms.)

17.15 *Draw the structure of the dinucleotide that is formed from two molecules of AMP.*

First, we draw a molecule of AMP. Since there's no initial "d," the sugar is ribose (–OH group on the 2' carbon); the base is adenine (A); and there's one phosphate group (MP = monophosphate). Then we draw two molecules in position for the condensation reaction to occur between the phosphate group on one AMP molecule and the –OH group on the 3' carbon of the sugar (ribose) of the other AMP molecule

NOTE:

We rearranged the =O, $-O^-$, and –OH groups on the phosphates to make this more convenient):

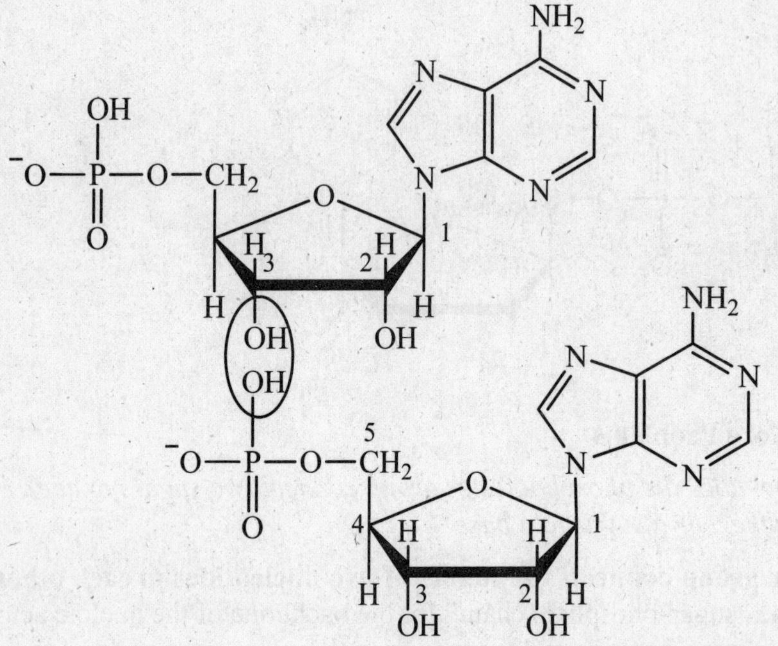

NH2 N N OH ⁻O—P—O—CH2 N N O O H 3 2 H 1 H H OH OH NH2 OH N N 5 ⁻O—P—O—CH2 N N O O 4 H H 1 H 3 2 H OH OH

Then we show the condensed molecule (after the removal of a molecule of H_2O) with the two AMP molecules linked together:

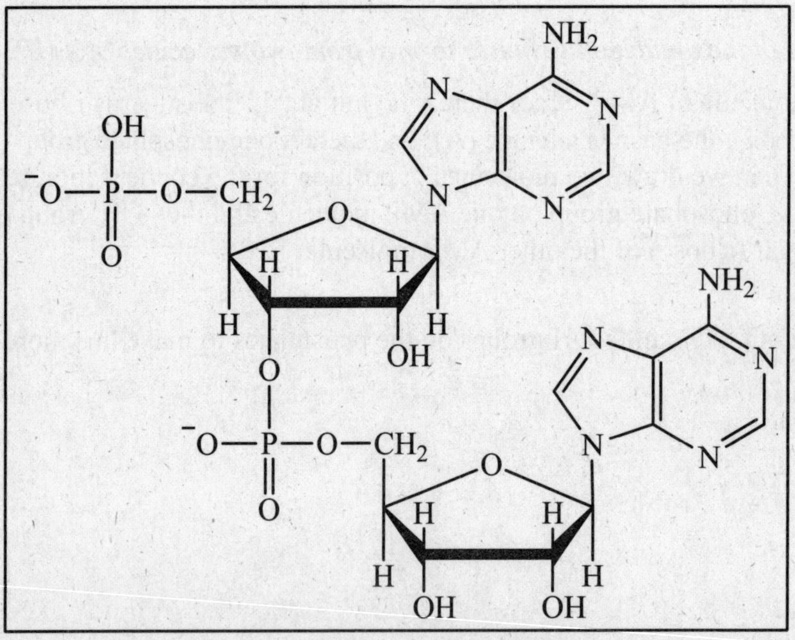

NH2 N N OH ⁻O—P—O—CH2 N N O O H H H H OH NH2 O N N ⁻O—P—O—CH2 N N O O H H H H OH OH

17.17 *Which base forms a stable base pair with each of the following bases in a molecule of DNA?*

Again, if your instructor wants you to know this, there's no way around memorizing this: A with T in DNA or U in RNA, C with G in both.

a) G always pairs up with **C**

b) A always pairs up with **T** in DNA

17.19 *A DNA strand has the base sequence AGTGGC.*

a) *What base is at the 5' end of this strand?*

DNA sequences are always written starting with the 5' end on the left, so **A, adenine, is the base at the 5' end of the strand.**

b) *What base is at the 5' end of the complementary strand?*

G, guanine: The 5' end of the complementary strand pairs up with the 3' end of this strand (by convention, always shown on the right), which has a C (cytosine).

c) *Write the base sequence of the complementary strand.*

Remember this is DNA, so A pairs with T (in RNA, A pairs with U).

original strand: (5' end) A – G – T – G – G – C (3' end)

complementary strand: (3' end) T – C – A – C – C – G (5' end)

Since we always write DNA sequences starting with the 5' end on the left, we must reverse the order of the nucleotides. The correct sequence for the complementary strand is **GCCACT**.

17.21 a) *What is a histone?*

Histones are positively charged proteins that provide a way for DNA strands to arrange themselves neatly and compactly into coils to form chromosomes. Like spools holding thread, histones help keep the long DNA strands tidy.

b) *Are histones positively charged, negatively charged, or electrically neutral?*

Histones are positively charged, so that they can attract the negatively charged phosphate groups in the DNA backbone. If you can remember that the phosphate groups are negative it makes it easy to remember that the histones need to be positive.

17.23 *Where is the DNA located in a eukaryotic cell?*

In eukaryotes, most of the DNA is located in the nucleus of the cell, with a small amount in the mitochondria.

Solutions to Section 17.3 Core Problems

17.25 *What is the function of DNA in an organism?*

DNA stores, in an easily reproducible form, the instructions for building every cell in an organism. The sequence of bases in DNA is a set of instructions for building proteins. The proteins are responsible for carrying out all cell functions.

17.27 a) *How many chromosomes are there in a typical cell in the human body?*

Humans have 46 chromosomes in most cells (23 in sex cells, none in red blood cells).

b) *How many of these are autosomal chromosomes?*

44 of the 46 are autosomal chromosomes (22 pairs).

17.29 *What is an activated nucleotide?*

An activated nucleotide is a nucleoside triphosphate (sugar + base + 3 phosphate groups).

17.31 *What is the function of each of the following proteins in DNA replication?*

a) primase **Primase builds short sections of RNA (using A, U, G and C) along a strand of DNA, using nucleotides in their activated (triphosphate) forms.** A hydrolysis reaction removes two of the phosphate groups from the nucleotides; the energy obtained from the hydrolysis reaction is used to carry out the condensation reactions that build the complementary chain. (These short sections of RNA are called primers. Looking at the parts of the word, the "-ase" ending indicates an enzyme, and the "prim-" part is a reminder that it builds primers.)

b) DNA ligase **When all the portions of the new strand are in place, DNA ligase replaces each U with a T and attaches (through condensation) the ends of the pieces of the strand to form one complete, finished strand of DNA.**

17.33 *What is the function of the proofreader protein in DNA replication?*

The proofreader protein, as the name indicates, checks the strand for errors–mismatches in the A–T and G–C pairings–and replaces any wrong nucleotides with the correct ones. (Actually the proofreader protein catches about 99% of the errors.)

Solutions to Section 17.4 Core Problems

17.35 *Which of the following is the best description of transcription?*

In transcription, a cell unwinds a section of DNA and separates the two strands, builds an RNA complement of the section, releases the RNA strand and puts the strands of the DNA molecule back together.

a) A cell makes a piece of RNA that is the complement of a piece of DNA. **This is the definition of transcription.**

b) A cell makes a piece of DNA that is the complement of a piece of RNA.

c) A cell makes a piece of RNA that is the complement of another piece of RNA.

17.37 *Which of the following could a cell use to build RNA?*

The process of building RNA, like that of building DNA (or, for that matter, building any other molecule in the body from smaller molecules), requires energy that is derived from the hydrolysis of a high-energy molecule. RNA is built from the activated RNA nucleotides (nucleoside triphosphates, each made of sugar + base + three phosphate groups). **ATP is the only activated nucleotide (triphosphate) listed.**

a) adenosine *b) AMP* *c) ADP* ***d) ATP***

17.39 *Which DNA strand actually binds to the RNA nucleotides during transcription, the coding strand or the template strand?*

The template strand provides the template for the RNA strand and actually binds to the RNA nucleotides during transcription.

17.41 *The coding strand in a section of DNA has the base sequence AACCTTG. If this section of DNA is transcribed, what will be the base sequence of the resulting RNA?*

The RNA has the same sequence as the coding strand of the DNA, except that we must replace T with U. (Look at the coding strand and the RNA strand in Figure 17.13 to see why; both strands have to fit the same template strand, but DNA uses T where RNA uses U.)

coding DNA strand:	(5' end) A – A – C – C – T – T – G (3' end)
template DNA strand:	(3' end) T – T – G – G – A – A – C (5' end)
RNA strand:	(5' end) **A – A – C – C – U – U – G** (3' end)

Remember that we always report sections of nucleic acids with the 5' end first (this happened naturally in this problem but it's good to keep track).

17.43 *Describe how the initial transcript of an rRNA molecule is processed.*

The initial transcript may contain, in one strand, several pieces of one or more types of RNA and extra nucleotides or chains of nucleotides (see Figure 17.14)**. Enzymes cut apart the initial transcript into pieces, break down excess RNA fragments into nucleotides, add a CCA sequence to the 3' end of each tRNA segment.**

17.45 *What is an intron, and what happens to introns during mRNA processing?*

Introns are sections of excess RNA in the initial transcript, which are removed and broken down into individual nucleotides during mRNA processing.

17.47 *What is the 5' cap in a mRNA molecule, and what is its function?*

The 5' cap in mRNA is a guanine nucleotide and a phosphate group. This cap identifies the molecule as messenger RNA.

Solutions to Section 17.5 Core Problems

You'll be using Tables 17.3 and 13.1 a lot throughout this chapter, so put a bookmark or a paper clip on each page so you can find them more easily. You might also consider making a photocopy of each table for convenient use.

17.49 *In translation, what molecule contains the coded instructions?*

mRNA contains the code for protein synthesis.

17.51 *How many bases are there in one codon?*

A sequence of **three bases** is a codon, and provides the code for one amino acid.

17.53 *What is the function of a stop codon?*

A stop codon signals the end of the protein sequence. It is a three-base sequence that doesn't code for an amino acid, and therefore the molecule ends.

17.55 *A piece of mRNA has the base sequence AAUCGCUUA. What is the amino acid sequence of the corresponding polypeptide?*

Split the mRNA sequence into three-base groups (codons): AAU – CGC – UUA. Then look in Table 17.3 for the translation: AAU = Asn, CGC = Arg, and UUA = Leu. The amino acid sequence is **Asn – Arg – Leu.**

17.57 *Using Table 17.3, answer the following questions:*

a) *How many different codons correspond to aspartic acid (Asp)?*

Two: GAC and GAU.

b) *Give one amino acid that has only one codon.*

Trp (UGG) **and Met** (AUG which is also the start codon) each correspond to only one codon.

17.59 *A piece of a mRNA molecule has the following base sequence:*

CAAUGUUGGCAUACG

This mRNA contains the code for the beginning of a polypeptide. What are the first four amino acids in this polypeptide?

If this mRNA contains the code for the beginning of a polypeptide, it must contain a start codon (AUG), so we look for this codon in the base sequence:

CA<u>AUG</u>UUGGCAUACG

To translate the mRNA, we need to divide up the remaining bases into 3-base sequences, or codons, beginning with the start codon and proceeding left to right. The cell ignores the bases before (to the left of) the start codon.

CA – <u>AUG</u> – <u>UUG</u> – <u>GCA</u> – <u>UAC</u> – G

Now we can use the genetic code to write the amino acid sequence. AUG codes for Met (methionine), UUG codes for Leu (leucine), GCA codes for Ala (alanine), and UAC codes for Tyr (tyrosine), so the first four amino acids are **Met – Leu – Ala – Tyr**. (We don't know what codon the ending G might have been part of.)

17.61 *A piece of a mRNA molecule has the following base sequence:*

GGCAGAGAGACUGACUA

This mRNA contains the code for the end of a polypeptide. What are the last three amino acids in this polypeptide?

If this mRNA contains the code for the end of a polypeptide, it must contain a stop (nonsense) codon (UAA, UAG or UGA) so we look for one of these codons in the base sequence:

GGCAGAGAGAC<u>UGA</u>CUA

To translate the mRNA, we need to divide up the bases into 3-base codons, working backward (right to left) from the stop codon. The cell ignores the bases after (to the right of) the stop codon.

GG – CAG – AGA – GAC – UGA – CUA

Now we can use the genetic code to write the amino acid sequence. (We don't know what codon the initial GG belongs to.) CAG codes for Gln (glutamine), AGA codes for Arg (arginine), GAC codes for Asp (aspartic acid), and UGA is the stop codon (it doesn't code for any amino acid). The last three amino acids are **Gln – Arg – Asp.**

17.63 *The coding strand of a DNA molecule contains the following sequence of bases:*
ACATCGTTC

a) *What is the base sequence in the mRNA when this DNA is transcribed?*

The RNA has the same sequence as the coding strand of the DNA, except that we must replace T with U. (Look at the coding strand and the RNA strand in Figure 17.13 to see why; both strands have to fit the same template strand, but DNA uses T where RNA uses U.)

coding DNA strand:	(5' end) A – C – A – T – C – G – T – T – C (3' end)
RNA strand:	**(5' end) A – C – A – U – C – G – U – U – C (3' end)**

b) *What is the amino acid sequence in the polypeptide when the mRNA is translated?*

The problem doesn't tell us that this sequence contains the code for the beginning or end of a protein (and in any case it doesn't contain the start sequence, AUG, or one of the stop sequences, UAA, UAG or UGA.) Therefore we just separate the sequence, as given, into three-base sections:

ACA – UCG – UUC

From Table 17.3, ACA codes for Thr (threonine), UCG codes for Ser (Serine), and UUC codes for Phe (phenylalanine). The amino acid sequence is therefore:

Thr – Ser – Phe

Solutions to Section 17.6 Core Problems

17.65 *Which type of RNA (mRNA, rRNA or tRNA) fits each of the following descriptions?*

a) *This type of RNA binds directly to amino acids.*

tRNA (see Figure 17.19)

b) *This type of RNA contains codons.*

mRNA (see Figure 17.18)

c) *This type of RNA helps catalyze protein synthesis.*

rRNA (ribosomes are the catalysts for translation, and are clusters of protein and rRNA.)

17.67 *What is the function of the ribosomes in a cell?*

Ribosomes are large clusters of proteins and ribosomal RNA that act as the catalysts for translation. The ribosomes bind to mRNA molecules and use the base sequence of the mRNA to make polypeptides.

17.69 *What is an anticodon, and what type of RNA contains an anticodon?*

See Figure 17.19. **An anticodon is a three-base sequence on a tRNA molecule that binds to mRNA during translation.**

17.71 *How many codons fit into a ribosome?*

The ribosome spans two codons at a time, so that two tRNA molecules can fit inside the ribosome side-by-side. See Figure 17.21.

17.73 *How many molecules of ATP must be broken down in order to make the bond between tRNA and an amino acid?*

See Figure 17.20. **Adding one amino acid to a tRNA molecule costs the equivalent of two molecules of ATP** (in the form of GTP).

17.75 *There are four different codons that correspond to the amino acid threonine (ACA, ACC, ACG and ACU), but there is only one tRNA that can bond to threonine. Why aren't there four different tRNA's that bond to threonine?*

tRNA anticodons don't always have to be an exact complement (in the third base) to the mRNA codons. Therefore the same tRNA molecule, one that binds to threonine, can bind to more than one codon (a total of four in this case) in an mRNA molecule. Since there are only 20 common amino acids, the body doesn't have to make a separate tRNA for every one of the 61 codons.

Solutions to Section 17.7 Core Problems

17.77 *What happens in each of the following types of mutations?*

a) *substitution* **In substitution mutations, one base in the DNA chain is replaced with different base.**

b) *deletion* **In deletion mutations, one or more consecutive bases are left out of (deleted from) the strand.**

17.79 *Explain why the addition of one base pair to a DNA molecule produces a frameshift.*

In a "frameshift," the cutoffs for the three-base codons are shifted one place over on the chain, which makes a completely new sequence of codons that will produce a different amino acid sequence. When a base pair is added to the DNA molecule (an addition mutation) or omitted from it (a deletion mutation), the three-base counts that make up codons shift one place in the chain.

17.81 *Classify each of the following substitutions as a silent mutation, a missense mutation, or a nonsense mutation. (The base sequences of the DNA coding strand are given.)*

See Table 17.4. A silent mutation gives a codon that codes for the same amino acid as the original codon, so there is no change in the resulting amino acid sequence; a missense mutation produces a codon that codes for a different protein than the original codon; and a nonsense mutation converts a codon that codes for an amino acid into a stop codon.

a) $GAA \rightarrow GAC$

These coding sequences and the corresponding RNA sequences are the same (there are no U/T substitutions to make). GAA codes for Glu, while GAC codes for Asp, so this is a **missense mutation.**

b) $GAA \rightarrow GAG$

Again, the DNA coding sequences and the RNA sequences are the same (no U/T substitutions). GAA and GAG both code for Glu, so this is a **silent mutation.**

c) $GAA \rightarrow TAA$

The corresponding RNA sequence change (exchanging T for U) is:

$$GAA \rightarrow UAA$$

UAA is a stop codon (or nonsense codon, one that doesn't code for any amino acid), so this is a **nonsense mutation**.

17.83 *Which of the following is the most likely to be harmful to a cell?*

a) Replacement of one base pair by another in the DNA

The replacement of one base pair, a substitution mutation, often affects only one amino acid in the resulting protein (except in the case of a nonsense mutation.) Replacements often have little or no effect, since the substitution could be silent or make little difference in the secondary and higher structures of the amino acid (although replacements can have dramatic effects in some cases, especially if they cause nonsense mutations.)

b) Addition of one base pair in the DNA

Addition (and deletion) mutations usually have a large effect on the function of a cell, because the frameshift causes a completely different amino acid sequence to be formed and the resulting protein is usually inactive.

17.85 *Which of the following substitution mutations is most likely to be harmful to a cell, and which is least likely to be harmful, based on their effect on the corresponding polypeptides? You will need to refer to tables 17.4 and 13.1 as you answer this problem, and explain your answers. (The base sequences of the DNA coding strand are given.)*

a) $ACA \rightarrow TCA$

The corresponding RNA sequence change (exchanging T for U) is:

$$ACA \rightarrow UCA$$

ACA codes for Thr (threonine), while UCA codes for Ser (serine). In the structures of these two amino acids (Table 13.1), we see that they are both neutral, hydrophilic amino acids, with the only difference between them being an extra methyl group in the side chain of serine. **It's likely that this mutation will have a relatively small effect on the function of the protein.**

b) ACA → ACG The corresponding RNA sequence change is the same (there are no T/U substitutions to make). ACA and ACG both code for Thr (threonine), so this is a silent mutation and **will have no effect on the corresponding polypeptide**.

c) ACA → CCA The corresponding RNA sequence change is the same (there are no T/U substitutions to make). ACA codes for Thr (threonine), while CCA codes for Pro (proline). In the structures of these two amino acids (Table 13.1), we see that threonine is a hydrophilic amino acid, while proline is hydrophobic. Hydrophilic amino acids tend to appear on the exterior of folded proteins, while hydrophobic side chains tend to be found in the interior, so this will likely change the shape of the polypeptide. Furthermore, you may remember that proline has a unique role as the amino acid that causes beta turns, the U-turns that are responsible for the folded structure of beta sheets (Figure 13.7). **Due to both changes in structure, it's very likely that this mutation will have a dramatic effect on the function of the protein.**

17.87 *a) Write the amino acid sequence that corresponds to the following DNA coding strand:*

AAACATAAGTTG

Since this is the sequence of the coding strand, the corresponding RNA sequence (exchanging T for U) is:

AAACAUAAGUUG

We break the sequence into 3-base codons and look up the corresponding amino acids in Table 17.3:

AAA – CAU – AAG – UUG

Lys – His – Lys – Leu

b) A mutation replaces the fifth base in this DNA with a G. What will be the amino acid sequence when this mutated DNA is transcribed and the resulting mRNA is translated?

The mutation will give the DNA sequence:

AAAC<u>G</u>TAAGTTG

and the corresponding RNA sequence (exchanging T for U):

AAACGUAAGUUG

We break the sequence into 3-base codons and translate into an amino acid sequence:

AAA – CGU – AAG – UUG

Lys – Arg – Lys – Leu

Since the substitution produces a codon that codes for a different amino acid (Arg instead of His), this is a missense mutation.

c) *A mutation adds a "T" between the fifth and sixth bases in the original DNA (so the sequence is now AAACATTAAGTTG). What will be the amino acid sequence when this mutated DNA is transcribed and the resulting mRNA is translated?*

Again, exchange T for U to get the RNA sequence:

AAACAUUAAGUUG

Divide into 3-base codons and translate:

AAA – CAU – UAA – GUU – G

Lys – His – STOP

The third codon in the sequence is now UAA, a stop codon, so the amino acid sequence encoded in this RNA segment, rather than being 4 amino acids (and presumably continuing on with the rest of the chain), will be only two amino acids long, then stop. Bases that follow the stop codon are ignored, so it doesn't matter that GUU codes for Val. The polypeptide chain will be shorter than that described by the correct sequence.

17.89 *What is a mutagen?*

A mutagen is a chemical compound that causes damage to DNA.

17.91 *A mutation in one cell in a developing embryo is more harmful than a mutation in one cell in an adult. Explain.*

Embryos contain only a small number of cells, each of which will multiply into a very large number of cells. If one cell in an embryo has damaged DNA, it may not be able to multiply, and even if it can, all of the cells in the growing fetus that come from that cell will also carry the mutation. However, an adult has many cells in each organ, so a damaged cell is unlikely to affect the function of the organ. The adult body can also afford to destroy a damaged cell.

17.93 *Many mutagens cause cancer, including all of the chemicals and radiation types in Table 17.6. Explain.*

If there is enough damage to the parts of a DNA molecule that control cell division, the cell, instead of ceasing to function, can grow out of control. Then uncontrolled cell division produces a mass of cells with the same mutations–a cancer.

17.95 *What is an autosomal genetic disorder, and under what circumstances will a person suffer from this type of disorder?*

The autosomal chromosomes are the ones that occur in pairs, all of our chromosomes except the sex chromosomes (X and Y). Human egg and sperm cells only contain 23 chromosomes, one from each autosomal pair plus one sex chromosome. A mutation in the DNA of an egg or sperm cell will (assuming that cell is involved in a successful fertilization) be passed on to every cell in the offspring.

Since the autosomal chromosomes occur in pairs, if one of the two chromosomes contains a mutation that would affect the function of the encoded protein, the other chromosome will allow cells to function normally. **However, if both chromosomes in a pair carry the same mutation, the cell won't be able to produce the proper protein**

that the sequence should be producing. An offspring with an autosomal genetic order suffers from the inability to produce a certain needed protein because of two defective genes, one in each of an autosomal pair.

17.97 *If a woman inherits a defective X chromosome from one of her parents, she will normally suffer no ill effects. Why is this?*

Female humans have two X chromosomes, so if one of them is defective, the other will usually be able to make the correct version of the affected protein. In females, the effect of a mutation in an X chromosome follows essentially the same rules as a mutation in an autosomal chromosome.

Solutions to Concept Questions

* indicates more challenging problems.

17.99 *Describe the backbone of a nucleic acid molecule and how the bases are attached to the backbone.*

The backbone of a nucleic acid molecule is an alternating chain of sugar molecules (ribose or deoxyribose) and phosphate ions, connected by phosphodiester groups formed during condensation reactions. The bases are connected to the chain by attaching to the 1' carbon. See Figures 17.2 and 17.3.

17.101 a) *Describe the double-helix structure of DNA.*

See Figure 17.7. **A "helix" is a stretched-out coil, and the "double-helix" describes two long strands that coil together.** It can also be described as a twisted ladder.

b) *What type of attractive force holds the two DNA strands next to each other?*

The two DNA strands are attracted to each other by hydrogen bonds between the complementary base pairs on each strand.

17.103 *Describe the function of DNA in a living organism.*

DNA carries the instructions for synthesizing all of the proteins that will carry out cell functions. It is a blueprint for building and operating an organism.

17.105 *Describe the steps that occur during DNA replication.*

(1) The two strands of DNA are separated from each other by an enzyme called helicase.

(2) A second enzyme, primase, begins assembling short sections of RNA (primers) along each strand, by linking together the complementary RNA nucleotides.

(3) Another enzyme, DNA polymerase, starts filling in the gaps between the RNA primers with DNA nucleotides.

(4) A final enzyme, DNA ligase, exchanges the U bases for Ts (replaces RNA with DNA) and seals the gaps between the segments, to complete the finished strand of DNA.

17.107 *How is the information in DNA used to build a living organism?*

The base sequences in DNA provide a code for the production of proteins, which carry out all of the processes of building, operating and replicating cells.

17.109 *How is mRNA modified after it is made by a cell?*

See Figure 17.15. **mRNA segments in the initial transcript are capped with a guanine nucleotide on the 5' end. A section of chain is removed from the 3' end, and a chain of adenosine nucleotides (poly-A tail) is added. Finally, sections of excess RNA called introns are removed from the RNA chain and broken down into individual nucleotides. The net result is the mature mRNA.**

17.111 *Humans make 49 different types of tRNA, each with its own anticodon. However, there are 61 different codons that correspond to amino acids. Why don't we need 61 different types of tRNA?*

tRNA anticodons don't always have to be an exact complement (in the third base) to the mRNA codons . Therefore the same tRNA molecule may be able to bind to more than one codon in an mRNA molecule.

17.113 a) *Which is more likely to be harmful to a cell, an error that occurs during replication or an error that occurs during transcription? Explain your answer.*

Replication is the production of an entire duplicate strand of DNA, which only happens during cell division. Replication is like making a complete copy of a set of blueprints, without making anything from them (or indeed even looking to see what they're for!). The transcription process copies only the portions of the blueprint that will actually be used; these sections of code are copied into the form of RNA, which will eventually produce the proteins that carry out the cell's functions.

Since much of our DNA contains no actual information that ever gets put to use ("junk DNA"), an error in replication may have no effect at all. In addition, since most of our DNA occurs in autosomal chromosome pairs–two versions of DNA instructions for the same processes–even an error in the replication of an important section of DNA may have no effect, if the other chromosome in the cell has a working set of instructions. (An error could have consequences for the daughter cell, though, if the damage causes there to be no working copy of a gene.)

An error during transcription, on the other hand, could give rise to proteins that would have the wrong structure, and might not function properly–or could even be harmful.

b) *Which is more likely to be harmful to a cell, an error that occurs during transcription or an error that occurs during translation? Explain your answer.*

A transcription error is more likely to be harmful. Transcription produces an RNA molecule that will produce many copies of the same protein. An error in transcription may produce many faulty copies of the protein, and could keep the cell from being able to function.

Translation is the process by which the instructions in nucleic acids are actually used to produce protein molecules. An error in translation would give one protein with an incorrect primary structure, but the other proteins produced by the RNA molecules would presumably be operational and cell function would likely be able to continue normally.

17.115 *Which is more likely to be harmful to a cell, a mutation in an intron or a mutation in an exon? Explain your answer.*

Introns are sections of RNA that are cut out of the initial transcripts and do not appear in the finished mRNA molecule. Exons are the sections of RNA that are kept to form the finished mRNA. **A mutation in an intron may have no effect at all on the mRNA molecule, but a mutation in an exon may (if it causes a codon to code for the wrong amino acid).**

17.117 *If the mutation in Problem 17.116 occurs in a bacterium, how damaging is it likely to be? Explain.*

Chapter 15 review! The mutation in Problem 17.116 "makes one of the enzymes of the citric acid cycle inactive." In eukaryotes (including plants and animals), the citric acid cycle takes place in the mitochondria. **Bacteria do not have mitochondria (though some bacteria do perform the citric acid cycle in their cytoplasm), and derive most of their energy from fermentation processes rather than from the citric acid cycle. Therefore the mutation described may have little or no effect on the function of a bacterium.**

17.119 *How are mutations related to biological evolution?*

While most mutations have relatively little effect, and those that do have an effect are usually harmful rather than helpful, occasionally a mutation will have an effect that makes an organism more successful. Without mutation, evolution doesn't happen–new genes would never occur.

Solutions to Summary and Challenge Problems

* indicates more challenging problems.

17.121 a) *List the four bases that occur in DNA.*

A (adenine), C (cytosine), G (guanine), and T (thymine).

b) *What base occurs in RNA, but not in DNA?*

U (uracil) occurs only in RNA, while T (thymine) occurs only in DNA.

c) *What are the complementary base pairs in DNA?*

A pairs with T, C pairs with G.

17.123 *a) Identify the bases and the sugar in the following nucleic acid molecule:*

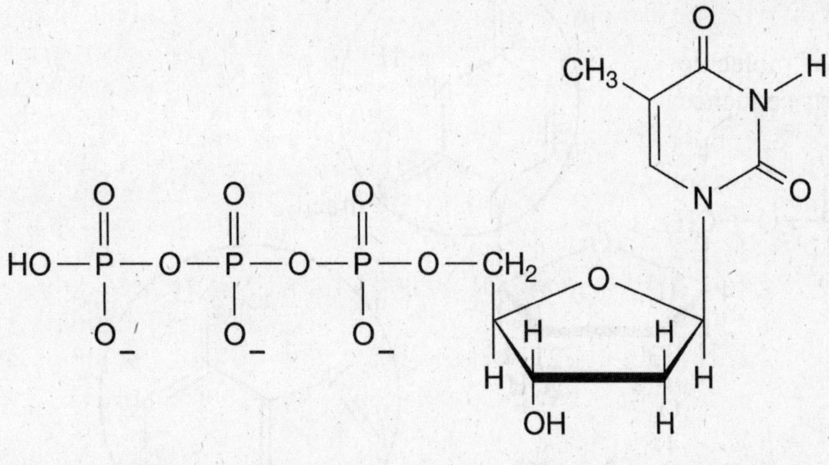

The bases are circled in the picture. Looking at Table 17.1, we identify them, in order from top to bottom, as **uracil (U), adenine (A) and guanine (G)**. The sugar in the molecule has an –OH group on the 2' carbon, so it is **ribose** (deoxyribose, the other option, has two H atoms on the 2' carbon, no –OH.)

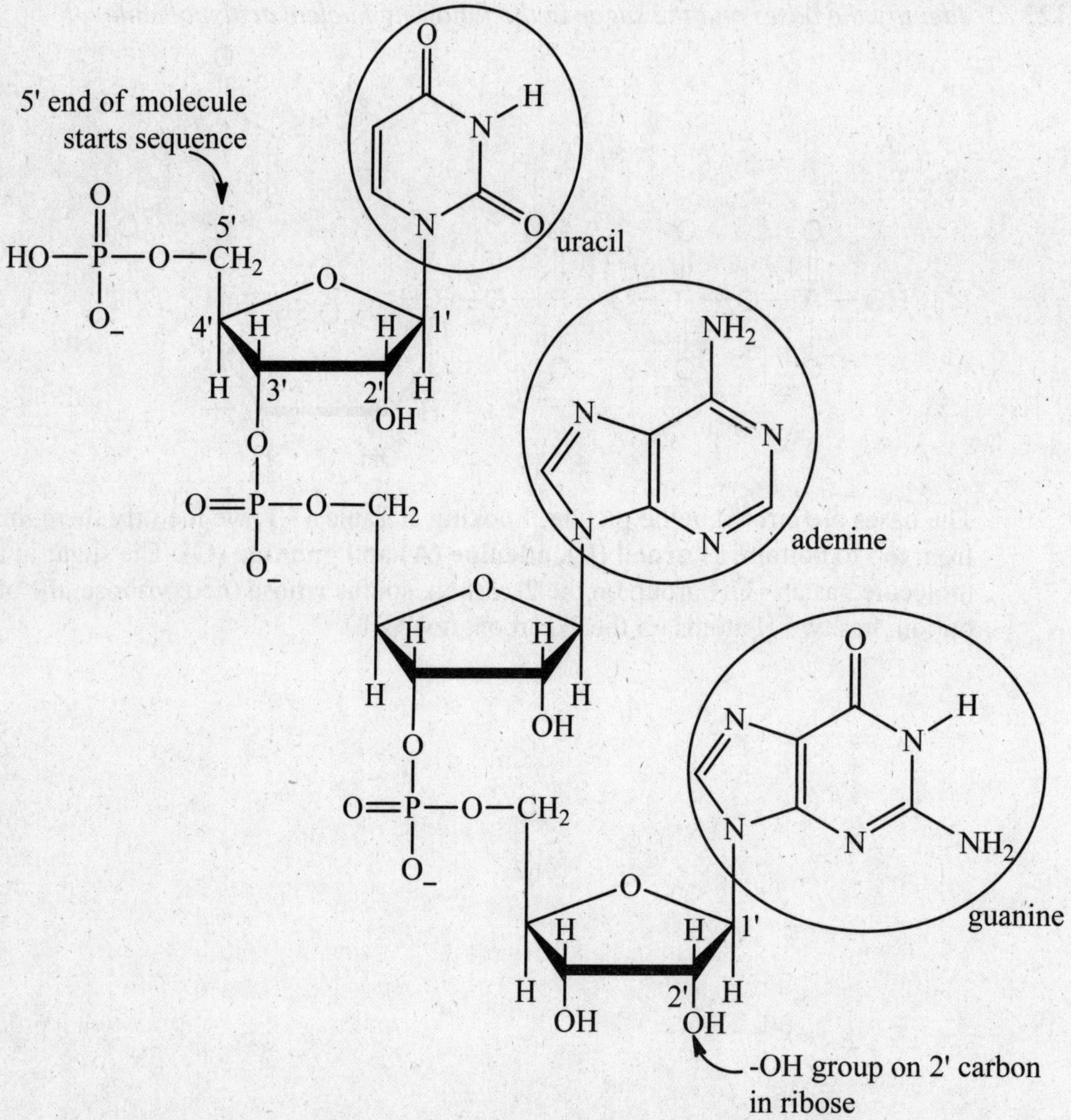

5' end of molecule starts sequence

uracil

adenine

guanine

-OH group on 2' carbon in ribose

b) *Is this a molecule of DNA or a molecule of RNA? How can you tell?*

It's a molecule of RNA. The sugar is ribose (DNA has deoxyribose) **and there is a U (uracil) base** (U only occurs in RNA, T only occurs in DNA.)

c) *Write the base sequence for this molecule. (Be sure to write the sequence in the correct direction.)*

Base sequences are always written starting with the 5' end, the end of the molecule with the phosphate ion on the 5' carbon of the sugar. In this molecule, the 5' end is at the top of the picture, so the base sequence is **UAG**.

17.125 *The diagram below shows two possible ways in which DNA could be replicated. In this diagram, the heavy blue lines represent the original DNA strands and the thin green lines represent the new DNA strands. Which of these is correct?*

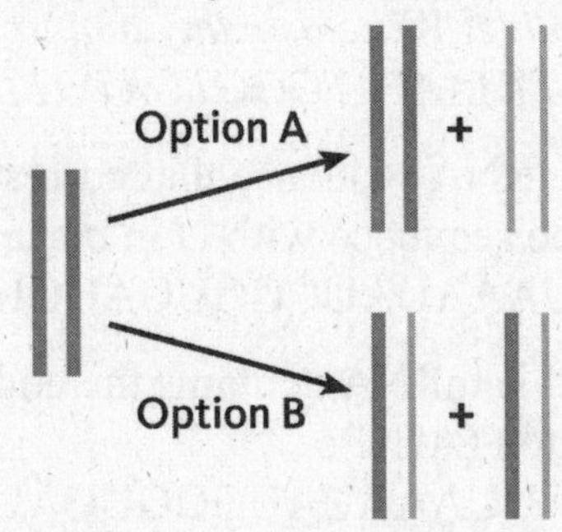

In DNA replication, the two strands of the double helix are separated and each is used as a pattern to build a new strand. The two old strands end up separated from each other and paired up with the new strands. Therefore **option B is a better representation of DNA replication**.

17.127 *A piece of DNA has the following base sequence (starting from the 5′ end):*

AATCAAGGC

a) *If this DNA is the coding strand, what is the base sequence in the template strand, and what is the base sequence in the corresponding mRNA? Be sure to write the base sequences in the correct direction.*

coding DNA strand (given): (5′ end) *A – A – T – C – A – A – G – G – C* (3′ end)
template DNA strand: (3′ end) T – T – A – G – T – T – C – C – G (5′ end)
RNA strand: (5′ end) A – A – U – C – A – A – G – G – C (3′ end)

Since nucleic acid sequences are always written starting with the 5′ end, **we write the template DNA sequence as GCCTTGATT and the RNA sequence as AAUCAAGGC.**

NOTE:
The RNA strand and the coding strand have the same sequence, with U in RNA substituted for T in DNA.

b) *If this DNA is the template strand, what is the base sequence in the coding strand, and what is the base sequence in the corresponding mRNA? Be sure to write the base sequences in the correct direction.*

coding DNA strand: (3′ end) T – T – A – G – T – T – C – C – G (5′ end)
template DNA strand (given): (5′ end) *A – A – T – C – A – A – G – G – C* (3′ end)
RNA strand: (3′ end) U – U – A – G – U – U – C – C – G (5′ end)

Both the coding DNA strand and the RNA strand are starting with the 3′ end as shown, so we reverse them to **write the coding DNA sequence as GCCTTGATT and the RNA sequence as GCCUUGAUU**. Again, the RNA strand and the coding strand have the same sequence, with U in RNA substituted for T in DNA.

17.129 *The following DNA sequence comes from a coding strand, and includes the code for the beginning of cytochrome C, a protein that plays an important role in the electron transport chain. Using this information, determine the amino acid sequence for the beginning of this polypeptide. Write as many amino acids as you can.*

ATTAAATATGGGTGATGTTGAGA

We first need to write the RNA sequence that corresponds to this coding DNA sequence, which means rewriting the sequence with U in place of T:

AUUAAAUAUGGGUGAUGUUGAGA

Since it is specified that this mRNA contains the code for the beginning of a polypeptide, it must contain a start codon (AUG):

AUUAAAUAUGGGUGAUGUUGAGA

To translate the mRNA, we need to divide up the bases into 3-base codons, beginning with the start codon. The cell ignores the bases before (to the left of) the start codon. Then we can use the genetic code to write the amino acid sequence:

AUUAAAU – AUG – GGU – GAU – GUU – GAG – A
Met – Gly – Asp – Val – Glu

We write the final amino acid sequence as **Met – Gly – Asp – Val – Glu** (methionine, glycine, aspartic acid, valine, glutamic acid). We don't know what codon the ending A might have been part of.

*17.131 *Chemists can make mRNA with a random sequence of A's and C's (for example, AAACACCACCCCAC...). When this mRNA is translated, the resulting polypeptide is built entirely from six of the twenty amino acids. Which six are they?*

Look in Table 17.3 and eliminate all codons that include U and G. We can start by covering up the bottom two rows and the right two columns, then eliminating the bottom two entries in each of the four boxes that remain, which leaves the codons that include only A and C. The amino acids that remain are **Lys, Ans, Thr, Gln, His and Pro.**

*17.133 *DNA samples that are extracted from cells usually contain large amounts of Mg^{2+} ions bound to the DNA. Explain why magnesium ions are so strongly attracted to DNA.*

The phosphate groups in the DNA backbone carry negative charges, which would be attracted to the positive charge of Mg^{2+} ions.

*17.135 *Single-stranded RNA can form base-paired regions. The following pictures show two ways in which this can happen. Which of these is more likely, and why?*

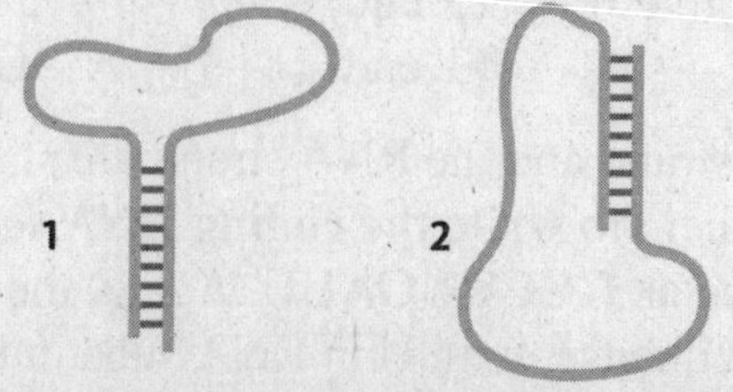

Nucleic acid strands always bind such that one strand's 5' → 3' direction lines up with the other strand's 3' → 5' direction (Figure 17.6). In picture 1, the 5' and 3' ends are matched up in the preferred orientation, but in picture 2, both strands have the 5' → 3' direction at the same time. **Picture 1 represents the likely arrangement.** (If you're having trouble seeing this, label one end of each strand as 5', and draw an arrow along the entire molecule.)

*17.137 *Would you expect Chargaff's Rule (see Problem 17.136) to apply to RNA samples? Why or why not?*

Chargaff's rule applies to DNA because DNA always appears as pairs of complementary strands. **Since RNA appears as single strands, there's no particular reason for the bases, or any particular combination of bases, to have the same total frequency. The bases do not occur in pairs. The rule, therefore, does not apply to RNA.**

*17.139 *A section of a mRNA molecule contains the base sequence below:*

AUUGUAUUUCCA

a) *If this entire section of mRNA is translated, what is the amino acid sequence of the resulting section of polypeptide?*

Split the mRNA sequence into groups of three bases, then look in Table 17.3 for the translation:

AUU – GUA – UUU – CCA
Ile – Val – Phe – Pro

b) *Is this section of the polypeptide hydrophilic, or is it hydrophobic?*

Use Table 13.1 to identify the amino acids. **Isoleucine, valine, phenylalanine, and proline are all hydrophobic.**

c) *Would you expect this section of the polypeptide to be on the inside or the outside of the folded protein?*

Chapter 13 review–hydrophobic side chains tend to appear on the **inside** of globular proteins (since the outside of the protein is in contact with an aqueous solution).

17.141 *A section of the coding strand of a DNA molecule contains the base sequence below:*

AAACGCAGTTTAGAGCT

a) *If this section is transcribed and the resulting mRNA is translated, what will be the amino acid sequence of the resulting polypeptide? (Assume that the entire mRNA is translated.)*

The mRNA sequence will be the same, except with U in place of T:

AAACGCAGUUUAGAGCU

We split the sequence into 3-base codons and translate to amino acids:

AAA – CGC – AGU – UUU – AGA – GCU
Lys – Arg – Ser – Phe – Arg – Ala

b) *A mutation changes the fourth base in the DNA from C to T. What will be the amino acid sequence that corresponds to this mutated DNA?*

This is a substitution that results in a missense mutation.
New coding sequence: AAATGCAGTTTTAGAGCT
new mRNA sequence: AAA – UGC – AGU – UUU – AGA – GCU
new amino acid sequence: **Lys – Cys – Ser – Phe – Arg – Ala**

c) *A different mutation adds an extra T between the third and fourth bases of the original DNA (so the sequence is now AAATCGC...). What will be the amino acid sequence that corresponds to this mutated DNA?*

This is an addition mutation and results in a frameshift.
New coding sequence: AAATCGCAGTTTTAGAGCT
new mRNA sequence: AAA – UCG – CAG – UUU – UAG – AGC – U
new amino acid sequence: **Lys – Ser – Gln – Phe – STOP**

The bases after the Stop codon are ignored, and the peptide ends.

d) *Which of these two mutations will probably have the larger impact on the activity of the protein, and why?*

The second (addition) mutation will probably have a greater effect on the activity of the protein than the first (substitution) mutation, because it causes the peptide chain to be terminated early as well as changing the amino acid sequence. This usually results in an inactive protein, and this type of mutation can be lethal to the cell. The substitution mutation leaves the protein the same length, but replaces an arginine (a large, basic amino acid) with a cysteine (a small, neutral amino acid), so it may or may not have a significant effect on the function of the protein (depending on how critical the affected amino acid is to the shape and function of the protein.)

*17.143 *Blood cannot clot unless it contains a variety of proteins called coagulation factors. One of these, called coagulation factor VIII, contains 2409 amino acids. The gene for coagulation factor VIII is located on the X chromosome. Mutations in this gene can cause hemophilia, a genetic disorder characterized by internal bleeding and slow wound healing. The following two mutations are known to occur in the gene for coagulation factor VIII.*

1) codon 80: GTT → GAT (produces severe hemophilia)
2) codon 85: GTC → GAC (produces mild hemophilia)

a) *What effect will each of these mutations have on the primary structure of coagulation factor VIII?*

The GTT → GAT substitution, when transcribed into mRNA, becomes a GUU → GAU substitution. The GTC → GAC substitution, when transcribed into mRNA, becomes a GUC → GAC substitution. In both cases, the substitution replaces a Val (valine, a hydrophobic amino acid that has no net charge at physiological pH) with an Asp (aspartic acid, a hydrophilic, acidic amino acid that has a negative charge at physiological pH). This could cause the affected region of the protein to change shape, because while the original amino acid would usually be folded into the inside of a protein (where hydrophobic side chains are usually found), the new amino acid will tend to stay on the outside of the protein (where hydrophilic side chains usually appear, in contact with the aqueous solution.)

b) *One of these mutations affects an amino acid in the active site of coagulation factor VIII. Which one is it, and how can you tell?*

Since both substitutions have the same effect on the primary structure of the protein, but the GTT → GAT substitution in codon 80 has a much greater effect, it is likely that the codon 80 substitution affects an amino acid in the active site of the protein. Changes to the active site of a protein have a much greater effect on protein function than changes to other areas away from the active site.

*17.145 a) *Is it possible to have an addition mutation within the coding portion of a gene that does not produce a frameshift? If so, how?*

Here are some ways that an addition mutation could happen without producing a frameshift:

(1) **If an addition mutation creates a Stop codon (UAA, UAG or UGA) at the site of the addition, the protein simply terminates and the remaining codons (which would otherwise be frameshifted) are not read at all.**

(2) **If an entire three-base codon were added at once, then the mRNA would add an extra amino acid to the protein chain, but there would be no frameshift.**

(3) **If an addition occurs at the beginning of the coding portion, it may not produce a frameshift.** For example, here's an original mRNA sequence that includes the code for the beginning of a protein, with the start codon underlined:

AAU<u>AUG</u> – GGU – GAU – GUU – GAG – A…

If an insertion of an A (in boldface) occurs after the 4th base in the sequence, then the new sequence becomes:

AAUA<u>**A**UG</u> – GGU – GAU – GUU – GAG – A…

NOTE:

Any other insertion in the start codon would either cause a frameshift, or cause the sequence not to be read at all (because the start codon would be disrupted).

(4) **Certain additions happening near the end of a protein could happen without causing a frameshift.** For example, an addition (in **bold**) that turned a UAG stop codon into UA**A**G, UA**G**G, or U**G**AG would still give a stop codon. (There are other scenarios where this works out, for example UAA becoming U**G**AA, UA**G**A or UA**A**A.)

b) *Is it possible to have a substitution mutation within the coding portion of a gene that produces a frameshift? If so, how?*

A substitution in the start codon could produce a frameshift. For example, using the same original RNA sequence in (3) above:

AAU<u>AUG</u> – GGU – GAU – GUU – GAG – A…

Substituting the fourth base with a G (in bold) moves the start codon and produces a frameshift for the rest of the sequence:

A**AUG** – UGG – GUG – AUG – UUG – AGA…

17.147 *XP (see Problem 17.146) is equally likely to occur in males and in females. Based on this, is the gene for the DNA repair enzyme located on a sex chromosome or on an autosomal chromosome? How can you tell?*

See Figure 17.23. **Genetic disorders that rise from mutations on one of the sex chromosomes are almost always sex-linked–that is, they occur much more often in one sex (usually males) than the other. Genetic disorders that have equal prevalence in males and females are usually autosomal. This mutation is almost certainly located on an autosomal chromosome.**

17.149 *Gene C is on Charles's X chromosome. (Charles is a man, so he has only one X chromosome.)*

a) *Could Charles's daughter Wilhelmina inherit gene C? If so, how?*

See Figure 17.23. **Since Charles's daughter will inherit one of her X chromosomes from him, Wilhelmina will inherit Gene C.** (The only way for Charles to produce a daughter is to pass on his X chromosome.)

b) *Could Charles's son Aloysius inherit gene C? If so, how?*

Any child of Charles who inherits his X chromosome will be female. **The only way for Charles to have a son is to pass on the Y chromosome instead of the X. Aloysius cannot inherit Gene C.**